Bhima R. Vijayendran (Ed.)
BioProducts

Also of interest

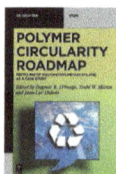

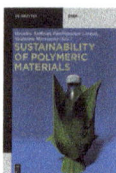

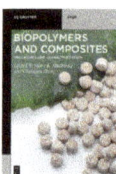

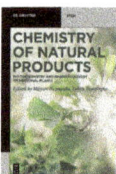

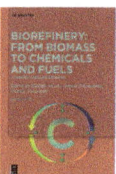

BioProducts

Green Materials for an Emerging Circular
and Sustainable Economy

Edited by
Bhima R. Vijayendran

DE GRUYTER

Editor
Bhima R. Vijayendran
Redwood Innovation Partners LLC
7040 Avenida Encinas
Carslbad, CA 92011
USA

ISBN 978-3-11-079121-1
e-ISBN (PDF) 978-3-11-079122-8
e-ISBN (EPUB) 978-3-11-079148-8

Library of Congress Control Number: 2023932362

Bibliographic information published by the Deutsche Nationalbibliothek
The Deutsche Nationalbibliothek lists this publication in the Deutsche Nationalbibliografie;
detailed bibliographic data are available on the Internet at http://dnb.dnb.de.

© 2023 Walter de Gruyter GmbH, Berlin/Boston
Cover image: Peach_iStock/iStock/Getty Images Plus
Typesetting: Integra Software Services Pvt. Ltd.
Printing and binding: CPI books GmbH, Leck

www.degruyter.com

Preface

I was approached by a representative of De Gruyter in early February of 2022 to develop a book on Bioproducts for the Emerging Green and Sustainable Economy, covering this vast subject from discovery research to commercialization. My initial reaction was this is a daunting task given the broad scope and global nature of the field. I reached out to a few of my fellow practitioners in the field that I have known and worked with for the last two decades. There was some skepticism from a few of them. However, the overall feedback was quite positive and encouraging. I got back to the representative and submitted an outline for the book which was accepted. The focus of the outline was on bioproducts, excluding biofuels and bioenergy, derived from biomass and or produced by biological processing and/or conventional catalytic thermal chemical processes.

To manage a reasonably credible treatment of the enormous scope, I steered toward recruiting researchers and active stakeholders in the field that I have worked with closely. I also tried to reach out to my expert colleagues in Europe, Asia, Canada, South America, and India, as the field of Bioproducts is truly global in nature. To my pleasant surprise, most of them enthusiastically volunteered to submit a chapter on a topic directly aligned with the scope of the book. So, we have 26 chapters covering major all industrial sectors and applications in the book because of their contributions. The book is organized into six sections.

The first section has an introduction and an overview along with general discussion on biomass sources, process schemes, and techno-economic analysis useful in project selection. Also included is a primer for budding entrepreneurs with mostly R&D background, who need financial tools and awareness for raising needed funding.

The second section is focused on early-stage R&D projects covering the diverse areas of biosurfactants, bio-lubricants, microbial-derived bioproducts, novel formaldehyde-free binders from furanic building blocks, bio-solvents, and value-added products from lignin.

The third section covers emerging synthetic biology and biomanufacturing to develop and commercialize high-value products in myriad applications such as nutraceuticals, food additives, and others. Due to the great interest and attendant activities in this area, it was well beyond the scope of this book to cover this particularly important topic comprehensively.

The fourth section covers a few recently commercialized bioproducts – microbial biosurfactants, bioplastcizers for polyvinyl chloride (PVC), soy-based additives for the construction industry, and emerging green hydrogen. There are many other new bioproducts for cosmetic, personal care, chemical intermediates, etc. that have been commercialized. However, inclusion of all of them was outside the scope of the book. Again, as noted above, the selection of the few highlighted new products in this section was based on the editor's direct involvement in their product development.

https://doi.org/10.1515/9783110791228-202

The fifth section is focused on bioplastics and biopolymers, a vast and growing field covering myriad industries and applications. The subject is covered by the contributing authors from their perspectives and viewpoints. One challenge is avoiding repetition and coverage of closely related subject matter. A serious attempt has been made to minimize this issue.

The sixth and concluding section covers biocomposites, a growing field with applications in building and construction, auto and transportation, turbine blades for wind power generation, and biomedical devices and implants.

I owe my deep sense of appreciation and thanks to all the contributors. They have been extremely generous with their time and made a sincere effort to keep a tight schedule.

I also would like to thank Dr. Blaine Metting, a former colleague at Battelle, for his support and guidance. My thanks also to Dr. Roger Wyse of Spruce Capital, Mr. Andy Shaffer of Shafer Biz Associates, Dr. Damiano Beccaria of Eastman Chemicals, and my good friend Mr. Freddy Shleihs of Fraunhofer IKTS for their support.

One person that I owe a great thanks to is Ms. Ria Sengbusch of De Gruyter for her timely help and feedback during the entire book project.

I also would like to thank my grandkids – Isla, Ian, Molly, and Gus – representing the next generation, for helping me select the book cover.

Overall, this has been a fulfilling and interesting experience over the last twelve months. It was great to connect with former colleagues and coworkers in developing the book. Finally, the responsibility for any inadvertent errors and omission of certain topics and their treatment is entirely on my shoulders. I would love to hear from the readers.

Bhima R. Vijayendran
Carlsbad, California, USA
June 2023

Contents

Section I: **Introductory chapters**

Section II: **Some early-stage R&D bioproducts**

Section III: **Synthetic biology**

Section VI: **Sustainable green biocomposites**

Some concluding remarks

After 26 chapters and about 600 pages some concluding remarks and observations are in order.

There has been substantial funding and innovation in this space. Consumer awareness and active support for sustainable green products are on the rise. However, commercial adoption has been slow and uneven.

Today, scientists and policymakers generally agree that climate change is an existential issue worldwide. Consumers and companies offering products to serve their customers in myriad markets, and applications are in tune with this assessment and starting to address them in a concerted manner. One standard economic policy prescription for cutting greenhouse emissions that drive climate change is adopting a carbon tax. This gives consumers an incentive to use less fossil-fuel-based products and an incentive to produce improved low-carbon technologies leading to electric vehicles, green power, and sustainable bioproducts.

Incentives for capturing greenhouse gases from current fossil-based products is a pragmatic, possibly interim, step (sometimes called blue products) in the transition to the ultimate goal of total green products. Another trend is an increased focus on recycle and reuse of products at the end of their useful life. Yet another one, mostly at the consumer level, especially with generation Z demographics, is reduced consumption. This is partially reflected in a 2021 book by J. B. McKinnon, entitled *The Day the World Stops Shopping: How Ending Consumerism Saves the Environment and Ourselves*. There is increasing awareness that life in a lower-consuming society really can be better with less stress and more time for personal growth and enjoyment. One of the challenges is balancing economic growth, a major engine for capitalism, against lower conspicuous consumption.

In the basic science and technology fronts, including R&D and manufacturing, of bioproducts rapid advances are taking place, and they are expected to keep accelerating facilitating market adoption of bioproducts.

And thus overall, prospects are bright. It is not going to be a linear path to the goal line. By the way, it is going to cost more for all.

Fight On!

https://doi.org/10.1515/9783110791228-204

List of contributing authors

Bhima R. Vijayendran
Redwood Innovation Partners LLC
Carlsbad
CA 92011
USA
bhima@redwdinnov.com

Ram Lalgudi
Aries Science and Technology LLC
Columbus
OH 45701
USA
lalgudir@ariesst.com

Michael A. Schultz
Indianapolis, IN
mike@ptisolns.com

John McArdle
Redwood Innovation Partners LLC
Boston
Massachusetts
USA
johnm@redwdinnov.com

Herman P. Benecke
Columbus, OH
beneckefl@gmail.com

Daniel B. Garbark
Lead Researcher
Battelle Memorial Institute
Columbus, OH
garbarkd@battelle.org

Claudia Creighton
Carbon Nexus at the Institute for Frontier
Materials
Deakin University
75 Pidgons Rd
Waurn Ponds
Victoria 3216
Australia

Nguyen Duc Le
Carbon Nexus at the Institute for Frontier
Materials
Deakin University
75 Pidgons Rd
Waurn Ponds
Victoria 3216
Australia

Huma Khan
Carbon Nexus at the Institute for Frontier
Materials
Deakin University
75 Pidgons Rd
Waurn Ponds
Victoria 3216
Australia

Srinivas Nunna
Carbon Nexus at the Institute for Frontier
Materials
Deakin University
75 Pidgons Rd
Waurn Ponds
Victoria 3216
Australia

Russell J. Varley
Carbon Nexus at the Institute for Frontier
Materials
Deakin University
75 Pidgons Rd
Waurn Ponds
Victoria 3216
Australia
russell.varley@deakin.edu.au

Cai Li Song
Petronas Research Sdn Bhd
Lot 3288 & 3289
Off Jalan Ayer Itam
Kawasan Institusi Bangi
43000 Bangi, Selangor
Malaysia
song.caili@petronas.com

https://doi.org/10.1515/9783110791228-205

Mohamad Fakhrul Ridhwan Samsudin
PETRONAS Research Sdn Bhd
Lot 3288 & 3289
Off Jalan Ayer Hitam
Kawasan Institusi Bangi
43000 Kajang, Selangor
Malaysia

Nur Amalina Samsudin
PETRONAS Research Sdn Bhd
Lot 3288 & 3289
Off Jalan Ayer Hitam
Kawasan Institusi Bangi
43000 Kajang, Selangor
Malaysia

Wong Mee Kee
PETRONAS Research Sdn Bhd
Lot 3288 & 3289
Off Jalan Ayer Hitam
Kawasan Institusi Bangi
43000 Kajang, Selangor
Malaysia

M. Syamzari B. Rafeen
PETRONAS Research Sdn Bhd
Lot 3288 & 3289
Off Jalan Ayer Hitam
Kawasan Institusi Bangi
43000 Kajang, Selangor
Malaysia
syamzari_rafeen@petronas.com.my
+6019-7344278

Nur Liyana Ismail
PETRONAS Research Sdn. Bhd.
Lot 3288 & 3289
Off Jalan Ayer Itam
Kawasan Institusi Bangi
43000 Kajang, Selangor
Darul Ehsan
Malaysia

Siti Fatihah Salleh
PETRONAS Research Sdn. Bhd.
Lot 3288 & 3289
Off Jalan Ayer Itam
Kawasan Institusi Bangi
43000 Kajang, Selangor
Darul Ehsan
Malaysia

Jofry Othman
PETRONAS Research Sdn. Bhd.
Lot 3288 & 3289
Off Jalan Ayer Itam
Kawasan Institusi Bangi
43000 Kajang, Selangor
Darul Ehsan
Malaysia

Sara Shahruddin
PETRONAS Research Sdn. Bhd.
Lot 3288 & 3289
Off Jalan Ayer Itam
Kawasan Institusi Bangi
43000 Kajang, Selangor
Darul Ehsan
Malaysia

Michelle Young
Lead Technologist
Carollo Engineers, Inc.
390 Interlocken Crescent
Suite 800
Broomfield
CO 80021
USA
myoung@carollo.com

Rachel Yoho
George Mason University
4400 University Dr.,
Fairfax VA 22030
USA
ryoho@gmu.edu

Beth Bannerman
Amyris, Inc.
San Francisco, CA
USA
bannerman@amyris.com

Sunil Chandran
Amyris, Inc.
San Francisco, CA
USA
chandran@amyris.com

Crispinus Omumasaba
Utilization of Carbon Dioxide Institute Co. Ltd.
2-4-32 Aomi, Koto
Tokyo 135-0064
Japan

Alain A. Vertés
Utilization of Carbon Dioxide Institute Co. Ltd.
2-4-32 Aomi, Koto
Tokyo 135-0064
Japan

F. Blaine Metting
Utilization of Carbon Dioxide Institute Co. Ltd.
2-4-32 Aomi, Koto
Tokyo 135-0064
Japan
And
Pacific Northwest National Laboratory, Retired
Tel. +81-3-6435-1150; Fax +81-3-6435-1188

Hideaki Yukawa
Utilization of Carbon Dioxide Institute Co. Ltd.
2-4-32 Aomi, Koto
Tokyo 135-0064
Japan
hyukawa@co2.co.jp

Vikas Kumar
Aquaculture Research Institute
Department of Animal, Veterinary and Food
Sciences
University of Idaho
Moscow, ID 83844-2330
USA

Anisa Mitra
Department of Zoology
Sundarban Hazi Desarat College
Pathankhali, South 24
Parganas, 743611
India

Dan Derr
Integrity BioChem
USA
danderr99@msn.com

Jacyr Quadros
Innoleics Serv. e Cons. Ltda.
São Paulo
Brazil
jqudros@innoleics.com

Gunnar Lynum
SMD Products Company, Inc. & Strategic Market
Development, LLC
P.O. Box 1634
Manchester, MO 63021
USA

Steven Lynum
SMD Products Company, Inc. & Strategic Market
Development, LLC
P.O. Box 1634
Manchester, MO 63021
USA

Jonathan Cristiani
Black and Veatch Corporation
11401 Lamar Avenue
Overland Park
KS 66211
USA
+1 919-463-3043
CristianiJM@BV.com

Grace Dearnley
Black and Veatch Corporation
11401 Lamar Avenue
Overland Park
KS 66211
USA

Justin Distler
Black and Veatch Corporation
11401 Lamar Avenue
Overland Park
KS 66211
USA

Andrew Doerflinger
Black and Veatch Corporation
11401 Lamar Avenue
Overland Park
KS 66211
USA

Vincent Mazzoni
Black and Veatch Corporation
11401 Lamar Avenue
Overland Park
KS 66211
USA

Ashok Adur
Everest International Consulting (LLC)
Jacksonville
FL 32256
USA
aadur@outlook.com

Pramod Kumbhar
Praj Matrix-R&D Center (Division of Praj
Industries Ltd.)
403/403/1098, Urawade
Pirangut, Mulshi
Pune 412115
India
pramodkumbhar@praj.net

Anand Ghosalkar
Praj Matrix-R&D Center (Division of Praj
Industries Ltd.)
403/403/1098, Urawade
Pirangut, Mulshi
Pune 412115
India

Yogesh Nevare
Praj Matrix-R&D Center (Division of Praj
Industries Ltd.)
403/403/1098, Urawade
Pirangut, Mulshi
Pune 412115
India

Sandeep Kulkarni
KoolEarth Solutions, Inc.
Alpharetta, 30005 GA
USA
skulkarni@koolearthsolutions.com

Ray Bergstra
TerraVerdae Bioworks Inc.
Edmonton
Canada
rjbergstra@terraverdae.com

Camille Sobrian Saltman
csaltman@incarenewtech.com
+250-300-5254

David Saltman
dsaltman@incarenewtech.com
+250-300-5253

Robert C. Joyce
Innovative Plastics and Molding, Inc,
Toledo, OH
USA
robert@fibertuff.us

Section I: **Introductory chapters**

Bhima R. Vijayendran

Chapter 1
Introduction and an overview

1.1 Introduction

Since humans settled down almost 10,000 years ago, agriculture and plant-derived products have played a significant role in all aspects of their lives and economies. This changed dramatically in the last two centuries with the increased use of fossil-derived products taking a major share of the economy due to advances in basic science, technology, engineering, and manufacturing. Many benefits have resulted in this transition based on fossil sources. One of the major impacts was the relatively inexpensive energy and products derived from them that propelled much of our way and quality of life.

Due to the indiscriminate, addictive, and uncontrolled reliance on nonrenewable fossil sources, the progress has come with significant cost, namely, damage to the environment with attendant pollution, greenhouse gas emissions, and global warming posing serious challenges for continued sustainability. For instance, carbon dioxide concentration in the atmosphere has risen from 280 parts per million (ppm) before the industrial era to about 420 ppm today, and it is currently rising at more than 2 ppm per year [1]. In practical terms, one ppm of carbon dioxide in the atmosphere is about 7.8 billion metric tons. There are other greenhouse gases such as methane with even more potent global warming effects. It is estimated that the average methane level in 2021 was 1,896 parts per billion (ppb), around 162% greater than the preindustrial levels [1]. One can easily relate to the scope and magnitude of the problem. Fortunately, there is also an opportunity to make use of these polluting emissions as feedstocks for renewable and sustainable biofuels and bioproducts and reduce the current reliance on polluting fossil sources.

In the last quarter century or so, these environmental and sustainability issues have received attention from all segments of the society on a global scale. Many passionate, dedicated, and motivated individuals, entrepreneurs, start-ups, established businesses, governments, leading research and development (R&D) organizations, and other stakeholders have been working diligently in solving many of these issues.

One approach that reduces greenhouse gas emissions is the adoption of cost-effective bioproducts – products derived from renewable resources processed by conventional catalytic thermochemical processes and/or fossil sources processed by emerging bioprocessing techniques including fermentation and synthetic biology. The

Bhima R. Vijayendran, Redwood Innovation Partners LLC, Carlsbad 92011, CA,
e-mail: bhima@redwdinnov.com

https://doi.org/10.1515/9783110791228-001

global economy with supportive public policies and consumer activism is poised for transitioning into a circular one with proactive initiatives to mitigate waste, greenhouse gas emissions, reuse and recycle of resources in a profitable and sustainable manner.

The book highlights some of these developments: from discovery, lab feasibility, scale-up, and eventual commercialization of bioproducts. Given the very vast scope of the field, their global nature, topics, and contributors are selected based on personal direct involvement of the editor in R&D in the field, commercialization of bioproducts, global relationships with several contributors in the field and a passion to make a difference in advancing bioproducts and contribute toward a sustainable and circular economy.

The book covers the challenges and frustrations of introducing new bioproducts despite committed efforts and investments. Fortunately, the overall environment for sustainable bioproducts is going through a sea change as consumers are starting to pay attention to greenhouse emissions and global warming. Businesses are taking note of this change in consumer behavior and are responding to meet their expectations with sustainable and environmentally friendly offerings.

1.2 Some significant developments over the last quarter century

Several major developments in bioproducts related to advances in R&D, processing, funding, commercialization, consumer awareness perception and acceptance, and public policy and legislative actions have occurred. Some of these are covered in this chapter.

1.2.1 An evolutionary process

The progress in the field has been an evolutionary one with cycles and phases [2]. The initial focus was based on the expectation that green bioproducts would have a premium and the consumers would be willing to pay a higher price. This was a misguided one, a lesson learned, somewhat slowly, with serious consequences in the commercialization of early bioproducts. The second cycle or wave was focused on fungible or direct petroleum substitute products. Several fungible bioproducts were developed, and their impact on reducing carbon footprint reduction has been reported [3]. One example is bio-monoethylene glycol (MEG) by India Glycols Ltd. Bio-MEG is projected to avoid 407 kg of CO_2/ton of the bioproduct used in polyester soft drink bottles (see Chapter 9 for details on bio-MEG). Another focus has been on drop-in replacements such as bio-ethylene and bioglycols. One more approach has been to develop functionally equivalent bioproducts such as polyesters based on 2,5-furandicarboxylic acid. A few of these are covered in Chapter 2. However, commercial impact of such fungible, direct replacement, and functional equivalent bioproducts has been somewhat muted to date.

A more recent trend has been to focus on higher value but smaller volume functional molecules of interest in nutraceuticals, cosmetics, pharma, and others (see Chapter 12). Another successful approach has been to develop molecules with desirable target functionality, not readily achievable by fossil-based products. A commercially successful example of this is 1,3-propanediol (PDO) from sugar-based feedstock and fermentation using engineered *E. coli* organism (see Chapter 3 for details).

A promising emerging trend is to use greenhouse gases and industrial wastes as feedstocks to produce higher value bioproducts (see Chapter 2).

1.2.2 Product and process: R&D

Significant advances have been made in basic sciences – synthetic and systems biology, analytical tools including data analytics/machine Artificial Intelligence learning, and instrumentation – thermochemical catalyst technologies and processing, enzyme design and bioprocessing, and manufacturing at scale. Such advances have occurred due to significant and persistent investments from various stakeholders. In the United States, major investments by the US Department of Energy (DOE), the US Department of Agriculture (USDA), and the National Science Foundation (NSF) are worth noting. It is also worth mentioning that similar sustained efforts have been happening in Europe, Asia, and elsewhere. A few of the US-based initiatives and programs are covered in the following two sections.

1.2.2.1 Unites States Department of Energy (DOE) initiatives

The US DOE has been a major driver in the renewable and sustainable fields for decades. A few specific ones are listed further:

– The DOE National Laboratories such as Pacific Northwest National Labs, National Renewable Energy Labs, Oak Ridge National Laboratory, and Idaho National Labs have been active in advancing various aspects of bioproduct development. Their websites have full details of their research activities in the field. It is worth noting that Battelle Memorial Institute (BMI) based in Columbus, OH, an independent nonprofit R&D organization that manages or comanages many of these DOE labs, has been involved in bioproducts since the mid-1980s. See more on this later.
– DOE is actively supporting innovative technology and product development in the field through periodic solicitations related to improved catalysts, scale-up and manufacturing at scale, use of industrial wastes such as CO_2 as feedstock to make bioproducts, enzyme design, and bioprocessing (https://eere-Exchange,enrgy.gov/).
– DOE has also been funding regional lab facilities and trained staff to facilitate start-up companies in the field with their product development and scale-up needs (http://abpdu.lbl.gov/).

– A good example of strategic leadership is the Roadmap effort of DOE in identifying top 12 target molecules for the emerging bioproducts industry in mid-2000s [4]. See Chapter 2 for further details.

1.2.2.2 United States Department of Agriculture (USDA) initiatives

USDA has been an active funding source for much of the impactful work in the field. A few efforts are highlighted here:

– USDA has several regional lab facilities such as in Illinois, Pennsylvania, and California where dedicated scientists are involved in all aspects of bioproduct R&D. A couple of specific examples are the work on xanthan gum, an extracellular microbial polysaccharide polymer, and novel estolide compound-based fatty acid chemistry. Further details of these products are given in Chapter 2.
– USDA has also pioneered in facilitating commercialization of bioproducts through their USDA-Certified Biobased Product Label Program started with 2002 Farm Bill and Reauthorized in 2018 [5, 6]. Several hundred products from numerous companies from all the 50 states covering diverse products from paints, industrial products, cosmetics, cleaners, and so on are listed in the program and the list is growing rapidly. The construction products described in Chapter 17 and plasticizer additives covered in Chapter 16 are some examples.

1.2.2.3 Farmer-based organizations

There are several farmer-based organizations at the national and state levels that are active in funding research and product development derived from crops. An illustrative example is the National Soybean Checkoff program that supports funding for bioproducts development at the national level (United Soybean Board, USB) and soybean-growing state associations such as the Ohio Soybean Council (OSC). They have been supporting bioproducts R&D since the checkoff program started almost 25 years ago. More recently, the OSC has started a laboratory, Airable Labs,(https://www.airableresearchlab/com/), dedicated to developing soy-based industrial bioproducts.

1.2.2.4 A few other organizations

There are several active organizations in Europe, Canada, and Australia that are also supporting bioproducts' R&D activities and commercialization. A few of them are listed:

– Nova Institute in Germany – www.bio-based.eu
– EuropaBio – Industrial Biotechnology Enabling a Circular BioEconomy in Europe – www.eurobio.org

- Cosun Biobased Products – www.cosunbiobased.com
- BIOTECCanada – www.Biotech.ca
- Ontario East Economic Development Commission – www.ontarioeast.ca
- Queensland University of Technology – www.qui-edu.au/research/ctcb

1.3 Role of bioproducts

The role and importance of bioproducts that are not fuel and/or energy related – the focus of this book – in the evolving sustainable green circular economy are worth reviewing. As shown in Chapter 2, only about 10% of the nonrenewable carbon from fossil sources is used for chemical production. Thus, the bulk of polluting greenhouse gas emissions is from the energy sector for use in transportation and power generation industries. Yet, in a typical fossil industry composed of energy and chemical products, the output from this 10% of polluting carbon provides the needed profitability as measured by return on investment for the entire petrochemical enterprise due to the higher value of chemical products in the marketplace. In other words, the petrorefinery complex becomes a profitable venture as every carbon atom from fossil resources is converted to higher value and profitable outcomes. It is expected that in the emerging bioeconomy, a similar biorefinery model would evolve and thus make bioindustry an economically sustainable business. Furthermore, eliminating even part of the 10% of polluting fossil carbon-based products by bioproducts should have a significant impact on reducing greenhouse emissions. Use of fossil fuel combustion products as bioproduct feedstocks is especially attractive as higher value products are created while capturing greenhouse gases.

There are other benefits as well. Bioproducts such as bioplastics have the potential to be biodegradable and offer additional benefit in improved waste management. The option for recycling is also an attractive feature. Issues, challenges, and opportunities in this topic are covered in Chapters 19 and 21.

1.4 A few takeaways from the quarter century of development in bioproducts

One of the questions that practitioners in the field are asked is what has been accomplished. It is a fair challenge from two perspectives. One is to get a realistic gut check and the other is to build from lessons learned to guide and redirect future efforts, as needed. However, answers to this question are not simple and straightforward. As mentioned earlier, the activities over the last quarter century have been quite diverse, covering a lot of ground. Perhaps, it would be helpful to break this analysis into a few tangible and measurable parts. An attempt is made in the following section.

1.4.1 Advances in basic science relevant to bioproducts

Even a casual review of the literature would clearly show the tremendous advances made in this area, especially in areas of synthetic and systems biology and their impact on industrial biotechnology-derived bioproducts.

It is often said that the twenty-first century is the century of synthetic biology, as late nineteenth and much of early twentieth centuries were for synthetic chemistry that propelled the growth of fossil-based energy and petrochemical industries. Much is expected from the disruptive synthetic biology: technology-innovative approaches for engineering new biological processes or redesigning existing ones for purpose-driven outcomes. Synthetic biology has been described as the heart of the emerging bioeconomy, capable of delivering novel solutions to global healthcare, agriculture, manufacturing, and environmental challenges [7]. Please visit the Synbiobeta website for a comprehensive overview of the activities in the field (www. Synbiobeta.com).

A recent report [8] highlights some of the advances and their potential impact on the bioeconomy:

- Breakthrough advances in gene sequencing, tools, and techniques to manipulate metabolic pathways of interest in myriad fields including developing industrial and consumer products.
- Database powered by AlphaFold algorithm's ability to predict over 200 million protein structures from over 10 million species. It is expected that the expanded database will aid countless researchers in their work and open new avenues of scientific discovery [9]. One potential area of interest in bioproducts is the design of enzymes that can decompose plastic pollution into harmless substances.
- Lower capex biomanufacturing facilities to produce high-value bioproducts with more benign environmental footprint.
- Improved next-generation pesticides and crop control agents with improved safety, efficacy, and handling.
- Enzymes that improve efficiency and reduce energy use in traditional industries such as pulp and paper, textiles, and food processing.
- Cosmetics and personal care items made from sustainable bioproduced components that do not use ingredients from threatened and endangered species (see Chapter 12).

1.4.2 Deployment of improved and cost-effective manufacturing at scale

In this area, there are several success stories, and a few are listed further for illustrative purposes:
- Commercial-scale production of several diols including propylene glycol and 1,3-butylene glycols. See Chapter 3 for details.
- Commercial-scale production of 1,4-butanediol for nylon intermediates and other polymers. See Chapters 19 and 20 for details.
- Full-scale commercialization of PDO for improved polyester polymer. See Chapter 3 for details.
- Continued global expansion of polylactic acid (PLA) manufacturing, including the recent announcement by ADM and LG Chemicals [10].
- Announcement by Avantium and Covation, a Du Pont spin out, to produce furanic-based polymers. See Chapter 5 for details.

This list goes on. However, still the market penetrations of bioplastics and biopolymers are small.

1.4.3 Commercialization challenges

Commercialization has certainly been an uphill battle and continues to be one, for many reasons. A few of them are explored later. The obvious ones are higher cost, undeveloped/nonexisting supply chains, unproven performance and associated risks, poor compatibility with existing industry infrastructure, and sunk investments. All these issues are formidable, leading often to the premature death of a new bioproduct, especially if the new product is a replacement or substitute product for a well-established fossil product in a relatively mature market. Lack of consumer and market pull for many of these failed products have been a common theme, as bioproducts do not have any subsidy unlike the biofuels.

It is probably worth looking at a few bioproducts that have been highly successful in the marketplace. Some like the first two – guar and xanthan gum – have been commercial for decades.

1.4.3.1 Guar

A nonionic high-molecular-weight galactomannan water-soluble polymer is derived from the plant source. See Chapter 20 for details. Guar has myriad uses in diverse industrial and food applications. The major application that has contributed to its growth is in oil and gas well fracturing. Due to its structure and functionality such as high viscosity, easy cross-link ability, and breakdown with enzymes in the downhole environment, guar is uniquely suited in the fracking application. Low cost and reliable supply from India and Pakistan are also contributing factors. The first plant to make derivatized propylene-oxide-modified guar in the mid-1970s after the first oil shock had a capacity of 10 million pounds/year. This quickly grew to 100 million pounds by the end of that oil shock period. Today, guar consumption is approaching a billion dollar in sales and growing at 7.9% per year.

1.4.3.2 Xanthan gum

Xanthan gum is an extracellular anionic water-soluble polymer produced by fermentation (see Chapter 20 for details). It has excellent viscosifying and thickening properties of value in myriad food and industrial applications. No other polymer has commensurate properties and is hence used in large quantities even though the cost is relatively high. The market size is estimated at 960 million dollars/year growing at 6% per year.

1.4.3.3 Wood plastic composite (WPC)

A very successful wood biocomposite product introduced about 25 years ago for the building and construction industry. Its main use is in decking, as replacement for pressure-treated lumber. This product is cost-competitive and compatible with the existing plastic manufacturing process, infrastructure, and supply chain. Further, the product is more durable, aesthetically pleasing, and the manufacturers could provide 25 or longer lifetime warranty. Also, it does not have the issue of treated lumber with toxic chemicals and their leaching. Current sales of WPC are in several billion dollars/year and growing at a high rate. See Chapter 24 for details.

1.4.3.4 Polylactic acid (PLA)

PLA has been studied and under development for a long time due to its inherent biodegradability. However, it has some serious deficiencies in its properties and processing in the existing equipment. It has a long and circuitous history. Early work started in the late 1980s at BMI funded by Du Pont and Dainippon Inc. The work led to over 30 patents, where the main development was the stereoselective vapor-phase cyclization of L-lactic acid mixtures directly to L-lactide that effectively underwent ring-opening polymerization to L-polylactic acid. Unfortunately, the sponsors lost interest and did not move forward with the technology developed at BMI. The patents were acquired by Chronopol, a venture arm of Coors Beer company. They tried for a few years and the patents were acquired by Cargill, which later formed a joint venture with Dow to further develop and commercialize PLA. The venture exited PLA development and Cargill took on the task and challenge of further developing and commercializing PLA. Today, PLA is a multimillion pound/year commercial product with several players including Total Energies-Corbion (www.totalenergies-corbion.com), and the recent ADM-LG Chemicals announcement [10] with manufacturing plants spread across the globe (see Chapters 19, 20 and 21 for details).

The story of PLA shows challenges in introducing a new product, be it bio or otherwise, that has marginal properties and a tenuous value proposition, namely, biodegradability of unproven status in practical real-world environments. Further, there is a need for patience and the ability to weather the inevitable valley of death with financial and strategic commitments.

1.4.3.5 1,3-Propanediol (PDO)

This has been mentioned earlier and in Chapter 3. In this case, the fermentation approach using an engineered organism and low-cost sugars, as feedstock, led to cost-effective manufacturing of the monomer of value in the polyester and other industrial markets. Having commitment and support from a major company, namely, DuPont with extensive knowledge of the end use markets, customers and supply chain were factors in its successful commercialization. The estimated current global sales of bio-PDO are 250 million dollars/year.

1.4.3.6 Some key takeaways

A few lessons learned from the commercialization initiatives are worth noting, even though they may be obvious at least in hindsight. Some of them are highlighted:
- Commercialization of any new product including bioproducts is not for the faint of heart – it requires sustained financial resources and commitment.
- Having a clear and easy to translate value proposition to the end user in the marketplace is critical. Having a lower price as the main value proposition is a bit problematic, especially in mature and well-established replacement applications and markets. Relying on somewhat hard to quantify environmental benefits will not be sufficient but could be beneficial if price is equivalent.
- Bioproducts with enhanced functionality providing high-value performance attributes in final products/applications have a better shot.
- Compatibility with the existing infrastructure and supply chains and/or the ability to develop one facilitates commercialization and reduces external hurdles.
- As noted in a recent Barron's article (July 12, 2022), the transition to bio needs to hit a Trifecta:

- The sources that should replace fossil need to be simultaneously clean, affordable, and reliable.
- The transition, especially for energy, will require vast amounts of resources (new technology and infrastructure developments, manufacturing, supply chain, etc.) that may impact sustainability of renewables.
- The transition will require new partnerships, alliances, and business models.

Many bioproducts currently under development including those described in this book have not achieved this Trifecta, but many are getting closer. It will neither be easy nor a linear path but overall, many are on an encouraging trajectory.

1.5 Public awareness and policy advances related to bioproducts

One of the significant and major developments over the last decade or so is the increased awareness of the environmental movement and sensitivity of the harmful effects of global warming that threatens our future. Consumers are becoming more proactive in their purchasing behaviors, and the leading manufacturers of products are starting to pay increased attention in their product development and marketing strategies. Several major brand name companies in consumer products and others are partnering with raw material and chemical intermediate suppliers to drive the supply chains toward bioproducts.

At the public policy and legislative levels, there is a global trend to encourage products including bioproducts that address and mitigate greenhouse gas emissions and global warming. In some cases, tax and other financial incentives such as carbon credits are being tried to accelerate the transition to a green and sustainable economy. A very positive and encouraging recent development is the enactment of Inflation Reduction Act (IRA) passed by the US Congress in August 2022. Over $390 billion is allocated for renewables, including bioproducts and manufacturing.

The investment and financial community closely watched these developments and started to view this area of green and sustainable products and markets favorably as a long-term attractive option.

Overall, the public perception and support for bioeconomy are very bullish.

1.6 Some trends for bioproducts

A few major trends emerging in the field are listed:
- As synthetic biology with its rich and vast potential matures, more and more industrial chemical products/intermediates will be produced through biological

routes, perhaps, as a hybrid approach with a traditional thermochemical catalytic process.

–   Increased use of waste feedstock as carbon sources by taking advantage of potential negative raw material cost in the overall production cost of bioproducts. This is important as over 50%, in some cases even more, of manufacturing cost is from raw materials.

–   A stepwise progressive transition to green products seems to be evolving. This involves moving from current gray fossil polluting sources to blue (where the CO_2 is captured from the conventional fossil industry processes) and then to green, where no fossil carbon is used. A good example is current H_2 production (gray hydrogen) from steam reforming of methane that releases massive amount of CO_2 and moving to blue hydrogen which captures the polluting CO_2 derived from fossil sources, and finally making green hydrogen from water electrolyzers using solar and wind energy. The blue and green hydrogen can be used to make products such as blue and green ammonia (see Chapter 18 for details). It is estimated that global production of ammonia accounts for about 1% of greenhouse emissions. As hydrogen is used in numerous chemical processes, one can easily envision similar transitioning with other chemical products.

–   As discussed earlier under commercialization, there is increased attention to revisit business models in developing and commercializing bioproducts:

    –   The first one is to build capabilities and tools that can be deployed across promising and attractive products through strategic partnerships. A recent example is the acquisition of Zymergen, a strong biotech company, with product focus by Ginkgo Bioworks, a powerhouse in the field.

    –   The second is to develop product and process technologies with a strong patent portfolio and then license them for large-scale manufacture and strategic partnerships with potential end users.

    –   The third is to focus on low-volume specialty/fine chemical markets and high-value luxury niche applications.

–   Now there is recognition that it is important to incorporate how science and engineering affects citizens' safety and health. This has given rise to "science and technology studies" or variations thereof, and European Innovation and Patent regimes, considering these aspects in the development phase. NAS and NSF are active as well, as are funding organizations, for example, Technology in the Public Interest Program Strategy – MacArthur Foundation (macfound.org).

–   Transitioning to bioproducts, including fuels and energy, is not going to be a linear path but a bumpy and a painful one, costing a lot more than we are all used to.

In summary, prospects portend well for bioproducts.

References

[1] Spinard R. Increase in Atmospheric Methane Set Another Record During 2021. National Oceanic and Atmospheric Administration, U.S. Department of Commerce; April 7, 2022.

[2] Vijayendran BR. A commentary bio products from bio refineries-trends, challenges and opportunities. J Bus Chem 2010;7:3.

[3] www.icca.chem.org;ryan_baldwin@americanchemistryco.

[4] Top Value-Added Chemicals from Biomass Volume 1- Results of Screening for Potential Candidates from Sugars and Synthesis Gas. Northwest National Laboratory (PNNL), National; Energy Laboratory (NREL). Office of Biomass Program (EERE); Werpy, T and Petersen, G., Series editors, 2002.

[5] help@usdabioproducts.net.

[6] www.Biopreferredgov/BioPreferred/.

[7] El-Karoui M, Hoyos-Flight M, Fletcher L. Future trends in synthetic biology – a report. Front Bioeng Biotechnol 2019.

[8] The US Bioeconomy: Charting a course for a Resilient and Competitive Future. http://www.HTTPS://DOI.ORG/10.55879/D2HRS72WC.

[9] C&ENews. DeepMind releases Structure predictions for nearly every known protein, Aug 4, 2022

[10] C&ENews. LG Chem and ADM detail lactic acid plants, August 22, 2022.

Bhima R. Vijayendran, Ram Lalgudi
Chapter 2
Feedstock for bioproducts

2.1 Introduction

Feedstock in the development and commercialization of bioproducts is derived from various sources. Useful and economically viable ones have several critical features such as low cost, reliable supply chain, amenable to industrial-scale conversion processes and functionality that can impart properties and performance in the finished bioproducts of value in the marketplace. Historically renewable biomass from agriculture and forest products has been the main stay of feedstock to produce chemical intermediates and products. Over the last 100 years or so, the feedstock picture has changed and shifted to nonrenewable fossil-based sources. The main drivers are abundant supply and low cost and advances in process engineering including catalyst technology that propelled its dramatic growth. Today, only about less than 10% of global industrial chemical market is based on renewable feedstock sources. As noted in Chapter 1, the picture is changing again due to numerous factors such as increased awareness and concerns from greenhouse gas emissions from fossil sources, progress in science and technology related to renewable-based feedstocks, process advances and catalyst innovations, and regulatory and public policy pressures.

In this chapter, an overview of feedstock for bioproducts is covered. Specifically, the following sources of feedstock are reviewed:
Plant seed oils
Sugars
Carbohydrates
Lignin
Waste and coproducts
Miscellaneous

2.1.1 Feedstock overview

An overview of fossil-derived and bioderived feedstock is shown in Figures 2.1 and 2.2. Figure 2.1 highlights a few characteristics of fossil-based and bio-based feedstocks. Less than 10% of fossil oil and gas production is used in the manufacture of petrochemical products.

Bhima R. Vijayendran, Redwood Innovation Partners LLC, Carlsbad, CA 92011,
e-mail: bhima@redwdinnov.com
Ram Lalgudi, Aries Science and Technology, LLC, Columbus, OH 45701, e-mail: lalgudir@ariesst.com

https://doi.org/10.1515/9783110791228-002

Features	Fossil	Bio
Resource Distribution	A few regions	More distributed
Reserves	Finite	Potentially infinite
Sustainability	Increasingly challenging	Manageable
Current Global Use	Over 35 billion barrels/year	Less than 6-10 billion boe*/year
Production Technology	Well established	Emerging
Infrastructure	Well established	Emerging
Price	About $100/ barrel	About $20/ boe**
Chemistry	Simple hydrocarbon	Complex oxygenated hydrocarbons
Conversion Processes	Catalytic thermochemical-well established	Emerging
Environmental Foot Print	High	Low

*boe = barrel of oil equivalent
**$60/ on of dry biomass

Figure 2.1: Some features of fossil and biofeedstock.

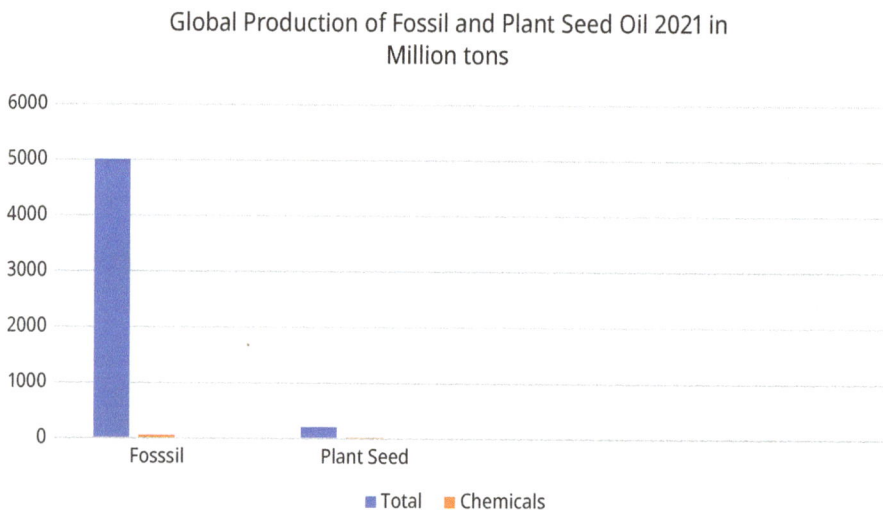

Figure 2.2: Indicative relative scale of global fossil and plant oil production and utilization in 2022 [1].

2.1.2 Plant seed oils

Global production of vegetable and plant oils is over 200 million tons/year [1]. There are many sources of plant oils from various parts of the world. The main oils of commercial interest are soybean, palm, sunflower, canola and castor oil. There are some specialty oils such as *Vernonia*, cashew nut and emerging *Camelina* oil.

A few of these are shown in Figures 2.3–2.8.

Figure 2.3: Soybean (courtesy of Merideth at https://unsplash.com/).

Figure 2.4: Palm (courtesy of Szewczyk at https://unsplash@com/).

Figure 2.5: Sunflower (courtesy of Bonnie Kittel at https://unsplash.com).

Figure 2.6: Castor seed (courtesy of CDC at https://unsplash.com/).

Figure 2.7: *Camelina* (courtesy of Christina Eynck at Canada Government).

Figure 2.8: Canola (courtesy of Michael Ohey Sekara at http://unsplash.com/).

2.2 Composition of plant oils

Chemical composition of the three significant seed oils, namely, soybean, palm and sunflower, is shown in Table 2.1.

Table 2.1: Triglyceride compositions and functionality [2].

Triglyceride	Soybean oil	Palm oil	Sunflower
Stearic	6.5%	5.4	4.2
Palmitic	12.2	39.6	5.6
Oleic	29.1	42.9	26.2
Linoleic	44.1	8.9	64.1
Linolenic	2.7		
Higher chain length	<1%	<0.5%	<1%

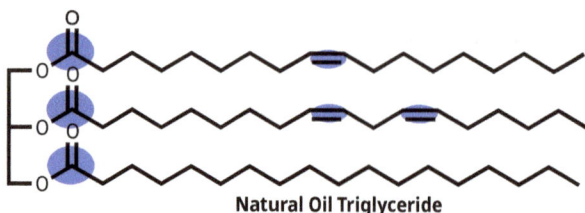

Natural Oil Triglyceride

Figure 2.9: Functionality.

2.2.1 Functionality modifications

Plant oils are quite versatile in their functionality being amenable to a myriad of chemical modifications and functionalization using cost-effective industrial processes and existing infrastructure (Figure 2.9).

Various chemical modifications are extensively covered in the literature [3–7]. A few of them are highlighted further. A schematic of this is shown in Figure 2.10.

An area worth noting is the use of plant breeding and recently introduced genetically modified soybean oil with high oleic acid content. Having a high-level oleic acid and concomitant reduction in saturated fatty acids imparts useful functionality for chemical conversion to several industrial bioproducts such as bioplasticizers, biolubricants, surfactant, and coatings [8].

Hydrogenated vegetable oil

Hydrogenation

OH

OH

Azelaic acid   Ozonolysis   Epoxidation

OH

Pelargonic acid

Triglyceride (vegetable oil)   Epoxidized vegetable oil

Transesterification

Fatty acid methyl ester (biodiesel)

Hydrogenation and   Hydrogenation and
Self metathesis   Cross metathesis

1-Octadecene

9-Octadecenedioicacid 1,18-dimethyl ester

9-Octadecene   Methyl 9-decenoate

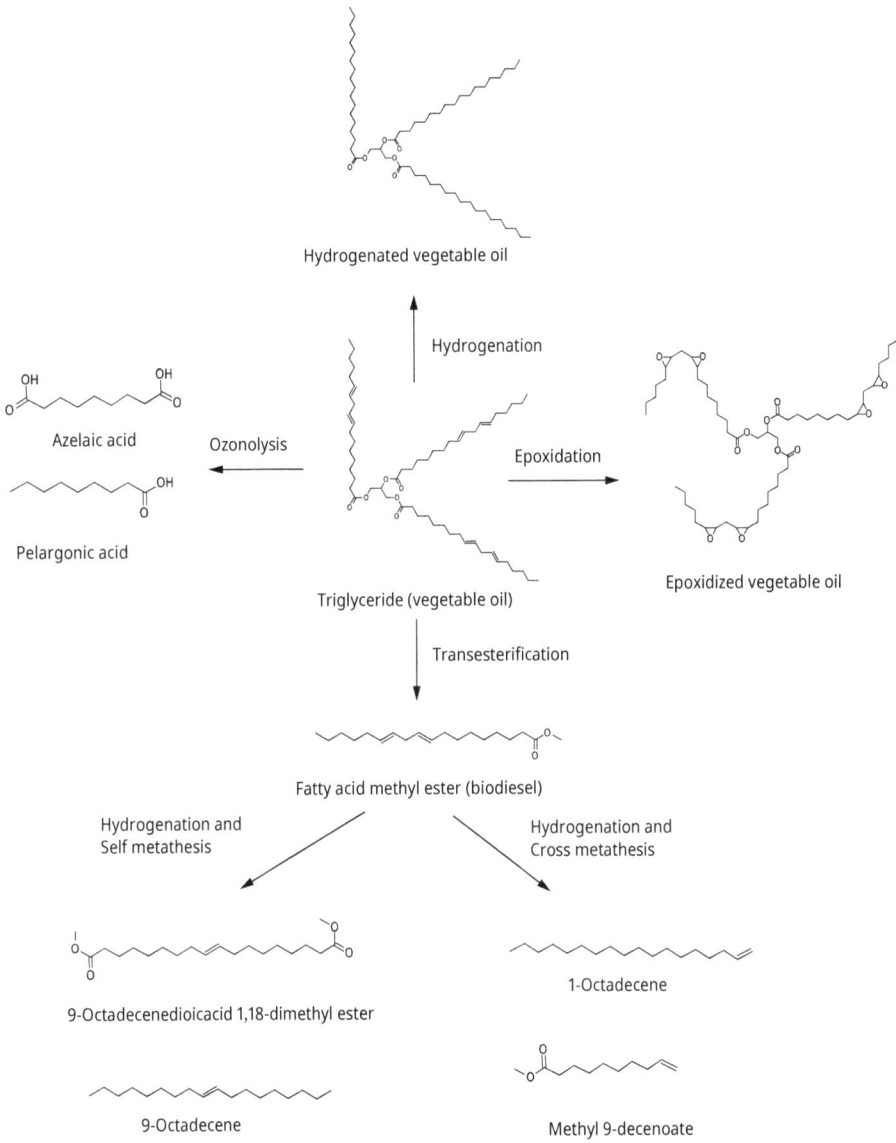

Figure 2.10: Overview of triglyceride functionalization and chemistry [3–7].

Myriad of useful and commercial industrial products such as plasticizers, surfactants, lubricants, coatings and adhesives, polymers, and plastics have been successfully developed from plant oils. Chapters 6, 8, 10, 15 and 16 cover some of these bioproducts.

2.2.2 Specialty oils

A few specialty oils are also of interest and value commercially. They are covered further.

2.2.2.1 Castor oil

Castor is a nonfood industrial crop mainly cultivated in Africa, South America, and India. It grows well with high oil content in tropical regions with warm climate conditions, typically below 38 °C. Frosting weather is known to adversely affect the plant growth. Attempts to grow and process castor in the United States have not been successful due to unfavorable climate for castor plantation, high labor cost, undesirable oil yield and the process safety associated with the ricin toxicity, among others [9].

It is a vegetable oil obtained by pressing the seeds of the castor oil plant (*Ricinus communis* L.), that is, castor oil is a nondrying oil with an unique triglyceride derived from 12-hydroxy-9-*cis*-octadecenoic acid or commonly referred to as ricinoleic acid (RA; Figure 2.11). The high content of RA triglyceride in castor oil allows them to be directly used as polyol for making coatings [10], foams [11] and surfactants [12–14].

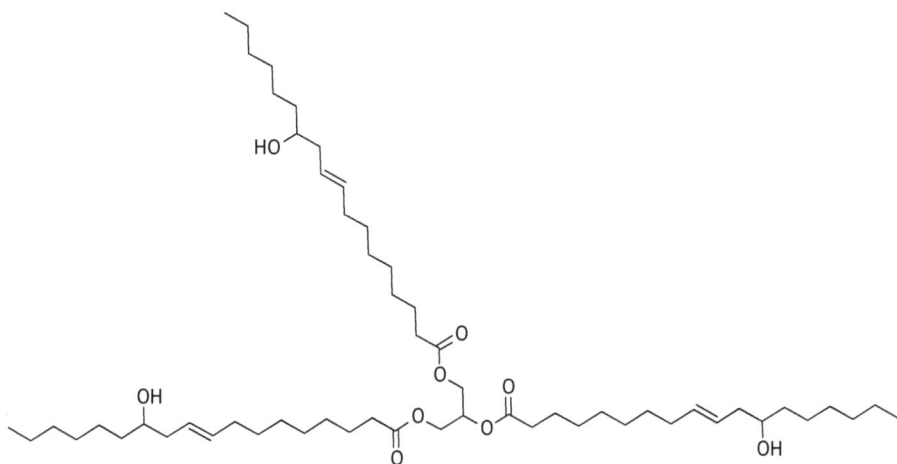

Figure 2.11: Chemical structure of castor oil.

2.2.2.2 *Vernonia* oil

Vernonia is an industrial crop that can grow as weeds in fields or in woodlands of Africa under wide agroecological conditions [15]. *Vernonia* crop has naturally occurring epoxy fatty acids in the triglyceride (Figure 2.12), which makes them an economically attractive candidate for plasticizer [16]. Typically, plasticizers are generally derived from epoxidation of soybean oil. Furthermore, the epoxide ring in the *Vernonia* oil can be hydroxylated, and the resulting polyol can be used to make biobased polyurethane coating [17, 18]. The wide industrial application of neat *Vernonia* oil has created significant interest in domestication of *Vernonia* as an industrial oil seed crop in the United States [19]. *Vernonia* oil is still not available at scale for industrial use, and significant improvement in agronomics is needed for growing *Vernonia* crop in the United States.

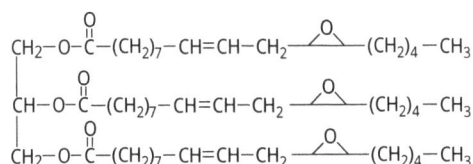

Figure 2.12: Idealized representations of *Vernonia* oil (structure adopted from Trumbo et al. [6]).

2.2.2.3 Cashew oil

Cashew nutshell liquid (CNSL) is a dark brown viscous liquid which is a very important agricultural byproduct of cashew nut production. Natural CNSL is a mixture of phenolic compounds [20] (Figure 2.13) with aliphatic side chains, and typically comprises 70% anacardic acid, 5% cardanol and 18% cardol. Solvent treatment followed by vacuum distillation is the most used extraction method to produce CNSL [21].

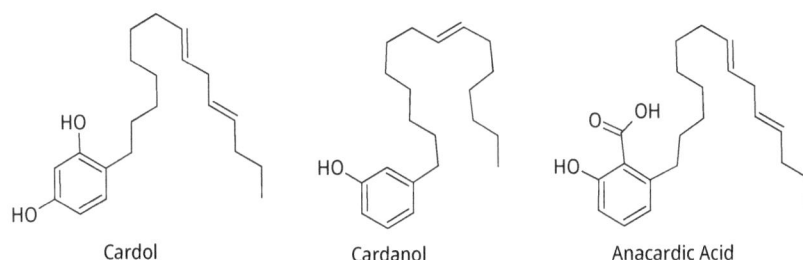

Cardol   Cardanol   Anacardic Acid

Figure 2.13: Typical chemical composition of oils extracted from cashew nuts.

CNSL finds several industrial uses, owing to its attractive chemical structure. The phenolic moiety has been utilized to prevent oxidation [22] and corrosion [23] and to produce thermosetting resins by reacting with formaldehyde [24]. The hydrophobic structure of cardanol resembles petroleum-derived nonylphenol. Ethoxylation of cardanol produces a biobased nonionic surfactant that can replace toxic petroleum-derived nonylphenol-based nonionic surfactants [25].

2.2.3 Biotechnology advances

Advances in synthetic biology and industrial biotechnology have started to play a bigger role in bio-derived feedstocks.

Plant biotechnology has played and continues to play a bigger role in crop and plant-based feedstocks. The impacts have been in several areas such as increased yield, lower input costs, improved drought and insect resistance. One area of particular interest from feedstock perspective is trait modifications and increased functionality of value in myriad bioproduct applications.

2.3 Sugars and carbohydrates

Simple C_5 and C_6 sugars (Figure 2.14) and complex ones provide a reliable source for renewable feedstock for bioproducts.

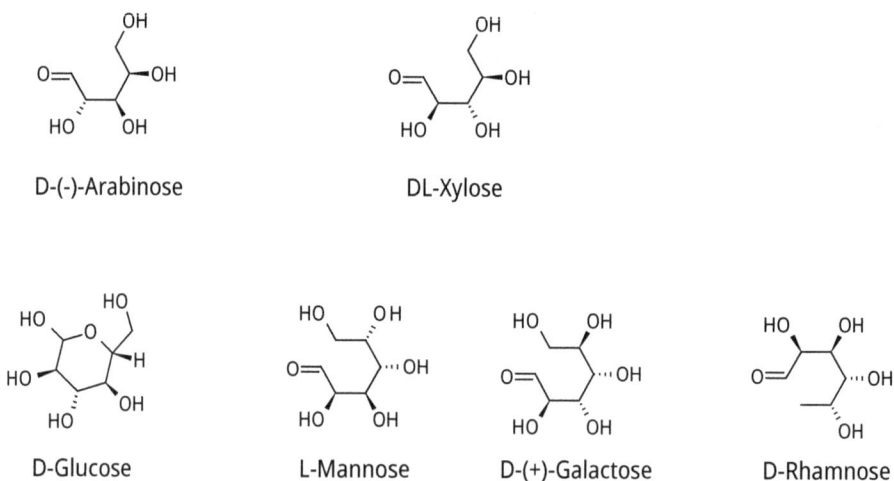

Figure 2.14: Structures of simple sugars.

A study by the Department of Energy (DOE) () [26] highlighted top 11 chemicals that could be derived from sugars through conventional catalytic thermochemical and emerging bioprocessing schemes

Some of these target molecules such as 1,3-propanediol, propylene glycol and 1,3-butanediol have moved on to full commercial scale. Others such as acrylic acid and glucaric acid are still in the development stage. A few are covered in Chapter 9 and 19.

Sugar and carbohydrate-based feedstocks have been converted to functional C_4 intermediates such as succinic acid as platform intermediates for further modifications to useful chemical products.

A couple based on succinic acid and furans are shown later from reference [26] (Figures 2.15 and 2.16).

Chapter 5 on furanic-based fiberglass binders has some recent updates on the furanic platform.

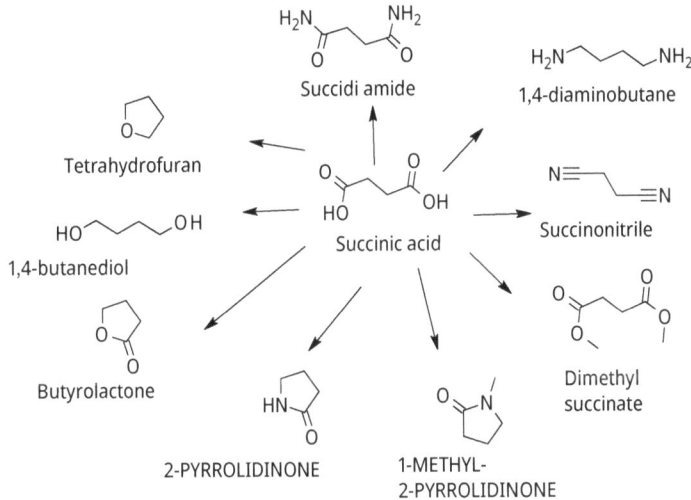

Figure 2.15: Succinic acid platform [26].

Some recent works have successfully converted simple sugars to high-value chemical intermediates such as fatty acid and hydroxy long-chain molecules using engineered *E. coli* in a fermentation process [27]. Start-up companies are developing high-value plant oils for food and nutraceutical applications with engineered organisms in bioreactors, touted as zero-farmed oils.

Furans also demonstrate the potential of a platform molecule

Derived from biomass, 5-hydroxymethylfurfural (HMF) can be converted into many types of compounds now obtained from petroleum sources.

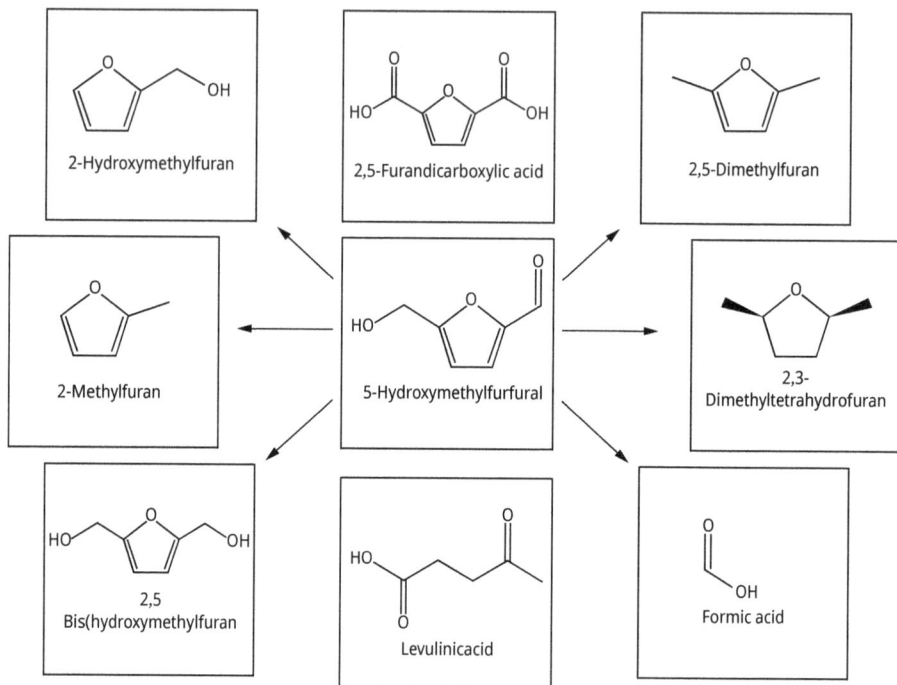

2-Hydroxymethylfuran

2,5-Furandicarboxylic acid

2,5-Dimethylfuran

2-Methylfuran

5-Hydroxymethylfurfural

2,3-Dimethyltetrahydrofuran

2,5 Bis(hydroxymethylfuran

Levulinicacid

Formic acid

Figure 2.16: Furanic platform [26].

2.4 Lignin

A ubiquitous biomass source is available in large quantity having a complex condensed polyphenolic structure. Lignin is a natural phenolic-based polymer that is abundant in the cell walls of plants, especially in land-based plants. It provides mechanical support for plant growth, as well as assisting in the long-distance transport of water in xylem, resisting a variety of stresses, and providing protection against pathogens. The chemical structure of native lignin comprises three p-hydroxycinnamyl alcohol monomers, referred to as monolignols. These are p-coumaryl alcohol (p-hydroxyphenyl, H unit), coniferyl alcohol (guaicyl, G unit) and sinapyl alcohol (syringyl, S unit), which are characterized by their degree of methoxyl substitution at the third and fifth carbon (C_3 and C_5) positions of the aromatic ring, that is, p-coumaryl alcohol has no methoxyl group, whereas coniferyl alcohol has a methoxyl group at C_3 position, and sinapyl alcohol has a methoxyl group at both C_3 and C_5 positions. The most predominant intermonomer linkage within lignin gives rise to its complex three-dimensional amorphous network structure called phenyl-propane β-aryl ether (β-O-4). Other interunit linkages that have been identified to date are biphenyl and dibenzodioxocin (5-5), phyenylcoumarin (β-5), 1,2-diaryl propane (β-1), phenylpropane α-aryl ether (α-O-4), diaryl ether (4-O-5) and β–β-linked structures (β–β). See Chapter 7 for additional details on lignin structure and chemistry.

Despite lignin's low cost and huge supply, its use in bioproducts has been extremely limited despite numerous attempts and substantial R&D investments by large companies, academic institutions and others. This is mainly due to the rather complicated and involved process in deconstructing lignin structure to useful building blocks at scale in an economic way.

A few examples of lignin-based bioproducts are in lignin-derived carbon fiber chapter 7.

2.5 Miscellaneous

Glycerin, a coproduct in the manufacture of soaps and biodiesel manufacture, shows promise as a cost-effective building block for several industrial products such as polyols [28], propylene glycol and epichlorohydrin (Figure 2.17).

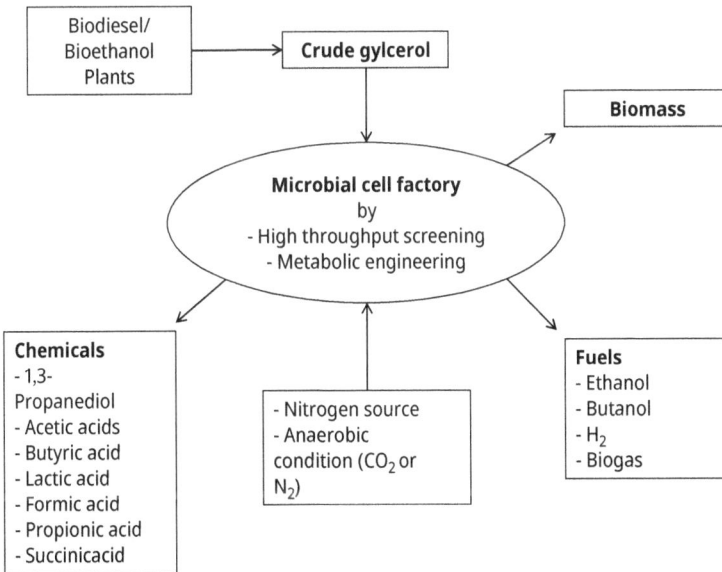

Figure 2.17: Glycerin platform.

2.6 Waste-derived feedstock

There is increased interest in converting industrial waste such as flue gas emissions, garbage and plastics to useful products. A recent article in *C&E News* (June 2, 2002) highlights the production of syngas from trash.

Genetically modified *E. coli* has been used to convert flue gas into isopropanol [29].

Some recent work done with algae shows the production of algal oils similar in composition to canola oil from flue gas feedstock in a photobioreactor [30].

There are other initiatives to convert methane from waste dump sites to useful chemicals using engineered organisms in a fermentation unit.

These and other efforts to valorize waste streams (see *C&E News* June 2, 2022) provide a lot of benefits to address greenhouse emissions as well as waste issues by converting them to useful products. In addition, if there is carbon credit, the economics for such schemes become extremely attractive and viable.

This brief survey clearly shows that there are several attractive and commercially viable renewable feedstocks for the emerging sustainable green economy.

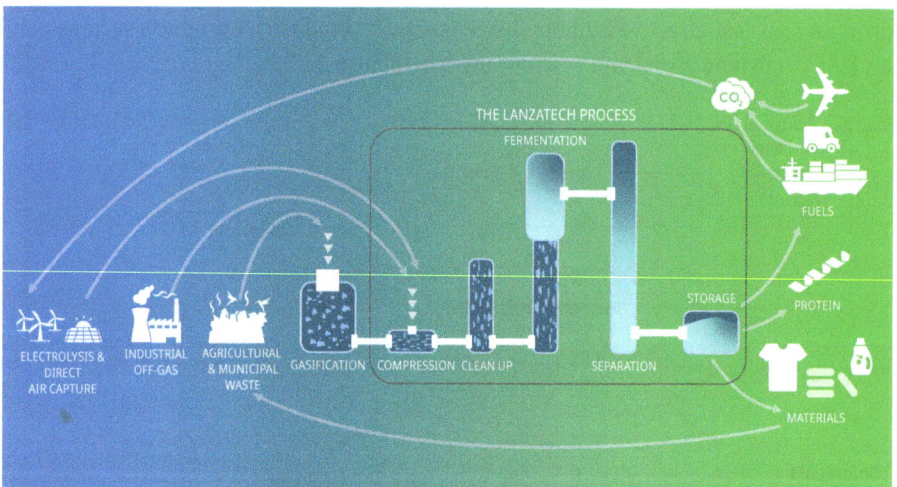

Figure 2.18: LanzaTech's Process Courtesy of Lanzatech.

References

[1]   Statista Research Department, Feb 11, 2022, and https://Www.Ees/Org.

[2]   Alves AQ, da Silva VAJ, Góes AJS, Silva MS, de Oliveira GG, Bastos IVGA, de Castro Neto AG, Alves AJ. The fatty acid composition of vegetable oils and their potential use in wound care. Adv Skin Wound Care 2019;32(8).

[3]   Armylisas AHN, Hazirah MFS, Yeong SK, Hazimah AH. Modification of olefinic double bonds of unsaturated fatty acids and other vegetable oil derivatives via epoxidation: A review. Grasas Y Aceites 2017;68:174.

[4]   Tang Q. Bio-based epoxy resin from epoxidized soybean oil. In: Chen Y, editor. Soybean, Qian Li, Translator; Huizhi Gao, Series Ed., Rijeka: IntechOpen; 2019p. Ch 8. https://doi.org/10.5772/intechopen.81544.

[5]   Graiver D, Patil M, Narayan R. Recent advances in ozonation of vegetable oils. Recent Pat Mater Sci 2010;3(3):203–18. https://doi.org/10.2174/1874464811003030203.

[6]   Marvey BB. Sunflower-based feedstocks in nonfood applications: Perspectives from olefin metathesis. Int J Mol Sci 2008;9(8). https://doi.org/10.3390/ijms9081393.

[7]   Petrović ZS, Javni I, Ionescu M. Biological oils as precursors to novel polymeric materials. J Renewable Mater 2013;1:167–86.

[8]   Heggs R. Chapter 10 – Industrial Uses Of High-Oleic Oils. In: Flider FJ editor, High Oleic Oils, AOCS Press; 2022, pp. 245–59. https://doi.org/10.1016/B978-0-12-822912-5.00009-5.

[9]   Patel VR, Dumancas GG, Kasi Viswanath LC, Maples R, Subong BJJ. Castor oil: Properties, uses, and optimization of processing parameters in commercial production. Lipid Insights 2016;9:1–12. https://doi.org/10.4137/LPI.S40233.

[10]   Thakur S, Karak N. Castor oil-based hyperbranched polyurethanes as advanced surface coating materials. Prog Org Coat 2013;76(1):157–64. https://doi.org/10.1016/j.porgcoat.2012.09.001.

[11]   Wang HJ, Rong MZ, Zhang MQ, Hu J, Chen HW, Czigány T. Biodegradable foam plastics based on castor oil. Biomacromolecules 2008;9(2):615–23. https://doi.org/10.1021/bm7009152.

[12]   Zhou J, Sun Y, Zhu K, Di Serio M, Zhang Y, Sun J, Wu H, Ding L, Liang H. Influence of sulfonic acid group on the performance of castor oil acid based methyl ester ethoxylate sulfonate. Null 2018;39 (12):1693–98. https://doi.org/10.1080/01932691.2018.1461641.

[13]   Zhang Q, Sun Y, Zhi L, Zhang Y, Di Serio M. Properties of ethoxylated castor oil acid methyl esters prepared by ethoxylation over an alkaline catalyst. J Surfactants Deterg 2015;18(2):365–70. https://doi.org/10.1007/s11743-014-1657-3.

[14]   Zhang -Q-Q, Cai B-X, Xu W-J, Gang H-Z, Liu J-F, Yang S-Z, Mu B-Z. Novel zwitterionic surfactant derived from castor oil and its performance evaluation for oil recovery. Colloids Surf A Physicochem Eng Asp 2015;483:87–95.https://doi.org/10.1016/j.colsurfa.2015.05.060.

[15]   Baye T, Becker HC. Exploration of *Vernonia galamensis* in Ethiopia, and variation in fatty acid composition of seed oil. Genet Resour Crop Evol 2005;52(7):805–11. https://doi.org/10.1007/s10722-003-6086-5.

[16]   Krewson CF, Riser GR, Scott WE. Euphorbia and vernonia seed oil products as plasticizer-stabilizers for polyvinyl chloride. J Am Oil Chem Soc 1966;43(6):377–79. https://doi.org/10.1007/BF02646792.

[17]   Trumbo DL, Rudelich JC, Mote BE. Applications of Vernonia oil in coatings. In: Janick J editor, Perspectives on New Crops and New Uses, Alexandria, VA: ASHS Press; 1999, pp. 267–71.

[18]   Dirlikov S, Frischinger I, Islam MS, Lepkowski TJ. Vernonia oil: a new reactive monomer. In: Gebelein CG, editor. Biotechnology and Polymers, Boston, MA: Springer; 1991.

[19]   Thompson AE, Dierig DA, Johnson ER, Dahlquist GH, Kleiman R. Germplasm development of Vernonia galamensis as a new industrial oilseed crop. Ind Crops Prod 1994;3(3):185–200. https://doi.org/10.1016/0926-6690(94)90066-3.

[20] Morais SM, Silva KA, Araujo H, Vieira IGP, Alves DR, Fontenelle ROS, Silva AMS. Anacardic acid constituents from cashew nut shell liquid: NMR characterization and the effect of unsaturation on its biological activities. Pharmaceuticals 2017;10(1). https://doi.org/10.3390/ph10010031.

[21] Nyirenda J, Zombe K, Kalaba G, Siabbamba C, Mukela I. Exhaustive valorization of cashew nut shell waste as a potential bioresource material. Sci Rep 2021;11(1):11986. https://doi.org/10.1038/s41598-021-91571-y.

[22] Andradedos de T. J. A. S.Araújo BQ, Citó AMDGL, da Silva J, Saffi J, Richter MF, Ferraz ADBF. Antioxidant properties and chemical composition of technical cashew nut shell liquid (TCNSL). Food Chem 2011;126(3):1044–48. https://doi.org/10.1016/j.foodchem.2010.11.122.

[23] Ghosh T, Karak N. Cashew nut shell liquid terminated self-healable polyurethane as an effective anticorrosive coating with biodegradable attribute. Prog Org Coat 2020;139:105472. https://doi.org/10.1016/j.porgcoat.2019.105472.

[24] Chuayjuljit S, Rattanametangkool P, Potiyaraj P. Preparation of cardanol–formaldehyde resins from cashew nut shell liquid for the reinforcement of natural rubber. J Appl Polym Sci 2007;104 (3):1997–2002. https://doi.org/10.1002/app.25915.

[25] Roy A, Fajardie P, Lepoittevin B, Baudoux J, Lapinte V, Caillol S, Briou B. CNSL, a promising building blocks for sustainable molecular design of surfactants: A critical review. Molecules 2022;27(4). https://doi.org/10.3390/molecules27041443.

[26] Top Value Added Chemicals from Biomass Volume I – Results of Screening for Potential Candidates from Sugars and Synthesis Gas Produced by the Staff at Pacific Northwest National Laboratory (PNNL), National Renewable Energy Laboratory (NREL), Office of Biomass Program (EERE); Werpy, T., Petersen, G., Series Eds., 2004.

[27] Del Cardyre SB, Brubaker S, Keasling JD. Modified Organisms and Uses Therefor, U.S. Pat 8,535,916, Feb 13, 2007.

[28] Benecke HP, Vijayendran BR, Garbark, Mitchell KP. Low cost and highly reactive polyols: A co-product of the emerging biorefinery economy. Clean- Air Water 36(8):694–99.

[29] Jewett M. Carbon -negative production of acetone and isopropanol by gas fermentation at industrial pilot scale. Nat Biotechnol 2022. doi: 10:1038/S41587777-021-01195-w. Www.Nature.Com/Articles/S441587-021-01195-w.

[30] Vijayendran BR, Randall M, Schmid E. Algal Oil Based Bio-Lubricants. U.S. Pat 9,458,407 Issued Oct 4, 2016.

Michael A. Schultz

Chapter 3
Process scale-up for bioproducts: enabling the emerging circular economy

Abstract: This chapter provides guidelines and best practices for scale-up of bioproducts and bioprocessing technology, with the aim of reducing the time, cost, and risk of scale-up. Experimental planning, modeling, and technoeconomics along with early-stage engineering are critical to this effort. Similarities and differences in the scale-up of bioproducts compared to the scale-up of more conventional thermochemical processes using petroleum-based feedstocks will be presented, along with a case study and commercial success stories for the scale-up of bioproducts.

3.1 Introduction

We have seen a growth in bioproducts over the last number of years [1]. The drivers for this growth include an increasing focus on sustainability, with the goal to reduce emissions of CO_2 and other air and water pollutants, and to enable circularity through reuse of carbon and carbon-based products. Some cases offer the potential for a lower cost of carbon, through use of waste feedstocks from industrial gaseous emissions, end-of-life plastic, and waste biomass. In addition, in some cases, the bioproduct is actually a better product than the petroleum-based version, coming with a cheaper, safer processing route and performance advantages over traditional materials.

While fossil fuels still dominate, bioprocessing represents an opportunity to be a key piece of the puzzle to reduce CO_2 emissions from fossil-based products. Market transformation is only possible by continuing to push boundaries, including:

- Growth in installed capacity – Production of biodiesel, renewable diesel, and ethanol [2] is increasing dramatically [3]. Sustainable Aviation Fuel (SAF) is an emerging alternative fuel [4]. Markets, supply chains, equipment manufacture, and technology know-how in this space are following suit.
- Emergence of tools to support industrial bioprocessing – Advanced tools are emerging, mirroring the past advances in thermochemistry such as advanced analytics, high-throughput screening, and genetic tools for custom microbes.
- Scale-up methodologies – These are critical to advancing existing technologies and developing the next generation.

Michael A. Schultz, Managing Director, PTI Global Solutions LLC, Indianapolis, IN,
e-mail: mike@ptisolns.com

https://doi.org/10.1515/9783110791228-003

3.2 Challenges with scale-up of emerging bioprocessing technology

Scale-up of emerging technology for production of bioproducts faces many hurdles along the way. One of these is the need to reduce technology risk through measured approaches of increasing scale, starting at the lab or bench scale (perhaps with several lab scales from milliliter quantities on plates or in shake flasks, to 1 L CSTRS and then 3–10 L CSTRS), moving to a pilot, then demonstration, and finally commercial. Of course, not only are the volumes increasing but the time to build each scale, and the capital cost associated with each scale increases dramatically as well (see Figure 3.1).

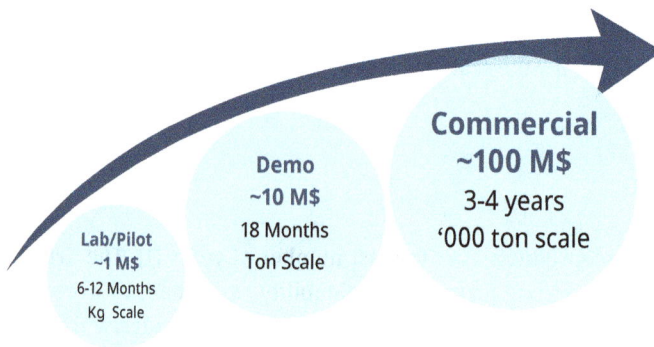

Figure 3.1: Typical scale-up stages.

Often a key challenge is to significantly *reduce* cost even while scaling up. Finally, because the first commercial scale is typically small relative to the conventional technology, it is difficult to compete on a unit cost basis with incumbent products because of the higher capital cost factor.

Bioprocessing also brings scale-up challenges that are unique to this area relative to conventional fuel and chemical technology:

- **Lack of established data and models:** For example, bacterial growth kinetics and flux models.
- **New separation challenges:** CO_2 from air and intercellular bioproducts.
- **New optimization criteria:** Carbon footprint and other environmental, social, and governance and life cycle assessment (LCA) metrics.
- **Bio-based catalysis:** Biological catalysts, whether using enzymes or whole cells, require different handling than conventional heterogeneous catalysis using zeolites and other inorganics.

3.3 Commercial-scale project development: bioprocessing technology

The project development cycle for an established process technology is well known with an initial conceptual design phase to define the project, develop a block flow diagram, and generate a cost estimate that is typically ±50%. Basic engineering, detailed engineering, procurement and construction, and start-up then follow.

The conceptual design phase for an established technology can generally be completed in 2–4 months. However, for a new bioprocessing technology, much more time is needed to get this right. To do this, creative process engineering should be brought into the picture as early as possible, even during discovery research and development (R&D). In fact, by starting with the conceptual design, or process concept, this framework can be used to drive new technology development, scale-up, and commercialization. This process concept is not set in stone and, in fact, should be reviewed and updated throughout the scale-up effort.

Ultimately, this approach must result in a design to move into the basic engineering phase while reducing technical risk and optimizing economics. The process concept is established to drive the scale-up effort, not just inform it. This then enables the innovation to be directed to areas that create the greatest value from breakthrough and disruptive ideas, by identifying challenges early, failing fast when it is cheaper and quicker, and making sure efforts are focused on solving commercially relevant problems.

The process concept is developed, and frequently revised, through a combination of creative process engineering, multiscale experimental data, and modeling and analysis.

Creative process engineering: The flow scheme is developed, the material balance is estimated, and key process design decisions are identified to establish the best process flowsheet for the technology.

Modeling and analysis: A good model can save time and resources in the lab. Coupled with the right analysis, the scale-up team can prioritize objectives in the lab, pilot, and demo units. As a cautionary note, useful models are more important than perfect models.

Experimental data: The right data is needed to prove out breakthrough ideas, secure partners and investors, and develop engineering data for equipment design. Multiscale data is critical to this effort, and with good planning, multiple assets and external resources can be leveraged.

The key benefits to this approach are:
- Prioritization of R&D to de-risk and optimize a new technology
- Identification of cost reduction opportunities throughout the scale-up effort
- Anticipation of engineering needs as early as possible

In this way, the team can reduce risk and optimize economics of the new design, while efficiently managing the time, cost, and deployment of limited technical and financial resources of its efforts.

3.4 Key element no. 1: creative process engineering

Creative process engineering (see Figure 3.2) is critical for success for first-of-its-kind technology. Creative process engineers understand the commercial plant design and can also accommodate the ambiguity that is common with any new technology. This creativity enables the engineers working closely with the science experts to develop the process concept, establish the material balance, and make key process design decisions to set the framework for the evolving novel technology. These decisions fall into the following categories:

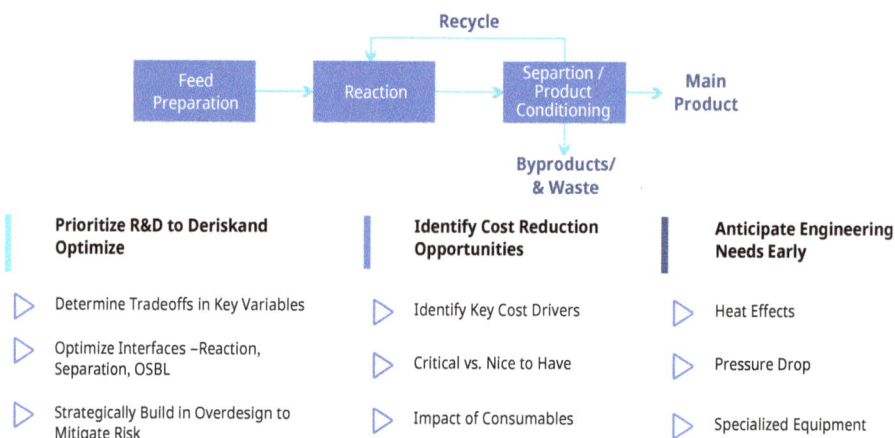

Figure 3.2: Creative process engineering.

– **Product requirements**: product quality, waste versus byproduct, and batch versus continuous
– **Catalyst**: Composition, biological versus thermochemical, size/shape, and heterogeneous versus homogeneous
– **Major unit operations**: reactor concept, feedstock processing, and separation processes
– **Major equipment**: standard or custom, and pump/exchanger/compressor type
– **Design conditions**: temperature, pressure, and product specifications

The challenge of translating discoveries from the lab into viable process flowsheets has been described by Douglas [5] to require assumptions that fix (1) parts of the process flowsheet, (2) some of the design variables, and (3) connections to the environment. Douglas estimates that more than 1 million process flowsheets can be generated just from the varied assumptions associated with the first process flowsheet. Clearly, it is not feasible to evaluate all of these alternatives. The good news is that the number of alternatives can be quickly reduced to a more manageable number but good engineering judgment is needed to make decisions with relatively little information. This is where the *creative* aspect of process engineering is critical.

In practicality, it is best to identify the reactor concept and separation scheme that are the best options, and then build the flow scheme around these. Often, this is a case of screening out the "bad options," resulting in several process concepts that make sense for further evaluation.

Separations for bioprocesses (often called downstream processing) bring particular challenges as compared to conventional chemical and petrochemical processes, especially for bioproducts with high-value and/or strict purification requirements (nutraceuticals, food, monomers, etc.). Typically, product isolation and purification require several processing steps, which add cost to the flowsheet, and the potential for product loss in each step. This is often underestimated at the planning stage, leading to added cost and development time. Scoping these needs out early and evaluating optimization with the reaction section can increase the chances of scale-up success.

The following represents a process design algorithm that has proven to be successful in the scale-up of new bioprocessing technology:

- Determine system requirements
- Size reactor system
- Size supporting equipment (pumps, heat exchangers, etc.)
- Integrate separation system
- Develop cost model
- Evaluate trade-offs and set targets for further data generation

3.5 Key element no. 2: efficient experimentation

Experimental data is clearly the lifeblood of any new technology. Getting data to prove out an invention can be the key to obtaining an important patent, generating early-stage investment, and securing key partnerships. However, generating data is expensive and time-consuming, particularly as scale increases, making it critical to ensure that the right data is generated to make the best use of available resources.

The entire scale-up effort can be viewed as one integrated data-gathering exercise, with the overall goal of generating the necessary data to define the commercial process design. Along the way, data is also needed to demonstrate a reduction in technical risk and allow optimization of the process economics. This is a bit of a different mindset from trying to prove out a "result" at each scale (e.g., proving conversion of raw materials A and B into product C with the desired efficiency X in the lab, then the lab–pilot, the pilot, and finally the demo).

Upfront planning can ensure that the right data is gathered. This planning effort will yield a scale-up plan with experiments designed to generate the necessary design data and identify the parameters that have the greatest impact on economics and technical risk. In fact, the product of this effort is data rather than the bioproduct itself.

A key aspect of this early-stage planning is decoupling these parameters, understanding that "science parameters" such as reaction kinetics and separation factors can, and should, be explored at the lab stage. Conversely, a lab-scale effort to evaluate issues related to heat and mass transfer or pressure drop will be a futile effort at best, leading to inconclusive or even incorrect results and is best done at a larger scale. This decoupling is illustrated in Table 3.1.

Table 3.1: Decoupling scale-up parameters.

	Chemistry/Biology parameters	Engineering parameters
	Process inputs (temperature, pressure), Kinetics, Contaminants, etc	Heat Transfer, Mass Transfer, Hydrodynamics, Fouling, Etc.
Lab Scale	X	
Lab-Pilot Prototype	X	X
Pilot Scale		X
Large Demo/Commercial		X

Multiscale data is beneficial for many additional reasons:
- Model development: Data at multiple scales enables generation of robust models for process development and equipment design.
- Troubleshooting: The smaller lab and pilot rigs can be instrumental to troubleshooting challenges in the larger units. If possible, it is worth the investment to keep these smaller units operating in support of the larger scale operations.
- Continuous improvement: This is often needed while scaling a new technology to meet aggressive timelines and cost targets. These improvements can be identified and scaled in parallel to ensure that the first commercial unit has the benefit of the learning from several generations of technology improvements that are identified and de-risked in multiscale operations.

3.6 Key element no. 3: modeling and analysis

Modeling and analysis is used to set targets for economic and sustainability performance, encapsulate experimental data into engineering models, and design process
equipment. However, it is critical to recognize the limitations of models. British statis-
tician George Box liked to say that all models are *wrong*, but some are *useful* [6]. For
scale-up purposes, models should be useful tools to support process development,
scale-up, and design, rather than exact replications of the system in question.

An effective ("useful") modeling effort typically starts off with something simple,
adding more detail as scale-up progresses. A simple overall mass balance using a
spreadsheet is a great place to start. Additional detail is later added to this simple
model, with additional types of models developed depending on the requirements. Examples of additional types of useful models include:

- Kinetic models for chemical and biological reaction systems
- Reactor design models for common reactor types, such as packed bed, trickle
  flow, fluidized bed, and external loop
- Phase equilibrium models to support design of separation systems
- LCA models for sustainability analysis
- Technoeconomic models for economic analysis
- Process simulation models for flowsheet and equipment design

The level of detail needed is driven by the requirements of the task at hand (see
Figure 3.3).

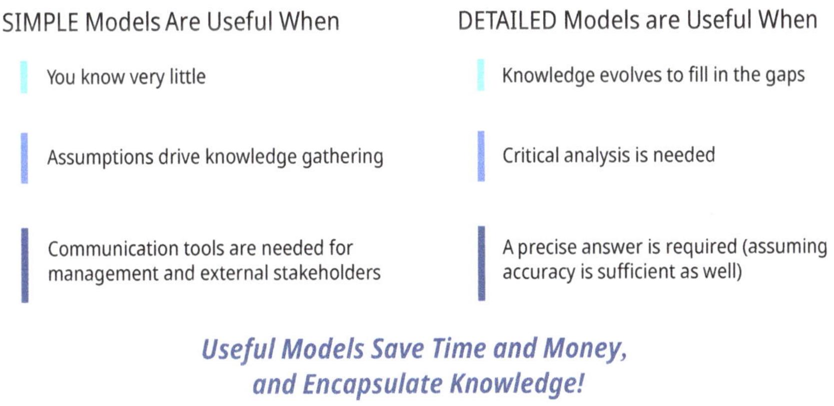

SIMPLE Models Are Useful When   DETAILED Models are Useful When

| You know very little | Knowledge evolves to fill in the gaps |

| Assumptions drive knowledge gathering | Critical analysis is needed |

| Communication tools are needed for management and external stakeholders | A precise answer is required (assuming accuracy is sufficient as well) |

Useful Models Save Time and Money,
and Encapsulate Knowledge!

Figure 3.3: Useful models for scale-up.

Where data does not exist, or is inconclusive, assumptions can be used to establish a
working model. The criticality of these assumptions to the system in question is evaluated by exploring sensitivities. If the answer is "very critical," this result can be used to

inform the upcoming experimental activities. This interplay between engineering design, modeling, and experimentation is quite important. When modeling is done in a vacuum, with little or no interaction with experimentalists, the results are often a very beautiful model with limited value. Similarly, some experimentalists insist that it is impossible to model their system and find no value in the results from a model. The reality is that a useful model can, and should, complement experimentation to reduce the time and cost of scale-up, providing insight as to when additional data is needed to enhance understanding. A great model can also produce results and understanding that may be too time-consuming, costly, or just not possible through additional experimentation. The models can also direct future opportunities for experimental programs.

The models should then be refined as more data are collected. These data should be generated at multiple scales to enhance the robustness and utility of the model.

3.7 Process safety and operation: design issues to consider during bioprocess scale-up

The field of safety in R&D facilities and manufacturing plants is necessarily quite broad. Much of this is driven by regulations at the local and national levels, but best practices have developed in the process industries with the aim of designing and operating lab-, pilot-, and commercial-scale facilities in a manner that minimizes risk of the workers and the surrounding environment from both expected and unexpected process conditions. The best run facilities, in regard to worker safety, have management commitment to provide a safe workspace, and necessary work practices, procedures, and equipment to follow these best practices. In turn, the workers are obligated to follow these practices, wear appropriate safety equipment, ensure their work area is free of hazards, and ensure that their fellow workers are safe as well.

First, some definitions (many taken from the AIChE Center for Chemical Process Safety (CCPS) Glossary [7], noted in quotes with CCPS) are:

- Fault tree analysis (FTA) *(CCPS)*: "A method used to analyze graphically the failure logic of a given event, to identify various failure scenarios (called cut-sets), and to support the probabilistic estimation of the frequency of the event."
- Failure mode and effect analysis (FMEA) *(CCPS)*: "A hazard identification technique in which all known failure modes of components or features of a system are considered in turn and undesired outcomes are noted."
- Hazard identification (HAZID): *(CCPS)*: "Part of the Hazards Identification and Risk Analysis (HIRA) method in which the material and energy hazards of the process, along with the siting and layout of the facility, are identified so that a risk analysis can be performed on potential incident scenarios."
- Hazard and operability study (HAZOP): *(CCPS)*: "A systematic qualitative technique to identify process hazards and potential operating problems using a series

of guide words to study process deviations. A HAZOP is used to question every part of a process to discover what deviations from the intention of the design can occur and what their causes and consequences may be. This is done systematically by applying suitable guidewords. This is a systematic detailed review technique, for both batch and continuous plants, which can be applied to new or existing processes to identify hazards."

- Layers of protection analysis (LOPA) (*CCPS*): "The extension of a HAZOP Study to include aspects of a LOPA, including selecting identified scenarios for further analysis; evaluating the initiating event frequency, consequence severity and effectiveness of Independent Protection Layers (IPLs) on an order-of-magnitude basis; considering enabling conditions and/or conditional modifiers as appropriate when evaluating scenario risk; and comparing the calculated scenario risk to a risk goal to determine the adequacy of existing risk control measures."

- Management of change (MOC) (*CCPS*): "A management system to identify, review, and approve all modifications to equipment, procedures, raw materials, and processing conditions, other than replacement in kind, prior to implementation to help ensure that changes to processes are properly analyzed (for example, for potential adverse impacts), documented, and communicated to employees affected."

- Pre-start-up safety review (PSSR) (*CCPS*): "A systematic and thorough check of a process prior to the introduction of a highly hazardous chemical to a process. The PSSR must confirm the following: Construction and equipment are in accordance with design specifications; Safety, operating, maintenance, and emergency procedures are in place and are adequate; A process hazard analysis has been performed for new facilities and recommendations have been resolved or implemented before startup, and modified facilities meet the management of change requirements; and training of each employee involved in operating a process has been completed."

- Process hazard analysis (PHA): This is a generic term for any number of hazard assessments that can be carried out for a process unit. In some cases, the term "PHA" is used interchangeably with "HAZOP." It is important to make sure that if the term PHA is used in an organization that there is alignment on what type of assessment is to be carried out.

- Root cause analysis (RCA) (*CCPS*): "A formal investigation method that attempts to identify and address the management system failures that led to an incident. These root causes often are the causes, or potential causes, of other seemingly unrelated incidents. Identifies the underlying reasons the event was allowed to occur so that workable corrective actions can be implemented to help prevent recurrence of the event (or occurrence of similar events)."

- What-if analysis (*CCPS*): "A scenario-based hazard evaluation procedure using a brainstorming approach in which typically a team that includes one or more persons familiar with the subject process asks questions or voices concerns about what could go wrong, what consequences could ensue, and whether the existing safeguards are adequate."

The number of assessments defined earlier can lead to confusion and an "alphabet soup" of acronyms that cloud the overall goal of an inherently safe process unit. In this section, we review the types of assessments that can be carried out at various stages of new engineering projects. Please refer to the abovementioned definitions for the various hazard assessment techniques identified later. In addition, references at the end of this document contain a more comprehensive description of which type of assessment tool to use for a particular scenario.

– Research/bench scale
  – Typically, at this scale we are screening different reaction chemistries. We can identify potential raw materials, catalysts, solvents, intermediates, products, and byproducts, and use material safety data sheets for each to identify major hazards. We can also assess process conditions to identify high pressure or temperature hazards. We should avoid, however, the temptation to start going too far into process design at this stage. As an example, the hazard review (such as a FTA) could identify a particularly hazardous solvent. The technical personnel assigned to the project may then wish to perform additional experiments to evaluate alternative solvents. Those solvents may be less effective chemically, but with a lower hazard rating may lead to a more inherently safe process design. The goal is to generate sufficient data that the design engineer can use carry out trade-offs and propose the best design.

– Conceptual design (FEL-1)
  – At this point, we are able to develop a process flow diagram (PFD), preliminary heat and material balance, and preliminary equipment list. It is useful at this time to carry out a HAZID. The goal is to understand potential safety hazards and adjust the PFD if needed, or perhaps to generate process alternatives for further evaluation. We also would like to identify flammable raw materials, intermediates, and solvents which may require the design of hazardous classification areas and special considerations for explosion-proof equipment, all of which can add cost to the design.

– Basic engineering (FEL-2)
  – In this stage, the piping and instrumentation diagram (P&ID) is developed, along with the detailed heat and material balance and control scheme. At this point, we can carry out the HAZOP. For the HAZOP to be effective, the P&ID should be complete from a process perspective, with control schemes defined and operating instructions developed. The HAZOP is best executed with an independent facilitator, and key disciplines represented from process chemistry/biology, design engineering, plant engineering, operations, and analytical. Recommendations from the HAZOP may result in additional design changes to the P&ID. If this is the first time in the design process that safety hazards have been reviewed, it is possible, in fact likely, that somewhat major design

changes are required that can impact both the cost and project schedule. To prevent this, it is worth taking the time earlier in the design process to review hazards as discussed earlier. Note that the overall goal is not to simply complete a HAZOP document – the assessment should inform design decisions and future safety reviews and instructions. If HAZOP is completed and put on the shelf never to be reviewed again, it has not been useful.

– Start-up/commissioning
  – The pre-start-up safety review (PSSR) is a critical piece of any precommissioning activities. During the PSSR, the team (ideally the same team that was part of the HAZOP) reviews that the plant was built as designed, safety/operating/maintenance/emergency procedures are in place, and that *all* recommendations from the HAZOP have been implemented and resolved. The plant "owner," meaning the engineer or other individual responsible for plant start-up, should take ownership of this review to ensure that the review is complete, and the plant is ready for start-up.

– Post start-up, existing facilities
  – Process safety considerations do not stop once the plant has started up. A working "Management of Change" (MOC) process is important to determine how the plant will respond to changes introduced that are outside the scope of the original HAZOP. These changes may include introduction of new raw materials, changing operating conditions, or mechanical changes to the process itself. A robust MOC process is used to determine what type of safety review is warranted for a particular change, ranging from a "what-if" analysis to a revised HAZOP. Finally, root cause and fault tree analyses are used to determine the causes of a process failure that occurs during operation.

This section touches on key elements of process safety but is by no means exhaustive. The reader is referred to additional references which are much more comprehensive [8, 9].

3.8 A case study

This chapter has presented some guidelines for practical scale-up of bioprocessing technology and the bioproducts produced by these technologies. These concepts will be illustrated with a case study. Consider the conversion of lignin to a surfactant product using an aqueous caustic reaction chemistry.

Section 3.4 presented an algorithm to develop and optimize the process flow scheme:

- Determine system requirements
- Size reactor system
- Size supporting equipment (pumps and heat exchangers)
- Integrate separation system
- Develop cost model
- Evaluate trade-offs and set targets for further data generation

The block flow for this process is initially developed, as shown in Figure 3.4. By identifying the key processing steps in a block flow, first, before advancing to more detail in a PFD, the key reaction and separation steps are clearly identified, and an overall mass balance can be prepared.

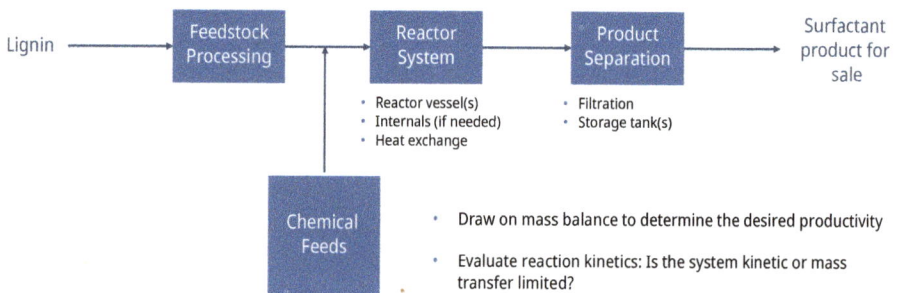

Figure 3.4: Block flow for new bioprocess.

The next step is to size the reactor system. Figure 3.5 shows a graph of conversion versus residence time. Other parameters such as temperature and feedstock properties may be explored using the design of experiments.

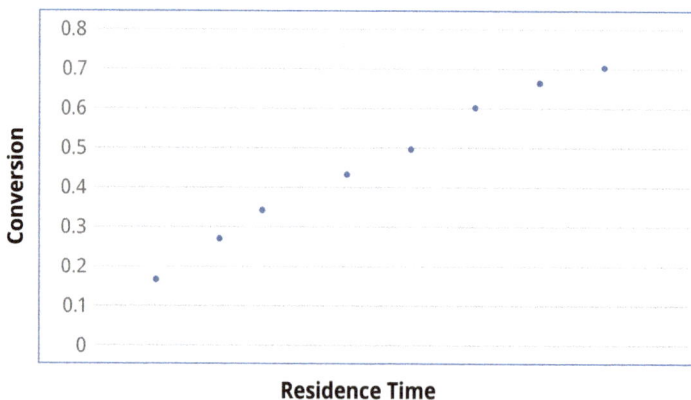

Figure 3.5: Conversion versus residence time.

The separation system can then be integrated to the flow scheme, so that the reaction/separation trade-offs can be explored and optimized. For instance, operating at a very high conversion may lead to a greater yield of nonselective reaction byproducts, increasing the cost of the separation system. It may be better to operate at a lower conversion, reducing the load on the separation system. By carrying out the reaction experiments described earlier, these trade-offs can be explored along with the separation scheme.

A technoeconomic model is developed to enable these trade-offs to be explored. For instance, Figure 3.6 shows the cost performance of two different cases relative to the overall cost targets.

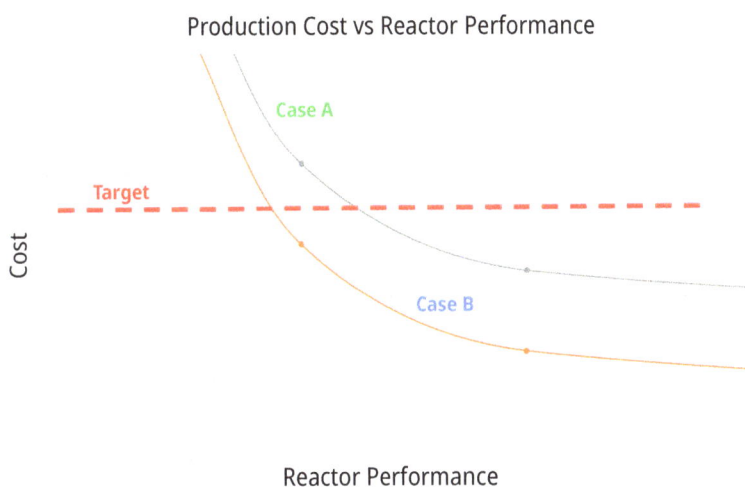

Figure 3.6: Techoeconomic trade-offs for new bioprocess.

3.9 Commercial success stories

In recent history, a number of bioproducts, whether using bio-based feedstocks, a biological catalyst, or both, have achieved commercial success. A few examples are given further, meant to be illustrative rather than exhaustive, as of Q2 2022:

- 1,3-Propanediol (PDO). PDO has been made conventionally by hydration of acrolein or hydroformylation of ethylene oxide. More recently, a bioprocessing-based route has been developed by modifying *E. coli* to enable production by fermentation of feedstocks such as glucose and glycerol. Commercial producers now include Covation Bio (formally DuPont Biomaterials) with a capacity of 77,000 tons/year, along with others, such as Shenghong Group Holdings, Zhangjiangang, and Metabolic Explorer, bringing capacity onstream in the near future [10]. The PDO is then converted by Covation Bio to produce a range of bioproducts under the trade names Susterra® and Zemea® [11].

– Polylactic acid (PLA). PLA was produced by Natureworks, a joint venture established by Cargill and Dow Chemical, with the first commercial production coming onstream in 2001. Natureworks now is a joint venture between PTT Global Chemical and Cargill. Lactic acid is produced by fermentation, and the PLA is then produced through a thermochemical condensation reaction. Currently, Natureworks has a production capacity of 150,000 tons/year at their Nebraska facility, with plans to build out additional capacity in Thailand [12]. Other manufacturers include Total Corbion (75,000 tons/year), WeForYou (75,000 tons/year), and several manufacturers in China, Japan, and Europe with capacities of 1,000–15,000 tons/year [13]. PLA is used to make biodegradable plates, cups, cutlery, and containers, along with a variety of films, fibers, and 3D printing applications.

– Genomatica has commercialized a fermentation-based route to produce 1,4-butanediol (1,4-BDO). Like PDO, 1,4-BDO has been produced historically through thermochemical routes using fossil-based feedstocks. As a technology licensor, Genomatica now has licensed technology for two commercial plants, for a total global capacity of more than 100,000 tons/year [14]. 1,4-BDO is used in downstream applications for conventional products such as polyester and polyurethanes.

– Global ethanol production in 2021 was 27 billion gallons (~80 million tons) [15], primarily used for blending into fossil-based gasoline. Virtually, this uses sugarcane or corn as a starting material, and is known as first-generation ethanol. To further reduce the carbon intensity of this alternative fuel, and to avoid competition with food sources, the technology for producing ethanol from alternate feedstocks (second-generation, 2G, ethanol) is starting to come on stream. Examples include:
    – Raizen (Brazil, 10 million gallons/year, 30,000 tons/year) [16]
    – Clariant (Romainia, 17 million gallons/year, 50,000 tons/year) [17]
    – Praj (India, 250,000 gallons/year, 740 tons/year) [18]
    – In addition, LanzaTech has commercialized a plant in China to convert steel mill off-gas into ethanol. The Shougang plant has a capacity of 16 million gallons/year, or 46,000 tons/year [19]

3.10 Conclusions

It is critical to "start with the end in mind" using a Technology Concept (or Process Concept) that is used as a framework to drive new technology development, scale-up, and commercialization (see Figure 3.7). This technology concept is not set in stone, and, in fact, should be reviewed and updated as we progress throughout the scale-up effort. We establish the technology concept to drive the scale-up effort, not just inform it. This then enables us to direct the innovation to create the ***greatest value from***

breakthrough and disruptive ideas, as we identify challenges early, fail fast when it is cheaper and quicker, and make sure our efforts are focused on solving commercially relevant problems.

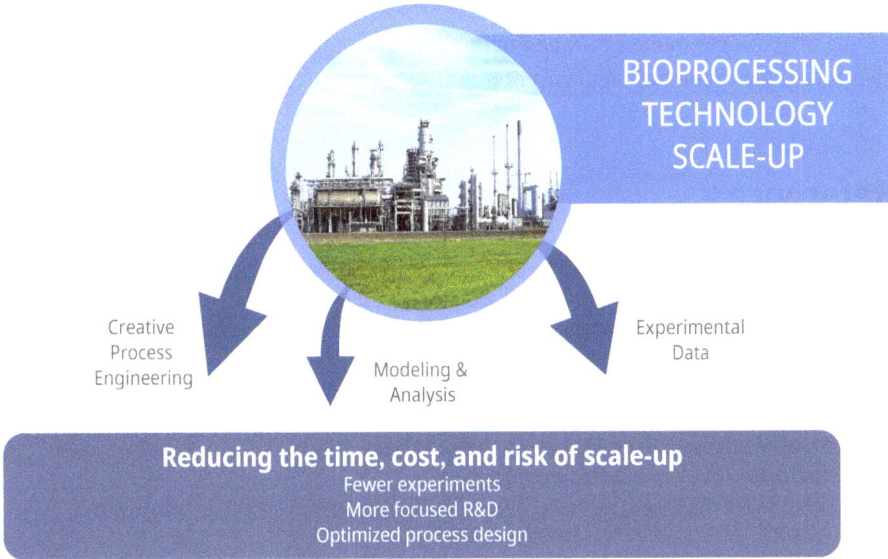

Figure 3.7: Reducing the time, cost, and risk.

The technology concept is developed, and iteratively revised, through a combination of *creative process engineering, multi-scale experimental data, and modeling and analysis.*

Creative process engineering: The flow scheme is developed, the material balance is estimated, and key process design decisions are identified so that we can establish the best process flowsheet for the technology.

Modeling and analysis: A good model can save time and resources in the lab. Coupled with the right analysis, this can be used to prioritize objectives in the lab, pilot, and demo units. Cautionary note: useful models are more important than perfect models!

Experimental data: We need the right data to prove out breakthrough ideas, secure partners and investors, and develop engineering data for equipment design. Multiscale data is critical to this effort, and with good planning, multiple assets and external resources can be leveraged.

The key benefits to this approach are:
- Prioritization of R&D to de-risk and optimize a new technology
- Identification of cost reduction opportunities throughout the scale-up effort
- Anticipation of engineering needs as early as possible

In this way, we can *reduce risk and optimize economics* of our new design, while efficiently managing the time and cost of our efforts.

References

[1] Global Bioproduct Market Size, Share & Growth Analysis Report (bccresearch.com).
[2] https://ethanolrfa.org/resources/ethanol-biorefinery-locations.
[3] Global biofuel production in 2019 and forecast to 2025 – Charts – Data & Statistics – IEA.
[4] Sustainable Aviation Fuels (SAF) (icao.int).
[5] Douglas JM. A hierarchical decision procedure for process synthesis. AIChE J 1985;31(3):353–62.
[6] Box GEP. Robustness in the Strategy of Scientific Model Building. In: Launer RL, Wilkinson GN, editors. Robustness in Statistics, Academic Press; 1979, pp. 201–36.
[7] https://www.aiche.org/ccps/resources/glossary.
[8] Guidelines for Risk Based Process Safety, by the AIChE Center for Chemical Process Safety (Guidelines for Risk Based Process Safety | AIChE).
[9] https://www.process-improvement-institute.com/_downloads/Selection_of_Hazard_Evaluation_Techniques.pdf.
[10] https://biorrefineria.blogspot.com/2021/06/Biobased-1-3-propanediol-PDO.html.
[11] https://covationbiopdo.com/.
[12] https://www.natureworksllc.com/About-NatureWorks.
[13] https://www.gianeco.com/en/faq-detail/1/37/where-to-buy-pla-.
[14] https://www.genomatica.com/news-content/second-commercial-plant-will-expand-capacity-to-over-100000-tons.
[15] https://ethanolrfa.org/markets-and-statistics/annual-ethanol-production.
[16] https://www.reuters.com/business/energy/brazils-razen-build-second-cellulosic-ethanol-plant-filing-2021-06-25/.
[17] https://www.clariant.com/en/Corporate/News/2022/06/Clariant-produces-first-commercial-sunliquid-cellulosic-ethanol-at-new-plant-in-Podari-Romania.
[18] https://www.praj.net/businesslines/2g-ethanol.
[19] https://www.chemengonline.com/lanzatech-starts-up-waste-to-ethanol-plant-in-china/#:~:text=With%20a%20capacity%20of%2046%2C000,fuel%2C%20diesel%20and%20household%20products.

John McArdle

Chapter 4
Bioproduct projects
Financial and investment analysis tools

4.1 Overview

This chapter provides a brief introduction to financial and investment analytical methods that are used by early-stage companies to attract investment funding. These analytical methods can be used by entrepreneurs, researchers, and students involved in the development of novel green material bioproducts.

4.2 Background

Differentiated technologies and products with high value are very important in any early-stage company. Equally, if not, more critical is the ability to secure needed funding to realize the potential of the company with a compelling economic return financial story.

It is not always possible for early-stage companies to fully achieve their potential for growth by self-financing or operating cash flow. Accordingly, outside funding is often required. However, early-stage companies do not often have access to the funding sources that are available to more established companies.

For example, traditional banks are not willing to lend money to companies that do not have an established line of credit or sufficient collateral, both of which can be in short supply with early-stage companies. Accordingly, these companies often rely on private "angel" investors and venture capitalists (VCs) for the funding needed to support growth.

Investors do not fund early-stage companies because they intend to operate such companies. Instead, they provide funding because of their expectation of high financial return with the sale of the company.

Investor returns are primarily, but not exclusively, based on the valuation and ownership share of the early-stage company that are negotiated by the investor and principals of the company at the time of the funding.

John McArdle, Redwood Innovation Partners, LLC, Boston, MA, USA, e-mail: johnm@redwdinnov.com

https://doi.org/10.1515/9783110791228-004

4.3 Investment analysis

Investment analysis is used to identify attractive funding opportunities that are believed to provide a reasonable chance of high financial reward relative to their financial risk. Investment analysis can also be used to rank the attractiveness of individual investment opportunities under capital constraints.

Payback, internal rate of return (IRR), and net present value (NPV) are methods of investment analysis. Payback and IRR methods have shortcomings that can lead to less-than-optimal investment decisions. When properly applied, NPV will reliably identify attractive investment opportunities.

Attractive investment opportunities in early-stage companies are not always obvious. First, there can be significant uncertainty in projecting sales revenues and operating costs. Second, different methods of investment analysis, in some scenarios, provide unhelpful or contradictory results.

Table 4.1 shows projected cash flows for two hypothetical investments. Funding is $1.0 million for each investment with different cash flows projected over 4 years. Investment A is more attractive based on a shorter payback period and a higher IRR. However, investment B is more attractive based on a higher NPV @ 10% discount factor.

Table 4.1: Investment analysis methods.

Year	Investment $, out	Cash flow $, in			
	0	1	2	3	4
Investment A	1,000,000	1,000,000	500,000	100,000	50,000
Investment B	1,000,000	500,000	500,000	500,000	500,000
Investment analysis method	Payback year	IRR %	NPV @ 10% $		
Investment A	1	42	432,000		
Investment B	2	35	584,000		

It is apparent from these contradictory results that the output from different methods of investment analyses should be carefully interpreted to properly identify optimal investment opportunities.

4.3.1 Net present value

NPV is the sum of an initial investment plus the discounted future cash flows that are projected to arise from such investment over a defined period of time. Investments that have positive NPVs are attractive. Conversely, investments that have negative NPVs are not attractive.

Projected future cash flows are adjusted to present values by the discount rate. Discount rate represents the financial return that can be expected from an alternative investment that has an equivalent risk to that of the investment being considered. For this reason, the discount rate is also referred to as the opportunity cost of capital.

Higher and lower discount rates are used for NPV calculations with more risky and less risky investment opportunities, respectively. The long-term, pre-tax financial return of a well-diversified portfolio of common stock has been about 10% per year including capital gains and dividends. This historical stock market return provides a useful benchmark that can be used to select the discount rates for other investment opportunities taking into account the risk of such investments.

Investments in early-stage companies often have higher financial risks than an investment in a diversified portfolio of common stocks. Accordingly, investors target higher financial returns for these companies than can be expected based on the historical performance of a diversified portfolio of common stocks.

4.4 Investors

Investors recognize the risk of investing in early-stage companies. In fact, they expect that the majority of their investments made in these companies will have minimal or negative financial return.

Investors often invest in a portfolio of early-stage companies with each company in the portfolio believed to have the potential for a high financial return. In this manner, the high financial returns from a few successful investments will balance out the majority of those that have minimal or no return. In theory, this will provide an overall average financial return for the investment portfolio that is higher than can be expected from other less risky investments including, for example, a well-diversified portfolio of common stocks.

Financial return for a private investor in an early-stage company is realized with the sale of the company to a third party or an initial public stock offering. Typically, the investment period for private investors is 5–8 years.

4.5 Company valuation

Company valuation is the amount of money that transfers from the buyer to the seller with the sale of a company. Until the actual sale of a company, the calculation of company valuation is an estimate.

Investors funding early-stage companies negotiate the company valuation and percentage of company ownership with the principals of the company at the time of a funding round. These parameters establish the basis for the investor's financial return with the sale of the company as well as the compensation for the ideas and sweat equity of the founders.

Valuation of an early-stage company is primarily based on the expectation of company growth and profitability. Company valuation must increase during the funding period if the investor is to achieve a positive return on their investment.

Investors take into account many factors when evaluating an early-stage company. These include the current financial performance of the company as well as the expertise and track record of the management team, size of the market opportunity, strength of the competition, commercial readiness of the product, and maturity of the sales channels.

Early-stage, pre-revenue companies have minimum or no-sales revenue and partly developed products and sales channels. For these companies, valuation is based on a more subjective analysis of the company's potential to grow into a profitable business.

Early-stage, post-revenue companies have some sales revenue and more developed products and sales channels. For these companies, valuation is based on a more quantitative assessment of the company's projected growth and profitability.

4.5.1 Pre-revenue company valuation calculation method

Valuations of pre-revenue companies can be based on the reported average valuation of other previously funded pre-revenue companies in a similar business sector including, for example, industry, application, and region. These reported valuations are adjusted for the "target" pre-revenue company using weighted scores of valuation parameters that are considered by the investors to be most critical to achieve significant company growth and profitability.

Table 4.2 illustrates a valuation method that uses team qualifications, potential market size, competitive advantage, and prototype performance as the valuation parameters of a "target" company. Depending on investor preferences, other valuation parameters, as well as different parameter weightings, can be applied.

In this example, the calculated valuation of the target company is $5.5 million compared to an average valuation of $5.0 million for other previously funded pre-revenue companies in a similar business sector. The higher valuation of the target company reflects a higher total weighted score of the individual valuation parameters of this company compared to that of the previously funded companies.

Company valuations with the valuation parameters and weightings used in this example range from a maximum of $7.5 million with all strong 1.5 scores, to a minimum of $2.5 million with all weak 0.5 scores. Investors can rate any of these parameters as

Table 4.2: Pre-revenue company valuation [1].

Average Company Valuation	5,000,000			
Target Company Valuation	**5,500,000**			
Score:	0.5	Weak		
	1.0	Average		
	1.5	Strong		
Company Score Parameter	**Weight**	**Score**	**Weighted Score**	**Value Contribution**
Team Qualifications	0.40	1.5	0.60	3,000,000
Potential Market Size	0.30	1.0	0.30	1,500,000
Competitive Advantage	0.20	0.5	0.10	500,000
Prototype Performance	0.10	1.0	0.10	500,000
Total	**1.00**		**1.10**	**5,500,000**

zero. However, a zero rating in any of these critical valuation parameters will probably eliminate the company from funding consideration.

4.5.2 Post-revenue company valuation calculation method

Valuation of an early-stage, post-revenue company can be based on the projected financial performance of the company taking into account sales revenue, factory cost, indirect costs, and taxes.

Table 4.3 shows the projected financial performance of a hypothetical post-revenue company. Investor funding of $5.0 million is provided to purchase new factory equipment. This funding enables the company to increase manufacturing and associated sales revenue while at the same time maintaining positive cash flows.

Net sales revenues are based on projected unit sales, price, and factory cost. Economies of scale and increased competition are assumed over time to reduce unit factory cost and reduce sales price, respectively.

Sales, general, and administrative (SG&A), research and development (R&D), engineering, and net working capital are calculated using multipliers based on total sales revenue. Depreciation of factory equipment is based on 7-year straight-line schedule. Land is not depreciated.

NPV is about $1.7 million at 25% discount factor over a 5-year period. Sales revenue, net income, and cash flow are increasing after the initial 3-year period indicating a successful ongoing business operation.

Company valuation is estimated at $120 million based on a 15× EBITDA ($8.0 MM × 15) at 5 years [2]. The financial return for the investor is based on this valuation along with ownership share.

NVP does not necessarily correlate to an attractive investment opportunity because investors are less focused on the financial performance and more focused on the

Table 4.3: Post-revenue company valuation.[1,2]

Investment Analysis					
Net Present Value	$	1,667,000			
Discount Rate	%	25.0			
SG&A, % total revenue	%	20.0	Minimum	$	200,000
R&D, % sales revenue	%	5.0	Minimum	$	200,000
Engineering, % sales revenue	%	5.0	Minimum	$	200,000
Working Capital, % incremental revenue	%	20.0			
Tax	%	20.0			
Total Investment	$	5,000,000			
Company Value, Multiple of EBITDA	#	15.0			
Company Market Value @ Year 5	$	120,000,000			

Year		1	2	3	4	5
Investment	$	5,000,000	–	–	–	–
Cost of New Factory Equipment	$	3,500,000	–	–	–	–
Cost of New Land	$	250,000	–	–	–	–
Unit Sales	#	500	1,000	2,500	5,000	10,000
Unit Sales Price	$/#	5,000	5,000	4,500	4,250	4,000
Unit Factory Cost	$/#	4,000	3,000	2,500	2,250	2,000
Unit Sales Price less Unit Factory Cost	$	1,000	2,000	2,000	2,000	2,000
Gross Margin	%	20	40	44	47	50
Total Sales Revenue	$	2,500,000	5,000,000	11,250,000	21,250,000	40,000,000
Total Factory Cost	$	2,000,000	3,000,000	6,250,000	11,250,000	20,000,000
Net Sales Revenue	$	500,000	2,000,000	5,000,000	10,000,000	20,000,000
SG&A	$	500,000	1,000,000	2,250,000	4,250,000	8,000,000
R&D	$	200,000	250,000	563,000	1,063,000	2,000,000
Engineering	$	200,000	250,000	563,000	1,063,000	2,000,000

1 Appendix 1 for accounting terms.
2 Appendix 2 for accounting factors.

Table 4.3 (continued)

Year		1	2	3	4	5
EBITDA	$	(400,000)	500,000	1,624,000	3,624,000	8,000,000
Depreciation	$	-	500,000	500,000	500,000	500,000
Taxable Income	$	(400,000)	–	1,124,000	3,124,000	7,500,000
Tax	$	-	-	225,000	625,000	1,500,000
Net Income After Tax	$	(400,000)	–	899,000	2,499,000	6,000,000
Change in Net Working Capital	$	500,000	500,000	1,250,000	2,000,000	3,750,000
Cash Flow	$	350,000	–	149,000	999,000	2,750,000
Present Value	$	280,000	–	76,000	409,000	901,000
Incremental Present Value	$	280,000	280,000	356,000	765,000	1,666,000

valuation of a company, although these two items can be closely correlated. Investors only realize attractive financial returns with companies that over time achieve increased earnings because this translates into higher company valuations.

4.5.3 Investor target return

Investors target 6× to 10× return multiples on their funding for early-stage companies over an investment period of 5–8 years. Investors in earlier funding rounds often target higher return multiples of up to 30× to account for any possible dilution of their ownership share resulting from subsequent funding rounds.

Table 4.4 shows that IRR is 25.0–58.4% for this range of return multiples and investment periods. IRR increases with higher return multiples and shorter investment periods. NPV is based on a $1.0 million investment at 25% discount rate. NPV is zero at the lowest return multiple (6×) and longest investment period (8 years). NPV, under this analysis, is by definition zero at 25% IRR.

High target returns are required for each individual company in an investment portfolio of early-stage companies so that in the event that the majority of these companies return minimal or no return, as expected, the overall "risk-adjusted" return of the portfolio remains attractive compared to other investment options.

Table 4.4: Investor return matrix [3].

Investment		$1,000,000		
Minimum target return		25.0%		
Years to exit	Analysis method	Investment return multiple		
		6×	8×	10×
5	IRR	43.1%	51.5%	58.4%
	NPV	$966,000	$1,621,000	$2,277,000
6	IRR	34.8%	41.4%	46.8%
	NPV	$573,000	$1,097,000	$1,621,000
7	IRR	29.1%	34.6%	38.9%
	NPV	$258,000	$678,000	$1,097,000
8	IRR	25.0%	29.7%	33.3%
	NPV	$0	$342,000	$678,000

4.5.4 Investor ownership

Table 4.3 shows an investment of $5 million in a hypothetical early-stage company. Table 4.4 shows that investors in early-stage companies target 6× to 10× multiple on their investment in 5–8 years. Accordingly, investors will expect to receive $30 million (6 × $5 million) to $50 million (10 × $5 million) from the sale of the company.

Projected valuation of this company in 5 years is $120 million based on 15× EBITDA multiple. Investors will likely attempt to negotiate not less than 25% ownership share in this company ($30 million/$120 million). This ownership share and company valuation will provide investors with 43.1% IRR in 5 years.

If this company has the same $120 million valuation but is instead sold in 8 years, this 25% ownership share will provide the investor with a (minimum) acceptable 25% IRR. This lower return is directly correlated to the time value of money at 25% discount rate for the longer investment period.

4.5.5 Dilution

Company valuation and ownership share negotiated at the time of a funding round establish the basis for investor return upon sale of the company. However, as shown in this section, the impact of future funding rounds can also play a significant role.

Table 4.3 shows that the $5.0 million funding for the hypothetical early-stage company provides sufficient cash to support company growth during the initial 3 years of operation. In later years, cash from operations is sufficient to support the company growth. Under this scenario, no further funding round is required.

Series A Round valuation and ownership shares are shown below for one conceptual scenario. Many other scenarios are possible. Share price increases from $1.00/share to $12.00/share with $120 million company valuation (year 5). Series A investor payout is $36 million. Series A investor return is 7.2×.

Series A round

Pre-money valuation:	$10 million (negotiated)
Series A funding:	$5 million
Post-money valuation:	$15 million
Issued shares:	10 million
Pre-money share price:	$1.00/share ($10 million/10 million shares)
Series A ownership:	30% (negotiated)
Series A investor shares:	3.0 million (0.30 × 10 million shares)
Company valuation:	$120 million @ Year 5
Share price:	$12.00/share ($120 million/10 million shares)
Series A investor payout:	$36 million (3.0 million shares × $12.00/share)
Series A investor return:	7.2× ($36 million/$5 million)

Multiple funding rounds can occur for rapidly growing companies. This additional funding can be used to purchase additional manufacturing equipment and cover increased working capital requirements.

However, later funding rounds reduce the ownership share of the investors of earlier funding rounds. Ownership "dilution" can result in lower returns for the earlier investors than would otherwise be expected.

Series B round valuation and ownership shares are shown below for one conceptual scenario. Many other scenarios are possible. Series B round funding dilutes ownership share of Series A investors from 30% to 24%. Share price increases from $1.00/share to $9.60/share, instead of $12.00/share, over 5 years. Series A investor payout has been reduced from $36 million to $29 million and return reduced from 7.2× to 5.8×. Series B investor payout is $24 million with a return of 4.8×.

Series B round

Series B funding:	$ 5 million
Series B ownership:	20% (negotiated)
Series B investor shares:	2.5 million (2.5 million/(10 million + 2.5 million) = 0.20)
Post-money issued shares:	12.5 million (10 million + 2.5 million)
Series A ownership:	24% (3 million shares/12.5 million shares)
Company valuation:	$120 million @ Year 5
Share price:	$9.60/share ($120 million/12.5 million shares)

Series A investor payout:	$29 million (3.0 million shares × $9.60/share)
Series A investor return:	5.8× ($29 million/5.0 million)

Series B investor payout:	$24 million (2.5 million shares × $9.60/share)
Series B investor return:	4.8× ($24 million/$5 million)

Series B investors may be willing at the time of Series B round to accept a lower return multiple than the earlier Series A investors if the perceived risk of the company is lower, and the investment period is shorter than for the Series A round.

4.5.6 Divergence

For the Series A round, with no later funding rounds, the 12× ratio of share price increase from $1.00/share to $12.00/share is exactly the same as the 12× ratio of company valuation increase from $10 million to $120 million. There is no "divergence" between the ratio of the increase of company share price and ratio increase of company valuation.

For the Series B round, the 9.6× ratio of share price increase from $1.00/share to $9.60/share is less than (and diverges from) the 12× ratio of company valuation increase from $10 million to $120 million. In this scenario, the divergence ratio is 1.25 (12×/9.6×). Industry surveys indicate that divergence ratios can be 3× to 5× for early funding rounds.

Private investors use various strategies to attempt to remediate the negative impact on their investment returns arising from later funding rounds. For example, investors can increase the target return on their investment by negotiating combinations of higher ownership share and lower company valuation. Other contract terms can be negotiated at the time of the funding including warrants to purchase additional shares, dividends, management structure, and liquidation preferences.

4.6 Summary

Early-stage companies, including green material bioproduct companies, often require funding from "angel" or VC investors to achieve commercial success. It is important that the principals of these companies have an understanding of the financial and investment analytical methods described in this chapter that are used by investors to project their financial return in early-stage companies.

Appendix 1: definitions for investment analysis

1. *Discount rate*
   Percentage that is used to discount cash for each year of the investment period. Accounts for the time value of money. Discount rate is also referred to as the opportunity cost of capital.

2. *Investment*
   Funding provided by the private investor or VC with expectation of repayment, with a profitable return, with the sale of company. Third-party investment is a positive cash flow.

3. *Cost of new factory equipment*
   Cost of new factory equipment is a negative cash flow. Factory equipment is depreciated.

4. *Cost of new land*
   Cost of new land is a negative cash flow. New land cost is not depreciated.

5. *Unit sales*
   Number of units that are sold each year.

6. *Unit sales price*
   Price of units that are sold in each year.

7. *Unit factory cost*
   Materials, factory labor, and other factory operating costs.

8. *Unit sales price less unit factory cost*
   Difference between unit sales price and unit factory cost.

9. *Gross margin*
   Percentage ratio of unit sales price less unit factory cost and unit sales price.

10. *Total sales revenue*
    Number of units sold in each year multiplied by the unit sales price.

11. *Total factory cost*
    Number of units sold in each year multiplied by the unit factory cost.

12. *Net sales revenue*
    Difference between total sales revenue and total factory cost.

13. *SG&A*
    Expense for SG&A. Industry benchmark, 20% of total revenue.

14. *R&D*
    Expense for R&D. Industry benchmark of 5% total revenue.

15. *Engineering*

    Expense for engineering. Industry benchmark of 5% of total revenue.

16. *EBITDA*

    Earnings before interest, tax, depreciation, and amortization. Net sales revenue, less SG&A, less R&D, and less engineering.

17. *Depreciation*

    Tax shield based on 7-year straight-line depreciation of cost of new factory equipment. Depreciation is a positive cash flow.

18. *Taxable income*

    EBITDA less depreciation.

19. *Tax*

    Corporate federal, state, and local tax as percentage of taxable income.

20. *Net income after tax*

    Taxable income, less tax. Net income after tax can be a negative or positive cash flow.

21. *Change in net working capital*

    Net working capital is current assets less current liabilities. Industry benchmark of 20% of change in year-to-year total sales revenue. Change in net working capital is a negative cash flow.

22. *Cash flow*

    Third-party investment plus depreciation plus net income after tax, less cost of new factory equipment, less cost of new land, less change in net working capital.

23. *Present value*

    Discounted cash flow for each year during the five investment horizon.

24. *Incremental present value*

    Accumulated present value for each year during the five investment horizon.

25. *Net present value (NPV)*

    Accumulated present value over the 5-year investment horizon.

Appendix 2: accounting factors

Net working capital

Net working capital is current assets less current liabilities. Current assets are cash, accounts receivable, and inventory. Current liabilities are short-term notes and payable accounts, taxes, and interest. Investment analysis considers an increase in net working capital as a negative cash flow and a decrease in net working capital as a positive cash flow.

Companies will often have an increase in net working capital with growing sales revenue. This can impede the ability of a company to grow because of cash flow constraints.

Based on industry benchmarks, a change in net working capital is estimated as 20% of the year-to-year difference in total sales revenue.

Change in the net working capital is assumed to be only a function of the annual increase or decrease in total sales revenue and does not take into account any change in company efficiency of managing net working capital.

Sales, general, and administrative (SG&A)

SG&A includes all commercial expenses of operations incurred during the regular course of business. SG&A includes advertising, commissions, marketing, selling, and other administrative expenses.

Based on industry benchmarks, SG&A is estimated at 20% of total sales revenue. However, higher percentages of SG&A are assumed during the first few years of rapid company growth when there is a relatively higher amount of SG&A required relative to total sales revenue.

Research and development (R&D) and engineering

R&D and engineering include all expenses that are incurred to develop, design, and improve a product.

Based on industry benchmarks, R&D and engineering are each estimated at 5% of total sales revenue. However, higher percentage R&D and engineering are assumed during the first few years of rapid company growth when there is a relatively higher amount of R&D and engineering required relative to the total sales revenue.

References

[1]   Adapted from Scorecard Valuation Methodology (Rev 2019), Frontier Angels, Bill Payne. www.angel capitalassociation.org/blog/scorecard-valuation-methodology-rev-2019-establishing-the-valuation-of-pre-revenue-start-up-companies/
[2]   www.eval.tech/valuation-multiples-by-industry.
[3]   Adapted from Investment Valuations of Seed and Early-Stage Ventures, Luis Villalobos, 2007. www.entrepreneurship.org/articles/2007/07/investment-valuations-of-seed-and-earlystage-ventures

Section II: **Some early-stage R&D bioproducts**

Herman P. Benecke

Chapter 5
Formaldehyde-free binders

5.1 Background

Typical US homes may contain insulation in their attic and other locations consisting of glass fiber mats that are bound together by a coating consisting of mixed phenol/formaldehyde (PF) and urea/formaldehyde (UF) resins. Glass fibers are bound to each other on a porous continuous conveyer belt by evaporating aqueous solutions of mixed PF/UF binders at temperatures up to 400°F within air-blown ovens where evaporation of water occurs initially and the resin is at this temperature for about 20 s. A vacuum is applied to the bottom of the belt so a continuous glass fiber mat is collected on the conveyor belt. During this drying process, the binder collects where glass fibers contact each other, thereby generating a cross-linked and semirigid matrix. Furniture as well as molded media and structural board found in homes may be prepared from wood fibers that use PF/UF binders to cement these fibers together.

However, these binders pose significant health risks since the UF component is known to slowly release formaldehyde [1], which has been classified as a known carcinogen by the State of California [2] and the World Health Organization [3]. Allergic contact dermatitis is also caused by formaldehyde exposure [1].

5.2 Early formaldehyde-free binders

Companies producing PF/UF binders responded to these health hazards by developing alternate formaldehyde-free binders. Early work involved the esterification reactions at about 400°F of aqueous solutions containing glass fibers, the polymeric carboxylic acid polyacrylic acid (PAA), polyols (compounds containing two or more hydroxyl groups) such as glycerin [4] or triethanolamine [5] when using sodium hypophosphite as the esterification catalyst. These processes generate polyester networks that also effectively encapsulate glass fibers as the water evaporates. These binder systems do not evolve formaldehyde but major deficiencies still exist. One is that their curing times are relatively long, leading to decreased production of cured glass fiber mats requiring increased production lines and increased capital and plant operating costs. These binders also have relatively high acidities due to the required high concentration of carboxylic acid functionality that leads to erosion of binder curing equipment or the need for expensive corrosion-resistant steel.

Herman P. Benecke, Columbus, OH, e-mail: beneckefl@gmail.com

https://doi.org/10.1515/9783110791228-005

5.3 Improved formaldehyde-free binders

An improved approach has been developed that involves the synthesis of polyols that serve as one component of an alternate and improved binder system [6–8]. These polyols are partially derived from 2,5-furandicarboyxlic acid (FDCA), a bio-based and renewable substance that can be derived from fructose which is in oversupply partly due to the popularity of noncaloric sweeteners. As shown in Figure 5.1, one approach to prepare FDCA involves the initial acid-catalyzed cyclization of fructose to 5-hydroxymethylfurfural (HMF) followed by its oxidation to FDCA.

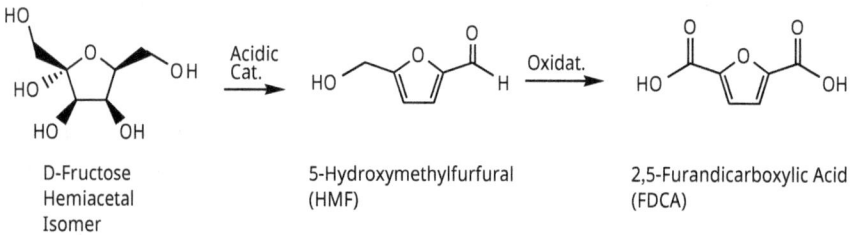

Figure 5.1: Conversion of D-fructose to 5-hydroxymethylfurfural to 2,5-furandicarboyxlic acid.

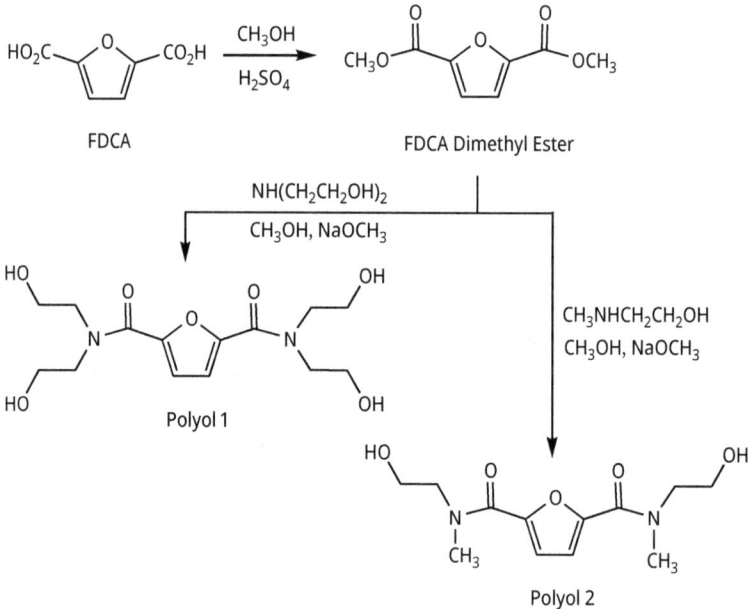

Figure 5.2: Conversion of FDCA to polyols 1 and 2.

The preparation of two representative next-generation polyols is shown in Figure 5.2, wherein FDCA is first esterified with methanol to produce FDCA dimethyl ester. This ester is amidified with diethanolamine in methanol using sodium methoxide as catalyst to produce polyol 1 or with *N*-methyl-*N*-ethanolamine with the same reactive system to produce polyol 2. The main reason for producing and testing polyols 1 and 2 is that they incorporate the tertiary beta-hydroxyethyl amide structure whose primary hydroxyl groups undergo esterification reactions faster than nonfunctionalized primary alcohols [9]. As described later, the enhanced esterification rates of these hydroxyl groups incorporated in tertiary beta-hydroxyethyl amide functionality provided desired faster cure rates.

5.4 Cure times for reaction of PAA with glycerin versus polyol 1

5.4.1 Testing procedure and stroke/cure tests

To obtain comparative standards, aqueous binder test solutions containing glycerin and PAA with the esterification catalyst sodium hypophosphite were prepared. Aqueous test solutions containing polyol 1 and PAA with and without this catalyst were also prepared. All solutions had the same carboxylic acid to hydroxyl mole ratios (1.65:1), the same final water content (69.1%), and the same weight ratios of catalyst to reactants. The pH of fully formulated binder solutions was determined before and after curing. One purpose of these tests was to determine the comparative times needed for curing of binders based on glycerin/PAA versus polyol 1/PAA.

5.4.2 Stroke/cure tests

Binder compositions were evaluated in a preliminary manner by placing equivalent quantities at time zero on a hot plate maintained at 180°C. These samples were stroked back and forth in a regular fashion with a spatula in "stroke/cure" tests, and the times needed to first obtain a viable and self-supporting string were noted as well as the slightly longer times needed for the entire mass to solidify into a tight mass. These later times were recorded as the actual cure times.

5.4.3 Catalyzed cure times of glycerin with PAA

The average cure times obtained in seven experiments for the curing of glycerin with PAA in the presence of the esterification catalyst sodium hypophosphite was 102 s with a standard deviation of 3.3 s. The initial percentage of PAA in these mixtures was 22.3%, and the initial pH values of each solution were in the range of 1–2.

These experiments were performed to determine the variability involved in the stroke/cure tests and also to determine the cure times for the reaction of glycerin with PAA as a standard to surpass in the development of improved processes.

5.4.4 Catalyzed and uncatalyzed reactions of polyol 1 with PAA

The average cure time of two experiments involving curing polyol 1 with PAA in the presence of the catalyst sodium hypophosphite was 84 s. To maintain the same carboxylic acid to hydroxyl mole ratios in the reaction of polyol 1 with PAA compared to the reaction of glycerin with PAA, the initial weight percentage of PAA in the polyol 1 reaction had been decreased to 17.1% while the weight percent of polyol 1 was increased. This compositional shift is due to the higher molecular weight of the tetraol polyol 1 (330.3 g/mole) where 82.6 g is needed to provide 1 mole of hydroxyl groups, while only 30.7 g of the lower molecular weight triol glycerin (92.1 g/mole) is needed to provide 1 mole of hydroxyl groups. Thus, PAA is more dilute in its reaction with polyol 1 reaction, which causes the initial pH values of polyol 1-based solutions to be in the 2–3 pH range (less acidic than glycerin).

When polyol 1 was esterified twice with PAA, but without the esterification catalyst, the average cure time was 87 s and the pH was the same as when performed with the esterification catalyst. Determination of these very similar cure times of polyol 1 with PAA, with and without the catalyst sodium hypophosphite, and given the standard deviation of 3.3 s in determining the catalyzed cure times of glycerin with PAA with the same stroke/cure test, indicates that these two cure times are essentially the same. Thus, use of this catalyst was redundant and did not provide meaningful catalytic acceleration.

5.4.5 Implication of cure time differences between glycerin and polyol 1

Comparison of the cure time of glycerin with PAA in the presence of esterification catalyst sodium hypophosphite (102 s) versus the cure time of polyol 1 with PAA in the absence of catalyst (87 s) implies that the hydroxyl groups of polyol 1 are activated and had access to an efficient esterification mechanism that enhanced their rate of esterification (see later) while glycerin did not.

5.4.6 Mechanism involved in esterification of polyol 1 with PAA

As shown in Figure 5.3, the presumed cure mechanism for the enhanced esterification rate of activated polyol 1 (and presumably activated polyol 2) with PAA is that in the presence of a carboxylic acid, the tertiary beta-hydroxyethyl amide functionality undergoes rapid cyclization to a hydroxy-substituted oxazolidine [9]. This intermediate undergoes the shown reaction to convert the carboxylic acid to a negatively charged carboxylate anion that is paired with a positively charged resonance-stabilized oxazolinium cation that efficiently react with each other to generate polyol 1 or 2 esterified with any carboxylic acid such as PAA.

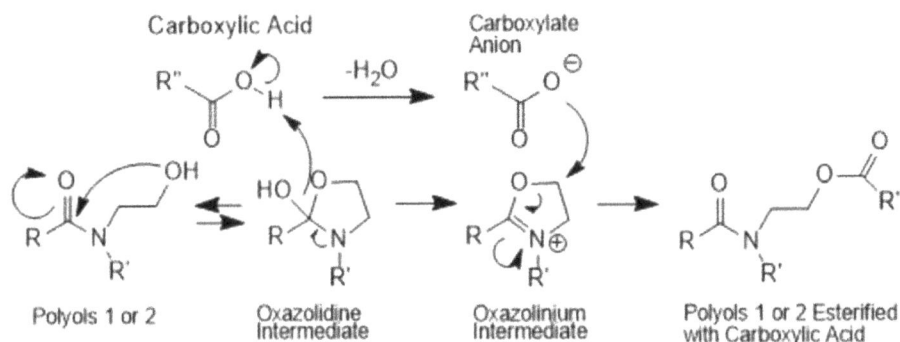

Figure 5.3: Potential mechanism providing accelerated esterification of tertiary beta-hydroxyethyl amide primary hydroxyl groups.

5.4.7 Implication of testing results of activated polyol 1/PAA-based binder and expected properties

The reduced cure times of polyol 1/PAA-based binder compared to the cure times of the glycerin/PAA-based binder indicate that the polyol 1-based fiberglass insulation production time rate for the polyol 1-based binder would provide a cure time that is 85% of the glycerin/PAA cure time without use of an esterification catalyst in a binder production plant. Not needing an esterification catalyst should lead to very reproducible binder properties with no need for exact replication of esterification catalyst concentrations as needed for ester-based binders incorporating glycerin or other similar nonactivated polyols.

That the polyol 1-based binder required appreciably less PAA than the glycerin/PAA-based binder should provide an economic advantage since PAA is a petroleum-based product whose cost will mirror the potential price increases of crude oil. Another advantage of the polyol 1/PAA binder system is that pH values before and during curing were higher (less acidic) than the glycerin/PAA system which should cause decreased production line corrosion and repair and maintenance expenses.

Blending polyol 1 and polyol 2 (that has half the hydroxyl groups compared to polyol 1) would allow adjustment of binder cross-linking that will appreciably influence the viscosity of binder reaction solutions and the rigidity of cured binder mats.

The presence of furanic rings within the cured binder should impart reduced flammability since it is known that furanic compounds are excellent char-formers [10]. Char formation is thought to prevent the spreading of fire and/or cellular degradation because the char has enough integrity to stay in place, thus reducing the heat transfer from the fire to nonburned material.

5.5 Preparation and testing of FDCA/polyol 1 oligomers as binder components

5.5.1 Rationale for preparing and evaluating polyol 1/FDCA oligomers

Another related-type binder system was obtained by pre-esterifying polyol 1 with the diacid FDCA to obtain derived oligomers. Polyol 1 was found to undergo efficient esterification with FDCA to form an oligomer of the type shown in Figure 5.4 [6–8]:

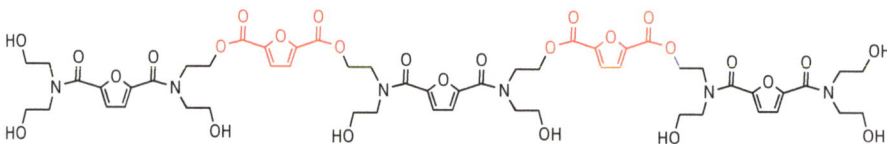

Figure 5.4: Potential oligomer from esterification of polyol 1 with FDCA when using a 3:2 molar ratio.

As an aid to understand the oligomer structure shown above, polyol 1-derived components are colored black while the FDCA-derived components are colored red. This 3:2 oligomer would be prepared by heating this molar ratio of polyol 1 and FDCA, small amounts of sodium hypophosphite catalyst and water in a flask and slowly increasing the reaction temperature over time while passing an inert gas such as nitrogen or argon through the flask to sweep out water to drive the esterification reactions to completion [8]. The structures of isolated oligomers were ascertained primarily by nuclear magnetic resonance (NMR) spectroscopy and infrared (IR) spectroscopy. Stroke/cure testing was performed to obtain cure times.

The oligomeric structure shown above was arbitrarily drawn in a linear fashion with a limited molecular weight due to drawing space restrictions, while branching, cross-linking of molecular chains, as well as variations in chain lengths would also have been expected. The molecular weights of these oligomers would be influenced by the molar ratio of FDCA to polyol 1, and the closer this ratio is to 1:1, the higher the expected oligomer molecular weights.

Use of these types of previously prepared oligomers represents a type of pre-curing that was expected would decrease the binder curing times when esterified with PAA compared to the curing times involved in direct esterification of polyol 1 with PAA as previously described. Efficient esterification of the activated oligomeric primary alcohol sites with PAA during curing should rapidly induce cross-linking and desired rigidity increases.

The presence of multiple hydroxyl groups in these oligomers leads to their having high water solubility caused by the water-solubilizing effect of hydroxyl groups due to hydrogen bonding effects. Binder candidates are required to have high water solu-bilities since they would be applied to fiberglass mats from relatively concentrated aqueous solutions.

5.5.2 Determining polyol 1/FDCA oligomer cure times with PAA

Oligomer cure times for reacting with PAA were determined by stroke/cure tests on a hot plate maintained at 180°C. As previously described, the moles of carboxylic acid provided by PAA relative to the moles of hydroxyl groups provided by different oligomers were maintained very close to 1.65:1, and these compositions had the same water percentage of 69.1% and also the standard weight ratio of catalyst to reactants.

5.5.3 Cure times for reaction of polyol 1:FDCA oligomers with PAA

Stroke/cure tests were performed by reacting aqueous solutions of oligomers of polyol 1:FDCA having molar ratios ranging from 1:1 to 1:0.6 with PAA. The initial PAA percen-tages were in the 12–15% range and thus would provide lower corrosion potential than the polyol 1/PAA as well as the glycerin/PAA curing systems.

Some test mixtures had insoluble components and only those test results obtained from solutions that were completely homogeneous are listed in Table 5.1. As men-tioned in Section 5.4.4 concerning the curing of polyol 1 with PAA, the fact that cure times were almost the same with and without the catalyst sodium hypophosphite in-dicates that the use of this catalyst was redundant and had minimal catalytic effect. Thus, if sodium hypophosphite had been omitted, the same results should have been obtained.

It can be seen that oligomers 2 and 3 with polyol 1:FDCA compositions of 1:0.7 and 1:0.6 had the shortest cure times of 57 s. Continued development and testing of the polyol 1:FDCA oligomer approach would focus on these two oligomer compositions.

Table 5.1: Stroke/cure test results for reaction of polyol 1:FDCA oligomers with PAA.

Oligomer number	Oligomer molar composition		Cure time (s)
	Polyol 1	**FDCA**	
1	1	0.8	83
2	1	0.7	57
3	1	0.6	57

5.6 Overall comparison of cure times for reactions of glycerin, polyol 1, and polyol 1/oligomers with PAA

Table 5.2 provides a comparison of the cure times of glycerin, polyol 1, and polyol 1/FDCA oligomers when cured with PAA. It is apparent that polyol 1:FDCA-based oligomers 2 and 3 have by far the shortest cure times, and these results validate the advantage of this type of pre-curing. Further advantages of this oligomer approach are the reduced use of petroleum-derived PAA in terms of decreased PAA expense and equipment corrosion and consequential equipment repair or replacement.

Table 5.2: Stroke/cure test results of binder compositions with PAA at 180°C.

Samples	Cure times (s)	Percent of glycerin cure time
Glycerin	102	–
Polyol 1	87	85
Polyol 1: FDCA Oligomers 2 and 3	57	56
	57	56

Additionally, an oligomer of polyol 1 and malic acid (2-hydroxybutanedioic acid) had a 46 s cure time when cured with PAA but malic acid is not being promoted since it is not commercially available at scale at this time [8].

5.7 FDCA availability

The relatively low availability and high price of FDCA have provided a severe economic obstacle to the commercialization of products incorporating FDCA and the polyol 1-based compounds and oligomers described in this chapter. However, Avantium N.V., a Dutch producer of numerous bio-based products, announced in a press release on April 7, 2021, that it had plans to open the world's first FDCA plant in 2023. The planned FDCA flagship plant is set to economically produce 5 kton per annum of FDCA, the key building block for the 100% fossil-free, recyclable polymer polyethylene furanoate that is expected to replace high-volume petroleum-based polyethylene terephthalate (PET). Avantium's plans to produce a wide variety of bio-based and renewable products have been published [11].

DuPont recently sold its biomaterial business to China's Huafon Group, and the resulting Covation Biomaterials plans to work with Huafon to produce furan dicarboxylic acid methyl ester that is an intermediate shown in Figure 5.2 for producing polyol 1 and polyol 2 [12].

5.8 Potential development and marketing activities for formaldehyde-free binders

Continued development of formaldehyde-free binders will primarily depend on the economic production of FDCA, diethanolamine and N-methyl-N-ethanolamine which are the central components of polyol 1, polyol 2, and polyol 1/FDCA oligomers, respectively. As indicated earlier, the availability of FDCA depends on Avantium, Covation/Huafon, or another company successfully opening an FDCA plant.

A more accurate method to determine cure times could be developed by using solid-state NMR spectroscopy or IR spectroscopy to determine cure times to replace useful but visually-judged stroke/cure tests. Tensile strength testing methods need to be developed to determine bonding strength between glass fibers or wood particles bonded together by the described formaldehyde-free binders.

US patents 7638592, 8309676, and 9290698 will remain in effect until 2026, 2029, and 2033, respectively, to allow interested companies time to recoup continued research cost before patent expiration. Expiration dates were calculated by adding 17 years to the issuance date.

5.9 Summary

The properties and main features of proposed binders that completely avoid the release of formaldehyde and have curing times with PAA that are appreciably shorter than that of glycerin and other nonactivated polyols are as follows:
- Polyol 1/FDCA oligomers have been developed with cure times with PAA that are appreciably less than polyol 1/PAA-based binders and glycerin/PAA-based binders.

- Polyol 1/FDCA oligomers require reduced use of PAA compared to polyol 1/PAA and glycerin/PAA systems to help moderate future increased prices of petroleum.
- Polyol 1-FDCA oligomers have higher pH values (less acidic) when cured with PAA than polyol 1/PAA and glycerin/PAA-based binders that would lead to decreased production-line corrosion and repair expenses.
- The fact that polyol 1/FDCA oligomers and polyol 1-based binders do not require esterification catalysts for reaction with PAA should lead to very reproducible binder properties with no need for exact replication of catalyst concentrations as needed for binders based on glycerin or similar nonactivated polyols.
- Use of polyol 2 that has only two hydroxyl groups per molecule in combination with polyol 1-based systems should lead to adjustable reaction viscosities and mat rigidities.
- Use of furan-containing polyols in various binder approaches should lead to reduced binder flammability.
- Biobased and renewable fructose is an essential and available starting material needed to synthesize FDCA and Polyols 1 and 2.

References

[1] Latorre N, Silvestre JF, Monteagudo AF. Allergic contact dermatitis caused by formaldehyde and formaldehyde releasers. Actas Dermosifiliogr 2011;102(2):86–97.

[2] State of California Environmental Protection Agency, Office of Environmental Health Hazard Assessment. Safe drinking water and toxic enforcement act of 1986-chemicals known to the state to cause cancer or reproductive toxicity, September 29, 2006.

[3] World Health Organization (WHO). International agency for research on cancer monographs on evaluation of carcinogenic risks for humans by formaldehyde.

[4] Strauss CR. Fibrous glass binders, US Patent 5318990.

[5] Arkens CT. Curable aqueous composition and use as fiberglass non-woven binder, US Patent 5763524.

[6] Benecke HP, Garbark DB, Kawczak AW, Clingerman MC. Formaldehyde free binders, US Patent 7638592.

[7] Benecke HP, Garbark DB, Kawczak AW, Clingerman MC. Formaldehyde free binders, US Patent 8309676.

[8] Benecke HP, Garbark DB, Kawczak AW, Clingerman MC. Formaldehyde free binders, US Patent 8378056 (abandoned).

[9] Wicks ZW Jr., Appelt MR, Soleim JC. Reaction of N-(2-hydroxyethyl) amido compounds. J Coat Technol 1985;57(726):51–61.

[10] Benecke HP, Garbark DB. Biobased polyols for potential use as flame retardants in polyurethane and polyester applications, US Patent 9,290,698.

[11] Reisch MS. Avantium tries for an encore. Chem Eng News 2018;11:24–25.

[12] McCoy M. DuPont biomaterials emerges as Covation. Chem Eng News 2022;10.

Herman P. Benecke, Daniel B. Garbark
Chapter 6
Biobased lubricants and hydraulic fluids

6.1 Background and advantages of soybean oil

Vegetable oils such as soybean oil (SBO) are composed of the triol glycerin that is esterified with three fatty acids to generate triglycerides that have very favorable lubricant properties. This lubricity is based on their excellent film-forming properties due to the strong association of their ester carbonyl groups with metal surfaces. Other advantages of SBO include its high fume point and flash point of approximately 200 °C and 300 °C, respectively [1]. These performance properties and the fact that SBO is biobased, sustainable, and biodegradable has led to its increased use as lubricants and hydraulic fluids over the last 15–20 years [2].

SBO is the major vegetable oil harvested in the USA with a 2021 production rate of about 27 million tons. Many varieties of SBO are grown in the USA, which are characterized by the relative proportion of the unsaturated fatty acids (oleic acid), the polyunsaturated fatty acids (linoleic and linolenic acids), and the saturated fatty acids (stearic and palmitic acids). Based mainly on genetic modification and a minor amount of crop breeding, SBO production has been shifting toward higher oleic SBO due to the enhancement of its lubricant properties with increased oleic acid and other properties as described in this chapter. The fatty acid compositions of these SBO grades are shown in Table 6.1, where we refer to the original starting composition as "commodity"-grade SBO.

Table 6.1: Percentages of oleic, polyunsaturated fatty acids, and saturated fatty acids in US soybean oils.

Description	% Oleic	% Polyunsat.	% Saturated
Commodity	22	61	16
Low saturates	22	69	9
Mid oleic	53	32	15
High oleic	77	11	12
90% oleic	90.5	1.5	8.0

Herman P. Benecke, Columbus, OH, e-mail: beneckefl@gmail.com
Daniel B. Garbark, Columbus, OH, e-mail: garbarkd@battelle.org

https://doi.org/10.1515/9783110791228-006

6.2 Performance limitations of SBO

A serious liability of commodity-grade SBO is the oxidative instability of its allylic methylene groups that are adjacent to the lone double bond in oleic acid and especially the double allylic methylene groups flanked by two double bonds as found in linoleic and linolenic acids. Hydrogen atoms on these methylene groups are readily abstracted by molecular oxygen (a diradical) at normal and especially at elevated temperatures. Hydrogen abstraction is enabled as the resulting radicals are stabilized by resonance with one double bond as found in oleic acid and by two double bonds as found in linoleic and linolenic acids. These carbon radicals initiate fatty acid bridging processes as well as scission leading to nondesired deposits and volatilization [3].

Other SBO liabilities when functioning as lubricants include the hydrolysis of ester linkages, relatively low viscosities and relatively high pour points (temperature below which the oil does not flow). These liabilities and solutions are addressed by different chemical modifications described in this chapter.

One tactic used to address the oxidative instability of SBO had been to employ heavy metal-based antioxidants but these materials can be harmful since about 50% of used motor oil worldwide is dumped in the environment [3]. As mentioned, mainly genetic modification and some crop breeding have been used to reduce the relative proportion of linoleic and linolenic acids in SBO that are most reactive toward oxygen and also to reduce saturated fatty acids to help reduce oil pour points. However, the oxidative and hydrolytic stabilities of relatively high oleic SBOs achieved by these methods still do not measure up to the properties provided by petroleum-based lubricants and hydraulic fluids that have minimal oxidative and hydrolysis issues.

6.3 Chemical modifications to enhance the oxidative stability of SBO

Two related SBO molecular modifications were developed to address the oxidative instability that involves adding specific substituents to the original double-bonded carbon atoms in SBO's unsaturated fatty acids, thus removing double bonds that are the cause of their oxidative vulnerability [1, 3]. In one approach, two ester groups are added across each double bond, and in the second approach, one ester group and one hydrogen atom are added across each double bond. Both approaches are initiated by fully epoxidizing all SBO double bonds to their epoxidized derivatives by reaction with formic acid and hydrogen peroxide.

6.3.1 Preparation of triglyceride vicinal diesters

Figure 6.1 illustrates the reaction of epoxidized linoleic acid with carboxylic acid an-
hydrides in the presence of the base potassium carbonate. This process leads to positioning two vicinal (adjacent) ester groups along the fatty acid at each original
olefinic carbon atom to form "diesters." These modifications shown with epoxidized
linoleic acid will also be taking place with oleic acid and epoxidized linolenic acid so
that essentially no double bonds remain to activate methylene group oxidation.

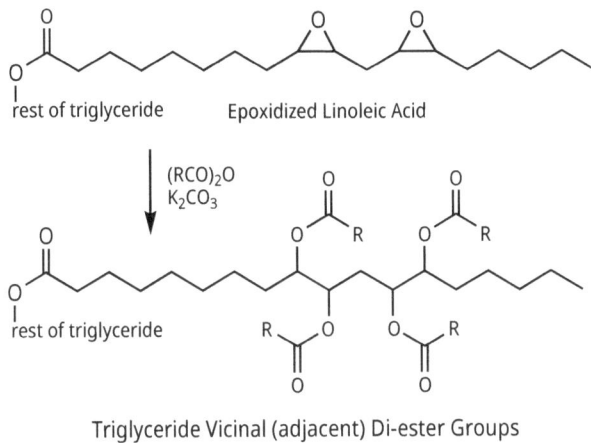

Figure 6.1: Preparation of triglyceride vicinal diesters.

When the term "diester" or a more specific term such as "dihexanoate" is used, we
are referring to the vicinal diesters positioned at each end of all original double
bonds as implied in Figure 6.1. The size of "R" groups may be varied to achieve variations in viscosities and physical properties to influence their lubricant and hydraulic
fluid applicability.

6.3.2 Preparation of triglyceride monoesters

Figure 6.2 shows a two-step reaction sequence starting with epoxidized linoleic acid
that involves the hydrogenolysis of each epoxide ring to deposit a hydroxyl group and
a hydrogen atom at either end of the original double bond. One of four possible regioisomers are shown in Figure 6.2 but other regioisomers (where hydroxyl and hydrogen groups are switched) would also be generated. The hydroxyl groups are
typically acylated with acid chlorides to generate "monoester" functionality positioned along the fatty acids at their original double-bonded carbon atom sites.

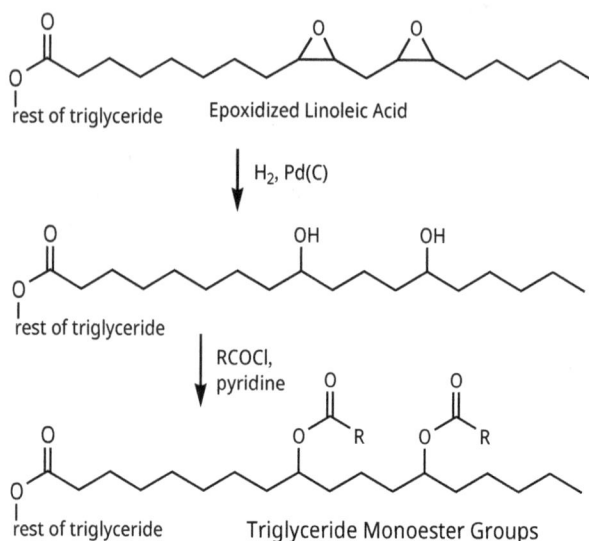

Figure 6.2: Preparation of triglyceride monoesters.

The size of "R" groups may also be varied to influence their lubricant and hydraulic fluid applicability.

6.4 Determining oxidative resistance

Two tests were used to determine the oxidative resistance of SBO and the ester derivatives described earlier when exposed to specified high temperatures and heated air flows for specified time periods to determine their suitability for use as lubricants and hydraulic fluids.

6.4.1 Noack test

The Noack test (described in ASTM D5800) determines the loss by volatilization of samples heated at 250 °C for 60 min to initiate oxidative processes with a constant flow of heated air over the samples that also remove volatile components. This test data is relevant for combustion engine oil candidates since it measures the weight loss of lighter components formed by oxidative degradation that would increase the viscosity of remaining oil, thus contributing to poor engine circulation and reduced lubricity. In general, the Noack test for engine oils requires volatility losses less than 15% for certification for use as an engine oil. Multiple Noack tests were run with a variety of derivatized SBO diester and monoester samples and all received passing scores, thus helping qualify their potential use as motor oils. Hydraulic fluid candidates may also be characterized by this test.

6.4.2 Penn. state microoxidation test

In the Penn. state microoxidation test, specified amounts of samples are placed on stainless steel coupons and heated at 180 °C, and specified volumes of air are passed over the sample for specified times. Samples are weighed afterward to determine the amount of volatilized sample and washed with the nonpolar solvent tetrahydrofuran to dissolve the oil and leaving behind potential deposits that are weighed to determine their quantity [3].

6.5 Performance evaluation of SBO diesters and monoesters

In a type of test described later, SBO lubricities and SBO triglyceride diesters and monoesters were determined by measuring the scars incurred in four-ball wear tests as described in ASTM D4172. In these tests, stainless steel balls are rotated past each other under high load while in contact with lubricant candidates.

Seal swelling is not desired in lubricant and hydraulic systems. A testing approach to determine the extent of seal swelling of lubricant and hydraulic oil candidates was developed as described further [3].

6.5.1 SBO ester derivative lubricities determined with four-ball wear tests

Four-ball wear tests were performed with commodity-grade SBO and commodity-grade triglyceride diester and monoester derivatives [3]. These tests were performed with stainless steel balls at 18 kg load at 75 °C, 1,200 rpm, and for 60 min, and the results are shown in Table 6.2. ZDDP (zinc dialkyl dithiophosphate) is a wear additive and its use was evaluated at 1% with SBO and modified SBO samples. In four-ball wear tests, the lower the scar diameter, the higher the lubricant's lubricity.

As shown in Table 6.2, nonmodified SBO had the highest scar diameter that was reduced to about half with 1% ZDDP. The best performing modified oil was SBO dihexanoate *without any ZDDP*, and its scar diameter was equal to or less than that obtained from a fully formulated SHC-134 gear oil and an SAE 10W-30 motor oil. The next highest performer was SBO monoisobutyrate that when formulated with 1% ZDDP had a scar diameter only slightly larger than the scar diameters obtained with

Table 6.2: Four-ball wear tests of commodity-grade SBO, diester, monoester, and commercial oils at 18 kg load.

Description:	SBO	Dihexanoate	Monoisobutyrate	SHC-134	SAE 10W-30
Wear scar (mm)	0.62	0.26	0.32	0.26	0.27
Wear scar +1% ZDDP (mm)	0.32	0.28	0.28	–	–

the commercial oils. The scar diameters of four other modified oils when modified with 1% ZDDP were shown to be only slightly higher than those of the commercial oils. In summary, these test results show that certain commodity-grade SBO diesters and monoesters had significantly smaller wear scars than commodity-grade SBO, and their scar diameters were equal to or comparable to those of a fully additized gear oil and commercial motor oil [3].

The reduced scar diameters displayed by the behavior of SBO ester derivatives reflects enhanced lubricities provided by the added ester groups positioned along the triglyceride fatty acids. These extra ester groups presumably provide extra bonding with metal surfaces in addition to the bonding provided by the three triglyceride ester carbonyl groups.

Four-ball wear tests were also performed under 40 kg load in which diesters and monoesters did not perform as well as commercial oils. However, addition of 1% ZDDP provided performances equal to or nearly equal to fully additized commercial oils. Continued research could evaluate the use of SBO diesters and monoester of grades other than commodity to determine if suitable performance can be obtained in 40 kg four-ball wear tests without ZDDP.

6.5.2 EPDM seal swelling in the presence of lubricants and hydraulic fluids

Seal swelling is a problem in many lubricant and hydraulic fluid applications, and the seal material EPDM (ethylene propylene diene, M class, hardness) is commonly used in these applications.

EPDM pieces were weighed and immersed in commodity-grade-derived SBO diesters and monoesters for 24 h at 68 °C, thoroughly dried, and then weighed again to determine any weight gains or losses [3]. These tests determined that 8.9% of commodity-grade SBO was absorbed in EPDM while zero or nil quantities of the diesters SBO triglyceride dihexanoate, SBO triglyceride-bis-2-ethylhexanoate, and SBO triglyceride-bis-2-ethyl butyrate were absorbed in EPDM. Close to 1% of two commodity-grade SBO triglyceride monoesters were absorbed in EPDM.

In summary, these tests indicated that EPDM was essentially resistant to the commodity-grade SBO triglyceride diesters tested. Further testing would determine the magnitude of swelling over extended times to determine the potential long-term absorption of these SBO derivatives in EPDM. Nitrile rubber was essentially unchanged when exposed to both nonmodified and modified commodity-grade SBO.

6.6 Utility of SBO triglyceride diester and monoester lubricants

6.6.1 SBO triglyceride dihexanoates

Table 6.3: Viscosities of SBO triglyceride dihexanoate lubricants from various SBO grades.

Description	90% oleic	High oleic	Mid-oleic	Commod.	Low saturates	SAE 50
Viscosity (cSt)						
At 40 °C	210	272	349	398	601	222
At 100 °C	23.5	28.8	33.6	36.2	47.6	20.4
VI	138	141	137	134	132	107
Pour point	−23	−24	−21	−21	−21	−18

Data shown in Table 6.3 indicate that the ISO values [viscosity values in centistokes (cSt) at 40 °C] of SBO triglyceride dihexanoates steadily increase between 90% oleic and the low saturate SBO grade [1]. As given in Table 6.1, this viscosity shift is accompanied by a corresponding general reduction in the concentration of oleic acid (containing one double bond) and an increase in the combined concentrations of linoleic and linolenic acids (containing two and three double bonds, respectively). Thus, this compositional shift provides an increase in the number of triglyceride fatty acid double bonds as one moves from left to right in Table 6.1. Given that each double bond is functionalized by two hexanoate ester groups, the increase in ISO viscosity is primarily explained by the increased contribution of these dihexanoate ester groups to the molecular weights of triglyceride fatty acids as they undergo this compositional shift. However, increased amounts of saturated fatty acids also contribute to higher ISO viscosities.

It can be seen that the SBO triglyceride dihexanoate lubricant prepared from 90% oleic acid with an ISO value of 210 is in the range of SAE 50 motor oil that is used in high-performance race cars.

Viscosity index (VI) is a measure of a fluid's viscosity change with the change in temperature, and higher VIs are desirable since they undergo significantly lower viscosity changes with temperature changes. Note in Table 6.3 that the VIs for all SBO triglyceride dihexanoates are slightly higher than that of the commercial SAE 50 oil shown.

The higher viscosity SBO triglyceride dihexanoates shown in Table 6.3 could be used as base oils for grease applications and gear oils. It has been found that lower molecular weight diesters have higher ISO viscosities than higher molecular weight diesters. Thus, the ISO viscosity of commodity-grade SBO triglyceride dihexanoate is 398 (pour point: −21 °C) while the ISO viscosity of commodity-grade SBO triglyceride dipentanoate is 422 cSt (pour point: −21 °C). The ability to readily achieve a variety of high-viscosity grease base oils presents the opportunity to use decreased amounts of expensive thickeners such as polyureas.

6.6.2 SBO triglyceride monohexanoates

Commodity-grade SBO was also converted to SBO triglyceride monohexanoate (ISO 136) and monoisobutyrate (ISO 222) that could be used for SAE 40 and SAE 50 motor oils, respectively. This comparison illustrates that monoester derivatives have lower ISO viscosities than comparable diester derivatives when functionalized with the same ester due to their lower overall molecular weights.

6.7 Glycerin replacement (backbone modification) in SBO triglycerides with dialcohols and monoalcohols

One approach for overcoming high viscosities found in triglyceride derivatives is to hydrolyze a desired SBO-grade sample and re-esterify the recovered fatty acids with a dialcohol (diol) or a monoalcohol (mono-ol) [1]. Figure 6.3 shows the esterification of high oleic acid with the diol 1,2-propylene glycol (1,2-PG) to generate 1,2-PG dioleate that would be epoxidized and then acylated with a carboxylic acid anhydride of choice to generate the product 1,2-PG dioleate dicarboxylate. Exchanging the alcohol in an ester is referred to as changing the ester "backbone" or performing a "backbone" modification.

Another approach to obtain the same product would involve the transesterification of high-oleic SBO with methanol to prepare high-oleic methyl oleate (a specific soy biodiesel) that would be efficiently transesterified with 1,2-PG to produce 1,2-PG dioleate. This intermediate would also undergo the last two reactions shown in Figure 6.3 to produce 1,2-PG dioleate dicarboxylate.

6.7.1 Utility of 1,2-PG disoyate dihexanoates

Note that the ISO viscosities of the SBO 1,2-PG disoyate dihexanoate samples shown in Table 6.4 are appreciably lower than the ISO viscosities of the corresponding grade SBO triglyceride dihexanoate samples in Table 6.3 due to their lower molecular

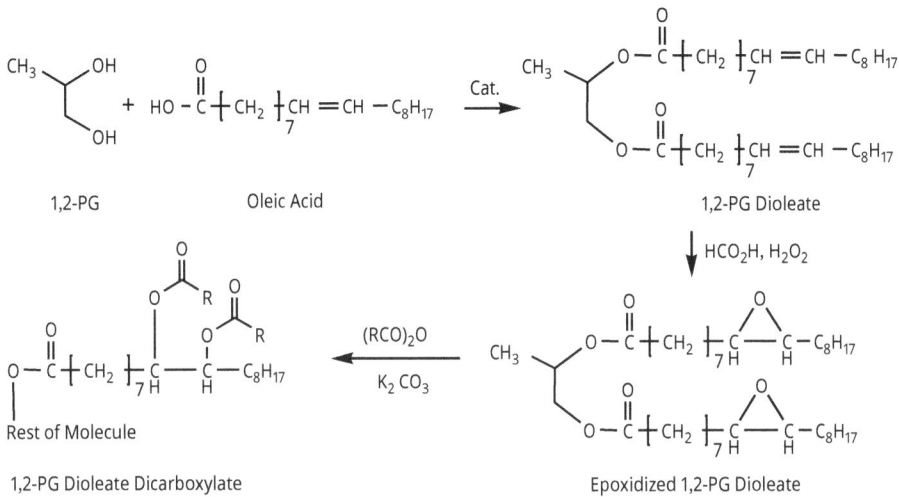

Figure 6.3: Re-esterification of fatty acids from SBO with 1,2-PG.

Table 6.4: Viscosity and pour point data for 1,2-PG disoyate dihexanoates for various SBO grades and select commercial motor oils.

Description	90% oleic	High oleic	Commodity	SAE 10W-30	SAE 10W-40	SAE 50
Viscosity (cSt)						
At 40 °C	78.0	99.3	232	69.7	95.4	222
At 100 °C	12.1	13.7	20.8	10.7	14	20.4
VI	153	139	105	141	151	107
Pour point (°C)	−34	−12	−12	–	−27	−18

weights [1]. Also note that as one moves from left to right in Table 6.4, the ISO viscosities of different grade SBOs increase significantly as the oleic acid concentration decreases while the linoleic plus linolenic acid concentrations increase (see Table 6.1). This ISO viscosity shift represents another example of viscosity dependence on the "oleic acid/(linoleic plus linolenic acid)" ratio.

The 1,2-PG disoyate dihexanoate sample from 90% oleic SBO with an ISO viscosity of 78 cSt is in the range of 10W-30 motor oil. The beneficial pour point of −34 °C is due to the relatively low percentage of saturated fatty acids in this sample (nominally 8%).

The 1,2-PG disoyate dihexanoate sample from high oleic SBO has an ISO viscosity of 99.3 cSt with a pour point of −12 °C. This sample is in the range of an ISO 100 hydraulic fluid as well as 10W-40 motor oil. High oleic SBO has a nominal saturate level of 12% that contributes to this sample's higher pour point. Another factor contributing to this high pour point is that the sum of linoleic and linolenic acids in high oleic SBO is 11% while it is only 2% in 90% oleic SBO (whose 1,2-PG disoyate dihexanoate's pour point is −34 °C as described earlier).

The 1,2-PG disoyate dihexanoate from commodity-grade SBO with an ISO viscosity of 232 is in fairly close range to SAE 50 motor oil. Contributing to its relatively high pour point of −12 °C is the relatively high saturate content of 16% while the sum of linoleic plus linolenic acid (61%) also contribute to this sample's high viscosity and pour points. Use of pour point depressants should also improve its low-temperature behavior.

A specialty 88% "oleic" oil (Emersol 233) containing 74% oleic, 14% palmitoleic (a 16-carbon, monounsaturated fatty acid like oleic acid), 4% linoleic, and 8% stearic acids was converted to 1,2-PG "Di-Emersol 233" dihexanoate. This product has an ISO viscosity of 97.2 with a VI of 149, and pour point of −39 °C, and is thus in the range of an ISO 100 hydraulic oil and 10W-40 motor oil. The low pour point of −39 °C is an attractive property for either use.

6.7.2 Utility of 2-butyl "oleate" diisobutyrate

Branched mono-ols such as 2-butanol can be esterified with different grades of soy fatty acids to produce diesters and monoesters with relatively low viscosities due to their relatively low overall molecular weights. Pioneer Plenish-brand SBO (oleic acid content 76%) was hydrolyzed to produce oleic acid that was esterified with 2-butanol, and this ester reacted with diisobutyric anhydride to generate 2-butyl "oleate" diisobutyrate. This product has an ISO viscosity of 25.0 with a VI of 109 and a pour point of <−40 °C, so it is expected to be an effective low-temperature hydraulic fluid.

6.8 Biodegradable SBO diesters and monoesters and potential uses

Commodity-grade SBO triglyceride dihexanoate and SBO triglyceride mixed mono (hexanoate/2-ethylhexanoate) were shown to be highly biodegradable when using ASTM 5864 [3]. That these diverse diester and monoester samples demonstrated biodegradability strongly suggests that the range of SBO ester derivatives is also biodegradable and this property will help mitigate irresponsible dumping and related environmental pollution. Given their adaptability to fit various viscosity and low-temperature profiles, specific SBO-based esters described herein could also fit well in outdoor applications, including two-stroke engine oils, open-gear lubricants, rail flange oils and greases (to limit friction on straight and especially curved rails), rock-drilling fluids, and transformer fluids (with sufficiently high dielectric strength).

6.9 Controlling hydrolysis rates of ester-based lubricants and hydraulic fluids

An issue in ester-based motor oils is their potential hydrolysis due to contact with water produced in combustion-driven engines. It is interesting that hydrolysis of ester-based motor oils should be nonexistent in battery-powered car engines due to the absence of "water of combustion" in their engines.

It is known in general that the rates of ester hydrolysis are retarded by steric crowding around the ester carbonyl group provided by alkyl substituents positioned alpha and beta to the alcohol hydroxyl functionality. Alcohols that provide some degree of steric crowding to reduce ester hydrolysis include the diols 1,2-PG, 2-methyl-1,3-propane diol, and the mono-ol, 2-butanol. Inspection of models of the triol 1,1,1-tris(hydroxymethyl) propane [also known as trimethylolpropane (TMP)] indicates that the ester carbonyl groups of TMP fatty acid esters, where each TMP alcohol group is esterified with a fatty acid, should experience significant steric hindrance toward hydrolysis.

Ester hydrolysis in hydraulic systems should be at a minimum since these are closed systems.

The hydrolytic stabilities of ester-based samples were obtained via ASTM D2619 by heating weighed samples in a tumbled vessel immersed in water heated at 93 °C for 48 h and then titrating with KOH to determine released carboxylic acid. A summary of an increase in acid number of these samples as well as other test data is given in Table 6.5.

Table 6.5: Backbone effects on hydrolytic stabilities, volatilities, and flash points.

Hydrolytic stability tests (ASTM D2619)		Noack % volatility ASTM 5800B	Flash point (°C) ASTM D92-05a
Lubricants: dihexanoates of commodity SBO	Change in acid number		
Triglyceride backbone	+1.34 mg KOH/g	6.1*	298
2-Methyl-1,3-propane diol backbone	+0.65 mg KOH/g	5.8	273
TMP backbone	+0.22 mg KOH/g	6.3	243

*Penn state microoxidation test: % weight loss at 180 °C for 60 min.

A passing score in the hydrolytic stability test is that not higher than 2.0 mg KOH/g is needed, so all substrates passed this test. It can be seen that replacing the glycerin backbone with 2-methyl-1,3-propane diol and TMP provided increased hydrolytic stability. The pendant methyl groups in 1,2-PG and 2-butanol fatty acid esters are also expected to provide some hydrolytic stability.

In the Noack test, engine oils require volatility losses less than 15% for certification; hence, the above mentioned samples pass this test for motor oil use given the above volatilization data.

The observed high flash points also appear to be reasonable for a number of uses.

6.10 Viscosity and oxidation stability comparisons between nonmodified SBO versus SBO diester and monoester derivatives

For comparison purposes, refined, bleached, and deodorized (RBD) commodity-grade samples of nonderivatized SBO were tested and determined to have an average ISO viscosity of 32.0 with an average VI of 228 [3]. These commodity-grade SBO samples and an 82% oleic SBO sample were tested by the Penn. state microoxidation test, and each sample demonstrated some deposit weight gain and appreciable evaporative loss indicating that oxidative degradation processes were active in both commodity-grade SBO and relatively high oleic acid SBO which indicates that it also has some oxidative vulnerability.

Importantly, the low ISO viscosities of these RBD SBO samples do not come close to matching the much higher viscosities of high-performance motor oils and gear oils and base oils for grease applications that are described in Section 6.6 for specified SBO triglyceride diesters and monoesters. These RBD SBO viscosities are also much lower than the viscosities described in Section 6.7.1 for 1,2-PG disoyate dihexanoate derivatives that fit the viscosity requirements of 10W-30, 10W-40, SAE 50 motor oils, and an ISO 100 hydraulic fluid.

This comparative data supports the practice of converting specific grades of SBO to their diester and monoester derivatives to generate a range of usable and oxidatively stable SBO-derived lubricants and hydraulic fluids that are not available from direct use of nonmodified SBO.

6.11 Continued development and commercialization activities

This chapter describes various SBO modifications that overcome the oxidative vulnerability of SBO while simultaneously increasing the inherently low SBO viscosities and producing numerous applications with useful properties. Companies currently producing lubricants, motor oils, greases, and hydraulic fluids from petroleum sources should be highly motivated in further evaluating the potential and cost-effectiveness of SBO diesters and monoesters described in this chapter. Their interest would be mainly driven by the apparent shift away from fossil fuels due to climate change considerations that may significantly reduce the availability of petroleum-based starting materials.

Continued research could focus on preparing lubricants and hydraulic fluid candidates that more closely matched the viscosities of commercialized materials to evaluate performance evaluations.

Data presented in Section 6.5.1 shows that SBO triglyceride dihexanoate in 18 kg load four-ball wear tests without the use of the zinc-containing wear additive ZDDP provided scar diameters that were lower than those obtained from fully additized commercial lubricants. However, when this test was repeated at 40 kg load, this SBO derivative did not perform as well as these commercial lubricants. Continued research could determine whether a range of 1,2-PG disoyate dicarboxylates would pass the 40 kg load test to determine if metal-containing wear additives were not needed which would provide environmental and cost advantages while also providing systems more resistant to hydrolytic cleavage.

Research could also be continued in evaluating the same type of 1,2-PG diester derivatives to potentially obtain more effective seal-resistant lubricants and hydraulic fluids compared to previously tested SBO triglyceride diesters (see Section 6.5.2).

US patents 8357643 and 9359572 will remain in effect until 2030 and 2033, respectively, thus allowing interested companies time to recoup continued research and development costs before their expiration.

6.12 Summary

The tendency of SBO to readily undergo oxidative cleavage and polymerization greatly inhibits SBO's direct use as a base oil for a variety of lubricants and hydraulic oils. Based on Noack and other test results, this liability is overcome by modification of SBO's double bonds by addition of two ester groups to form "diesters" or by addition of one ester group and a hydrogen atom across these double bonds to form monoesters [1, 3].

The attributes of diester and monoester derivatives across the range of SBO grades and compositions are as follows:
- A wide range of base oil viscosities are obtained from different grades of SBO that are characteristic of gear oils, greases, engine oils, and hydraulic fluids.
- Gear oils and greases maintain their triglyceride structures and higher viscosities.
- Engine oils and hydraulic fluids may require triglyceride backbone exchange with dialcohols such as 1,2-PG or monoalcohols such as 2-butanol to achieve needed decreased viscosities and decreased pour points.
- The ability to readily achieve high-viscosity greases presents the opportunity to avoid the use of expensive thickeners such as polyureas.
- An SBO triglyceride diester and a mixed monoester were shown to be highly biodegradable, and this general property will help mitigate irresponsible dumping.
- Based on their biodegradability, the described modified lubricants fit well with outdoor applications such as two-stroke engine oils, open-gear lubricants, rail flange oils and greases, rock-drilling fluids, and transformer fluids (with sufficiently high dielectric strength).

- In ester-modified SBO, viscosities decrease with increases in oleic acid content and decrease in linoleic and linolenic acids while viscosities increase with decreases in oleic acid concentration and increases in linoleic plus linolenic acids.
- Hydrolytic stability of diester-based lubricants and hydraulic fluids is increased by triglyceride backbone exchange with alcohols bearing substituents at their alpha and beta positions.
- Hydrolysis of diester and monoester lubricants would be minimal when used in electric motors due to the lack of "water of combustion" formed in these engines.
- Pour points move to desired lower values with decreasing saturated fatty acid contents.
- Hydraulic fluid candidates that fit the high-volume ISO 100 market (pour point: −39 °C) and also the low-temperature ISO 22 market (pour point: <−40 °C) have been prepared.
- Four-ball wear-test scar diameters (18 kg load) of certain SBO triglyceride diester and monoesters are significantly smaller than obtained from SBO (all without antiwear additives) and match those attained in commercial additized lubricants.
- Swelling of EPDM rubber by certain SBO diesters was significantly reduced (or was nil) compared to significant swelling in SBO which has positive implications in overcoming seal swelling.
- Due to its reduction in double bonds, higher oleic SBO requires only about two-thirds of the reagents needed to produce diester or monoester derivatives compared to commodity-grade SBO.

References

[1]   Benecke HP, Garbark DB, Vijayendran BR, Cafmeyer J. Modified vegetable oil lubricants. US Patent 9359572 (issued on Jun. 7, 2016).
[2]   Vijayendran BR. Biobased motor oils are ready for primetime. Ind Biotech 2014;10:1–5.
[3]   Benecke HP, Vijayendran BR, Cafmeyer J. Lubricants derived from plant and animal oils and fats. US Patent 8357643 (issued on Jan. 22, 2013).

Claudia Creighton, Nguyen Duc Le, Huma Khan, Srinivas Nunna,
Russell J. Varley*, Cai Li Song*

Chapter 7
An introduction to lignin-based carbon fibre

7.1 Background and history of carbon fibre

Carbon fibre possesses a unique combination of properties that sets it apart from most other fibres, such as high specific strength and modulus, low density, creep resistance, high thermal and electrical conductivity, excellent chemical resistance and high temperature resistance in inert atmospheres [1–4]. Translating these properties into a carbon fibre composite therefore has enabled the development of lightweight, high-strength and stiffness structural materials and components that have revolutionized a wide range of industries such as transportation, sporting goods, defence, wind energy, pressure vessels and construction [2, 5–7].

Since the discovery of carbon fibre more than 140 years ago, carbon fibre properties have continued to improve and evolve as improved manufacturing methods have been developed, and the knowledge of the underlying materials chemistry of the precursor fibre has increased. The first carbon fibre was made by Joseph W. Swan and Thomas A. Edison in 1880 by meticulously carbonizing cotton threads and used as a filament for an incandescent bulb [8–10]. Unfortunately, soon after its invention, tungsten replaced carbon fibre as a bulb filament [11]. However, in 1909, W.R. Whitney continued the research and showcased an effective way of graphitizing cellulosic fibres for bulb filament applications [12]. After 50 years, Roger Bacon and a group of scientists at Union Carbide Corporation, Ohio, USA, extensively researched pathways to produce high-performance carbon fibres from cellulose-based rayon precursor [12–14]. Their research efforts led to the commercialization of high-modulus rayon-based carbon fibres and found applications in the defence industry sector [15]. While researchers based in the USA were exploring rayon-based carbon fibres, in 1961, Akio Shindo from the Government Industrial Research Institute, Osaka, invented polyacrylonitrile (PAN)-based carbon fibres [16–18]. Shortly after this invention, in 1967, scientists R. Moreton, W. William and W. Johnson from Royal Aircraft Establishment (RAE) developed protocols to produce high-strength

---
*Corresponding author: Russell J. Varley, Carbon Nexus at the Institute for Frontier Materials,
Deakin University, 75 Pidgons Rd, Waurn Ponds, Victoria 3216, Australia, e-mail: russell.varley@deakin.edu.au
*Corresponding author: Cai Li Song, Petronas Research Sdn Bhd, Lot 3288 & 3289, Off Jalan Ayer Itam,
Kawasan Institusi Bangi, 43000 Bangi, Selangor, Malaysia, e-mail: song.caili@petronas.com
Claudia Creighton, Nguyen Duc Le, Huma Khan, Srinivas Nunna, Carbon Nexus at the Institute for
Frontier Materials, Deakin University, 75 Pidgons Rd, Waurn Ponds, Victoria 3216, Australia

https://doi.org/10.1515/9783110791228-007

and high-modulus carbon fibres from PAN-based precursor fibres with the application of tension during the thermal treatment process [19, 20]. Two years later, in 1969, carbon fibres were also produced from other sustainable precursors such as lignin, and this work was patented by Sugio Otani et al. [21], which was assigned to Nippon Kayaku Co Ltd. However, after few years of commercialization, in 1973, the production of lignin carbon fibres ceased due to its poor mechanical properties [22]. Nevertheless, due to their characteristic advantages, research on lignin–based carbon fibres have continued until today, primarily focussing upon improving its mechanical performance. In 1977, Leonard Sidney Singer from Union Carbide Corporation, New York, USA, invented high-modulus carbon fibres from mesophase pitch precursor [23]. Horikori et al. [24] from Sumitomo Chemical Company Ltd, Japan, published a patent in 1978 which describes the method to produce carbon fibres from polyethylene precursor source through sulphonation and subsequent carbonization process. As the history of making carbon fibre shows, over the past, scientists explored various precursor sources to produce carbon fibres; however, since the 1970s only PAN-based carbon fibres have seen an exponential market growth and remains by far the most commonly used precursor for carbon fibre today.

7.1.1 Market trends

In 2019, the global nameplate capacity of carbon fibre was 161,200 metric tonnes (MT). The carbon fibre market is dominated by PAN-based carbon fibre (representing almost 90% of the total carbon fibre production) [25]. By 2026, this figure is expected to increase to 180,400 MT [26]. In terms of value, this translates to roughly $3.5 billion [27]. The markets that are expected to drive the substantial increase in demand for carbon fibre over the next few years are primarily the wind, aerospace or defence, and the infrastructure and energy sectors (Table 7.1). In fact, the wind energy industry alone is already the largest consumer by volume of carbon fibre in the world, making up to 28,000 MT in 2021. Further driven by the rapid surge of onshore and offshore wind turbines, more so in China, this amount is expected to double to 54,000 MT in 5 years' time, with the USA being the largest supply of wind turbine blades. In contrast, in the aerospace industry, although the demand for composite-intensive aircrafts, such as the likes of Boeing 787 and Airbus A350, has suffered from the covid-19 pandemic, it is likely to return to pre-COVID-19 levels by 2025, with a market capacity of around 29,100 MT expected. In the aerospace sector, significant weight reduction can be achieved through the use of carbon fibre in manufacturing. According to a research from Toray Industries, if all passenger cars and aircraft in Japan adopted carbon fibre to reduce weight and improve fuel efficiency, 22 million tonnes of CO_2 could be saved [28].

By far, due to the complicated carbon fibre manufacturing process requiring some of the world's most advanced production technology, the global carbon fibre manufacturers are still mainly concentrated in Japan and the USA. Some big players on the PAN-based

Table 7.1: Summary of carbon fibre demand by its tow and geographic region (adopted with permission from [29] Copyright 2022 USDOE/OSTI).

	CAGR 2012–2018*	2018 demand (tonnes)*	Fibre tow	Region of greatest CF demand (% of worldwide demand)
Wind energy	11.5%	25,070	24K–50K	Europe (68%)
Aerospace	13.0%	18,667	Small tow	Europe (46%)
Automotive	20.5%	9,863	Large tow (>24K)	Europe (62%)
Pressure vessel	21.9%	7,225	12K–50K	North America (60%)
Overall	12.5%	92,802	–	–

*Average of estimates by Red and Zimm (2012), Lucintel (2012) and Industry Experts (2013).
 CAGR, compound annual growth rate.

carbon fibre market are Toray (Japan), Hexcel (USA), Mitsubishi Chemical Carbon Fibre & Composites (USA), Teijin Limited (Japan), SGL Carbon (Germany) and many more. Among all the carbon fibre manufacturers and suppliers, Toray dominates the carbon fibre supply chain with 54,870 MT of annual capacity, in both PAN-based regular tow and large tow carbon fibre [30]. Meanwhile, the pitch-based carbon fibre on the market started with Kureha (Japan), the first company in the world, to successfully develop an industrial manufacturing process for pitch-based carbon fibre in 1970, and has now been penetrated by companies like Cytec Industries, Inc. (USA), Hexcel Corporation (USA) and Nippon Graphite Corporation (Japan) [31, 32]. Tempering the figure of the estimated market capacity in the future is the knockdown effect, which is the potential carbon fibre shortfall between the nameplate capacity of existing carbon fibre facilities and the actual production capacity (125,150 MT) that is estimated to reach 55,250 MT in 2026, after taking into account all the recent expansion announcement by Zoltek (Toray Group, Japan), Hexcel (USA), Hyosung (South Korea) and DowAksa (a joint venture of DOW Chemical, USA, and AKSA, Turkey). Among the issues that hinder the growth and competitiveness of the carbon fibre industry to fill the shortfall, the most significant is the production costs that directly affect the price of carbon fibre. Besides, there is the issue with the cost of energy, primarily in the context of energy policy or strategy that would account for the energy savings contributed by the lightweight materials [29]. This deficit in production will start to show its sign as early as in 2023 or 2024, which could cause a market disequilibrium, and subsequently, a surge in price of the carbon fibre [33].

The demand for carbon fibre composites has grown over the years due to the need for light-weighting that is required to meet the stringent emission controls set by regulatory authorities around the world [34]. In 2018, it was estimated that the composite market revenue will increase from US $26 to 41.4 billion by 2025 [34]. Aligning with the predicted progress trend, market reports in 2018 [35] and 2019 [36] showcased an increase of 1.1% in global production capacity of carbon fibres in 2019 compared to 149.3 kilotonnes in 2018; at the same time, the compound annual growth rate of carbon fibre market demand from 2010 to 2018 is obtained as 11.45% [35]. The global demand for carbon fibre composites from 2010 to 2018 and the estimated values from 2019 to 2023 are shown in Figure 7.1 [36].

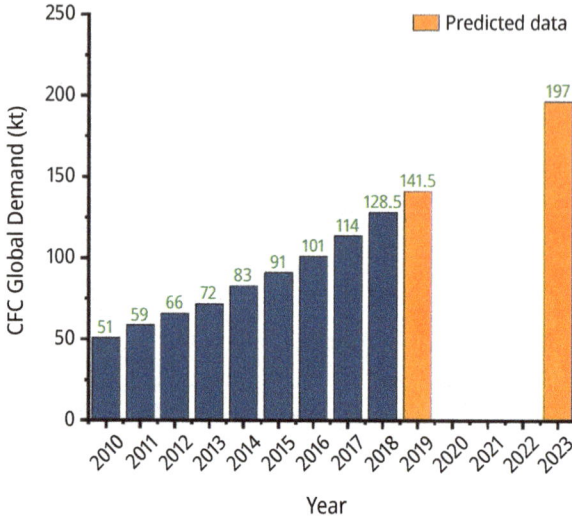

Figure 7.1: Carbon fibre composites' global demand from 2010 to 2018 and estimations from 2019 to 2023 [36].

7.1.2 Various grades of carbon fibres

A fibre with more than 92 wt% of carbon atoms is defined as carbon fibre [37]. The carbon content and its properties are strongly governed by the final carbonization temperature. Carbonizing fibres at temperatures up to 1,500 °C results in approximately 95 wt% carbon, while further increasing the treatment temperature to 3,000 °C leads to 99 wt% carbon content in fibres. Commonly, these fibres are described as graphite fibres and possess a highly aligned microstructure and corresponding high tensile modulus [37]. Carbon fibres with different combinations of mechanical properties can be manufactured by controlling the process parameters along the entire process chain, from the initial development of the precursor to carbonization. Various grades of carbon fibres that are on the commercial market are given in Table 7.2 and tend to be categorized according to their modulus. It is important to note, however, that these fibres are primarily produced from fossil fuel-derived precursors such as PAN and pitch in the case of ultra-high-modulus fibre.

Carbon fibres also come in various tow sizes, or number of individual fibre filaments, varying generally from 1k to 50k, although even higher tow sizes continue to be developed. With an increase in tow size, the cost per kg of carbon fibres reduces, but tends to

Table 7.2: Different grades of carbon fibres from PAN and pitch precursor types.

Carbon fibre grade	Tensile strength (GPa)	Tensile modulus (GPa)	Examples of commercial carbon fibres
Low modulus	1.0–2.5	50–229	XN-15-30S, XN-10-30S, XN-05-30S
Standard modulus	3.5–6.0	230–279	T300, T700S, PX35, IM6
Intermediate modulus	4.4–7.0	280–350	T800S, T1100S, IM9, IMS6L
High modulus	3.0–4.7	351–500	M46J, UMT430, HM63
Ultra-high modulus	2.5–4.0	501–950	M60J, UMT530, K13C2U, XN-80-60S

come with higher variability in properties and hence performance [38, 39]. Larger tow sizes, such as 48k and 50k, are mainly used for industrial and non-aerospace applications, and the smaller tow sizes are generally used for aerospace applications and stricter requirements with respect to variability in fibre specifications and performance [39].

7.1.3 Carbon fibre precursors

Carbon fibres from fossil fuel-derived precursors such as PAN dominate the industry with more than 90% of all carbon fibres currently being produced from PAN. There are many reasons for this, but its high carbon yield of about 50–60% and the long history and experience of working with PAN enable a wide variety of properties to be manufactured [1, 40–42]. However, the high cost associated with the production of PAN precursor remains high, accounting for more than 50% of the carbon fibre production cost. To date, this has impeded its wider proliferation into high-volume industries such as the automotive industry and others that do not have the stringent requirements of the aerospace sector [43]. To enter broader utilization of carbon fibre composite material in the automotive industry, for example, it has been proposed that carbon fibres that cost less than US \$15.4/kg whilst achieving a minimum tensile strength of 1.72 GPa and tensile modulus of 172 GPa [44] are required. In order to meet these requirements, potentially inexpensive alternative sustainable precursor materials such as lignin and cellulose are considered excellent candidates for carbon fibre production. While cellulose is the most widely available natural resource on the planet, lignin is the second most abundantly existing sustainable resource [45]. Cellulose is the most preferred bio-based precursor, in view of its availability and molecular orientation, but its poor carbon yield of only 10–30% after its carbonization inhibits its use in structural material applications [46]. Recently, Vocht et al. [47] developed novel stabilization protocols for cellulose precursor on lab scale to produce subsequent carbon fibres with a tensile strength of 2.8 GPa and a tensile modulus of 112 GPa. Results from this study provide an optimistic future for it to replace PAN-based fibres for specific mass market applications. Compared to cellulose, lignin possesses a high carbon content of 60–65% which is extremely advantageous and making it a potentially ideal sustainable precursor if its processing challenges such as long stabilization times during manufacture can be overcome [46]. The present state of carbon fiber development from renewable precursor sources is still in its infancy at the research scale, despite ongoing efforts to scale up the manufacture of lignin and cellulose for fibre production. At the moment, the properties of the resultant carbon fibres, cellulose, lignin and their blends are mostly inferior to carbon fibres produced from PAN and pitch-based precursor fibres. However, given the availability and potential cost of these bio-derived precursors if lignin carbon fibres with similar properties to PAN-based carbon fibre can be manufactured, the cost of carbon fibre could be cut in half [37]. Considering the economic and environmental benefits, current research is now refocusing on the development of bio-based precursors more explicitly using lignin as a primary source. The relative advantages and disadvantages of the most common alternative precursors and how they compare to PAN are given in Table 7.3.

Table 7.3: Benefits and limitations of using various renewable and non-renewable precursors for the production of carbon fibres [37, 40, 44–46, 48–52].

Precursor	Advantages	Disadvantages
PAN	High tensile strength, good compression properties, good carbon yield	High cost per kg, HCN emissions during thermal treatment
Pitch	High tensile modulus, high thermal conductivity, high carbonization yield	Expensive and complex purification process, low tensile strength
Polyethylene	Low cost	Harsh chemicals required for the stabilization process, low mechanical properties
Cellulose	Low raw material cost, high molecular orientation	Low carbon yield, high carbon fibre production cost
Lignin	Inexpensive, high carbon content, thermally stable, high aromatic structure, no HCN emission as by-product	Low tensile properties, disordered and complex microstructure, diversity of source, many purification steps are required prior to spinning, longer oxidation time

7.2 Synthesis and processing of lignin-based carbon fibre

7.2.1 Source, structure, properties and extraction method

Lignin is a natural phenolic-based polymer that is abundant in the cell walls of plants, especially in land-based plants. It provides mechanical support for plant growth, as well as assists in the long-distance transport of water in xylem, resisting a variety of stresses and providing protection against pathogens [53, 54]. The chemical structure of native lignin comprises three *p*-hydroxycinnamyl alcohol monomers, referred to as monolignols. These are *p*-coumaryl alcohol (*p*-hydroxyphenyl, H unit), coniferyl alcohol (guaicyl, G unit) and sinapyl alcohol (syringyl, S unit), which are characterized by their degree of methoxyl substitution at the third and fifth carbon (C_3 and C_5) positions of the aromatic ring, that is, *p*-coumaryl alcohol has no methoxyl group, whereas coniferyl alcohol has a methoxyl group at C_3 position, and sinapyl alcohol has a methoxyl group at both C_3 and C_5 positions, respectively [55]. The most predominant inter-monomer linkage within lignin gives rise to its complex three-dimensional amorphous network structure called phenylpropane β-aryl ether (β-O-4). Other inter-unit linkages that have been identified to date are biphenyl and dibenzodioxocin (5–5), phenylcoumarin (β-5), 1,2-diaryl propane (β-1), phenylpropane α-aryl ether (α-O-4), diaryl ether (4-O-5) and β–β-linked structures (β–β) (Figure 7.2) [56, 57].

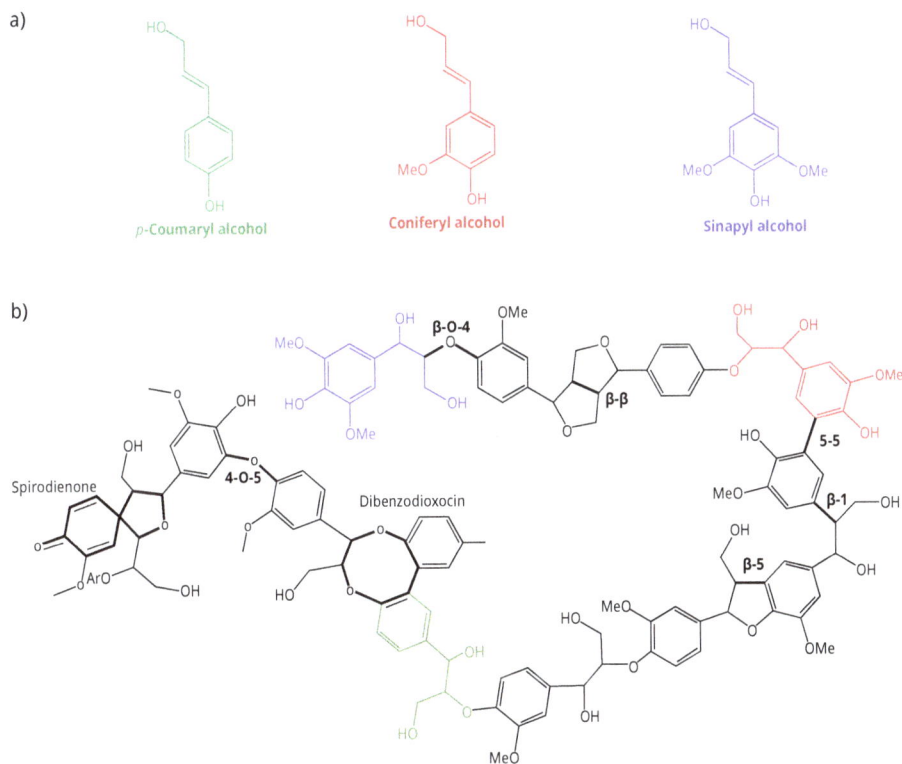

Figure 7.2: Molecular structures of (a) the monomers of lignin and (b) the lignin compound with the functional groups. Reprinted (adapted) with permission from [58]. Copyright 2022 American Chemical Society.

Lignin, being a highly branched and amorphous macromolecule, varies greatly in its composition depending on the plant species, maturity stage and environmental conditions [53, 50]. In general, hardwood (dicotyledonous angiosperm) lignin consists principally of G and S units and traces of H units, whilst softwood (gymnosperm) lignin is made up mostly of G units with low levels of H and S units. By contrast, lignin from grass (monocot) has comparable levels of G and S units, and more H units than hardwood and softwood lignins [54]. For all biomass types, these monomers are linked without defined repeating units [49].

However, before lignin can be used to make a precursor fibre, let alone a carbon fibre, it needs to be extracted from the plant. Whilst in principle it can be extracted from any plant, in practise, viable sources are required such as jute, hemp, flax, pine, cotton and wood pulp [59]. There are many different methods of extracting lignin from lignocellulosic feedstocks using a different chemical and physical processing routes, all of which play a critical role in controlling carbon yield, mechanical properties and hence economic viability.

To produce a value-added product, lignin is first separated from the biomass, and is commonly referred to as "technical lignin," which differs from native lignin in its physical properties, such as molecular structure and molecular weight, as well as chemical properties, such as solubility and hydrophobicity [60]. It is important to understand that the properties of the lignin are changed as a result of the processing conditions and chemical reactions performed during the isolation procedure. Technical lignin is a common by-product of the paper and pulping industry, whereby the qualitative and quantitative yields of cellulose have been optimized in a series of delignification processes [61]. In these extraction media, lignin is progressively broken down into smaller molecular weight fragments via the cleavage of bonds [62]. At present, the chemical pulping industry uses primarily sulphite, kraft or soda pulping process for separating lignin on an industrial scale from lignocellulosic feedstocks. Other popular fractionation methods include the organosolv treatment and steam explosion [63]. Some of these common technical lignins are described as follows.

7.2.1.1 Kraft lignin

Kraft pulping is the most dominant form of chemical pulping in the world producing about 130 million tonnes of pulp annually. In the kraft pulping process, lignocellulosic feedstock is digested at elevated temperature (145–170 °C) and pressure in a blend solution of sodium sulphide (Na_2S) and sodium hydroxide (NaOH), also known as the "white liquor," for a few hours. Under these conditions, the hydrosulphide and hydroxide anions react with the lignin polymers, by hydrolysing the β-O-4 bond and fracturing the lignin polymer into smaller fragments which readily dissolve in alkali solution, thus liberating the cellulose fibres initially bound with lignin. The dissolution of lignin in the spent pulping liquor gives its distinct darkish brown appearance, thereby giving rise to the name – the "black liquor" [64]. As high as 95% of the kraft lignin can effectively be recovered in the black liquor by acidification, with molecular mass ranging from 150 to 200,000 [65], the side chain structure of the kraft lignin experiences substantial structural changes, comprising a reduction in the interunit ether linkage and a concomitant increase in C–C bonds, thiol groups and phenolic hydroxyl groups [66]. Kraft lignin is hydrophobic and has a lower molecular mass than native lignin. Furthermore, it is almost insoluble in water and most solvents, but is highly soluble in alkaline media with pH above 11 [65]. In 2015, the production of kraft lignin reached about 630 kilotonnes/year, however, is still predominantly utilized in low-value applications, such as a combustion material for heat recovery, with little industrialization for high-value applications [67]. This limitation for high-value applications is largely attributed to the contamination of kraft lignin with carbohydrate molecules from hemicelluloses and fatty acids and the considerable amount of sulphur (1–3 wt%) covalently bound to kraft lignin in the form of thiols [65].

7.2.1.2 Lignosulphonate

Due to the limitations of kraft lignin, lignosulphonate is currently the main commercial source of lignin, accounting up to 90% of the total lignin market, with an annual global production of lignosulphonates reaching up to 1.8 million tonnes [65]. Lignosulphonate and kraft lignin have different properties, most notably its high solubility in water attributed obviously to the presence of the sulphonate group but also tends to have a similar or higher molecular weight to that of kraft lignin. The disadvantage of lignosulphonate is that it contains a high ash content. The extraction of lignosulphonates involves the digestion of lignocellulosic feedstock at a high temperature ranging from 130 to 180 C, in an aqueous solution of a sulphite (SO_3^{2-}) or bisulphite (HSO_3^-) salt of sodium, ammonium, magnesium or calcium. In this process, the most important reaction is the sulphonation of the aliphatic chains of lignin that lends the lignosulphonate its key characteristics.

7.2.1.3 Soda lignin

Soda lignin is obtained from the soda pulping process. Briefly, soda pulping is similar to the kraft process, except that sodium sulphide is excluded from the cooking medium. Soda lignin displays properties similar to kraft lignin, but unlike kraft lignin and lignosulphonates which are classified as sulphur-containing polymers that would give adverse environmental impact in their processing, soda lignin is sulphur free and has a higher purity, which makes its structure akin to native lignin when compared to other types of technical lignins [68]. Consequently, soda lignin is preferred in many applications such as tissue engineering, drug encapsulation, animal nutrition and also in polymer synthesis such as phenolic resin whereby a high purity and biocompatibility of lignin are required [69]. That said, soda pulping has limited use on easily pulped materials, for example, straws and bagasse, and still poses a great challenge when it comes to the digestion of most hardwoods. Owing to that, soda lignin from the softwood source can contain a significant amount of silicate and nitrogen, along with a higher sodium ion content, depending on the cooking procedure used.

7.2.1.4 Organosolv lignin

The organosolv pre-treatment process, commonly known as organo-solvation, is used to extract lignin by mixing organic solvents with an inorganic acid catalyst to break the lignin bonds. The latter functions to facilitate the breakdown of the β-ether linkages. Studies show that the presence of a catalyst allows the removal of 91% of the lignin, while its absence reaches only 14% [70]. Another similar technique is a mild organosolv extraction with alcohol which is carried out at a lower temperature of less than 150 °C. Ethanol in particular is among the most popular alcohols in this process to obtain organosolv lignin. When ethanol is used as the solvent, it is believed that it can be incorporated in the β-aryl ether units via α-alkoxylation, resulting in the β-O-4-rich lignin. This solvent incorporation is a crucial step in preventing undesired condensation reactions by trapping the reactive benzylic cations which is typically formed under acidic pre-treatment process, more so in harsh acidic environments [71]. The mild extraction method has shown to be able to produce clean lignin with excellent retention of a high amount of β-aryl ether content, and contains very low carbohydrate, ash and other non-lignin-derived impurities [71]. After all, there is still limited understanding of the actual relationship between the extraction method and on the lignin characteristics. The chemical structures of lignin and its properties have to be considered on a case-by-case basis as they are very much dependent on processing conditions.

7.2.2 Lignin fibre spinning and processing methods

Lignin, the second most abundant biopolymer after cellulose, has been investigated as a carbon fibre precursor since the 1960s primarily due to its vast abundance and bio-based origin [72, 73]. Successful use of lignin as an alternative precursor polymer is predicted to have a cost reduction of up to 50% in finished carbon fibre over conventional PAN polymer [43]. However, despite many studies published on investigating different routes for producing lignin-based carbon fibres [74–77], much further development is required to make lignin carbon fibre a commercial product.

Lignin-based precursor fibres have most frequently been produced using melt-spinning technique due to its affordability and scalability. Generally, to produce melt-spun lignin-based fibres, lignin powder is converted into pellets and extruded into fibres by heating them above the softening temperature (30–70 °C higher than the T_g of lignin) [78], followed by cooling in cold air before collecting the fibres [79]. The melt-spinning technique offers several advantages such as no solvent is required for fibre spinning, high-throughput rate and ease of the process. However, to process lignin fibre through melt-spinning, lignin should have a melting point well below its decomposition temperature, and its viscosity should remain stable during spinning [80]. A major drawback of melt-spinning lignin fibres is that it provides limited scope to improve the mechanical fibre properties as the fibre morphology is challenging to control. The fibre structure is governed by the removal of volatile components from lignin when it is heated above its glass transition temperature (T_g) during spinning, leading to the formation of defects and voids in the fibres [78]. Besides, depending on the source material, lignin may have varying low-molecular-weight fractions and sometimes no specific melting point due to its highly amorphous chemical structure. Lignin tends to decompose when heated at a temperature higher than 200–250 °C [81], making the processing window for melt extrusion particularly narrow. The presence of low-molecular-weight lignin is also reported to cause fibre fusion during the stabilization process due to its low T_g [82]. Many studies use melt-spinning to produce lignin-based carbon fibres obtained from various sources and have generally reported relatively low mechanical properties compared to PAN and pitch-based carbon fibres. For example, when carbon fibres were produced from melt-spun unmodified pyrolytic lignin, macro-voids were observed in the formed carbon fibres, resulting in a low tensile strength of 370 MPa and a modulus of 36 GPa [83].

The alternative wet-spinning technique is the industrially preferred technique for current commercial carbon fibre production from PAN polymer [84]. It offers greater scalability potential and a highly controlled coagulation process that minimizes the formation of defects in the fibres during spinning, leading to high-quality precursor fibre. However, the extremely low molecular weight and amorphous structure of lignin make it challenging to be converted into fibres through wet-spinning as the spinning dope requires a certain shear behaviour, which is generally achieved by blending with other linear, high-molecular-weight polymers such as PAN, polylactic acid (PLA) and polyvinyl alcohol (PVA) [85, 86]. A common variation of the wet-spinning method is the dry-jet or air-gap spinning which utilizes a spinneret mounted vertically, at an air gap of about 10–200 mm from the coagulation medium. When the spinning dope falls freely under gravity due to the air gap, it promotes molecular orientation prior to coagulation, leading to improvement in mechanical properties of fibres.

The electrospinning technique is used to produce continuous fibres at micro-metre to nano-metre scale. This technique was first used to form lignin-based fibres in 2007 [87]. In this process, a lignin-spinning solution is introduced through a voltage-driven jet nozzle to form thin fibre mats, which are solidified and collected by a grounded collector [88]. Successful electrospinning of lignin-based fibres requires modification of lignin or blending with other polymers to prevent electrospray which is often caused due to lignin's inability to form chain entanglements [89]. Since the electrospinning process results in fibre mats rather than single filaments, it remains a great challenge to separate and measure the properties of individual fibres [90]. Another significant drawback of the electrospinning process is the low scalability potential and low yield. Thus, fibres produced from the electrospinning method have been limited to applications in energy storage and filtration [91].

7.2.3 Blending of lignin with cellulose and other polymers

The challenges in the lignin fibre-spinning process arise from the complex structure of lignin, the lack of a reliable supply chain of raw lignin and insufficient understanding of the conversion mechanism of lignin precursor into carbon fibre [92]. Blending with cellulose, which is another abundant biopolymer, is a very attractive solution to produce renewable fibres [93]. Lignin–cellulose fibres have been prepared by wet-spinning method, where cellulose was dissolved in a suitable ionic liquid and added to a lignin/DMSO solution [94]. Blending cellulose with lignin greatly enhances the carbon yield of fibres when compared to cellulose alone, where significant mass loss occurs due to the release of various pyrolysis gases during the carbonization process [95]. Cellulose also grants better molecular orientation to the cellulose–lignin blend [96]. However, inadequate interaction between cellulose and lignin often results in the leaching of lignin during coagulation. Moreover, high costs of ionic liquids, which are the preferred solvents for cellulose, along with the constraints on lignin quantities that can be blended with cellulose, are some of the limitations hindering the widespread use of cellulose–lignin fibres.

As previously mentioned, most of the commercial carbon fibre is manufactured using PAN, which is a linear polymer of acrylonitrile and prepared by solution or suspension polymerization with the addition of some suitable comonomers which aid the spinnability and stabilization process and also improve the carbonization yield. On the other hand, lignin is a highly cross-linked aromatic network which is amorphous in nature and has many hydrophilic hydroxy groups in its structure [97]. These differences in the basic structure make it difficult to adapt standard spinning processes to lignin,; hence, it requires the establishment of innovative processes for lignin fibre spinning and its carbonization.

To overcome some of these limitations, lignin is often blended with PAN, which, thanks to its linear long-chain polymer structure, is likely to improve the processability of PAN/lignin blends. Softwood kraft lignin has been blended in equal ratios with PAN to produce wet-spun lignin-based carbon fibres with a tensile strength of 1.2 GPa and modulus of 148 GPa [74]. However, when blended with PAN, solution spinning of lignin presents a risk of uncontrolled leaching of lignin into the coagulation bath resulting in void formation and brittle fibres [98–100]. Other than PAN, lignin has been blended with other materials such as PVA [85], PLA [101], polyethylene oxide (PEO) [102] and graphene oxide [103] successfully to aid the spinnability and improve the mechanical properties of the fibres. While blending with various materials has shown to improve the mechanical properties, researchers are seeking solutions to overcome the intrinsic lack of well-defined molecular architecture and orientation in lignin that led to the poor mechanical properties of fibres based solely on lignin. The best carbon fibre properties reported in the literature have been achieved by altering the structure of a bio-oil-based pyrolysed lignin, modifying it to create a more linear polymer, followed by melt-spinning into fibre. Furthermore, the authors reported the application of higher tension during stabilization and carbonization process, which resulted in carbon fibres with high mechanical properties with a tensile strength of 1.7 GPa and a modulus of 182 GPa [92].

7.2.4 Stabilization and carbonization of lignin-based precursor

7.2.4.1 Stabilization of lignin-based precursor

Stabilization is an important step in the production of carbon fibres to make precursors thermally stable enough to withstand high temperatures applied during the carbonization process [104, 105]. Various chemical and physical pre-stabilization treatment methods for lignin have been proposed to address the challenges during carbon fibre manufacturing, including the enhancement of the carbon yield, reduction of stabilization time and prevention of fibre fusion.

7.2.4.1.1 Thermal stabilization of lignin-based precursor
For lignin-based carbon fibre production, thermal treatment is the most common method where the precursor is passed through a series of low- or moderate-temperature air ovens often between 200 and 350 °C [106–108]. Under thermal treatment conditions, lignin performs a significant chemical change via decomposition, rearrangement and cross-linking. During the process, the carbon content in the fibre increases while other elements are removed, resulting in the formation of a thermally stable structure, which prevents the destruction of fibres in the carbonization step. The thermal stabilization of lignin starts around 200 °C beginning with the cleavage of phenol monomer and methoxy groups (-OCH$_3$), which links with others via C–O–C bonds [108, 109], releasing formaldehyde, methane and water as volatile by-products [110–112]. At the same time, oxidation reaction introduces new carbonyl groups (ketones and aldehydes) to the molecular structure which further promotes condensation and rearrangement reactions [113, 114]. The predominant challenge of thermal oxidative stabilization of lignin is the long duration since a slow heating rate is required to prevent fibre fusion. For example, Baker et al. [115] used different heating rates (0.05, 0.025 and 0.01 °C/min) to stabilize organic purified lignin up to 250 °C. Zhang et al. [116] performed stabilization of acylated softwood lignin in air with rate of 0.2 °C/min to 220 °C. Norberg et al. [117]

used ultrafiltration softwood kraft lignin and achieved a higher heating rate of 4 °C/min but still needed to hold for 30 min at 250 °C.

7.2.4.1.2 Pre-stabilization treatment of lignin-based precursor

In addition to oxidative thermal stabilization, several physical and chemical methods have been studied to stabilize lignin fibre or speed up the stabilization. Physical agents (electron beam, plasma, UV, etc.) which can introduce cross-linking in a polymeric structure have been proposed to be effective as alternative methods for stabilization. Zhang [118] pre-treated softwood kraft lignin by UV for 15 min, resulting in a significant reduction in stabilization time from 40 to 4 h. Electron beam stabilization has been conducted by Seo et al. [119] on electrospun lignin–PAN fibres, leading to an increase of 20% in carbon yields accounting for the enhancement of cross-links in treated fibre. Plasma treatment and microwave are other possible methods tested with PAN fibre but have not been applied to lignin fibre yet [120].

Chemical stabilization is another method with the potential to increase the rate of stabilization of lignin. Since lignin is a polyphenolic resin, Lin et al. [121] pre-treated polyethylene glycol/lignin fibres with hexamethylenetetramine and hydrochloric acid as a catalyst to improve cross-linking through methylene groups. These treated fibres were able to stabilize at a higher heating rate of 2 °C/min, leading to reduced stabilization time. Guo et al. [122] used a boron nitride preceramic polymer as a cross-linking agent for lignin–phenol–formaldehyde resin blend. The blend was electrospun, and fibres were subsequently thermally stabilized at 250 °C for 2 h before carbonization. The treatment led to reduced stabilization time over the traditional method, and the obtained fibres retained their shape without fusion after being carbonized at 800 °C for 2 h. Cho et al. [123] added 5 wt% nanocrystalline cellulose into lignin dope. The cellulose-added electrospun lignin fibres showed good thermal stability at a high heating rate of 10 °C/min during carbonization up to 1,000 °C, indicating the assistance of cellulose in preventing lignin fusion. Le et al. [106, 124] have significantly reduced the stabilization time by 30% and increased the char yield by 50% for cellulose–lignin-blended fibres by using different catalysts (phosphoric acid and boric acid). All the above-mentioned works show a clear potential for future research in promoting the stabilization of lignin fibres via chemical treatment.

7.2.4.2 Carbonization of lignin-based stabilized fibres

Carbonization typically occurs at 400–1,600 °C in an inert atmosphere, converting stabilized fibre into carbon fibre [120, 125]. During this step, all elements other than carbon, such as nitrogen, oxygen and hydrogen, are removed from the structure [125, 126]. Carbon in lignin continues to condense and arrange to form polycyclic aromatic hydrocarbons [127, 128], expanding the number of aromatic rings in the molecular structure as temperature increases [129, 130] to form a carbon network. The carbon structure in lignin-based carbon fibre is particularly turbostratic (the sheets of carbon atoms are haphazardly folded, or crumpled, together) (Figure 7.3a), in that it contains more defects and is more disoriented than PAN-based carbon fibre [131]. Because of this poor orientation, lignin-derived carbon fibre has inferior mechanical properties to PAN-based carbon fibre, leading Baker et al. to suggest that higher performance lignin-based carbon fibre could be achieved by enhancing the graphitic orientation (Figure 7.3b and c).

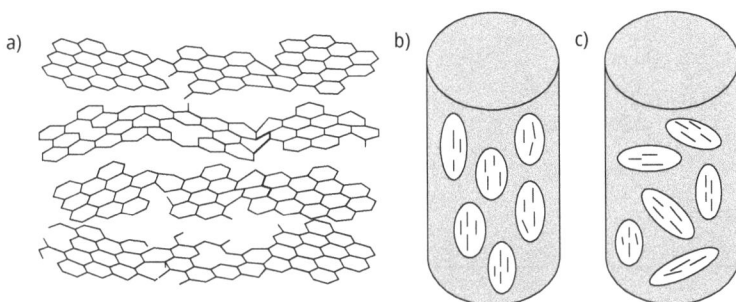

Figure 7.3: (a) Turbostratic carbon [108], (b) schematic representation of the orientation of graphite crystallites in PAN-based carbon fibre and (c) lignin-based carbon fibre [131] (reprinted (adapted) with permission, copyright 2016 Wiley Periodicals, Inc.).

7.3 Mechanical properties of lignin-based carbon fibres

This turbostratic microstructure, lacking in graphitic structure with increased defects [108, 132], is the major reason for the reduced mechanical properties of lignin-based carbon fibre compared with PAN- and pitch-based carbon fibres. Some examples include that of Jin et al. who reported a tensile strength and modulus of 1.39 GPa and 98 GPa, respectively, using a fractionated–solvated softwood kraft lignin [133]. Zhang et al. made carbon fibre from acylated softwood lignin and reported tensile strength and modulus of 1.05 GPa and 52 GPa, respectively [116]. Blends of lignin and PAN have been shown to have better mechanical strength with tensile strength and modulus of 1.72 GPa and 230 GPa, respectively [134]. Cellulose and lignin blends (70:30) have been widely studied, including the work by Bengtsson, who reported tensile strength and modulus of 1.17 GPa and 77 GPa, respectively [135]. Blends of lignin with other polymers such as PEO, polyethylene terephthalate and polypropylene have also been trialled somewhat successfully, but again display inferior mechanical properties due to the polymers' low thermal stability and resulting porous structure in the resulting carbon fibre [113, 136]. Despite this, Figure 7.4 summarizes the mechanical properties

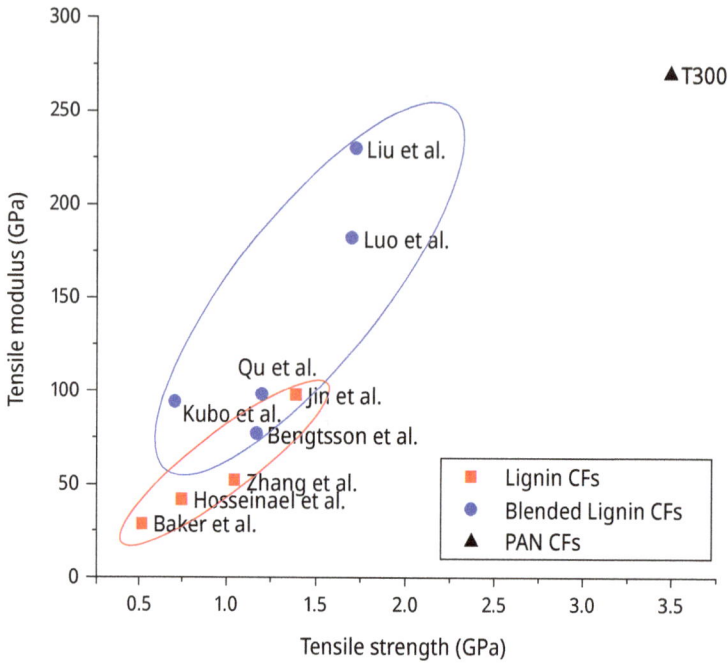

Figure 7.4: Mechanical properties of carbon fibres made from lignin and lignin blends in comparison with commercial Toray T300 carbon fibre.

of carbon fibres made from lignin and lignin blends using the commercial Toray T300 carbon fibre as a benchmark, illustrating the general improvement blended systems compared with lignin only.

Mechanical properties of lignin-based carbon fibre have been further improved via hot stretching during high-temperature graphitization above 2,000 °C, which is commonly used to enhance the mechanical properties of cellulose-based carbon fibre [120, 126]. The application of hot stretch is believed to refine the carbon structure by improving the graphitic levels and orientation of graphitic crystallites. However, the application of high-temperature graphitization of lignin-based carbon fibre may result in a lower carbon yield along with increased production cost [125]. In summary, the imperfect carbon structure of lignin-based carbon fibre is an open research question requiring further study to improve its mechanical properties.

7.4 Sustainable composites from bio-based carbon fibre

7.4.1 Applications and market demand

Lignin-based carbon fibre is an excellent and potentially inexpensive alternative to PAN-based carbon fibre particularly for general use composite materials. Unlike PAN-based carbon fibre whereby the manufacturing cost is highly dependent on oil price, with around 51% being precursor cost, 15% related to the oxidation process and 23% related to the carbonization process [43, 137], the extraction of lignin from biomass waste or pulp mills comes at a much lower cost (Table 7.4). Due to the high cost of PAN, carbon fibre is rather expensive, limiting its use to high-value, low-volume products. These include aerospace structural components, the body and brake discs of Formula 1 racing cars, pressure vessels and other defence-related products. The application of carbon fibre as structural composites is, in fact, dependent on its mechanical properties and cost. Replacing the PAN-based carbon fibre with lignin, in part or as a whole, offers the opportunity for it to be widely adopted in high-volume applications such as automotive, renewable energy and construction applications. In recent years, automakers have started to use carbon fibre composites more widely, as add-on parts or whole chassis elements, in order to reduce weight, including the likes of Audi, Mercedes-Benz, BMW in their i3 and i8 models, as well as Toyota that pioneered in the carbon fibre composites for the tailgate in its Prius PHV [138]. However, for this progress to continue, further reductions in cost and volume are required. For every 10% reduction in vehicle weight, it has been calculated that this would translate into an increase in vehicle fuel economy of about 6%, with a concomitant reduction in emissions. To place the potential increase in fuel economy into perspective, body-in-white modelling indicates that more than 60% of the steel in a vehicle could be replaced with carbon fibre-reinforced polymers without impacting vehicle crashworthiness [139]. According to the market research by Mordor Intelligence, the automotive carbon fibre market was valued at US $20 billion in 2020, and it is expected to grow up to US $33 billion in 2026, attributed to the expected rapid growth of electric vehicles [140].

Table 7.4: Cost, CO_2 and energy usage comparison of PAN and lignin-based carbon fibres (adapted from [141]).

Material	Precursor cost	Production cost	Total cost	CO_2 emissions	Total energy usage
Conventional PAN	$11.1/kg	$25.15/kg	$36.25/kg	31 kg/kg	704 MJ/kg
Textile-grade PAN	$4.4–13.2/kg	$12.25–25.4/kg	$16.65–38.6/kg	–	
Melt-spun PAN	$6.3/kg	$17.4/kg	$23.7/kg	–	
Lignin	$1.52/kg	$6.27/kg	$7.79/kg	24 kg/kg	554 MJ/kg

In addition to the excellent mechanical strength, durability and levity, there is another important property that carbon fibre possesses – electrical conductivity. This makes carbon fibre suitable to be used as an electrode in electrophysiological, electrochemical, biosensor and biofuel cell applications [142]. For example, traditionally, the negative electrode in lithium-ion battery is made up of copper, which makes up one-fourth of the total electrode weight but since the 1990s, carbon fibre has been employed as an alternative negative electrode in lithium-ion battery. This has the advantage of increasing efficiency by reducing the amount of non-active material in lithium-ion battery [143]. Lignin-derived carbon fibre is shown to be a promising precursor in several research studies. Some of the different forms studied includes fused electrospun carbon fibrous mats, fused lignin carbon fibre mats or as micro-sized particles in a mixture with pitch. A high charge capacity of around 400–500 mAh/g and a good cyclic stability at different current rates have been obtained. In particular, carbon (nano)fibre spun from the electrospinning method is expected to improve the charge and mass transport at the cathode, through its exceptional through-plane and in-plane electrical conductivity, and high porosity after the carbonization step [144]. Moreover, the porosity can be controlled by adjusting the spinning parameters, which allows the design and synthesis of fibre of controlled morphology and activity. With further modification, enhanced improvement on the performance of the lignin-based carbon fibre electrode can be obtained. For example, after thermal annealing of the carbon fibre in the presence of urea, the charge capacity could be improved by around 20% [145]. While the carbon fibre tow is generally used as anode, when the surface of carbon fibre was coated with lithium oxide ($LiFePO_4$), it can function as a positive electrode [146]. A specific capacity of 100 mAh/g was recorded for this type of cathode, as good as commercial positive electrodes that have a theoretical capacity at about 150 mAh/g.

Nowadays, research is underway to create structural carbon fibre composites that double as energy storage system, that is, the structural batteries (or solid-state batteries) made of carbon fibre-based electrodes separated by a glass fibre separator in a structural battery electrolyte matrix material. These types of structural battery composites by far have only achieved an energy density of 24 Wh/kg but despite their inability to store as much energy as lithium-ion batteries, they are highly attractive for use in vehicles and other applications as "mass-less" energy storage because the weight of the battery vanishes when it becomes part of the load-bearing structure [147]. The next-generation structural battery has fantastic potential in consumer technology. It could be quite possible within a few years to manufacture smartphones, laptops or electric bicycles that weigh half as much as today and are much more compact. Apart from the extensive use of carbon fibre in batteries, it has also been demonstrated that carbon fibre, especially the electrospun carbon nanofibres (CNFs), can be manufactured into support for the platinum catalyst in the proton exchange membrane fuel cell. The incorporation of small catalyst particles into the CNF web is known as the microfibrous entrapped catalyst (MFEC) [148]. The major advantage of this is the high surface area to volume of the CNFs, high dispersion and little catalyst aggregation, and consequently, the improved electrochemical active sites of the catalyst adhered on its surface as well as the synergism between the CNF and the immobilized catalytic species to achieve process intensification [144]. The particle size of MFECs is typically in the range of 100–200 µm, and the surface-area-to-volume ratio can reach up to 1×10^4 m^2/m^3 [148]. Depending on the catalytic species, alloys or bimetallic nanoparticles used to decorate the CNFs, either by electrospinning or doping in catalyst precursor, this leads to catalysts of various photo/electro/chemical catalytic properties. For example, CdS–TiO_2-doped CNFs display excellent photocatalytic activity for visible light hydrolytic dehydrogenation of ammonia borance; Au–TiO_2/CNFs for photodegradation of methyl orange, methylene blue and acid red 18 in wastewater under UV light irradiation [149]; Pd/CNFs for Heck's coupling reaction [150]; and last but not least, silver (Au)-doping CNFs in silver nitrate precursor could promote the reduction of 4-nitrophenol and benzylamine [151]. Noteworthy, the catalyst could be easily separated and recycled.

In short, lignin-based carbon fibre plays a paramount role in the transformation toward a more sustainable and affordable carbon fibre manufacturing. At the same time, there are challenges and issues ahead that needs to be overcome. Undoubtedly, future works looking into a combined economical and ecological approach of improving the performance of the lignin-based carbon fibre would make them promising as the next generation of carbon fibre in wider applications.

References

[1] Morris EA, et al. Early development of multifilament polyacrylonitrile-derived structural hollow carbon fibers from a segmented arc spinneret. Carbon 2021;178:223–32.

[2] Zan Gao JZ, Rajabpour S, Joshi K, Kowalik M, Brendan Croom YS, Zhang L, Bumgardner C, Brown KR, Diana Burden JWK, van Duin ACT, Zhigilei LV, Li X. Graphene reinforced carbon fibers. Sci Adv 2020;6(17):eaaz4191.

[3] Yang F, et al. Effect of amorphous carbon on the tensile behavior of polyacrylonitrile (PAN)-based carbon fibers. J Mater Sci 2019;54(11):8800–13.

[4] Ge Y, et al. The effects of chemical reaction on the microstructure and mechanical properties of polyacrylonitrile (PAN) precursor fibers. J Mater Sci 2019;54(19):12592–604.

[5] Groetsch T, et al. Gas emission study of the polyacrylonitrile-based continuous pilot-scale carbon fiber manufacturing process. Ind Eng Chem Res 2021.

[6] Nunna S, Setty M, Naebe M. Formation of skin-core in carbon fibre processing: a defect or an effect? Express Polym Lett 2019;13(2):146–58.

[7] Nunna S, et al. Time dependent structure and property evolution in fibres during continuous carbon fibre manufacturing. Materials (Basel) 2019;12(7).

[8] Dumanlı AG, Windle AH. Carbon fibres from cellulosic precursors: A review. J Mater Sci 2012;47(10):4236–50.

[9] Edison TA. Electric Lamp, USA: U.S.P. Office; Editor. 1880. pp. 1–3.

[10] Swan JW. Electric Lamp, U.S.P. Office; Editor. 1880. pp. 1–3.

[11] Park S-J. Carbon Fibers. In: Robert Hull CJ, Osgood RM, Parisi J, Seong T-Y, Uchida S-I, Wang ZM, editors. Springer Series in Materials Science, vol 210. Springer; 2015.

[12]   Ford CE, Mitchell CV. Fibrous Graphite, U.S.P. Office; Editor. 1963.

[13]   Tang MM, Bacon R. Carbonization of cellulose fibers-I low temperature pyrolysis. Carbon 1964;2:211–20.

[14]   Bacon R, Tang MM. Carbonization of cellulose fibers-II physical property sudy. Carbon 1964;2:221–25.

[15]   Gorss J. High Performance Carbon Fibers. 2003.

[16]   Cahn RW. Graphitic carbon copies. Nature 1991;349:97.

[17]   Swinbanks D. MITI makes its move. Nature 1990;348:186.

[18]   Nakamura O, et al. Study on the PAN carbon-fiber-innovation for modelling a successful R&D management. Synthesiology 2009;2(2):154–64.

[19]   Moreton R, Watt W, Johnson W. Carbon fibres of high strength and high breaking strain. Nature 1967;213(5077):690–91.

[20]   Watt W, Johnson W. New materials make their mark. Nature 1968;220:835.

[21]   Otani S, et al. Method for Producing Carbonized Lignin Fiber, USA: U.S.P. Office; Editor., 1969. pp. 1–5.

[22]   Souto F, Calado V, Pereira N. Lignin-based carbon fiber: A current overview. Mater Res Express 2018;5(7).

[23]   Singer LS. High Modulus, High Strength Carbon Fibers Produced From Mesophase Pitch, USA: U.S.P. Office; Editor. 1977, pp. 1–25.

[24]   Horikiri S, Iseki J, Minobe M. Process for Production of Carbon Fiber, U.S.P. Office; Editor. 1978.

[25]   Soulis S, et al. Impact of alternative stabilization strategies for the production of pan-based carbon fibers with high performance. Fibers 2020;8(6).

[26]   Composites world. Carbon fiber suppliers gear up for next-gen growth. 2020, [cited 2022 18 April]; Available from: https://www.compositesworld.com/articles/carbon-fiber-suppliers-gear-up-for-next-gen-growth.

[27]   Fernández L. Global market value of carbon fiber 2014–2026. 2021 [cited 2022 18 April]; Available from: https://www.statista.com/statistics/947941/global-market-value-carbon-fiber/.

[28]   The Manufacturer. Creating a sustainable future for carbon fibre in manufacturing. 2021, [cited 2022 18 April]; Available from: https://www.themanufacturer.com/articles/creating-a-sustainable-future-carbon-fibre/.

[29]   Das S, et al. Global Carbon Fiber Composites Supply Chain Competitiveness Analysis, United States; 2016.

[30]   Toray Industries. I. *Production Capacity*. 2021 [cited 2022 18 April]; Available from: https://www.toray.com/global/ir/management/man_010.html.

[31]   Ocean R. Pitch-based carbon fiber Market Scope and overview, Strategy, Revenue, Opportunity, Business Segment Overview and Key Trends 2022–2030. 2022, [cited 2022 18 April]; Available from: https://www.taiwannews.com.tw/en/news/4408935.

[32]   Nippon Graphite Fiber Co. L. What's Carbon fibers. 2022 [cited 2022 18 April]; Available from: https://www.ngfworld.com/en/fiber.html.

[33]   Composites World. The outlook for carbon fiber supply and demand. 2021, [cited 2022 18 April]; Available from: https://www.compositesworld.com/articles/the-outlook-for-carbon-fiber-supply-and-demand.

[34]   Zhang J, et al. Current status of carbon fibre and carbon fibre composites recycling. Compos Part B: Eng 2020;193.

[35]   Witten E, et al. Composites Market Report 2018. 2018, AVK. p. 1–59.

[36]   Sauer M. *Composites Market Report 2019*. 2019, AVK Carbon Composites e.V. p. 1–11.

[37]   Frank E, et al. Carbon fibers: Precursor systems, processing, structure, and properties. Angew Chem Int Ed Engl 2014;53(21):5262–98.

[38] Nunna S, et al. Development of a cost model for the production of carbon fibres. Heliyon 2019;5(10): e02698.

[39] Mirdehghan, Abolfazl S. Fibrous Polymeric Composites. In: Engineered Polymeric Fibrous Materials, 2021. pp. 1–58.

[40] Rahaman MSA, Ismail AF, Mustafa A. A review of heat treatment on polyacrylonitrile fiber. Polym Degrad Stab 2007;92(8):1421–32.

[41] Ghorpade RV, Lee S, Hong SC. Effect of different itaconic acid contents of poly(acrylonitrile-co-itaconic acid)s on their carbonization behaviors at elevated temperatures. Polym Degrad Stab 2020;181.

[42] Dér A, et al. Modelling and analysis of the energy intensity in polyacrylonitrile (PAN) precursor and carbon fibre manufacturing. J Cleaner Prod 2021;303.

[43] Mainka H, et al. Lignin – an alternative precursor for sustainable and cost-effective automotive carbon fiber. J Mater Res Technol 2015;4(3):283–96.

[44] Baker DA, Rials TG. Recent advances in low-cost carbon fiber manufacture from lignin. J Appl Polym Sci 2013;130(2):713–28.

[45] Dai Z, et al. High-strength lignin-based carbon fibers via a low-energy method. RSC Adv 2018;8(3):1218–24.

[46] Bengtsson A, et al. Carbon fibers from lignin-cellulose precursors: Effect of stabilization conditions. ACS Sustain Chem Eng 2019;7(9):8440–48.

[47] Vocht MP, et al. Preparation of cellulose-derived carbon fibers using a new reduced-pressure stabilization method. Ind Eng Chem Res 2022.

[48] Fang W, et al. Manufacture and application of lignin-based carbon fibers (LCFs) and lignin-based carbon nanofibers (LCNFs). Green Chem 2017;19(8):1794–827.

[49] Poursorkhabi V, et al. Processing, carbonization, and characterization of lignin based electrospun carbon fibers: a review. Front Energy Res 2020;8.

[50] Yongjian X, et al. Current overview of carbon fiber: toward green sustainable raw materials. Bioresources 2020;15(3):7234–59.

[51] Bajpai P. Raw Materials and Processes for the Production of Carbon Fiber. In: Carbon Fiber, 2021. pp. 13–50.

[52] Aldosari SM, Khan MA, Rahatekar S. Manufacturing pitch and polyethylene blends-based fibres as potential carbon fibre precursors. Polymers (Basel) 2021;13(9).

[53] Kang X, et al. Lignin-polysaccharide interactions in plant secondary cell walls revealed by solid-state NMR. Nat Commun 2019;10(1):347.

[54] Lourenço A, et al. Lignin composition and structure differs between xylem, phloem and phellem in Quercus suber l. Front Plant Sci 2016;7.

[55] José Borges Gomes F, et al. A review on lignin sources and uses. J Appl Biotechnol Bioeng 2020;100–05.

[56] Del Río JC, et al. Lignin monomers from beyond the canonical monolignol biosynthetic pathway: Another brick in the wall. ACS Sustain Chem Eng 2020;8(13):4997–5012.

[57] Froass PM, Ragauskas AJ, Jiang J-E. Chemical structure of residual lignin from kraft pulp. J Wood Chem Technol 1996;16(4):347–65.

[58] Zhang R, et al. Microbial ligninolysis: toward a bottom-up approach for lignin upgrading. Biochemistry 2019;58(11):1501–10.

[59] Watkins D, et al. Extraction and characterization of lignin from different biomass resources. J Mater Res Technol 2015;4(1):26–32.

[60] Ekielski A, Mishra PK. Lignin for bioeconomy: The present and future role of technical lignin. Int J Mol Sci 2021;22(1).

[61] Li T, Takkellapati S. The current and emerging sources of technical lignins and their applications. Biofuels Bioprod Biorefin 2018;12(5):756–87.

[62] Gratzl JS, Chen C-L. Chemistry of Pulping: Lignin Reactions. In: Lignin: Historical, Biological, and Materials Perspectives, American Chemical Society; 1999. pp. 392–421.

[63] McGrath JE, Hickner MA, Höfer R. 10.01 – Introduction: Polymers for a Sustainable Environment and Green Energy. In: Matyjaszewski K, Möller M, editors. Polymer Science: A Comprehensive Reference, Amsterdam: Elsevier; 2012. pp. 1–3.

[64] Cheremisinoff NP, Rosenfeld PE. Chapter 6 – Sources of Air Emissions from Pulp and Paper Mills. In: Cheremisinoff NP, Rosenfeld PE, editors. Handbook of Pollution Prevention and Cleaner Production, Oxford: William Andrew Publishing; 2010. pp. 179–259.

[65] Gellerstedt G, Henriksson G. Chapter 9 – Lignins: Major Sources, Structure and Properties. In: Belgacem MN, Gandini A, editors. Monomers, Polymers and Composites from Renewable Resources, Amsterdam: Elsevier; 2008. pp. 201–24.

[66] Fernández-Rodríguez J, et al. Chapter 7 – Lignin Separation and Fractionation by Ultrafiltration. In: Galanakis CM, editor. Separation of Functional Molecules in Food by Membrane Technology, Academic Press; 2019. pp. 229–65.

[67] Chen H. 3 – Lignocellulose Biorefinery Feedstock Engineering. In: Chen H, editor, Lignocellulose Biorefinery Engineering. Woodhead Publishing; 2015. pp. 37–86.

[68] Vishtal AG, Kraslawski A. Challenges in industrial applications of technical lignins. Bioresources 2011;6(3):2011.

[69] Witzler M, et al. Lignin-derived biomaterials for drug release and tissue engineering. Molecules (Basel, Switzerland) 2018;23(8):1885.

[70] Fernández-Rodríguez J, de Diego-Díaz B, Tapia-Martín ME. Chapter 7 – Biomethanization of Agricultural Lignocellulosic Wastes: Pretreatments. In: Tyagi V, Aboudi K, editors, Clean Energy and Resources Recovery. Elsevier; 2021. pp. 155–202.

[71] Zijlstra DS, et al. Mild organosolv lignin extraction with alcohols: The importance of benzylic alkoxylation. ACS Sustain Chem Eng 2020;8(13):5119–31.

[72] Kadla J, et al. Lignin-based carbon fibers for composite fiber applications. Carbon 2002;40(15):2913–20.

[73] Kubo S, Kadla J. Lignin-based carbon fibers: Effect of synthetic polymer blending on fiber properties. J Polym Environ 2005;13(2):97–105.

[74] Jin J, Ogale AA. Carbon fibers derived from wet-spinning of equi-component lignin/polyacrylonitrile blends. J Appl Polym Sci 2018;135(8):45903.

[75] Oroumei A, Fox B, Naebe M. Thermal and rheological characteristics of bio-based carbon fibre precursor derived from low molecular weight organosolv lignin. ACS Sustain Chem Eng 2015.

[76] Sudo K, Shimizu K. A new carbon fiber from lignin. J Appl Polym Sci 1992;44(1):127–34.

[77] Johnson D, Tomizuka I, Watanabe O. The fine structure of lignin-based carbon fibres. Carbon 1975;13(4):321–25.

[78] Qu W, et al. Towards producing high-quality lignin-based carbon fibers: a review of crucial factors affecting lignin properties and conversion techniques. Int J Biol Macromol 2021;189:768–84.

[79] Nordström Y, et al. A new softening agent for melt spinning of softwood kraft lignin. J Appl Polym Sci 2013;129(3):1274–79.

[80] Frank E, et al. Carbon fibers: Precursor systems, processing, structure, and properties. Angew Chem Int Ed 2014;53(21):5262–98.

[81] Jin Y, et al. Lignin-based high-performance fibers by textile spinning techniques. Materials 2021;14(12):3378.

[82] Qu W, et al. Potential of producing carbon fiber from biorefinery corn stover lignin with high ash content. J Appl Polym Sci 2018;135(4):45736.

[83] Qin W, Kadla JF. Carbon fibers based on pyrolytic lignin. J Appl Polym Sci 2012;126(S1):E204–E213.

[84] Mataram A, et al. Characterization and mechanical properties of polyacrylonitrile/silica composite fibers prepared via dry-jet wet spinning process. Mater Lett 2010;64(17):1875–78.

[85]   Föllmer M, et al. Wet-spinning and carbonization of lignin-polyvinyl alcohol precursor fibers. Adv
       Sustainable Syst 2019;3(11):1900082.
[86]   Shi X, et al. Stepwise fractionation extracted lignin for high strength lignin-based carbon fibers.
       New J Chem 2019;43(47):18868–75.
[87]   Lallave M, et al. Filled and hollow carbon nanofibers by coaxial electrospinning of Alcell lignin
       without binder polymers. Adv Mater 2007;19(23):4292–96.
[88]   Ma X, et al. Electrospun lignin-derived carbon nanofiber mats surface-decorated with MnO2
       nanowhiskers as binder-free supercapacitor electrodes with high performance. J Power Sources
       2016;325:541–48.
[89]   Fang W, et al. Manufacture and application of lignin-based carbon fibers (LCFs) and lignin-based
       carbon nanofibers (LCNFs). Green Chem 2017;19.
[90]   Poursorkhabi V, Mohanty AK, Misra M. Electrospinning of aqueous lignin/poly (ethylene oxide)
       complexes. J Appl Polym Sci 2015;132(2).
[91]   Poursorkhabi V, et al. Processing, carbonization, and characterization of lignin based electrospun
       carbon fibers: A review. Front Energy Res 2020;8:208.
[92]   Luo Y, et al. Enabling high-quality carbon fiber through transforming lignin into an orientable and
       melt-spinnable polymer. J Cleaner Prod 2021;307:127252.
[93]   Olsson C, Sjöholm E, Reimann A. Carbon fibres from precursors produced by dry-jet wet-spinning of
       kraft lignin blended with kraft pulps. Holzforschung 2017;71(4):275–83.
[94]   Byrne N, et al. Enhanced stabilization of cellulose-lignin hybrid filaments for carbon fiber
       production. Cellulose 2018;25(1):723–33.
[95]   Trogen M, et al. Cellulose-lignin composite fibres as precursors for carbon fibres. Part 1 –
       manufacturing and properties of precursor fibres. Carbohydr Polym 2021;252:117133.
[96]   Bengtsson A, et al. Carbon fibers from lignin–cellulose precursors: Effect of carbonization
       conditions. ACS Sustain Chem Eng 2020;8(17):6826–33.
[97]   Stewart D. Lignin as a base material for materials applications: chemistry, application and
       economics. Ind Crops Prod 2008;27(2):202–07.
[98]   Ingildeev D, et al. Novel cellulose/polymer blend fibers obtained using ionic liquids. 2012;297.
[99]   Song L, et al. Carbon fibers with low cost and uniform disordered structure derived from lignin/
       polyacrylonitrile composite precursors. Fibers Polym 2021;22(1):240–48.
[100] Dong X, et al. Polyacrylonitrile/lignin sulfonate blend fiber for low-cost carbon fiber. RSC Adv
       2015;5(53):42259–65.
[101] Wang Y, et al. Strong, ductile and biodegradable polylactic acid/lignin-containing cellulose
       nanofibril composites with improved thermal and barrier properties. Ind Crops Prod
       2021;171:113898.
[102] Jayaramudu T, et al. Adhesion properties of poly(ethylene oxide)-lignin blend for nanocellulose
       composites. Compos Part B: Eng 2019;156:43–50.
[103] Föllmer M, et al. Structuration of lignin-graphene oxide based carbon materials through liquid
       crystallinity. Carbon 2019;149:297–306.
[104] Frank E, et al. Carbon fibers: Precursor systems, processing, structure, and properties. Angew
       Chem – Int Ed 2014;53:5262–98.
[105] Le ND, et al. Understanding the influence of key parameters on the stabilisation of cellulose-lignin
       composite fibres. Cellulose 2021;28:911–19.
[106] Le N-D, et al. Chemically accelerated stabilization of a cellulose–lignin precursor as a route to high
       yield carbon fiber production. Biomacromolecules 2022.
[107] Norberg I. Carbon Fibres from Kraft Lignin, KTH Royal Institute of Technology; 2012. pp. 69.
[108] Ogale AA, Zhang M, Jin J. Recent advances in carbon fibers derived from biobased precursors. J Appl
       Polym Sci 2016;133.

[109] Le ND, et al. Effect of boric acid on the stabilisation of cellulose-lignin filaments as precursors for carbon fibres. Cellulose 2021;28:729–39.

[110] Braun JL, Holtman KM, Kadla JF. Lignin-based carbon fibers: oxidative thermostabilization of kraft lignin. Carbon 2005;43:385–94.

[111] Brodin I, et al. Oxidative stabilisation of kraft lignin for carbon fibre production. Holzforschung 2012;66.

[112] Beste A. Reaxff study of the oxidation of lignin model compounds for the most common linkages in softwood in view of carbon fiber production. J Phys Chem A 2014;118:803–14.

[113] Kadla JF, et al. Lignin-based carbon fibers for composite fiber applications. Carbon 2002;40:2913–20.

[114] Li Y, et al. Study on structure and thermal stability properties of lignin during thermostabilization and carbonization. Int J Biol Macromol 2013;62:663–69.

[115] Baker DA, Gallego NC, Baker FS. On the characterization and spinning of an organic-purified lignin toward the manufacture of low-cost carbon fiber. J Appl Polym Sci 2012;124:227–34.

[116] Zhang M, Jin J, Ogale A. Carbon fibers from UV-assisted stabilization of lignin-based precursors. Fibers 2015;3:184–96.

[117] Norberg I, et al. A new method for stabilizing softwood kraft lignin fibers for carbon fiber production. J Appl Polym Sci 2013;128:3824–30.

[118] Zhang M. Carbon Fibers Derived from Dry-Spinning of Modified Lignin Precursors. 2016. p. 168.

[119] Seo DK, et al. Preparation and Characterization of the Carbon Nanofiber Mat Produced from Electrospun Pan/lignin Precursors by Electron Beam Irradiation. In: Reviews on Advanced Materials Science, 2011. pp. 31–34.

[120] Choi D, Kil HS, Lee S. Fabrication of low-cost carbon fibers using economical precursors and advanced processing technologies. Carbon 2019;142:610–49.

[121] Lin J, et al. Improvement of mechanical properties of softwood lignin-based carbon fibers. J Wood Chem Technol 2014;34:111–21.

[122] Guo Z, et al. The production of lignin-phenol-formaldehyde resin derived carbon fibers stabilized by BN preceramic polymer. Mater Lett 2015;142:49–51.

[123] Cho M, et al. Skipping oxidative thermal stabilization for lignin-based carbon nanofibers. ACS Sustain Chem Eng 2018;6:6434–44.

[124] Le ND, et al. Cellulose-lignin composite fibers as precursors for carbon fibers: Part 2 – the impact of precursor properties on carbon fibers. Carbohydr Polym 2020;250:116918.

[125] Morgan P. Carbon Fibers and Their Composites. Carbon Fibers and Their Composites, 2005.

[126] Al Aiti M, et al. On the Morphology and Structure Formation of Carbon Fibers from Polymer Precursor Systems. In: Progress in Materials Science, Elsevier Ltd; 2018. pp. 477–551.

[127] Asmadi M, Kawamoto H, Saka S. Thermal reactivities of catechols/pyrogallols and cresols/xylenols as lignin pyrolysis intermediates. J Anal Appl Pyrolysis 2011;92:76–87.

[128] Cao J, et al. Study on carbonization of lignin by TG-FTIR and high-temperature carbonization reactor. Fuel Process Technol 2013;106:41–47.

[129] Zhou H, et al. Polycyclic aromatic hydrocarbon formation from the pyrolysis/gasification of lignin at different reaction conditions. Energy Fuels 2014;28:6371–79.

[130] Sharma RK, Hajaligol MR. Effect of Pyrolysis Conditions on the Formation of Polycyclic Aromatic Hydrocarbons (PAHs) from Polyphenolic Compounds. In: Journal of Analytical and Applied Pyrolysis, Elsevier; 2003. pp. 123–44.

[131] Baker FS. Low Cost Carbon Fiber from Renewable Resources, U.S Department of energy; 2010.

[132] Sagues W, et al. Are lignin-derived carbon fibers graphitic enough? Green Chem 2019;21.

[133] Jin J, et al. Carbon fibers derived from fractionated-solvated lignin precursors for enhanced mechanical performance. ACS Sustain Chem Eng 2018;6:14135–42.

[134] Liu HC, et al. Processing, structure, and properties of lignin- and CNT-incorporated polyacrylonitrile-based carbon fibers. ACS Sustain Chem Eng 2015;3:1943–54.

[135] Bengtsson A, et al. Carbon fibers from lignin-cellulose precursors: Effect of stabilization conditions. ACS Sustain Chem Eng 2019;7:8440–48.

[136] Kubo S, Kadla JF. Lignin-based carbon fibers: Effect of synthetic polymer blending on fiber properties. J Polym Environ 2005;13:97–105.

[137] Wang S, et al. Lignin-based carbon fibers: Formation, modification and potential applications. Green Energy Environ 2021.

[138] Just-auto Magazine. Lightweighting and Sustainability. June 2020 [cited 2022 18 April]; Sixth: [Available from: https://justauto.nridigital.com/just-auto_magazine_jun20/carbon_fibre_in_car_production_weighing_up_the_benefits.

[139] Ahmad H, et al. A review of carbon fiber materials in automotive industry. IOP Conf Ser: Mater Sci Eng 2020;971(3):032011.

[140] Mordor Intelligence. Automotive Carbon Fiber Market – Growth, Trends, Covid-19 Impact, and Forecast (2022 – 2027). 2022, [cited 2022 18 April]; Available from: https://www.mordorintelligence.com/industry-reports/automotive-carbon-fiber-composites-market.

[141] Chen MC-W. Commercial Viability Analysis of Lignin Based Carbon Fibre. 2014.

[142] Mohammadzadeh Kakhki R. A review to recent developments in modification of carbon fiber electrodes. Arab J Chem 2019;12(7):1783–94.

[143] Karl B, et al. Structural Positive Electrodes for Multifunctional Composite Materials. In: ICCM22 2019. Melbourne, VIC: Engineers Australia, 2019: 2272–2278.: In: ICCM22 2019, Melbourne, VIC: Engineers Australia; 2019. pp. 2272–78.

[144] Chan S, et al. Electrospun carbon nanofiber catalyst layers for polymer electrolyte membrane fuel cells: Structure and performance. J Power Sources 2018;392:239–50.

[145] Wang S-X, et al. Lignin-derived fused electrospun carbon fibrous mats as high performance anode materials for Lithium Ion batteries. ACS Appl Mater Interfaces 2013;5(23):12275–82.

[146] Hagberg J, et al. Lithium iron phosphate coated carbon fiber electrodes for structural lithium ion batteries. Compos Sci Technol 2018;162:235–43.

[147] Asp LE, et al. A structural battery and its multifunctional performance. Adv Energy Sustainability Res 2021;2(3):2000093.

[148] Guomin X, Guan W, Sun X. Filtration performance and application of activated carbon fiber enhanced microfibrous entrapped sorbent (ACF-MFES). Russ J Phys Chem A 2020;94(1):182–88.

[149] Gu Y, et al. One-step solvothermal synthesis of Au-TiO2 loaded electrospun carbon fibers to enhance photocatalytic activity. Vacuum 2016;130:1–6.

[150] Guo L, et al. Fabrication of palladium nanoparticles-loaded carbon nanofibers catalyst for the heck reaction. New J Chem 2013;37(12):4037–44.

[151] Yu B, et al. Preparation of electrospun Ag/g-C_3N_4 loaded composite carbon nanofibers for catalytic applications. Mater Res Express 2017;4(1):015603.

Nur Liyana Ismail, Siti Fatihah Salleh, Jofry Othman, Sara Shahruddin*

Chapter 8
Bio-based surfactant: overview, trend, and future outlook

8.1 Introduction

Surfactants, an abbreviation of "surface-active agents," are compounds that reduce water–oil, liquid–gas, and solid–liquid or solid–gas medium surfaces and interfacial tension [1, 2]. The surface energy is reduced by the presence of hydrophilic and hydrophobic sections of the same surfactant molecule, owing to preferred interactions at surfaces and interfaces. In aqueous solution, surfactant molecules arrange themselves at the interface, where the hydrophobic part is in the air (or oil) and the hydrophilic part is in water, while at high concentration or concentrations above the critical micelle concentration (CMC), surfactant molecules self-assemble into micelles (Figure 8.1). Not only are they widely used as cleaning agents, but other beneficial properties such as foaming, emulsification, and particle suspension also make surfactants known for their wetting ability and effectiveness as emulsifiers and stabilizers. Due to this characteristic, surfactants are found in a variety of products that we use every day, including food, pharmaceuticals, toiletries, detergents, automotive fluids, paints, and coatings [2]. Surfactants have steadily grown in popularity since their debut in the early twentieth century, and they are now among the most widely used synthetic compounds on the planet [3, 4]. The dynamics of the surfactant market are driven by the cost, variety, and availability of hydrophobes as well as the cost and complexity of attaching or creating hydrophilic head groups.

Petrochemical and renewable sources are the two primary feedstock groups used in the manufacture of surfactants [5, 6]. The development of petrochemical processing led to the acquisition of hydrophobic structures of surfactant molecules through polymerization of alkenes, such as ethylene or propylene. Although ethylene has been employed as a carbon chain building block, its increased applicability in industrial production resulted from the production of an intermediate or precursor, ethylene oxide [7]. Meanwhile, natural surfactants are usually derived from triglycerides found in vegetable oils or animal fats.

*Corresponding author: Sara Shahruddin, Petronas Research Sdn. Bhd., Lot 3288 & 3289, Off Jalan Ayer Itam, Kawasan Institusi Bangi, 43000 Kajang, Selangor Darul Ehsan, Malaysia, e-mail: sarashahruddin@petronas.com
Nur Liyana Ismail, Siti Fatihah Salleh, Jofry Othman, Petronas Research Sdn. Bhd., Lot 3288 & 3289, Off Jalan Ayer Itam, Kawasan Institusi Bangi, 43000 Kajang, Selangor Darul Ehsan, Malaysia

https://doi.org/10.1515/9783110791228-008

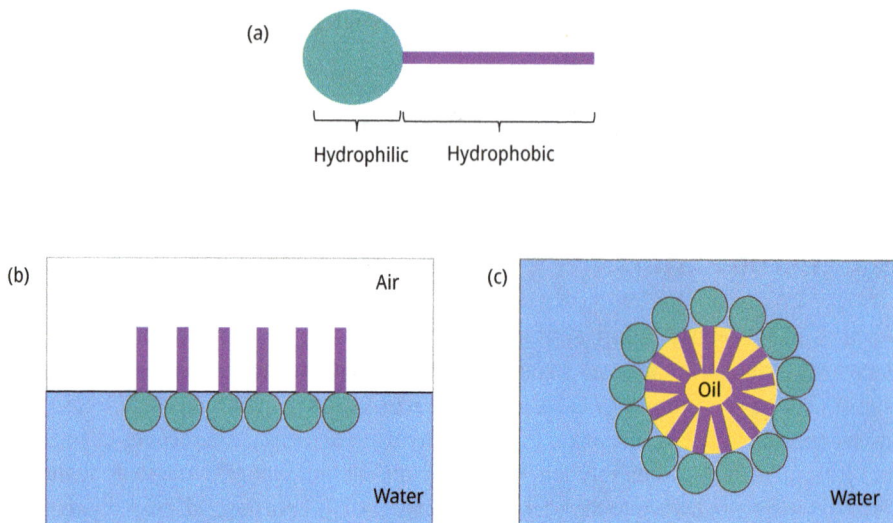

Figure 8.1: (a) Simplified surfactant molecule, (b) arrangement of surfactant monomers at the water surface, and (c) micelle formation above the critical micelle concentration (CMC).

The surfactant industry was focused on the saponification of oils and fats prior to petrochemical processing [8, 9]. Surfactants infiltrate water bodies after usage, where they can create issues if they remain for a long time, resulting in the build-up of potentially toxic or otherwise hazardous substances causing significant environmental concerns [10–12]. Synthetic surfactant-related water contamination has increased in recent years because of their widespread usage in domestic, agricultural, and other cleaning activities. This occurrence has caused global concern, forcing establishment of a series of new rules governing their usage and disposal [13, 14].

In addition, experts relate the production of petrochemical-based surfactants to the high net output of CO_2, a greenhouse gas linked to climate change and global warming. By switching to renewable feedstock, this rate can be minimized. A previous study shows that by using renewable resources instead of petrochemicals for surfactant synthesis would cut CO_2 emissions by 37% in the EU [15]. Aside from environmental concerns and regulations, growing consumer awareness and market pressures have prompted considerable R&D into bio-based surfactants as potential substitutes for synthetic surfactants. Industry experts highlighted the importance of moving toward surfactant production from renewable sources, designing for biodegradation at the end of their life and improving surfactant efficiencies in use. Innovation in these areas will lessen the market's dependency on fossil-derived feedstocks and reduce waste creation, while maximizing the value of surfactants in their current applications.

The term "bio-based surfactant" refers to the surfactant produced by a chemical or enzymatic process that uses renewable substrates as raw materials [16, 17]. According to ISO/DIS 21680, the definition of bio-based surfactant is a surfactant wholly or partly derived from biomass (based on biogenic carbon) [18]. Most applications need further processing of bio-based feedstocks to incorporate functional groups that can give the surfactant's functional characteristics, resulting in a variety of anionic, cationic, non-ionic, and amphoteric products. Many of these processes require the use of petroleum-based feedstocks or moieties that are not always environmentally friendly. The European Commission of Standardization has created categories for biosurfactants, including >95% completely bio-based, 50–94% majority bio-based, 5–49% minority bio-based, and 5% non-bio-based to assist in analyzing the bio-based surfactants' sustainability criteria (Table 8.1) [19].

Table 8.1: Bio-based surfactant classes according to CEN/TS 17035 [19].

Surfactant class	Bio-based carbon content, X% (m/m)
Wholly bio-based surfactant	≥ 95
Majority bio-based surfactant	$95 \geq X > 50$
Minority bio-based surfactant	$50 \geq X \geq 5$
Non-bio-based surfactant	$X < 5$

The hydrophobe, hydrophile, or both that derived from natural sources can be used in the production of bio-based surfactants. Plant oil fatty acids and animal fats are examples of natural hydrophobes, while glycerol, glucose, sucrose, and amino acids (aspartame, glutamic, lysine, arginine, alanine, and protein hydrolyzates) are the examples of natural hydrophiles. They can either be directly utilized from their original form or produced from complicated sources such as vegetable oil, sugarcane, sugar beets, and starch-producing crops. As for biosurfactants, they consist of hydrophilic sugar or peptide component and hydrophobic saturated or unsaturated fatty acid chains that are naturally produced by bacteria, yeast, and fungi. Hence, biosurfactant is classified as a wholly bio-based surfactant since all its raw materials are considered natural [20–22].

The hydrophobic part of bio-based and biosurfactant feedstock is mostly from fatty acyl groups. The fatty acyl groups are generally obtained from oilseeds in the form of triacylglycerol, but may also be derived from oleochemical by-products such as free fatty acid or phospholipids. Fatty acyl groups are generally utilized as lipophilic building blocks for surfactants in the form of free fatty acids or fatty acyl esters, which are produced via hydrolysis or alcoholysis of triacylglycerol [23, 24]. This fatty acyl group conjugates hydrophilic and lipophilic compounds via an ester bond. This bond makes the fatty acid-based surfactants suitable for foods, cosmetics, personal care, and pharmaceutical product applications, but not for laundry detergents since the ester bonds are unstable. More stable bonds such as ether, amides, and carbonate bonds can be produced by converting the fatty acid groups to fatty alcohols, fatty amines, or fatty acid chloride [25–27].

Algae are another potential renewable source of fatty acids. It has been an active research area in recent years due to its potential for high oil production per acre and the ability to leverage on non-arable soil [28–30]. Previously, Unilever has partnered with Sol-azyme, a microalgae firm, in the aim of finding a palm-oil-free replacement for its soaps and surfactants. Solazyme used the advantage of their intellectual property in the areas of recombinant DNA expression in algae and algae bioprocessing to create oils with spe-cific fatty acyl compositions [31]. Solazyme, later renamed as TerraVia, was acquired by Carbion in 2017 to focus on delivering innovative and high-value ingredients for food, personal care, and industrial applications [32]. Lignin has also been used as a feedstock in surfactant production due to its hydrophobic aromatic structure. Lignin-based surfac-tants are usually made by grafting hydrophilic groups or monomers onto the lignin to enhance its surface properties [33–35]. Extensive investigations are necessary to expedite the commercialization of lignin-based surfactants to the market since information on connecting performance and characteristics of lignin-based surfactants for their optimal usage is still lacking.

As for the hydrophilic part of biosurfactant, among the most significant renew-able feedstocks are vegetable oils (for glycerol), sugarcane and sugar beets (for su-crose), and starch-producing crops such as maize, wheat, potato, and tapioca (for glucose) [4, 23, 36]. The use of glycerol as an alternative hydrophilic building block to replace ethylene oxide in the synthesis of nonionic surfactants is a feasible option. The major glycerol-based surfactants on the market are ester-based mono- and digly-cerides, which are made by transesterifying triglycerides with excess glycerol and a base catalyst [4, 26, 37].

Carbohydrates such as sugar and sucrose are another useful biorefinery feedstocks that make up as surfactant hydrophiles. Table 8.2 provides an extensive look at the three main classes of carbohydrate-based surfactants: sucrose esters, alkyl polyglucosides (APG), and sorbitan esters. The discovery of sucrose monoesters, or long-chain fatty acid esters, was one of the first major achievements of the Sugar Research Foundation, which led to their use as nonionic surfactants, food additives, and emulsifiers [38]. The global sucrose ester market amounted to $71.9 M in 2018 and is expected to reach $137.85 M by 2027 [39]. However, selectivity in the synthesis of these esters remains a challenge where acylation with a single fatty acid can yield a wide range of isomers with varying degrees of substitution [40]. One of the solutions to tackle the selectivity problem is by using en-zyme catalysts such as lipases and proteases for regioselective sucrose ester production [41, 42]. Further improvement via lipase and protease protein engineering might increase the regioselectivity and yield of the catalyzed processes. The biotransformation of su-crose to sucrose esters, utilizing whole-cell fermentation methods might also give a new path to sucrose-based surfactant production.

Besides sucrose esters, glucose can also react directly with fatty alcohol in a glyco-sidation process to produce APGs. APG is a nonionic surfactant class with growing production and popularity. Among carbohydrate-based surfactants, APG is the most popular one and is abundantly produced for wide range of applications due to its ex-cellent dermatological properties, high stabilities toward oxidation and hydrolysis, as well as foaming and cleaning abilities [16]. APG's primary functions are cleansing,

Table 8.2: Carbohydrate-based surfactants (alkyl polyglucosides, sorbitan esters, and sucrose esters) that are fully biodegradable.

Carbohydrate-based surfactants	Raw materials	Product characterization	Synthesis	Products	Properties	Primary functions	Application
Alkyl polyglucoside	– Head (hydrophilic): monosaccharides from starch, glucose syrup, glucose monohydrate, water-free glucose, D-glucose, D-xylose – Tail (hydrophobic group): fatty alcohols	– Degree of polymerization (no. of glucose attached to the head) – Alkyl chain length of the fatty alcohol, branching	– Fischer glycosation process: acetalization of glucose in molar excess fatty alcohol (acid catalyzed) – Solvents (optional): Dimethyl-formamide, dimethylsulfoxide	– Alkyl mono-, di-, tri-, and oligoglycosides as mixtures of α- and β-anomers.	– HLB range: 11–14 (relatively hydrophilic) – Critical micelle concentration decreases as alkyl chain increases	– Cleansing agent – Emulsifying agent – Wetting agent – Degreasing agent – Solubilizing agent – Hydrotope – Foaming agent	Medium-chain $C_{12/14}$ – Personal care: cosmetics, bath products, cleansers and wipes, oral care products – Homecare: surface cleaners, dishwashing agents, laundry detergents Short-chain $C_{8/10}$ (or branched C_8) – Hard surface cleaners – Agrochemicals (substitute alkyl phenol ethoxylate) – Industrial and institutional cleaning products Other applications – Oil fields – Admixtures for cement, concrete, and plaster

(continued)

Table 8.2 (continued)

Carbohydrate-based surfactants	Raw materials	Product characterization	Synthesis	Products	Properties	Primary functions	Application
Sucrose esters	– Head (hydrophilic group): disaccharides such as sucrose and lactose – Tail (hydrophobic group): fatty acids from vegetable oils such as lauric acid, stearic acid, palmitic acid, and oleic acid	– Degree of esterification (no. of acyl groups attached to sucrose) – Alkyl chain length of fatty acid – Degree of unsaturation of alkyl chain	– Direct esterification with fatty acids, or transesterification with fatty acid methyl ester (base catalyzed) – Usually carried out with emulsifier such as fatty acid soap and/or sucrose esters, and solvents	– Mono-, di-, tri-, tetra-, or penta-sucrose esters	– HLB range: 1–16 – Monoesters: very hydrophilic, HLB = 16 – Diesters and triesters: more lipophilic	– Emulsifying agent – Wetting agent – Solubilizing agent – Foaming agent	Limited application potential: cater to specific emulsification needs only for: – Food additives – Personal care – Detergent – Stabilizer – Pesticide – SE products are good candidates as insecticides and possibly antimicrobial agents
Sorbitan esters	– Head (hydrophilic group): dehydrated sugar alcohols (sorbitan) – Tail (hydrophobic group): fatty acids from vegetable oils such as lauric acid, stearic acid, palmitic acid, and oleic acid	– Degree of esterification – Alkyl chain length of fatty acid – Degree of unsaturation of alkyl chain	– Intramolecular dehydration of sorbitol (acid catalyzed) and subsequent esterification with fatty acids (base catalyzed)	– Mono-, di-, or tri-sorbitan ester	– HLB range: 4–8 (relatively lipophilic) – Surface tension increases as alkyl chain increases	– Emulsifying agent – Stabilizing agent	– Emulsifier – Pharmaceuticals – Foods – Cosmetic products – Pesticides – Coatings – Explosives

emulsifying, wetting, degreasing, solubilizing, hydrotope, and foaming. Besides its superior foaming performance, it also has an upper hand with respect to its use in the cosmetic and oil industries. Performance properties of APG can be adapted to meet market requirement via precise control of average degree of polymerization and the length of chain alcohols.

In addition, glucose may also be chemically converted to sorbitan to produce sorbitan esters. Like sucrose esters, sorbitan esters are esters of fatty acids, only that the polar head group is derived from sugar alcohols instead of sugar. Sorbitan esters are used as lipophilic nonionic surfactants in multitude of applications, including personal care, pharmaceutical, household, as well as food and beverages. Commercially, sorbitan esters are commonly referred to as spans such as Span 20 (sorbitan monolaurate), Span 40 (sorbitan monopalmitate), Span 60 (sorbitan monostearate), and Span 80 (sorbitan monooleate). Aside from sorbitant esters, other glucose-derived products are sorbitol, sorbitan, *N*-methyl glucamine, and *O*-methylglucoside, as well as the enzymatically converted products such as amino, lactic, and citric acids, all of which can be leveraged to produce surfactants (Figure 8.2) [4].

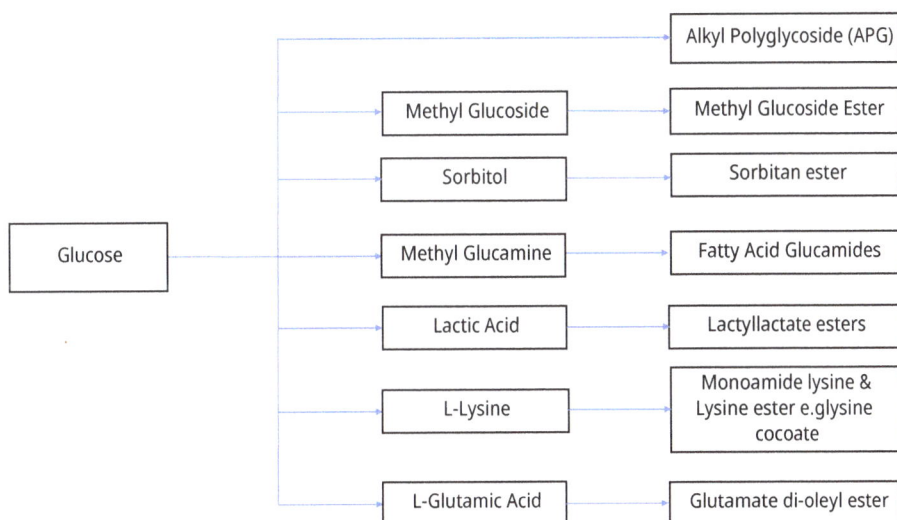

Figure 8.2: Simplified transformation pathway from glucose to several surfactant building blocks and surfactants.

Sugar-derived surfactants have a higher market demand than synthetic chemicals and surfactants due to their low toxicity, low cost, biodegradability, good cleaning and washing abilities, environmental compatibility, and high surface activity [43, 44]. However, if the demand for sugar surfactants grows in the long run, feedstock availability will become a concern. New methods that use bacteria and microorganisms to manufacture glucose are emerging; however, the issue of scalability has yet to be solved.

Aside from sugar-based surfactants, the development of new amino acid-based surfactants is on the rise, which may be influenced by advances in biotechnological amino acid synthesis. The two most produced amino acids in the market are L-glutamic acid and L-lysine. Besides that, alanine, aspartic acid, glycine, and arginine, as well as protein hydrolyzates are also used in the manufacture of some commercial surfactants [45–47]. In the meantime, sarcosine-based surfactants have been on the market for decades. Even though sarcosine is a naturally occurring molecule, it is mostly synthesized on a large scale by combining chloroacetic acid with N-methylamine [48–50]. Betaine, another naturally occurring molecule, is also synthesized in large scale using petrochemical-based trimethylamine and chloroacetic acid. Most betaine surfactants use an oleochemical hydrophobe precursor obtained from tropical oils as the bio-based component [51]. Glycine betaine is a promising biosurfactant that can be commercially extracted from brown algae and sugar beet molasses [52, 53].

Glycolipids are a type of complex carbohydrate that contains both a glycan and a lipid component. They are usually the main lipids of bacterial and fungal cell walls. In aqueous solution, glycolipids are amphiphilic substances that form stable micelles, and these molecules have the capacity to offer low interfacial tension [54, 55]. Rhamnolipids and sophorolipids are among the glycolipids that have been utilized the most as biosurfactants. Rhamnolipids are produced as one or two rhamnose sugar groups attached to one or two fatty acid chains by different bacterial species (i.e., *Pseudomonas aeruginosa*, *Pseudomonas chlororaphis*, and *Burkholderia pseudomallei*) [4, 56]. Aside from their favorable emulsifying, solubilizing, foaming, and antibacterial characteristics, the use of rhamnolipids is appealing due to their high production yields after relatively short incubation times [56]. Rhamnolipids are now available on a larger scale due to the optimized fermentation techniques and advanced extraction and concentration technologies. Sophorolipids, another extensively researched type of glycolipid, are biosynthesized by certain yeast strains like *Starmerella bombicola*, *Wickerhamiella domercqiae*, and *Candida batistae* from sophorose sugar and hydroxylated fatty acid. Sophorolipids are commercially used in dish and vegetable detergents, and in skin care formulations [57–60].

8.2 Recent progress in R&D and industrial production

Regulations on the environmental impact and hazardous chemicals are highly stringent, particularly in Europe and North America, which are the two largest markets for surfactants, especially in the home and personal care sectors. As a result, the surfactant industry is commencing to develop biosurfactants, which have lower levels of toxicity and a more environmentally friendly manufacturing process. Apart from complying with environmental regulations, the industry is seeing bio-based surfactants to achieve a sustainable competitive edge. The advent of biotechnology in the twenty-first century promoted the creation of novel bio-based and biosurfactants along with their enhanced commercial and economic viability. Extensive and significant R&D has also enabled high-quality and high-functionality bio-based surfactant formulations to evolve from the lab-scale to niche applications to commercial-scale production. Some of the bio-based surfactants that are commercially available in the market, their main manufacturers, and their applications are listed in Table 8.3.

Table 8.3: Commercially available bio-based surfactants, their manufacturers, and their applications.

Bio-based surfactants	Selected manufacturers	Fields of applications
Anionic		
Lignosulfonate	Vanderbilt Minerals, LLC	– Laundry
Methyl ester sulfonates	Huish Detergents, Inc., Lion Corp., Longkey, Stepan	– Food service and kitchen hygiene – Dishwashing
Anionic derivatives of alkyl polyglucoside	Cognis, Colonial Chemical	– Hard surface cleaning – Institutional cleaning and sanitation – Vehicle and transportation care
Nonionic		
Fatty alcohol alkoxylate	BASF, Dow	– Dishwashing
Fatty acid alkoxylate	BASF, Clariant, Croda	– Laundry
Alkyl polyglucoside	Croda, Esterchem, Huntsman	– Hard surface cleaning
Sorbitan ester	Akzo Nobel, BASF, Colonial Chemical, Dow, Huntsman	– Food service and kitchen hygiene – Institutional cleaning and
Alkanoyl-*N*-methylglucamide	BASF, Croda, Huntsman	sanitation – Vehicle care
Alkyl-ethoxylated mono- and diglycerides	Clariant, Kao, Kerry Ingredients and Flavors	– Personal care
Polyglycerol esters	BASF, Colonial Chemical, Hychem Corp., Kerry Ingredients and Flavors	
Amphoteric		
Cocamidopropyl betaine	BASF, Colonial Chemical, Stepan	– Food and beverage processing
Cocamidopropyl hydroxysultaine	Colonial Chemical, Stepan	– Personal care – Hard surface cleaning
Lauryl hydroxysultaine	Colonial Chemical, Stepan	
Glycolipid		

Table 8.3 (continued)

Bio-based surfactants	Selected manufacturers	Fields of applications
Sophorolipid	BASF, Clariant, Ecover, Evonik, MG Intobio Co. Ltd., Saraya Co. Ltd., Soliance	– Personal care – Vegetable liquid wash – Dish washing
Rhamnolipid	AGAE Technologies, BASF, Biotensidon GmbH, Clariant, Evonik, GlycoSurf, Henkel, Jeneil Biotech Inc., Logos Technology Rhamnolipid Companies Inc., TeeGene Biotech	
Amino acid surfactants		
Sodium cocoyl glutamate	Ajinomoto Co. Inc., Stepan, Zschimmer and Schwarz	Personal care
Sodium methyl cocoyl taurate	Clariant	
α-Acyl glutamate and sarcosinate	Schill + Seilacher	

In the current development of novel surfactants, there is a growing trend of utilizing nontraditional naturally occurring branching hydrophobic chains [61–63]. Nonionic surfactants based on twin-tail glycerol have been synthesized, and they have good oil-in-water and water-in-oil emulsifying characteristics [64]. Other structural analogues of glycerol-based surfactants have recently been created by employing heterogeneous interfacial acidic catalysts to direct etherify glycerol and dodecanol. These surfactants have been shown to be comparable with commercially available surfactants in terms of physicochemical assessment and detergency ability [65]. Another class of amphiphilic compounds with a glycerol backbone is bio-based dialkyl glycerol ethers. These compounds have good solvosurfactant characteristics and can function as solubilizers for hydrophobic dyes in aqueous media [66].

Natural edible flavor vanillin is used to create a cleavable vanillin-based polyoxyethylene nonionic surfactant. Because it contains cleavable acetal bonds which break down quickly under acidic circumstances, this environmentally beneficial surfactant is totally biodegradable in nature. The surfactant's surface activity, wettability, emulsifying, and foaming properties are on par to nonylphenol ethoxylate surfactants which are highly toxic to aquatic organisms and environment [67]. Several novel types of sustainable surfactants have been created in recent years by employing various types of terpenes, which are major components of essential oils derived from a variety of plants and flowers [68–71]. The terpenes were transformed to branched hydrophobic tail containing quaternary ammonium surfactants. Natural farnesol, a 15-carbon acyclic sesquiterpene alcohol found in neroli, lemon grass, tuberose, rose, citronella, and other plant species, was used to create a new form of terpene-based sustainable surfactant which have demonstrated excellent surfactant performance [71]. Under the trade name ECO-SURF, Dow Chemical Co. is now offering a range of sustainable oilseed-based nonionic surfactants. These surfactants are claimed to have minimal aquatic toxicity and are bio-degradable in nature, making them suitable candidates for paints and coatings, as well as home, industrial, and institutional cleansers and textiles [72].

TegraSurf, a sustainable water-based surfactant developed for energy, mining, agri-cultural, water treatment and other industrial applications, was released in July 2021 by Integrity BioChem, a technology-driven business producing next-generation biopolymers. TegraSurf is made of sustainable vegetal materials and is certified readily biodegradable by the OECD 301B guideline. After 90 days, it is no longer in the environment, making it safer and healthier for local population and allowing formulators to fulfill industry sus-tainability criteria [73]. BASF and Solazyme Inc. recently released Dehyton® AO 45, the first commercial microalgae-derived betaine surfactant made from microalgae oils as an alternative to conventional amidopropyl betaine surfactants [74]. Following the launch of sophorolipid-based surfactants in 2020, BASF formed an exclusive partnership with Holiferm Ltd. in the United Kingdom to focus on the development of glycolipids other than sophorolipids for personal and home care, as well as industrial uses [75].

Croda expanded its commercial-scale bio-based manufacturing capabilities and technology leadership in renewable raw materials by unveiling its 100% bio-based ethylene oxide production facility as an effort to make the world's products greener. Ethylene oxide is the key raw material used to produce surfactants. Croda's Atlas Point manufacturing plant in New Castle, Delaware, is the first of its type in the United States for the manufacture of 100% sustainable, 100% bio-based nonionic surfactants [76]. Ajinomoto is increasing to 60% of their global capacity for its Amisoft range of amino acid-based liquid surfactants by building a new plant in Brazil [77]. Sironix Renewables received $645,000 in investment from the University of Minnesota Discovery Capital In-vestment program and investors, as well as a $1.15 million grant from the US Depart-ment of Energy Advanced Manufacturing Office, to help them scale up their Eosix® production. The new renewable oleo-furan-based surfactants are 100% plant-based that offer unique and adjustable characteristics in a wide range of areas, including cleaning products, cleaners, cosmetics and personal care, agriculture and inks, and paint and coatings [78].

8.3 Industrial challenge on bio-based surfactant

This section covers the market performance, demand drivers, and growth prospects of biosurfactants. The market trend on bio-based and biosurfactants is discussed for the different geographic regions and in terms of changing market trends for biosurfactants in various application areas. Analysis of the industrial challenges of biosurfactants, which includes the growth-restraining factors and future opportunities, is provided.

8.3.1 The economy and market trend of bio-based surfactants

The worldwide surfactant industry, estimated to be worth $39 billion in 2019, is expected to expand at a rate of 2.6% per year over the following 5 years, reaching $46 billion in 2024. Surfactants are produced in a total of 17 million metric tons per year [79]. In the EU, of the 3 million metric tons of surfactants produced in 2019, roughly 50% were bio-based [80]. A market study by the Market Research Future [81] indicated that the global biosurfactant market value is around US $2.1 billion in 2020 and predicted it to reach US $2.8 billion by 2026, with a compound annual growth rate of over 5% from 2021 to 2026. The attractive performance of biosurfactants advances their high potential to substitute synthetic-based surfactants for drop-in applications and with unique properties that can overcome entry barriers for the emerging industrial areas.

Major types of biosurfactants such as sophorolipids, glycolipids, lipopeptides, polymeric biosurfactants, phospholipids, and fatty acids generally form the product demand application. Among them, sophorolipids provide the largest global market demand with detergents and industrial cleaning applications. The leading demand drivers for biosurfactants comprise a growing consumer preference, increasingly stringent regulatory requirements, and rising awareness toward eco-friendly alternatives. By being environmentally compatible and with low toxicity, many studies have considered biosurfactants as the next generation of industrial surfactants [82–84]. In terms of end-user applications, biosurfactants are finding usage in household detergents, industrial and institutional cleaners, cosmetics, and personal care within the major markets in Europe and North America [81]. Recently, they are gaining acceptance in the newer application areas such as in oil and gas as well as agricultural industries.

Furthermore, the increasing consumer awareness of biosurfactants' benefits and their wide range of application sectors form market drivers that increase their future growth potential. Higher growth of biosurfactants is seen in Asia-Pacific (APAC), especially in Southeast Asian countries, which have slightly different demand factors that involve the increasing purchasing power of mass consumers, growing concern on environmental issues, and the generation of harmful chemical by-products. In terms of APAC market segmentation, the major sales revenue for biosurfactants resides within the home care and personal care applications, as rising urbanization becomes the dominant factor for surfactant growth.

More importantly, a key growth enabler is in the innovative research on biosurfactants, especially when it can generate multifunctional and diversified products using renewable feedstock. This technological progress contributes to the desirable properties of biosurfactants to meet the changing consumer lifestyles in developing economies and consequently their increasing preference for usage in the end-user product formulation. As an example, within the home care detergent industry, the usage of biosurfactants as environmentally friendly products provides sustainable alternatives that are gaining a large market share [82, 85, 86].

The highest adoption of bio-based and biosurfactants is in Europe and North America, which dominates the bio-based surfactant market share in terms of revenue and volume. Increasingly stringent regulatory requirements enable a wider acceptance of biosurfactants in the place of synthetic surfactants. For example, the imposed government regulations such as CEN/TC-276 define the standards for surface-active agents and detergents to enhance the EU bio-based economy, and detergent regulation (EC) no. 648 that requires surfactants used in detergents to be biodegradable under aerobic conditions as per OECD 301 test series. In addition, the COVID-19 pandemic results in a sharp increase in the bio-based surfactant product demand for household detergents, personal care, and industrial cleaners due to the rising trend for sanitation.

8.3.2 The industrial challenges for bio-based surfactant

Bio-based surfactants are synthesized via chemical reaction which is usually carried out under harsh conditions. The use of hazardous solvents and toxic acid or base catalysts sometimes creates undesired waste or by-products that are detrimental to the environment. Enzymes have the potential to play a significant role in the production of numerous bio-based surfactants, although they are not currently used on a large basis. Enzymes provide several advantages over chemical processing, notably in terms of improving process sustainability. However, the main drawbacks of enzymes are their relatively higher price compared to chemical catalysts as well as their slower reaction speeds. Since energy costs are expected to rise, the need for production sustainability (lower operating energy, less waste, and safer operating condition) is critical.

Despite the growing demand for bio-based surfactants, several challenges exist, which restrain their further market growth and wider adoption. General customers tend to choose cost-effective surfactants. At present, the high cost of bio-based surfactants is the biggest hurdle in meeting the requirement of priced sensitive Asian customers. Increased sustainability of biosurfactant alone without significant higher performance is not well accepted, as the usual consumers will not be willing to pay a "green" premium for bio-based products.

Higher complexity and low-efficiency microbial fermentation process in biosurfactant manufacturing contribute to the high production cost and expensive capital cost investment. For example, the average price of sophorolipids is US $34/kg as compared to sodium dodecyl sulfate and amino acid surfactants that are priced at US $1–4/kg [87]. High purification cost of bio-based surfactants is also a big issue, as it accounts up to 60% the production cost, but this can be minimized for the case of biosurfactant application in crude forms, such as in an industrial environment [89]. However, for high-purity applications, improvement in downstream processing methods is needed to attain a competitive cost of production.

Furthermore, biosurfactant demand's reliance on the volatility and economic downturn of downstream end-user industries poses a serious risk to its long-term viability.

Industries that are applicable for biosurfactant applications such as oil and gas, enhanced oil recovery, food industry, construction, textiles, paints, pharmaceuticals, and detergents are known to be susceptible to general macroeconomic performance. In addition, the COVID-19 pandemic further leads to disruption in the end-user industrial demand and sustainability concern on the raw material supply. Opportunity exists in developing a new technology solution that utilizes a low-cost raw material such as industrial wastes for biosurfactant production. However, this needs to consider the overall production impact factors that include the availability, stability, and variability of each component [89].

8.4 Future outlook and prospect

The development of bio-based surfactants from renewable feedstocks is an attractive alternative to fossil-based surfactants with a significantly growing market attributed to their performance, biodegradability, biocompatibility, and nontoxicity [22, 33]. Additionally, advances in renewable technology, increased environmental concern, consumer awareness, and stringent regulatory requirements provide a continued push toward bio-based surfactant demand. Potential areas for use are growing fast and valuable outcomes depend on whether the bio-based surfactants can be customized for specific applications along with if they can be produced at a price that will make them attractive alternatives to the fossil-based surfactants.

At present, fossil-based surfactants are less expensive than bio-based surfactants [4, 93, 94]. However, this trend will likely be changed in future, thereby increasing the prospects of bio-based surfactants. Feedstocks and how the bio-based surfactants are produced are the two key factors governing final product costs [4, 36, 95, 96]. To use renewable feedstock in the industry, they should be cost-effective, available in large quantities, and can effectively be converted to value-added surfactants [96]. The surfactant design requires careful selection of the hydrophile and hydrophobe pairs so that they can be easily synthesized with minimum purification and provide the desired properties for the intended application [4, 16, 93, 97]. Triglycerides, fatty acid methyl esters, fatty alcohols, fatty acids and fatty amines are common examples of renewable hydrophobes used to produce bio-based surfactants. Meanwhile, sustainable hydrophilic head group can be designed using several molecules such as glycerol, carbohydrate feedstock such as sucrose, glucose, organic acids, and amino acids [4, 36, 95, 96].

Current production technologies are leveraged and further enhanced with green manufacturing principles to convert renewable feedstocks into valuable and new bio-based surfactants. Other than the aforementioned starting materials, alternative substrates such as agro-based industrial wastes or other suitable simple waste substrate is gaining a lot of research interest and can lead to significant cost reduction [98]. The use of biomass for surfactant manufacturing will help to further reduce greenhouse gas emissions. Besides that, as bio-based surfactants degrade, they only release back the quantitative amount of the carbon used by the plant to produce the surfactants [36]. Although the use of biomass appears to be a promising option, the lack of homogeneity and consistency in the feedstock is a serious issue that will result in inconsistency in the final products. Surfactant property and performance variations may result in unsatisfactory properties. As a result, waste material pretreatment is critical, and extensive testing will be required to minimize variations.

Besides feedstocks, catalyst design is crucial to ensure high selectivity of the processes by limiting the formation of by-products and to reduce the reaction times efficiently; with renewable materials, fundamental rethinking of catalyst development is necessary [99–101]. The other critical component toward achieving low product cost is the process required for optimal chemical conversion. Researchers are looking into equipment miniaturization such as continuous reactors to help reduce the raw material consumption and effluent production. Process intensification is another aspect that could help reduce the investment costs [100]. Research focusing on alternative or green solvents dedicated to the conversion of renewable feedstock to value-added products has led to several publications. Among those being researched include bio-based ionic liquids, deep eutectic solvents, bio-based solvents, CO_2-switchable solvents, and supercritical fluids [102–104].

Additionally, several operation and control factors provide important handles to minimize biosurfactant production costs. Batch cycle optimization on fermentation and purification process can reduce the idle time between batches and minimize chemical usage for equipment cleaning and energy use during sterilization. Productivity is the most important factor in the manufacturing economics of biosurfactant production at commercial scales [8]. Optimum batch sequencing campaign is critical to minimize start-up and shutdown frequency to lower the production downtime that improves productivity. Lastly, biosurfactant product development will need to fulfill time-consuming and expensive legislative requirements that restrain market growth [91], which will inevitably add a cost of compliance into the product development cost that is incurred by biosurfactant manufacturers. Other market entry requirements include biosurfactant products that are tested for long shelf life and the ability to maintain stable properties in the industrial environment [92].

8.5 Conclusions

In recent years, surfactant manufacturers have introduced a variety of new eco-friendly surfactant-based products to the market. Increased consumer awareness, combined with legislative demand for sustainable development, has resulted in the market introduction of numerous novel surfactant types based on renewable building blocks. Because of their improved biodegradability and low toxicity, these surfactants are a popular choice for innovative formulations in the industrial and consumer markets. However, the commercial production of bio-based surfactants faces numerous challenges that must be addressed in order for them to be economically viable, particularly the high production cost. As a result, low-cost biosurfactant production is critical for achieving an economically sustainable process and ensuring market continuity in the future. A balance should be struck between cost-effective formulations and efficient performance. Simultaneous design of bio-based surfactants for functional, economic, and environmental benefits will be required to ensure the long-term viability of business. Moving forward, continuous technological advancements using higher performance catalysts, optimized and integrated reaction processes using less expensive renewable feedstocks, and effective downstream purification are required in the future. It is hoped that these efforts will eventually lead to a greater use of bio-based surfactants, which will provide enormous benefits to consumers and the environment.

References

[1]   Rosen MJ, Kunjappu JT. Surfactants and Interfacial Phenomena. 4th edition. Hoboken, New Jersey: John Wiley & Sons; 2012. ISBN 978-1-118-22902-6.

[2]   Möbius D, Miller R, Fainerman VB. Surfactants: Chemistry, Interfacial Properties, Applications, Volume 13 of Studies in Interface Science. ISSN 1383-7303.

[3]   Traverso-Soto JM, González-Mazo E, Lara-Martín PA. Analysis of surfactants in environmental samples by chromatographic techniques, Chromatography – the Most Versatile Method of Chemical Analysis, Leonardo de Azevedo Calderon, IntechOpen, doi: 10.5772/48475. Available from: https://www.intechopen.com/chapters/40379.

[4]   Foley P, Beach ES, Zimmerman JB. Derivation and synthesis of renewable surfactants. Chem Soc Rev 2012, 41, 1499–518.

[5]   Jimoh AA, Lin J Biosurfactant: A new frontier for greener technology and environmental sustainability. Ecotoxicol Environ Saf 2019;184:109–607. https://doi.org/10.1016/j.ecoenv.2019.109607. PMid: 31505408.

[6]   Taddese T, Anderson RL, Bray DJ, et al. Recent advances in particle-based simulation of surfactants. Curr Opin Colloid Interface Sci 2020;48:137–48. https://doi.org/10.1016/j.cocis.2020.04.001.

[7]   Soler-illia GJDEAA, Sanchez C. Interactions between poly(ethylene oxide)-based surfactants and transition metal alkoxides: Their role in the templated construction of mesostructured hybrid organic-inorganic composites. New J Chem 2000;24(7):493–99. https://doi.org/10.1039/b002518f.

[8]   Zoller U, Sosis P, editors. Handbook of Detergents, Part F: Production. CRC Press; 2008 Nov 2008.

[9]   Holmberg K, Natural surfactants. Curr Opin Colloid Interface Sci 2001;6:148–59

[10] Deschênes L, Lafrance P, Villeneuve JP, Samson R. Adding sodium dodecyl sulfate and Pseudomonas aeruginosa UG2 biosurfactants inhibits polycyclic aromatic hydrocarbon biodegradation in a weathered creosote-contaminated soil. Appl Microbiol Biotechnol 1996;46(5):638–46.

[11] Takeda K, Sasaoka H, Sasa K, Hirai H, Hachiya K, Moriyama Y. Size and mobility of sodium dodecyl sulfate – bovine serum albumin complex as studied by dynamic light scattering and electrophoretic light scattering. J Colloid Interface Sci 1992;154(2):385–92.

[12] Rebello S, Asok AK, Mundayoor S, Jisha MS. Surfactants: Toxicity, remediation and green surfactants. Environ Chem Lett 2014;12(2):275–87.

[13] Siwayanan P, Bakar NA, Aziz R, Chelliapan S, Siwayanan P. Exploring Malaysian household consumers acceptance towards eco-friendly laundry detergent powders. Asian Soc Sci 2015;11(9):125–37.

[14] Farn RJ, editor. Chemistry and Technology of Surfactants. John Wiley & Sons; 2008 Apr 15.

[15] Patel M. Surfactants based on renewable raw materials: Carbon dioxide reduction potential and policies and measures for the European Union. J Ind Ecol 2003;7(3-4):47–62.

[16] Hayes DG, Smith GA Bio-based surfactants: Overview and industrial state of the art. Bio-based Surfactants 2019:3–8.

[17] Kandasamy R, Rajasekaran M, Venkatesan SK, Uddin M. New trends in the biomanufacturing of green surfactants: bio-based surfactants and biosurfactants. In: Next Generation Biomanufacturing Technologies. American Chemical Society. 2019, pp. 243–60

[18] ISO/DIS 21680(en) Surface active agents – Bio-based surfactants – Requirements and test methods {Internet]. 2020. Available from: https://www.iso.org/obp/ui/#iso:std:iso:21680:dis:ed-1:v1:en.

[19] Comite Europeen de Normalisation. CEN/TS 17035: 2017. Surface Active Agents – Bio-Based Surfactants – Requirements and Test Methods. Brussels: CEN; 2017.

[20] Md F. Biosurfactant: Production and application. J Pet Environ Biotechnol 2012;3(4):124.

[21] Van Renterghem L, Roelants SL, Baccile N, Uyttersprot K, Taelman MC, Everaert B, Mincke S, Ledegen S, Debrouwer S, Scholtens K, Stevens C. From lab to market: An integrated bioprocess design approach for new-to-nature biosurfactants produced by Starmerella Bombicola. Biotechnol Bioeng 2018;115(5):1195–206.

[22] Moldes AB, Rodríguez-López L, Rincón-Fontán M, López-Prieto A, Vecino X, Cruz JM. Synthetic and bio-derived surfactants versus microbial biosurfactants in the cosmetic industry: an overview. Int J Mol Sci 2021;22(5):2371.

[23] Hayes DG, Solaiman DK, Ashby RD. Bio-Based Surfactants: Synthesis, Properties, and Applications. Elsevier; 2019 Apr 30.

[24] Salimon J, Salih N, Yousif E. Industrial development and applications of plant oils and their bio-based oleochemicals. Arab J Chem 2012;5(2):135–45.

[25] Kreutzer UR Manufacture of fatty alcohols based on natural fats and oils. J Am Oil Chem Soc 1984;61(2):343–48.

[26] Biermann U, Bornscheuer U, Meier MA, Metzger JO, Schäfer HJ. Oils and fats as renewable raw materials in chemistry. Angew Chem Int Ed 2011;50(17):3854–71.

[27] Giraldo L, Camargo G, Tirano J, Moreno-Pirajan JC. Synthesis of fatty alcohols from oil palm using a catalyst of Ni-Cu supported onto zeolite. E J Chem 2010;7(4):1138–47.

[28] Jeon JY, Han Y, Kim YW, Lee YW, Hong S, Hwang IT. Feasibility of unsaturated fatty acid feedstocks as green alternatives in bio-oil refinery. Biofuel Bioprod Biorefin 2019;13(3):690–722.

[29] Hess SK, Lepetit B, Kroth PG, Mecking S. Production of chemicals from microalgae lipids–status and perspectives. Eur J Lipid Sci Technol 2018;120(1):1700152.

[30] De Luca M, Pappalardo I, Limongi AR, Viviano E, Radice RP, Todisco S, Martelli G, Infantino V, Vassallo A. Lipids from microalgae for cosmetic applications. Cosmetics 2021;8(2):52.

[31] Lerayer A. Solazyme:" unlocking the power of microalgae: A new source of sustainable and renewable oils". In: BMC Proceedings. 2014 Oct, Vol. 8, No. 4, pp. 1–2. BioMed Central.

[32] Innovative microalgae specialist TerraVia acquired by Corbion [Internet]. 2017. Available from: https://www.corbion.com/media/624357/20170729-tvia-completion-eng-final.pdf.

[33] Alwadani N, Fatehi P. Synthetic and lignin-based surfactants: Challenges and opportunities. Carb Res Conv 2018;1(2):126–38.

[34] Schmidt BV, Molinari V, Esposito D, Tauer K, Antonietti M. Lignin-based polymeric surfactants for emulsion polymerization. Polymer 2017;112:418–26.

[35] Zhou M, Wang W, Yang D, Qiu X. Preparation of a new lignin-based anionic/cationic surfactant and its solution behaviour. RSC Adv 2015;5(4):2441–48.

[36] Bhadani A, Kafle A, Ogura T, Akamatsu M, Sakai K, Sakai H, Abe M. Current perspective of sustainable surfactants based on renewable building blocks. Curr Opin Colloid Interface Sci 2020;45:124–35.

[37] Fan Z, Zhao Y, Preda F, Clacens JM, Shi H, Wang L, Feng X, De Campo F. Preparation of bio-based surfactants from glycerol and dodecanol by direct etherification. Green Chem 2015;17(2):882–92.

[38] Plat T, Linhardt RJ. Syntheses and applications of sucrose-based esters. J Surfactants Deterg 2001;4(4):415–21.

[39] Outlook for sucrose esters. Focus Surfactants 2020;2020(6):3.

[40] Queneau Y, Fitremann J, Trombotto S. The chemistry of unprotected sucrose: The selectivity issue. C R Chim 2004;7(2):177-88.t.

[41] Potier P, Bouchu A, Descotes G, Queneau Y. Lipase-catalysed selective synthesis of sucrose mixed diesters. Synthesis 2001;2001(03):0458–62.

[42] Cruces MA, Plou FJ, Ferrer M, Bernabé M, Ballesteros A. Improved synthesis of sucrose fatty acid monoesters. J Am Oil Chem Soc 2001;78(5):541–46.

[43] Jesus CF, Alves AA, Fiuza SM, Murtinho D, Antunes FE. Mini-review: Synthetic methods for the production of cationic sugar-based surfactants. J Mol Liq 2021:117389.

[44] Hill K, Rhode O. Sugar-based surfactants for consumer products and technical applications. Lipid/Fett 1999;101(1):25–33.

[45] Infante MR, Pérez L, Pinazo A, Clapés P, Morán MC, Angelet M, García MT, Vinardell MP. Amino acid-based surfactants. C R Chim 2004;7(6-7):583–92.

[46] Takehara M. Properties and applications of amino acid based surfactants. Colloids Surf 1989;38(1):149–67.

[47] Bordes R, Holmberg K. Amino acid-based surfactants–do they deserve more attention? Adv Colloid Interface Sci 2015;222:79–91.

[48] Baumann L. The preparation of sarcosine. J Biol Chem 1915;21(3):563–66.

[49] Cullum DC. The separation of sarcosine from methylaminediacetic acid. Analyst 1957;82(977):589–91.

[50] Morris ED, Bost JC. In Kirk-Othmer Encyclopedia of Chemical Technology, John Wiley & Sons, Inc; 2002.

[51] Clendennen SK, Boaz NW. Betaine Amphoteric Surfactants – Synthesis, Properties, and Applications. In Bio-Based Surfactants, AOCS Press; 2019 Jan 1, pp. 447–69.

[52] Goursaud F, Berchel M, Guilbot J, Legros N, Lemiègre L, Marcilloux J, Plusquellec D, Benvegnu T. Glycine betaine as a renewable raw material to "greener" new cationic surfactants. Green Chem 2008;10(3):310–20.

[53] Nsimba ZF, Paquot M, Mvumbi LG, Deleu M. Glycine betaine surfactant derivatives: synthesis methods and potentialities of use. Biotechnol Agron Soc Environ 2010;14(4):737–48.

[54] Bednarski W, Adamczak M, Tomasik J, Płaszczyk M. Application of oil refinery waste in the biosynthesis of glycolipids by yeast. Bioresour Technol 2004;95(1):15–18.

[55] Kakehi K, Suzuki S. Analysis of glycans; polysaccharide functional properties. Compr Glycosci 2007

[56] Irfan-Maqsood M, Seddiq-Shams M. Rhamnolipids: Well-characterized glycolipids with potential broad applicability as biosurfactants. Ind Biotechnol 2014;10(4):285–91.

[57] Van Bogaert IN, Zhang J, Soetaert W. Microbial synthesis of sophorolipids. Process Biochem 2011;46(4):821–33.

[58] Hirata Y, Ryu M, Oda Y, Igarashi K, Nagatsuka A, Furuta T, Sugiura M. Novel characteristics of sophorolipids, yeast glycolipid biosurfactants, as biodegradable low-foaming surfactants. J Biosci Bioeng 2009;108(2):142–46.

[59] Develter DW, Lauryssen LM. Properties and industrial applications of sophorolipids. Eur J Lipid Sci Technol 2010;112(6):628–38.

[60] Roelants S, Solaiman DK, Ashby RD, Lodens S, Van Renterghem L, Soetaert W. Production and applications of sophorolipids. Bio-based Surfactants 2019:65–119.

[61] Iskandar WF, Salim M, Hashim R, Zahid NI. Stability of cubic phase and curvature tuning in the lyotropic system of branched chain galactose-based glycolipid by amphiphilic additives. Colloids Surf A Physicochem Eng Asp 2021;623:126697.

[62] Kim JH, Oh YR, Hwang J, Kang J, Kim H, Jang YA, Lee SS, Hwang SY, Park J, Eom GT. Valorization of waste-cooking oil into sophorolipids and application of their methyl hydroxyl branched fatty acid derivatives to produce engineering bioplastics. Waste Manag 2021;124:195–202.

[63] Elsoud MM. Classification and Production of Microbial Surfactants. In: Microbial Biosurfactants, Singapore: Springer; 2021, pp. 65–89.

[64] Zhang L, Zhang X, Zhang P, Zhang Z, Liu S, Han B. Efficient emulsifying properties of glycerol-based surfactant. Colloid Surface Physicochem Eng Aspect 2018, 553:225–29.

[65] Fan Z, Zhao Y, Preda F, Clacens JM, Shi H, Wang L, Feng X, De Campob F. Preparation of bio-based surfactants from glycerol and dodecanol by direct etherification. Green Chem 2015,17, 882–92.

[66] Lebeuf R, Illous E, Dussenne C, Molinier V, Silva ED, Lemaire M, Aubry JM. ACS Sustainable Chem Eng 2016, 4:4815–23.

[67] Ding F, Zhou X, Wu Z, Xing Z. Synthesis of a cleavable vanillin-based polyoxyethylene surfactant and its pilot application in cotton fabric pretreatment. ACS Sustainable Chem Eng 2019, 7:5494–500.

[68] Ogunkunle T, Fadairo A, Rasouli V, Ling K, Oladepo A, Chukwuma O, Ayoo J. Microbial-derived bio-surfactant using neem oil as substrate and its suitability for enhanced oil recovery. J Pet Explor Prod 2021;11(2):627–38.

[69] Ma J, Gao J, Wang H, Lyu B, Gao D. Dissymmetry gemini sulfosuccinate surfactant from vegetable oil: A kind of environmentally friendly fat liquoring agent in the leather industry. ACS Sustainable Chem Eng 2017, 5:10693–701.

[70] Faßbach TA, Gaide T, Terhorst M, Behr A, Vorholt AJ. Renewable surfactants through the hydroaminomethylation of terpenes. Chem Cat Chem 2017, 9:1359–62.

[71] Bhadani A, Rane J, Veresmortean C, Banerjee S, John G. Bio-inspired surfactants capable of generating plant volatiles. Soft Matter 2015, 11:3076–82.

[72] ECOSURF™ SA surfactants- seed oil-based surfactants. Dow. 2008. (Internet) Available from: https://www.dow.com/documents/en-us/mark-prod-info/119/119-02222-01-ecosurf-sa-surfactants-seed-oilbased-surfactants.pdf.

[73] Integrity Biochem announces first-of-its-kind surfactant to improve industrial sustainability. Businesswire. 2021 (Internet). Available from: https://www.businesswire.com/news/home/20210728005149/en/Integrity-BioChem-Announces-First-of-its-Kind-Surfactant-to-Improve-Industrial-Sustainability.

[74] BASF and solazyme launch the first commercial microalgae-derived betaine surfactant. BASF. 2015 (Internet) Available from: https://www.basf.com/us/en/media/news-releases/2015/07/P-US-15-137.html.

[75] How bio-based surfactants are turning the world green. Cosmetics & Toiletries. 2020 (Internet). Available from: https://www.cosmeticsandtoiletries.com/formulating/category/natural/How-Bio-Based-Surfactants-are-Turning-the-World-Green–570209271.html.

[76] BASF strengthens its position in bio-surfactants for personal care, home care and industrial formulators with two distinct partnerships. BASF. 2021. Available from: https://www.basf.com/my/en/media/news-releases/global/2021/03/p-21-148.html.

[77] Ajinomoto to expand amino acid-based surfactants. Specialty Chemicals Magazine. 2019 (Internet). Available from: https://www.specchemonline.com/index.php/ajinomoto-expand-amino-acid-based-surfactants.

[78] Sironix renewables closes seed round to scale production of its plant-based surfactant for detergents, cleaners & shampoos. GlobeNewswire. 2020 (Internet). Available from: https://www.globenewswire.com/news-release/2020/09/16/2094579/0/en/Sironix-Renewables-Closes-Seed-Round-to-Scale-Production-of-its-Plant-based-Surfactant-for-Detergents-Cleaners-Shampoos.html.

[79] Assessing the sustainability and performance of green surfactants. IHS Markit. 2020 (Internet). Available from: https://ihsmarkit.com/research-analysis/assessing-sustainability-and-performance-of-green-surfactants.html.

[80] Bio-based chemicals – a 2020 update. IEA Bioenergy Task 42 report. 2020 (Internet). Available from: https://www.ieabioenergy.com/wp-content/uploads/2020/02/Bio-based-chemicals-a-2020-update-final-200213.pdf.

[81] Market Research Future. Available at, https://www.https://www.reportlinker.com/p06081965/Biosur factants-Market-Research-Report-by-Product-Type-by-Application-by-Region-Global-Forecast-to-Cumulative-Impact-of-COVID-19; 2021 [Accessed 2021-08-15].

[82] Rocha E Silva NMP, Meira HM, Almeida FCA, et al. Natural surfactants and their applications for heavy oil removal in industry. Sep Purif Rev 2019;48(4): 267–81. DOI: 10.1080/15422119.2018.1474477.

[83] Drakontis CE, Amin S. Biosurfactants: formulations, properties, and applications. Curr Opin Colloid Interface Sci 2020; 48:77–90. DOI: 10.1016/j.cocis.2020.03.013.

[84] Naughton PJ, Marchant R, Naughton V, et al. Microbial biosurfactants: current trends and applications in agricultural and biomedical industries. J Appl Microbiol 2019; 127(1): 12–28. doi: 10.1111/jam.14243.

[85] Almeeida DG, Sores da de Silva RCF, Luna JM, et al. Biosurfactants: promising molecules for petroleum biotechnology advances. Front Microbiol 2016;7:1718. doi: 10.3389/PMid:27843439.

[86] Silva RCFS, Almeida DG, Luna JM, et al. Applications of biosurfactants in the petroleum industry and the remediation of oil spills. Int J Mol Sci 2014;15(7): 12523–42. doi: 10.3390/ijms150712523.

[87] Rodriguez-Lopez L, Rincon-Fontan M, Vecino X, et al. Extraction, separation and characterization of lipopeptides and phospholipids from corn steep water. Sep Purif Technol 2020;248. doi: 10.1016/j.seppur.2020.117076.

[88] Roelants S, Van Renterghem L, Maes K, Evaraert B, Vanlerberghe B, Demaeseneire S, Soetaert W. Mirobial Biosurfactants: From Lab to Market. In: Press C, editor. Microbial Biosurfactants and their Environmental and Industrial Applications, 2019. doi: 10.1002/9781315271767.

[89] Singh P, Patil Y, Rale V. Biosurfactant production: Emerging trends and promising strategies. J Appl Microbiol 2019;126(1):2–13. doi: 10.1111/PMid:30066414.

[90] Santos DKF, Rufino RD, Luna JM, et al. Biosurfactants: Multifunctional biomolecules of the twenty-first century. Int J Mol Sci 2016 17:401. doi: 10.3390/ijms17030401.

[91] Natural Surfactants Market. Available at: https://www.marketsandmarkets.com/Market-Reports/nat ural-surfactant-market-25221394.html; 2018 [Accessed 2021-08-14].

[92] Souza EC, Vessoni-Penna TC, De Souza Oliveira RP. Biosurfactant-enhanced hydrocarbon bioremediation: an overview. Int Biodeterior Biodegrad 2014 89:88–94. doi: 10.1016/j.ibiod.2014.01.007.

[93] Geys R, Soetaert W, Van Bogaert I. Biotechnological opportunities in biosurfactant production. Curr Opin Biotechnol [Internet] 2014;30:66–72. Available from: http://dx.doi.org/10.1016/j.copbio.2014.06.002

[94] Marchant R, Banat IM. Microbial biosurfactants: Challenges and opportunities for future exploitation. Trends Biotechnol. 2012;30(11):558–65.

[95] Hayes DG. Fatty acids–Based Surfactants and their Uses. In: Fatty Acids [Internet], Elsevier Inc.; 2017. pp. 355–84. Available from: http://dx.doi.org/10.1016/B978-0-12-809521-8.00013-1.

[96] Bhadani A, Iwabata K, Sakai K, Koura S, Sakai H, Abe M. Sustainable oleic and stearic acid based biodegradable surfactants. RSC Adv 2017;7(17):10433–42.

[97] Farias CBB, Almeida FCG, Silva IA, Souza TC, Meira HM, Soares da Silva de RCF, et al. Production of green surfactants: Market prospects. Electron J Biotechnol [Internet] 2021;51:28–39. Available from: https://doi.org/10.1016/j.ejbt.2021.02.002.

[98]   Makkar RS, Cameotra SS, Banat IM. Advances in utilization of renewable substrates for biosurfactant production. AMB Express. 2011;1(1):1–19.

[99]   Marion P, Bernela B, Piccirilli A, Estrine B, Patouillard N, Guilbot J, et al. Sustainable chemistry: How to produce better and more from less? Green Chem 2017;19(21):4973–89.

[100]   Perathoner S, Centi G. Science and technology roadmap on catalysis for Europe. 2016.

[101]   Lange JP. Renewable feedstocks: the problem of catalyst deactivation and its mitigation. Angew Chem – Int Ed 2015;54(45):13187–97.

[102]   Gu Y, Jérôme F. Bio-based solvents: An emerging generation of fluids for the design of eco-efficient processes in catalysis and organic chemistry. Chem Soc Rev 2013;42(24):9550–70.

[103]   Jessop PG, Mercer SM, Heldebrant DJ. CO_2-triggered switchable solvents, surfactants, and other materials. Energy Environ Sci 2012;5(6):7240–53.

[104]   Eckert CA, Knutson BL, Debenedetti PG. Supercritical fluids as solvents for chemical and materials processing. Nature 1996;383(6598):313–18.

Mohamad Fakhrul Ridhwan Samsudin, Nur Amalina Samsudin,
Wong Mee Kee, M. Syamzari B. Rafeen*

Chapter 9
Sustainable ethylene glycol production: synthesis, properties, and future opportunities

9.1 Introduction

Why ethylene glycol (EG)? EG is the simplest diol having direct application as an automotive antifreeze, coolant, and as a precursor or building block chemical for the production of plastics such as polyethylene terephthalate (PET) and production of polyester resins for fiber [1, 2]. Other applications of EG are for film applications, defrosting and deicing aircraft, heat-transfer solutions for coolant for gas compressors, heating, ventilating, and air-conditioning, and water-based formulations such as adhesives, latex paints, and asphalt emulsions, manufacture of capacitors, and unsaturated polyester resins [3].

The birth of a molecule named "ethylene glycol" started with the work pioneered by Charles Adolphe Wurtz in 1859 where he treated 1,2-dibromoethane with silver acetate to produce EG diacetate (see Figure 9.1) [4]. The intermediate product, EG diacetate, was then hydrolyzed to EG. The final product, EG, also known as 1,2-ethanediol ($HOCH_2CH_2OH$), possesses the properties of a colorless, practically odorless, low viscosity, hygroscopic liquid of low volatility, and completely miscible with water and many organicliquids [3].

Since its discovery in 1859, EG was not produced industrially until World War I, when it was used as an intermediate for explosive (EG dinitrate) substituting glycerol in explosive industries and as coolant [1]. The production of EG back then was based on the chlorohydrin process developed also by Charles Adolphe Wurtz. The chlorohydrin process is the route for ethylene oxide (EO) production prior to further hydrolysis to produce EO where EO was prepared by elimination of hydrochloric acid from 2-chloroethanol (ethylene chlorohydrin) using potassium hydroxide solution [5, 6]. The chlorohydrin process enabled EG industrial production in 1914 which marked the start of steady growth for EG production [2, 3]. However, this process is no longer the intermediate pathway for

---

*Corresponding author: M. Syamzari B. Rafeen, Petronas Research Sdn Bhd, Lot 3288 & 3289, Off Jalan Ayer Hitam, Kawasan Institusi Bangi, 43000, Kajang, Selangor, Malaysia,
e-mail: syamzari_rafeen@petronas.com.my
Mohamad Fakhrul Ridhwan Samsudin, Nur Amalina Samsudin, Wong Mee Kee, Petronas Research Sdn Bhd, Lot 3288 & 3289, Off Jalan Ayer Hitam, Kawasan Institusi Bangi, 43000, Kajang, Selangor, Malaysia

https://doi.org/10.1515/9783110791228-009

Figure 9.1: Charles Adolphe Wurtz (1817–1884), French chemist who discovered ethylene glycol.

EG production due to pollution problems arising from the chlorine used being lost as calcium chloride and unwanted chlorine-containing by-products. Alternatively, direct oxidation was explored by Wurtz but not successful, until Lefort made a crucial discovery that the formation of EO from ethylene and oxygen is possible via a catalytic process where metallic silver is used as the catalyst [1, 7].

Direct oxidation has advantages over chlorohydrin process due to the elimination of the need for large volume of chlorine and absence of no chlorinated hydrocarbon by-products, simpler plant capex costs, and lower operating cost [2]. Direct oxidation technology can be divided into two categories depending on the source of the oxidizing agent which are the air-based process and the oxygen-based process. Union Carbide Corp. was the first to commercialize the air-based direct oxidation process in 1937 whilst Shell Oil Co. was the first to commercialize oxygen-based system in 1958 [1, 8].

Lefort contribution on the direct oxidation for the production of EO and followed by hydrolysis of EO to produce EG became and remain the dominant commercial source of EG production [5, 9]. EO hydrolysis can be performed with either acid or base catalyst system or uncatalyzed in neutral medium. In all ethylene hydrolysis process, the main by-product is diethylene glycol (DEG) whereas higher glycols, that is, triethylene glycol (TEG) and tetraethylene glycol are the remaining by-products [10, 11]. The large excess of water is used to minimize production of higher glycol homologues since production of higher homologues is inevitable because EO reacts more quickly with EGs than with water. Typically, a 20-fold molar excess is used to maximize production of EG [2]. In a typical process for EO hydrolysis, large excess of water from hydrolysis is removed in a series of multiple-effect evaporators and EG is refined using vacuum distillation [3].

EO hydrolysis is a well-established technology for the production of EG and remains to be the commercial source of EG. However, alternative synthesis routes are also being explored due to the constantly increasing demand for EG with methods such as the syngas route, biomass-derived glycerol, cellulose conversion, formaldehyde, and its derivatives conversion, and transesterification of ethylene carbonate [1].

According to recent market outlook for EG, it is expected to surpass the value of USD 61.7 billion by the end of 2031 with CAGR of 6.1% during the forecast period of 2021 till 2031 as compared to the global market value of USD 34.1 billion in 2022 [12]. This is driven by the market opportunities in the plastic and polyester industries as well as increasing use of MEG – expand in antifreeze formulation. In addition, the application of MEG as solvents and anti-icers for aviation is also driver for the growth in demand.

9.2 Synthesis of ethylene glycol

9.2.1 Hydration of ethylene oxide

The hydration of EO at elevated temperatures is the most established method for the production of MEG. EO is obtained from the oxidation of petroleum-based ethylene with air or oxygen. MEG further reacts with EO to form DEG, TEG, and polymerized EG in series. The aforesaid reactions are given in eqs. (9.1)–(9.3).

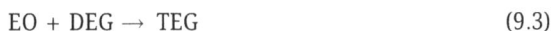

$$EO + H_2O \rightarrow MEG \tag{9.1}$$

$$EO + MEG \rightarrow DEG \tag{9.2}$$

$$EO + DEG \rightarrow TEG \tag{9.3}$$

By nature, EO has a higher tendency to react with MEG than water. The subsequent reactions that produce higher homologues are less desirable due to the economic value of the product [2].

In a typical thermal MEG production process scheme, EO reacts with water without catalysts at a temperature around 200 °C. The heavier glycols are typically suppressed by supplying excess water to the reaction. Increasing the molar ratio of water to EO has a significant effect on the selectivity of MEG. With mole ratio of water to EO above 20:1, 89% of MEG selectivity is achievable for noncatalyzed reaction [9]. About 8–10% of higher homologues, consisting of primarily DEG and TEG, are usually found in the reaction product. The excess water in the mixed EGs is removed through a series of evaporators. The dehydrated mixed EGs are then fed into distillation columns to separate the heavier EGs and purify MEG to the intended purity [13]. As the removal of water is an energy-intensive process, the higher water content that comes with the product increases the operation and capital expenses. Divided wall column, reactive distillation, and side reactor technology were among the techniques proposed to reduce the utility consumption for the dehydration of MEG [14–16].

Considering the separation issues from excessive water and heavy EG, a modified two-step reaction is proposed by Mitsubishi Chemical Corporation (MCC) [9]. In the initial step (eq. 9.4), ethylene is carbonated using carbon dioxide (CO_2) generated in the EO production process.

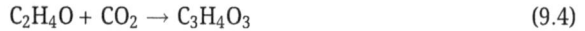

$$C_2H_4O + CO_2 \rightarrow C_3H_4O_3 \qquad (9.4)$$

The ethylene carbonate is then hydrolyzed to form MEG with CO_2 as by-product (eq. 9.5).

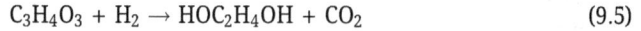

$$C_3H_4O_3 + H_2 \rightarrow HOC_2H_4OH + CO_2 \qquad (9.5)$$

The first step is catalyzed by phosphonium salt, which has demonstrated good stability and noncorrosive property. Attributing to its high catalytic activity for carbonation, phosphonium-based catalyst almost eliminates the possible dimerization reaction that produces DEG. Excellent EC conversion rate as high as 99.9999% is reported. Both carbonation and hydrolysis reactors can be operated at around 50 °C. A comparison of conventional and MCC's new MEG process is shown in Figure 9.2. Coupled with the low demand for water in the reactions, lower energy and capital investment are required in the purification section [9].

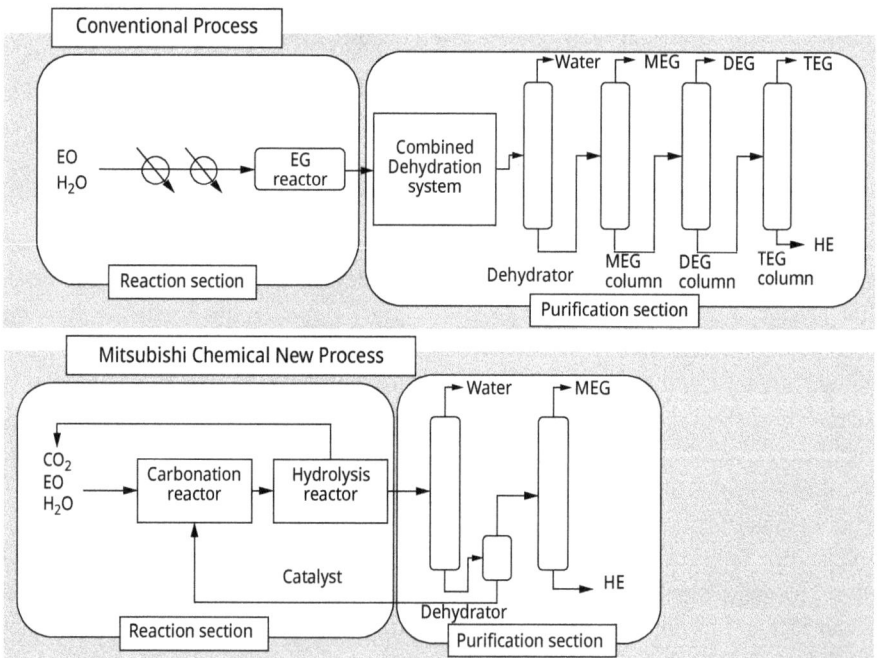

Figure 9.2: The conventional and new process of the ethylene glycol production (adapted from ref. [9] with permission).

This technology is exclusively licensed to Shell under the name OMEGA process in 2002 [17]. The OMEGA process integrates Shell's high-performance EO process, which enables a plant-wide process intensification that produces more effluent and saves energy as well as minimizes the capital cost. The technology is applied in commercial-scale plants ranging from 400 to 750 kiloton per year at various locations globally.

Another way to attain a higher MEG yield is by incorporating a catalyst in the reaction. Both homogeneous and heterogeneous catalysts are widely studied. Acids, soluble salts, cyclic amines, ion-exchange resins, zeolites, and ionic liquids (ILs) are among the catalysts tested to optimize selectivity, reduce the water requirement, and lower reaction temperature [18, 19]. Van Hal et al. [5] found that the pH of catalysts plays a central role in increasing the rate of hydration and maximizing MEG selectivity. Acid and bases of weaker strength are favored over their more concentrated counterparts, which promote oligomerization, hence generating more undesired coproduct. Compared to homogeneous catalytic systems, heterogeneous catalysts have the advantages of ease of separation and fewer corrosion issues [18].

Recent development shows successful demonstration of tandem catalysis. [CoIII (salen)] encapsulated in nanocages of mesoporous silica (FDU-12) by Li et al. [20] showed superior activity and catalytic performance. Owing to the novel cooperative activation effect, her work brought immobilization of homogeneous catalysts to the next level. Similar mini reactor concept was applied by Dai et al. [21] to prepare a thermally stable Lewis acid (Sn-H-SSZ-13) confined in zeolite cages that is capable of pushing the EO conversion and selectivity to 99%. The zeolite cage is designed in such a way that allows diffusion of EO, H_2O, and MEG through the pore windows, while the bigger molecules, such as DEG and TEG, are retained. The narrow space in the cage restricts the collision of two or more MEG molecules, hence minimizing the formation of undesirable by-products. The diffusion control mechanism of zeolite framework during EO hydration is depicted in Figure 9.3. The high-performance catalytic process also has lower energy requirement because of the near ambient reaction temperature and more environmentally benign in terms of wastewater produced.

Despite the exceptional performance of these new generation catalysts, the synthesis of the catalysts is intricate in view of the complicated structure of the catalysts. It is challenging to synthesize these novel catalysts in bulk for industrial-scale application in a cost-effective manner. A simpler approach was proposed by Ding et al. [15], where hydrolysis of EO is assisted with CO_2 pressure using binary catalysts. A significant increase in MEG yield and selectivity is reported compared to the tests performed under N_2 environment. One-pot two-step mechanism is elucidated, in which CO_2 participates in the reaction via the cycloaddition of EO to form ethylene carbonate intermediate. Subsequently, ethylene carbonate is hydrolyzed into MEG, while CO_2 is recovered. The process is catalyzed by IL and organic acid. At a low H_2O/EO ratio of 1.5:1 with CO_2 as a promoter and synergistic catalysis of IL and an organic base 97% selectivity is achieved.

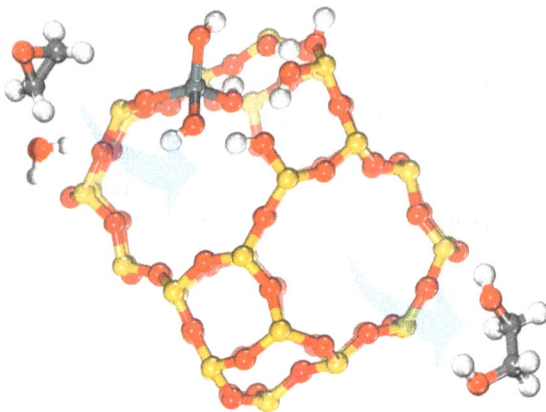

Figure 9.3: Schematic illustration of EO hydration to MEG via CHA zeolite (adapted from ref. [21] with permission).

9.2.2 Transesterification of ethylene carbonate and methanol

An alternative MEG synthesis method is transesterification of ethylene carbonate and methanol, in which MEG is coproduced with dimethyl carbonate (DMC). DMC is used in the automotive industry as a fuel additive and functions as carbonylation and methylation agents in chemical processes. Generally, ethylene carbonate is obtained from eq. (9.4). Then, EG reacts with methanol in the presence of a catalyst according to eq. (9.6) [22].

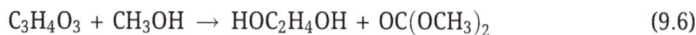

$$C_3H_4O_3 + CH_3OH \rightarrow HOC_2H_4OH + OC(OCH_3)_2 \qquad (9.6)$$

In the 1980s, a few patents related to the preparation of dialkyl carbonate via thermal catalytic reaction of glycol carbonate and alcohol were filed [23–25]. Glycols are collected as a by-product. A more refined process of transesterification of ethyl carbonate and methanol to produce MEG and DMC catalyzed by soluble salts of zirconium, titanium, and tin is patented by Texaco [24]. The reaction step [eq. (9.6)] to cosynthesize MEG and DMC is equilibrium-limited. Continual withdrawal of liquid product is favored to enhance the conversion beyond its equilibrium limitation [26]. A reactive distillation design was described by Buysch et al. [27]. Both reactants are fed counter currently into a gas–liquid contacting column. Ethylene carbonate is at the top, while methanol is in the vapor phase at the bottom. The heavier product, MEG, is removed from the column as the bottom product. DMC can be continuously collected from the top of the column.

Knifton and Duranleau [28] presented the outcome of six groups of homogeneous and heterogeneous catalysts including tertiary amine, acidic resins, ammonium-exchanged zeolites, and phosphine polymer. Base catalysts are reported to give better product yield, and the catalytic activity improved with increasing temperature and base concentration. More than 98 mol% selectivities are achieved under mild operating conditions. Bhanage et al. [22] studied the use of layered clay minerals for the transesterification of ethylene carbonate and methanol. Smectites containing magnesium and/or nickel are highly selective for MEG and suppress unwanted oligomerization reactions. The catalytic performance is subjected to the amount of alkali incorporated.

Given the increased attention to IL due to its superior thermal stability, low vapor pressure, and tunability, a series of ILs were investigated for simultaneous production of MEG and DMC. All ILs demonstrate high selectivity of MEG with the highest selectivity of 99.9% recorded by EMImCl (1-ethyl-3-methylimidazolium with chloride anion). Ethylene carbonate conversion increases with decreasing size of the cation of the IL salt as well as increasing temperature and CO_2 pressure [29]. For the ease of postreaction separation, researchers turn to immobilization of IL. Kim et al. [30] proved that an IL grafted to commercial silica (RImX-CS) can be used for three cycles without significant activity deterioration based on a mole of DMC over mole of RImX ratio. A longer alkyl chain length contributes to higher EC conversion because the bulkier alkyl chain extends the distance between anion and cation. The lower electrostatic bond increases the availability of anion. One-pot synthesis of MEG from supercritical CO_2, EO, and methanol was also attempted. Employing a catalyst system (potassium iodide supported on zinc oxide) that is active for both reaction steps [eqs. (9.4) and (9.6)], Chang et al. [31] successfully combined carbonation and esterification reactions as shown in Figure 9.4. EO conversion and MEG yield surpassed 99% and 57%, respectively, with by-products below 0.2%. Supercritical CO_2 plays a key role as a reactant and solvent in the reaction.

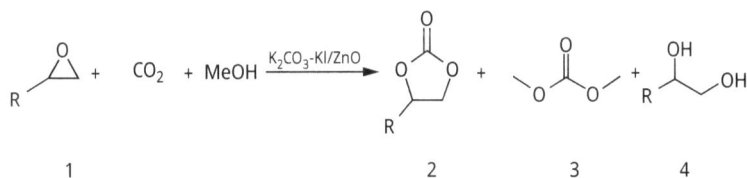

Figure 9.4: Schematic illustration of the one-pot synthesis process (adapted from ref. [31] with permission).

9.2.3 Bio-based synthesis process

In an effort to reduce dependence on petroleum, researchers have been exploring novel ways to produce EG from renewable resources. The valorization of biomass made up of cellulose, hemicellulose, and lignin received considerable attention in the past decade. Both cellulose and hemicellulose are composed of long chains of sugar. Upon breaking down this complex sugar, microorganisms, such as yeast, bacteria, and molds, can be used to convert the carbon-rich compounds into ethanol. The biochemical process proceeds slowly at low temperatures [32]. Subsequently, ethanol is dehydrated in the presence of phosphoric or activated alumina catalyst to form bio-ethylene [33]. Ethylene then goes through oxidation and hydration processes as explained earlier. Corn starch, sugarcane, and wheat starch are the major feedstock for global bioethanol production [34]. Owing to the abundance of sugarcane in Brazil, a commercial-scale bio-based EG plant with a capacity of 500 kiloton per annum is available [33].

Lippits and Nieuwenhuys [35] propose a one-pot conversion of bioethanol to EO, which combines the dehydrogenation, dehydration, and oxidation reactions in one reactor. With a prudent selection of catalysts and controlling the ethanol: O_2 ratio, the high selectivity of EO is achieved [36]. Taking into account the reaction kinetics and thermodynamics of separation, the feasibility of the direct conversion process was evaluated [37]. Simplifying the reaction path may improve the economic aspects of the production of bio-based EG.

Recently, catalyst specialists Haldor Topsoe and Braskem announced their collaboration to operate a demonstration plant employing a new sugar conversion technology [38]. The technology replaces the multistep sugar fermentation to bio-EG route with a simpler two-step process illustrated in Figure 9.5. Depolymerization of sugar was performed via pyrolysis, followed by catalytic hydrogenation of the C1–C3 oxygenates. In fact, the thermal cracking of sugar to obtain glycolaldehyde is documented in 2006 [39]. Glycolaldehyde is a versatile platform chemical that can be converted to EG and other numerous chemicals, including glycolic acid, methyl vinyl glycolate, and ethanol amines. Schandel et al. [40] studied hydrous thermolysis by spraying sugar

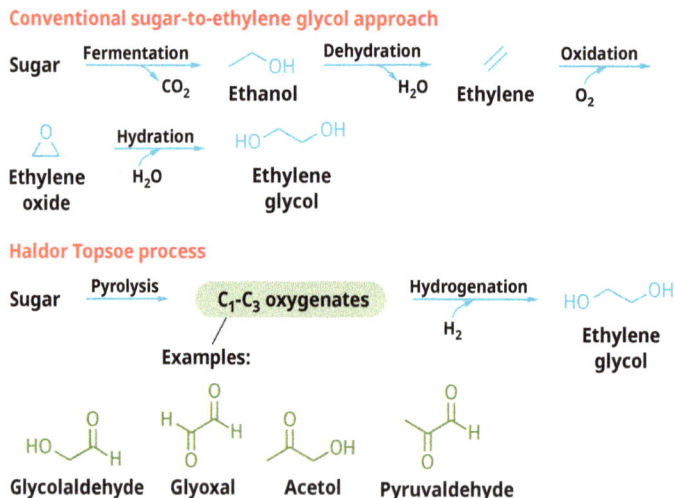

Figure 9.5: Comparison between conventional process and Haldor Topsoe process (adapted from ref. [38] with permission).

solutions (glucose, fructose, sucrose, and xylose) into a fluidized bed of glass beads. Their works show that when glucose is subjected to high heat (525 °C), the sugar rings are broken giving a yield as high as 73% of glycolaldehyde.

9.3 Properties of ethylene glycol

9.3.1 Physical properties

As mentioned earlier, EG is the simplest diol that was developed back in 1859 by Charles-Adolphe Wurtz [1]. Since the discovery of this simplest diol, the development of EG has started to bloom and received significant attention due to its multifunc-tional application. It is estimated that 70 million metric tons of EG will be produced globally in 2025 which is a 40% increase in production in comparison to the year 2020 [12, 17].

Generally, EG exists in a colorless and odorless form with a sweetish taste. It can be said that the physical properties of this EG are standing between ethyl alcohol (eth-anol) and glycerol. Despite the colorless properties, EG possesses a specific gravity of 1.1135 g/cm^3 at 20 °C, and it can be said that EG is heavier than ethanol (0.787 g/cm^3) and lighter than glycerol (1.129 g/cm^3) [8]. Furthermore, this simplest glycol is known to be less viscous than glycerol (1.412 Pa s) [2, 6]. In addition, the molecular weight of this diol is 62.07 g/mol and its corresponding boiling point and melting point are 197.3 and −12.9 °C, respectively [1]. Thereafter, Table 9.1 summarized the physical and chemical properties of the aforesaid diol.

Table 9.1: Physical and chemical properties of ethylene glycol (adapted from ref. [1] with permission).

Property	Ethylene glycol
Molecular weight	62.07 g/mol
Physical state	Liquid
Odor	Odorless
Boiling point at 101.3 kPa	197.3 °C
Melting point	−12.9 °C
Autoignition temperature	427.0 °C
Critical temperature	446.9 °C
Flash point, closed cup (Pensky-Martens closed cup ASTM D93)	126.7 °C
Flash point, open cup (Cleveland Open Cup ASTM D92)	137.8 °C
Specific gravity at 20 °C	1.1135 g/cm^3
Solubility	Miscible with water
Solubility of water in ethylene glycol at 20 °C	100 wt.%
Heat of combustion at 25 °C	−1053.0 kJ/g mol

Table 9.1 (continued)

Property	Ethylene glycol
Heat of vaporization at 1 atm	53.2 kJ/g mol
Critical specific volume	19.1×10^{-2} L/g mol
Lower explosive limit	3.2 vol%
Normal freezing point	13.0 °C
Onset of initial decomposition	240.0 °C
Refractive index at 25 °C	143.0×10^{-2}
Surface tension at 25 °C	48.0×10^{-3} N/m
Upper explosive limit	53.0 vol%
Vapor density (air = 1)	2.1
Vapor pressure at 20 °C	7.5 Pa
Viscosity at 20 °C	19.8×10^{-3} Pa s

9.3.2 Toxicity

Apart from being the simplest diols that are beneficial for multifarious applications, this diol possesses a toxicity behavior that is commonly available in many household and industrial products such as antifreeze, car wash fluids, detergents, cosmetics, paints, and vehicle brake fluids [2, 3]. Although the EG is odorless, however, exposure to this diol without proper personal protective equipment might cause significant morbidity and mortality to the user. The dangerous exposure could be originated from two situations: (i) accidental or (ii) intentional exposure. Both situations could cause the user to be treated with extensive medical treatments or minor treatment depending on the level of exposure.

As mentioned earlier, the EG has a sweet taste which can attract the children's attention while exploring their environment for unintentional consumption. Thus, if this unfortunate situation occurs, the gastrointestinal tract will absorb the EG upon ingestion and a series of alcohol and aldehyde dehydrogenase reactions which consequently lead to glycolic acid [41, 42]. These glycolic acids will cause the development of anion gap metabolic acidosis and further the development of acute renal injury. Thereinto, the consumption of EG as small as 120 ml is known to be enough for killing an average-sized man [43]. Figure 9.6 delineated the schematic of the metabolic toxicity of the EG accordingly.

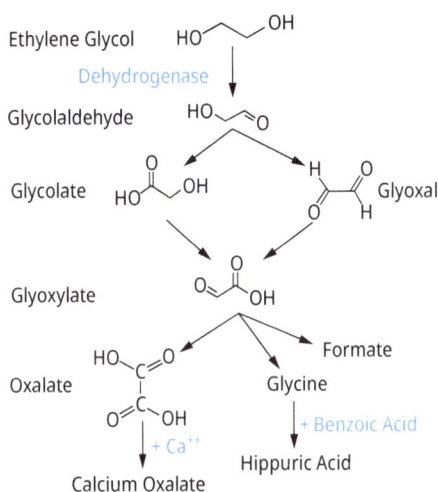

Figure 9.6: Toxicity pathways of ethylene glycol upon consumption (adapted from ref. [1] with permission).

9.4 Challenges and limitations

9.4.1 Production capacity

The traditional production of EG relies heavily on crude oil, which limits the production capacity of EG [44]. The hydration of EO is one of the common routes to produce EG, and it is a high energy-intensive process with several stream recycling to achieve the required reaction conversion. Since the precursor for EO is produced by the catalytic pyrolysis of petroleum, the increase in the EG demand is in conflict with the shortening supply of petroleum sources [45]. Despite the situation, little work has been done on the simulation of coal to EG process, and the research on analysis or optimization of the system performance is quite limited [46].

In their study, Yang et al. [47] compared three different routes to produce EG production which are oil-derived, coal-derived, and natural gas-derived from three angles which are (i) technical, (ii) economic, and (iii) environmental performances. They concluded that the coal-based route has a significant cost advantage; however, the trade-off of this process is the high energy consumption, low exergy efficiency, and environmental pollution in the form of wastewater discharge and CO_2 emission. The oil-derived route has the highest rate of return as this technology is considered the most mature among the three aforesaid technologies. In contrast, using a natural gas route, its technical and environmental evaluation shows better result; however, the total production cost and internal rate of return (IRR) are far worse.

9.4.2 Separation process for high-purity product

Conventionally, for the EG synthesis route, this glycol is produced in the form of mixed alcohols. The challenges often arise during the separation process to separate the EG from the mixed alcohols to achieve a high yield and purity of the desired product. Additionally, the EG and 1,2-butanediol (1,2-BDO) form an azeotropic mixture in nature as illustrated in Figure 9.7 [48]. With a close boiling point over the entire range of mixing composition, the separation of the two aforesaid alcohols poses challenges in obtaining high purity of EG as well as being very energy-consuming.

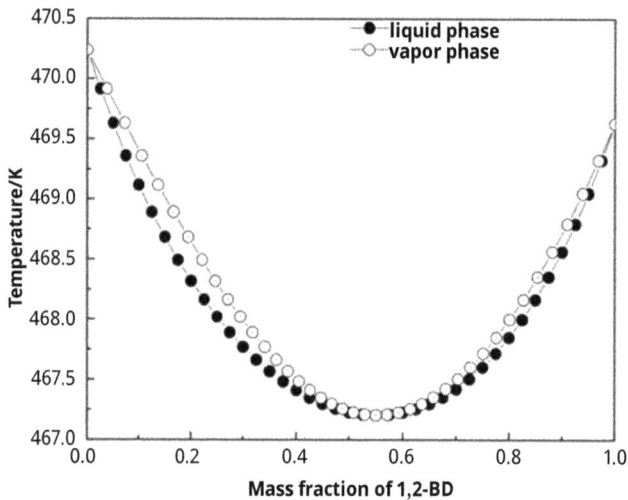

Figure 9.7: Binary phase diagram of ethylene glycol and 1,2-BDO (adapted from ref. [48] with permission).

In the coal-derived EG industry, the distillation process comprises five columns: two methanol recovery columns, a light component removal column, a 1,2-BDO removal column, and an EG refining column [46]. The limitations of this technology, however, are too glaring which include high energy consumption and low recovery yield while the quality of the product does not necessarily meet the intended specifications [45, 48]. The excessive amount of energy required, combined with the large reflux ratio and tray number to remove other glycols from EG, also makes the traditional separation process come with a painfully high CAPEX and OPEX [49, 50].

In a study by Wang et al. [51], the technology of the aldehyde-assisted chemical looping separation process (CSLP) was applied for the purpose using three different aldehydes: (i) formaldehyde (FA), (ii) acetaldehyde (AA), and (iii) prolyl aldehyde (PA). With the objective to optimize the technoeconomic outcome of the process while maximizing production yield and minimizing the environmental impact, this study found that PA-assisted CSLP gives the yield of EG with polyester-grade properties of 94.42%

and an IRR of 137.28%. Nevertheless, such a promising process suffers from a high-potential environmental impact due to liquid discharge. Meanwhile, AA-assisted CSLP has zero-liquid discharge, while FA-assisted CSLP consumes large energy consumption and has high liquid discharge. These three CSLPs studied show that the separation of EG from 1,2-BDO remains a challenge and is yet to be mitigated.

9.4.3 Release of hazardous gas

One of the basic technologies for producing EG from coal has been developed by Japan's Ube Industries, a process that harnesses syngas via a multiple-step reaction that involves the consumption and generation of methyl nitrite [52]. Although the coal-made EG is cost-attractive to polyester producers, this process comes with a challenge. The EG produces is of lower purity and the availability is not consistent due to the complexity of the plant. On top of that, the main drawback of the Ube process lies in the reliance on the methyl nitrite, a hazardous gas, as the intermediate material as shown in Figure 9.8.

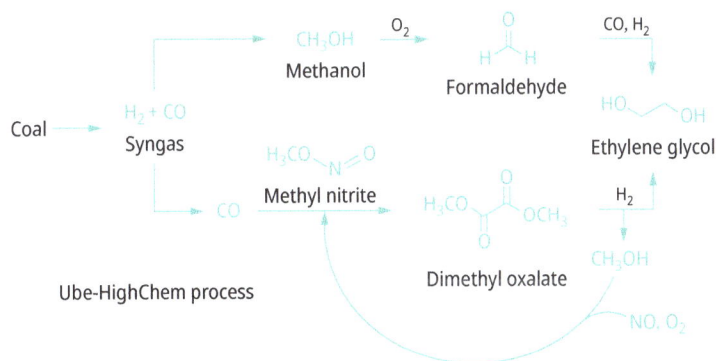

Figure 9.8: Binary phase diagram of ethylene glycol and 1,2-BDO (adapted from ref. [52]).

9.5 Current and future opportunities

In the conventional distillation route, the EG can be obtained at 95 wt.% purity with other glycols from the top of the distillation column and can be sold as a qualified product at a relatively lower price. In addition, there is also a potential to explore further the possible applications of EG with 95 wt.% purity since its production consumes much less energy [48].

As such, Yu and Chein [53] suggested an isothermal plug flow hydrogenation reactor and rearranging the reactors to further heightened the efficiency of the EG production via a hydrogenation process. In another related work published by Wei et al. [49], there are a few methodologies suggested to improve technology design related to this glycol production. This includes a CO coupling reactor with four series functional reactors to overcome the technological difficulties related to the coal-based syngas.

On the other hand, Yang et al. [46] suggest that it is important to perform a simulation to optimize the product yield as well as the system performance of the EG process derived from coal. They claimed that their simulated model is reasonable for such a process based on the increase in energy efficiency from 30.68 to 31.49%. Several key parameters were studied and optimized which include the mole ratio between CO and methyl nitrate, the pressure, and the catalyst height. Alternatively, the EG can be produced from xylitol by hydrogenolysis under heating in the presence of reducing agents and since xylitol is derived from biomass, this production route becomes favorable [54].

By studying the vapor-liquid equilibrium data of six binary systems consisting of EG, 1,2-BDO, 2,3-BDO, and 1,2-propylene glycol (1,2 PG), Li et al. [48] simulated in Aspen Plus four separation processes to obtain polymer-grade EG from the mixture components comparing the total annual costs and profits for each. The four processes are (1) conventional distillation, (2) azeotropic distillation, (3) distillation coupled with liquid–liquid extraction, and (4) reaction-assisted distillation. It is concluded that the economy depends mainly on the yield of the glycol, with the reaction-assisted distillation being the current best separation method among others. This method utilizes a chemical reaction to enhance the separation process. The reaction-assisted distillation was proven to pose great potential in obtaining a high yield of EG during separation [55].

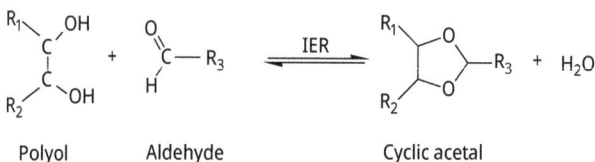

Polyol   Aldehyde   Cyclic acetal

Figure 9.9: Schematic illustration of cyclic acetal formation via polyol and aldehyde reaction (adapted from ref. [56] with permission).

In solving the problem related to the separation between EG and propylene glycol (PG) in an aqueous solution, Dhale et al. [56] suggested the use of a reactive distillation column. As illustrated in Figure 9.9, the acetaldehyde is used as a reactive agent and is combined with the polyols, subsequently forming a cyclic acetal. In light of this, the Amberlyst 15 cationic exchange resin is held in structured packing inside the column to catalyze the reaction. The glycols and the acetaldehyde are fed counter-currently into the column for mass transfer efficiency. This technology has been proven at a pilot scale with a 90% conversion rate achieved. Since this reaction is reversible, the downstream process involves the rapid process of hydrolysis of acetals to form high purity of EG and propylene glycol, while the acetaldehyde is being recycled into the process. This method has an advantage over the hydration of water which uses a lot of energy, while solvent extraction although is theoretically feasible, the affinity of glycols toward water makes it difficult to find suitable organic solvent to fulfill the separation requirement.

9.5.1 Emerging application of bio-based ethylene glycol

In the study carried out by India Glycols Limited, the comparison is made between the conventional production route of MEG and the production of bio-MEG from ethyl alcohol of sugarcane molasses-based on their life cycle assessment, respectively [57]. It is found that the production of 1 MT of MEG from petroleum emits 1,628 kg of CO_2. Meanwhile its bio-route counterpart only emits 1,221 kg of CO_2, a 25% reduction in the CO_2 emission over a MEG life cycle. This emission reduction is mainly contributed by the use of sugarcane molasses which can be cultivated again, hence reducing the carbon footprint of the overall bio-MEG production.

The production of MEG from sugar has become more established with Braskem [58] and Haldor Topsoe building their first demo-scale plant for the production of bio-MEG at the end of 2020. Although this technology is sustainable and has the potential to revolutionize the PET market, the source of feedstock tends to compete with the food production line. This is the same for other food-based feedstock such as corn, switchgrass, and wheat straw. Another challenge faced by bio-MEG production is that the cost and the price of bio-MEG are still higher compared to the fossil-based MEG. This wide price gap causes clients to take stock of the value of its use [59].

Bio-MEG can also be synthesized using the biotechnological route. The challenges of using this route predominantly lie in its yield which does not meet the theoretical value expected while the production titers are too low for commercial-scale production [60]. Comparing the two main biotechnological routes, the metabolic engineering of *Escherichia coli* shows a more promising result than the engineering yeast species.

9.5.2 Depolymerization of PET into ethylene glycol

Another interesting opportunity in this field is the depolymerization of PET. MEG is one of the main precursors in producing PET bottles. The extended period of time for PET to decompose naturally has caused the ever-rising environmental issue when these bottles end up in the landfills and in the oceans at the end of their use.

Realizing the opportunity to reverse the chemical process, a few companies including Eastman Chemical and Loop Industries have embarked on the pursuit to break down PET back into its building blocks, EG and dimethyl terephthalate (DMT) via chemical processes such as methanolysis and hydrolysis [61]. These two monomers are then condensed again to produce recycled PET, which has properties similar to virgin PET and is suitable for packaging applications including in stringent-requirement industries such as the food industry. Although the mechanical recycling of PET bottles is already at a satisfying rate, this emerging depolymerization chemical recycling technique will tackle the PET fiber that faces degradation challenges when recycled mechanically, thus accelerating the recovery of environmental issues caused by used plastics.

Eastman Chemical and Loop Industries are already on their way to building plants using this technology [61, 62]. According to Loop Industries, the depolymerizing PET route to make new PET saves more than 2 kg of CO_2 per kilogram of PET relative to the conventional route of making PET from fossil fuel. Fueled by this motivation, the CEO of Loop Industries, Daniel Solomita envisions building one depolymerization of PET plant in every country to recycle polyester resins and fibers as he sees this technology to be an essential infrastructure to have. At the moment, Loop is still at its first project partnered with PET maker Indorama Ventures to process 40,000 ton per year of postconsumer PET in Spartanburg, South Carolina. This project is supported by PepsiCo, Coca-Cola, and Danone who made commitments to buy more than half of the plant's output.

Meanwhile, Eastman has also started its modest production of EG and DMT via glycolysis of PET last year [61]. The company will commence a larger plant in Kingsport, Tennessee worth $250 million with an increase in efficiency in its methanolysis technology by the end of 2022. This plant will break down 100,000 MTPA of postconsumer PET waste, including from PET packaging and carpet fiber, into EG and DMT. These monomers will then be made into specialty polyesters, one of the products being refillable water bottles. Eastman claims that resins production using recycled content will decrease greenhouse gas production by 20–30% as compared to fossil-fuel-based feedstocks.

9.6 Conclusion

In conclusion, this chapter discussed the two main routes for synthesizing EG which are hydration of EO and transesterification of ethylene carbonate. The advantages and disadvantages of each synthesizing route were further detailed. Several related recent works on the synthesis of EG either via conventional or bio-based routes were summarized accordingly in this section. Subsequently, a detailed overview of the properties of the EG was presented which included the toxicity behavior of this glycol. The current challenges in terms of production capacity, a high-purity separation, and hazardous gas release are also deliberated in the subsequent section. Finally, this chapter offers a current perspective and potential future opportunities for the application of EG that suit the interests of the readers.

References

[1] Yue H, Zhao Y, Ma X, Gong J. Ethylene glycol: Properties, synthesis, and applications. Chem Soc Rev 2012;41:4218–44.

[2] Rebsdat S, Mayer D. Ethylene Glycol. In: Ullmann's Encycl. Ind. Chem., Weinhem: Wiley-VCH; vol 13, 2012, pp. 531–46.

[3] Kirk RE, Othmer DF. Kirk-Othmer Encyclopedia of Chemical Technology, 4th edition, Wiley; 1995.

[4] Wurtz A., Mémoire sur les glycols ou alcools diatomiques, Ann.Chim. Phys. [3], 55, 400–478, 1859.

[5] van Hal JW, Ledford JS, Zhang X. Investigation of three types of catalysts for the hydration of ethylene oxide (EO) to monoethylene glycol (MEG). Catal Today 2007;123:310–15.

[6] Kandasamy S, Samudrala SP, Bhattacharya S. The route towards sustainable production of ethylene glycol from a renewable resource, biodiesel waste: A review. Catal Sci Technol 2019;9:567–77.

[7] Lefort TE. France Patent Pat 729952 (to Societe Francaise de Catalyse Generalise), 27 March 1931.

[8] Dye RF. Ethylene glycols technology. Korean J Chem Eng 2001;18:571–79.

[9] Kawabe K. Development of highly selective process for mono-ethylene glycol production from ethylene oxide via ethylene carbonate using phosphonium salt catalyst. Catal Surv Asia 2010;14:111–15.

[10] Matar S, Hatch LF. Hydrocarbon intermediates. Chem Petrochem Process 2001;29–48.

[11] Matar S, Hatch LF. Chemicals based on ethylene. Chem Petrochem Process 2001;188–212.

[12] Research TM. Monoethylene Glycol Market to Expand at CAGR of 6.1% during Forecast Period, Notes TMR Study, (2022), [Online] https://www.prnewswire.com/news-releases/monoethyl.[Accessed 17 April 2022]

[13] Milligen HV, Milligen HV, Wells GJ. Enhancements in Ethylene Oxide/Ethylene Glycol manufacturing technology. 2021. [Online]. Available: https://catalysts.shell.com/en/ethylene-oxide-ethylene-glycol-production-white-paper. [Accessed 17 April 2022].

[14] Zhu F, Huang K, Wang S, Shan L, Zhu Q. Towards further internal heat integration in design of reactive distillation columns-part IV: application to a high-purity ethylene glycol reactive distillation column. Chem Eng Sci 2009;64:3498–509.

[15] Ding T, Zha J, Zhang D, Zhang J, Yuan H, Xia F, Gao G. CO_2 atmosphere enables efficient catalytic hydration of ethylene oxide by ionic liquids/organic bases at low water/epoxide ratios. Green Chem 2021;23:3386–91.

[16] Yan L, Witt PM, Edgar TF, Baldea M. Static and dynamic intensification of water-ethylene glycol separation using a dividing wall column. Ind Eng Chem Res 2021;60:3027–37.

[17] Harmsen J, Verkerk M. 15 shell OMEGA only monoethylene glycol advanced process. Process Intensif 2020;166–75.

[18] Zavelev DE, Tsodikov MV, Zhidomirov GM, Kozlovskii RA. Role of the surface hydroxyl groups of modified titanium oxide in catalytic ethylene oxide hydration. Kinet Catal 2011;52:659–71.

[19] Altiokka MR, Akyalçin S. Kinetics of the hydration of ethylene oxide in the presence of heterogeneous catalyst. Ind Eng Chem Res 2009;48:10840–44.

[20] Li B, Bai S, Wang X, Zhong M, Yang Q, Li C. Hydration of epoxides on [Co III(salen)] encapsulated in silica-based nanoreactors. Angew Chemie – Int Ed 2012;51:11517–21.

[21] Dai W, Wang C, Tang B, Wu G, Guan N, Xie Z, Hunger M, Li L. Lewis acid catalysis confined in zeolite cages as a strategy for sustainable heterogeneous hydration of epoxides. ACS Catal 2016;6:2955–64.

[22] Bhanage BM, Fujita SI, He Y, Ikushima Y, Shirai M, Torii K, Arai M. Concurrent synthesis of dimethyl carbonate and ethylene glycol via transesterification of ethylene carbonate and methanol using smectite catalysts containing Mg and/or Ni. Catal Letters 2002;83:137–41.

[23] Landscheidt H, Wolters E, Klausener A, Blank H, Birkenstock U. Process for the preparation of dialkyl carbonates. Bayer Aktiengesellschaft Patent 5,288,894, 1993.

[24] Knifton JF. Process for co-synthesis of ethylene glycol and DMC. Texaco Patent 4,661,609, 1987.

[25] Duranleau RG, Nieh ECY, Knifton JF. Process for production of ethylene glycol and dimethyl carbonate. Texaco Patent 4,691,041, 1986.

[26] Pacheco MA, Marshall CL. Review of dimethyl carbonate (DMC) manufacture and its characteristics as a fuel additive. Energy Fuels 1997;11:2–29. doi: 10.1021/ef9600974.

[27] H.-J. Buysch, A. Klausener, R. Langer and F.-J. Mais, "Process for the Continuous Preparation of Dialkyl Carbonates". Bayer Aktiengesellschaft Patent 5,231,212, 1993.

[28] Knifton JF, Duranleau RG. Ethylene glycol-dimethyl carbonate cogeneration. J Mol Catal 1991;67:389–99. doi: 10.1016/0304-5102(91)80051-4.

[29] Ju H-Y, Manju MD, Park D-W, Choe Y, Park S-W. Performance of ionic liquid as catalysts in the synthesis of dimethyl carbonate from ethylene carbonate and methanol. React Kinet Catal Lett 2007;90(1):3–9.

[30] Kim K-H, Kim D-W, Kim C-W, Koh J-C, Park D-W. Synthesis of dimethyl carbonate from transesterification of ethylene carbonate with methanol using immobilized ionic liquid on commercial silica. Korean J Chem Eng 2010;27(5):1441–45.

[31] Chang Y, Jiang T, Han B, Liu Z, Wu W, Gao L, Li J, Gao H, Zhao G, Huang J. One-pot synthesis of dimethyl carbonate and glycols from supercritical CO_2, ethylene oxide or propylene oxide, and methanol. Appl Catal A: Gen 2004;263(2):179–86.

[32] Cherubini F. The biorefinery concept: Using biomass instead of oil for producing energy and chemicals. Energy Convers Manage 2010;1412–21.

[33] Pang J, Zheng M, Sun R, Wang A, Wang X, Tao Z. Synthesis of ethylene glycol and terephthalic acid from biomass for producing PET. Green Chem 2016;342–59.

[34] Zamani AMA, Taherzadeh MJ. Bioethylene production from ethanol: a review and techno-economical evaluation. Chem Bio Eng Rev 2022;75–91.

[35] Lippits M, Nieuwenhuys B. Direct conversion of ethanol into ethylene oxide on gold-based catalysts: Effect of CeOx and Li_2O addition on the selectivity. J Catal 2010;274(2):142–49.

[36] Lippits M, Nieuwenhuys B. Direct conversion of ethanol into ethylene oxide on copper and silver nanoparticles: Effect of addition of CeO_x and Li_2O. Catal Today 2010;154(1):127–32.

[37] Ripamonti D, Tripodi A, Conte F, Robbiano A. Feasibility study and process design of a direct route from bioethanol to ethylene oxide. J Environ Chem Eng 2021;9(5):105969.

[38] Tullo A. New route planned to biobased ethylene glycol. C&EN Global Enterprise 2017;95(46):10.

[39] Majerski PA, Piskorz JK, Radlein DSAG. United States Patent US7094932B2, 2006.

[40] Schandel CB, Høj M, Osmundsen CM, Jensen AD, Taarning E. Thermal cracking of sugars for the production of glycolaldehyde and other small oxygenates. Chem Sus Chem 2020;13(4):688–92.

[41] Leth PM, Gregersen M. Ethylene glycol poisoning. Forensic Sci Int 2005;155:179–84.

[42] Lovrić M, Granić P, Čubrilo-Turek M, Lalić Z, Sertić J. Ethylene glycol poisoning. Forensic Sci Int 2007;170:213–15.

[43] Moore MM, Kanekar SG, Dhamija R. Ethylene glycol toxicity: chemistry, pathogenesis, and imaging. Radiol Case Rep 2008;3:122.

[44] Yang Q, Zhu S, Yang Q, Huang W, Yu P, Zhang D, Wang Z. Comparative techno-economic analysis of oil-based and coal-based ethylene glycol processes. Energy Convers Manag 2019;198:111814.

[45] Song H, Jin R, Kang M, Chen J. Progress in synthesis of ethylene glycol through C1 chemical industry routes. Cuihua Xuebao/Chin J Catal 2013;34:1035–50.

[46] Yang Q, Zhang D, Zhou H, Zhang C. Process simulation, analysis and optimization of a coal to ethylene glycol process. Energy 2018;155:521–34.

[47] Yang Q, Yang Q, Xu S, Zhu S, Zhang D. Technoeconomic and environmental analysis of ethylene glycol production from coal and natural gas compared with oil-based production. J Clean Prod 2020;273:123120.

[48] Li H, Zhao Z, Qin J, Wang R, Li X, Gao X. Reversible reaction-assisted intensification process for separating the azeotropic mixture of ethanediol and 1,2-butanediol: vapor-liquid equilibrium and economic evaluation. Ind Eng Chem Res 2018;57:5083–92.

[49] Wei R, Yan C, Yang A, Shen W, Li J. Improved process design and optimization of 200 kt/a ethylene glycol production using coal-based syngas. Chem Eng Res Des 2018;132:551–63.

[50] Chen YC, Hung SK, Lee HY, Chien IL. Energy-saving designs for separation of a close-boiling 1,2-propanediol and ethylene glycol mixture. Ind Eng Chem Res 2015;54:3828–43.

[51] Wang R, Na J, Li X, Li H, Gu S, Gao X. Techno-economic and environmental evaluation of aldehyde-assisted chemical looping separation technology in coal-based and biomass-based ethylene glycol purification. J Clean Prod 2021;313:127675.

[52] Tremblay J-F. Polyester made from coal? china is betting on it. C & EN Glob Enterp 2019;97:22–24.

[53] Yu BY, Chien IL. Design and optimization of dimethyl oxalate (DMO) hydrogenation process to produce ethylene glycol (EG). Chem Eng Res Des 2017;121:173–90.

[54] Delgado Arcaño Y, Valmaña García OD, Mandelli D, Carvalho WA, Magalhães Pontes LA. Xylitol: A review on the progress and challenges of its production by chemical route. Catal Today 2020;344:2–14.

[55] Li H, Huang W, Li X, Gao X. Application of the aldolization reaction in separating the mixture of ethylene glycol and 1,2-butanediol: Thermodynamics and new separation process. Ind Eng Chem Res 2016;55:9994–10003.

[56] Dhale AD, Myrant LK, Chopade SP, Jackson JE, Miller DJ. Propylene glycol and ethylene glycol recovery from aqueous solution via reactive distillation. Chem Eng Sci 2004;59:2881–90.

[57] India Glycols Limited. Reduction of Greenhouse Gas Emissions via Use of Chemical Products, International Council of Chemical Associations, 2016.

[58] Braskem. Newsletter Carbon Neutral Circular Economy. Braskem, 4 February 2021. [Online]. Available: https://www.braskem.com/newsletter-carbon-neutral-detalhe-en/first-demonstration-scale-of-renewable-bio-meg-production. [Accessed 17 May 2022].

[59] Green Chemicals Blog. What future for coal-to-meg and bio-meg in a low-cost brent world? Technon Orbichem, 11 June 2020. [Online]. Available: https://greenchemicalsblog.com/2020/06/11/what-future-for-coal-to-meg-and-bio-meg-in-a-low-cost-brent-world/. [Accessed 17 May 2022].

[60] Salusjärvi L, Havukainen S, Koivistoinen O, Toivari M. Biotechnological production of glycolic acid and ethylene glycol: current state and perspectives. Appl Microbiol Biotechnol 2019;103:2525–35.

[61] Hullo A. Companies are placing big bets on plastics recycling. Are the odds in their favor? Chemical & Engineering News, 11 October 2020. [Online]. Available: https://cen.acs.org/environment/sustainability/Companies-placing-big-bets-plastics/98/i39. [Accessed 13 May 2022].

[62] Tullo A. Eastman will build a $250 million plastics recycling plant. Chemical & Engineering News, 1 FEBRUARY 2021. [Online]. Available: https://cen.acs.org/environment/recycling/Eastman-build-250-million-plastics/99/web/2021/02. [Accessed 13 May 2022].

Daniel Garbark

Chapter 10
Soybean oil-based surfactants

10.1 Surfactants background and advantages of High Oleic Soybean Oil (HOSO)

Surfactants are defined as "surface-active agents" [1]. These ingredients, commonly used in detergents, have unique properties which allow the ingredient to be compatible with water and at the same time be attracted to organic materials such as grease and dirt. Surfactants are used in a variety of applications with the largest market being detergents. These materials perform a balancing act of being attracted to the grease and dirt (lipophilic) while also being attracted to water (hydrophilic). Water alone will not adequately clean the laundry. The lipophilic tail is attracted to the grease and dirt while the hydrophilic head group is attracted to the water. The surfactant suspends the grease and dirt in the water so that it can be rinsed away. Therefore, the surfactant performs the primary cleaning function for a laundry detergent.

Surfactants have changed significantly over the years. The earliest known surfactants consist of fatty acid soaps that still consist of a large portion of surfactants used. These soaps helped in lowering the surface tension of water allowing for appropriate cleaning of hands and clothing. However, with the advent of laundry machines, the fatty acid soaps were found to leave deposits on clothing. This led to the development of linear alkylbenzene sulfonate (LAS). With improved water miscibility LAS was a superior choice to traditional soaps. Surfactant development has continued to increase efficacy and selectivity leading to the development of nonionic surfactants and later cationic and amphoteric varieties. Overall, there are four categories of surfactants: anionic, nonionic, cationic, and amphoteric.

Surfactants developmental history consists of various feedstocks, including both petroleum and biobased. For improved water miscibility, surfactants focused on shorter chain lengths such as those based on lauric acid. Even though many were produced from petroleum sources, a shift to biobased options led to use of coconut fatty acids and palm kernel fatty acids both purified and as crude mixes. While these biobased feedstocks show promising performance, there is no US production. This led to increased consideration of feedstocks such as soybean oil (SBO). SBO is the largest vegetable oil harvested in the United States and with the advent of soy fatty acid varieties, some characteristics could be exploited for the production of equivalent performing surfactants

Daniel Garbark, Lead Researcher, Battelle Memorial Institute, Columbus, OH, e-mail: garbarkd@battelle.org

https://doi.org/10.1515/9783110791228-010

based on coconut fatty acids. This chapter presents recent work done in this area at Battelle [2,3]. With the introduction of high oleic soybean oil (HOSO), many reactions can be performed that could not previously be used with commodity SBO without interference from unwanted side reactions. For example, when reacting epoxidized SBO with hindered or low reactivity hydroxyls, significant portions of tetrahydrofuranyl (THF) ethers and cross-liked fatty acids can be formed. This can be seen in Figure 10.1.

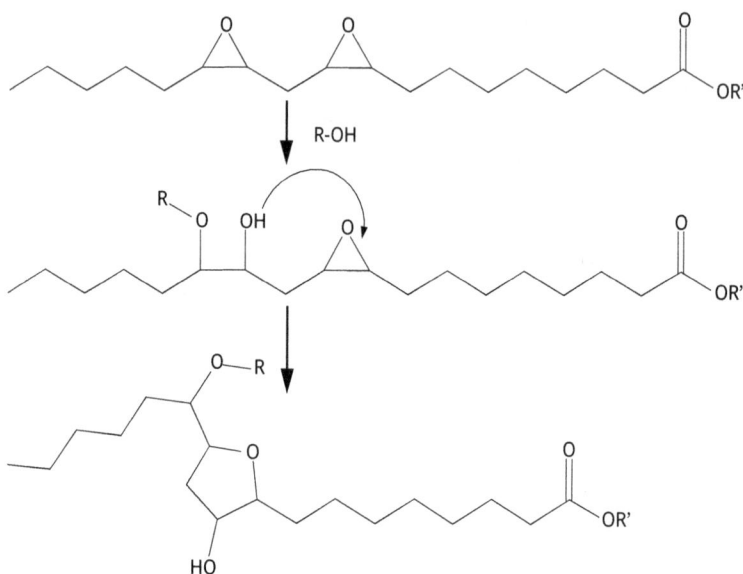

Figure 10.1: Reaction of epoxidized linoleate with hindered or low-activity alcohols.

The various levels of fatty acids in the assortment of SBOs can be found in Chapter 20. The reason for the THF ethers is due to the higher levels of linoleic and linolenic fatty acids (53%wt and 8%wt, respectively) in commodity SBO which contain greater levels of doubly allylic olefins. In using epoxidized HOSO, the same reaction occurs slowly and due to the lower amount of linoleic and linolenic acids (10%wt and 1%wt, respectively), the potential for THF ethers is significantly reduced. This allows for the use of various moieties when using HOSO as a building block to a variety of surfactants. Commodity SBO can be used in cases where the epoxide is reacted with highly active primary hydroxyls such as those from alkyl alcohols, polyether alcohols, and polyether alcohol monoalkyl ethers.

Biobased poly-acids have unique opportunity for use in a variety of applications such as chelation in laundry detergent builders. This can be seen in the use of trisodium citrate as a water softener in various detergent formulations. These softeners, which bind metal cations, also work to maintain detergent stability, help keep soil from redepositing during washing, and improve the cleaning efficiency of the detergent formulation.

Traditionally, materials used as detergent builders have included sodium triphosphate and ethylenediaminetetraacetic acid (EDTA). Regulations have pushed for reduction of phosphates used in detergents which resulted in greater interest in identifying and developing environmentally friendly alternatives. One group of surfactants produced from HOSO consists of active structural sites similar to EDTA but with a fatty acid inserted in the middle. This allows for dual functionality as the water softener maintains surfactant behavior through the remaining aliphatic tail.

Through designed chemical structure, epoxidized HOSO allows for an array of functionalities leading to targeted structure to stain determinations such as soy lactate performing improved make-up removal. The main improvement over the older fatty acid soaps is by having a hydrophilic group in the middle of the chain, one creates greater water miscibility while still maintaining surface activity through an aliphatic C8–C9 chain. In the cases where commodity SBO is used, the aliphatic tail is shortened to C5–C6. With the ability to vary side chains at the original olefin site, along with functional groups at the original fatty acid group, the hydrophilic–lipophilic balance (HLB) of the molecule can be more readily controlled. Selection of functional groups and reactants also leads to high biobased content and good biodegradability. In some cases, soy-based candidates were produced using hydroformylation and hydrogenation as an alternative to epoxidation. To date, 62 soy-based surfactant structures have been produced as potential replacements to those currently used in the marketplace. A few of these structures are described in this chapter.

10.2 Anionic surfactants

Anionic surfactants are the largest group of surfactants, accounting for approximately 40% of the world's surfactant production (2). They contain functional groups such as sulfonates, sulfates, carboxylates, or phosphates at the structural head. These products exhibit superior wetting and emulsifying properties and tend to be higher-foaming materials. Fatty acid soaps and alkylbenzene sulfonates fall into this category. Fatty acid soaps, the most common of surfactants, typically have a HLB of approximately 15, which means they function more as a cleaning agent than as a surfactant. Alkylbenzene sulfonates are petroleum-based and do not biodegrade under anaerobic conditions.

Anionic surfactants include structures such as sodium lauryl sulfate (SLS), sodium dodecylbenzene sulfonate (SDBS), and alkyl-ether sulfates such as sodium laureth sulfate. Other anionic surfactants consist of fluorinated materials such as perfluorooctanesulfonate and perfluorobutanesulfonate. Fluorinated surfactants have come under scrutiny lately due to their resilience in decomposition, sometimes described as forever chemicals.

The HOSO-based anionic functional variations were produced by reacting epoxidized HOSO with a hydroxy acid such as glycolic, lactic, malic, and citric acids. Their production was performed to increase hydrophilicity through increased carboxylate salt groups, while maintaining a stable ether bond with the hydroxy acid. The reaction scheme for the HOSO-lactate can be seen in Figure 10.2. The methyl ester of epoxidized HOSO is used for the reaction as it decreases the amount of side reactions and final intermediate viscosity. The ethyl or methyl ester of the hydroxy acid is used to protect the epoxide group from reacting more favorably with the carboxylic acid instead of the hydroxyl group. The final step not shown is the hydrolysis with sodium or potassium hydroxide and the removal of methanol and ethanol.

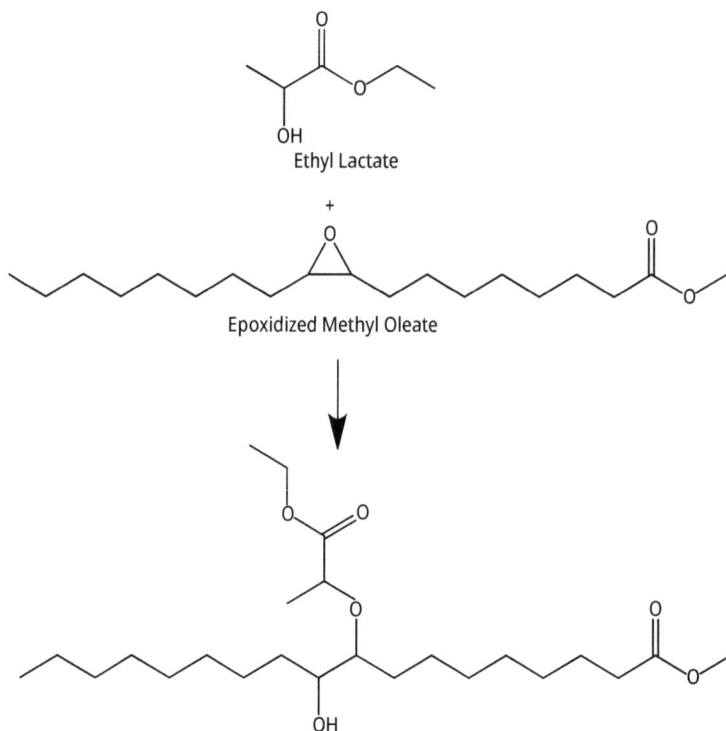

Figure 10.2: Reaction of ethyl lactate with epoxidized methyl HOSO.

An alternative reaction scheme allows for greater reaction speed to produce similar products more efficiently. The hydroxy acid esters can first be reacted through alkoxylation, preferably with ethylene or propylene oxide, to increase the reactivity of the hydroxy acid with the epoxidized SBO. The final structure produced then contains a polyalkoxyl ether chain between the hydroxy acid and the fatty acid.

When substituting the hydroxy acid-soy derivatives for SDBS, we observed that all formulations performed poorly in stain removal testing (ATSM D4265). Stain removal testing is performed in triplicate by utilizing reflectance of stain standards before and after one laundry cycle. What was observed was the citrate-soy candidate improved the removal of the stains from coffee, EMPA 112 (cocoa), EMPA 116 (blood, milk, and ink), and red wine. The lactate-soy candidate significantly improved the removal of makeup stains. We later found that by substituting the hydroxy acid-soy candidates for the non-ionic portion of the standard formulation allowed for great improvements in overall stain removal.

In using similar reaction schemes (ring opening of epoxide followed by secondary reactions), other anionic soy-based derivatives can be formed. For example, the epoxide group can easily be reacted with any polyethylene glycol (PEG) group. After the initial ring opening and purifying of the intermediate, the remaining primary hydroxyl group on the PEG ether has greater reactivity than the secondary and hindered hydroxyl at the original olefin site. Through reaction of the intermediate with chlorosulfonic acid, followed by purification and saponification, a structure containing both the sulfonate salt and carboxyl salt is produced. The final structure can be seen in Figure 10.3.

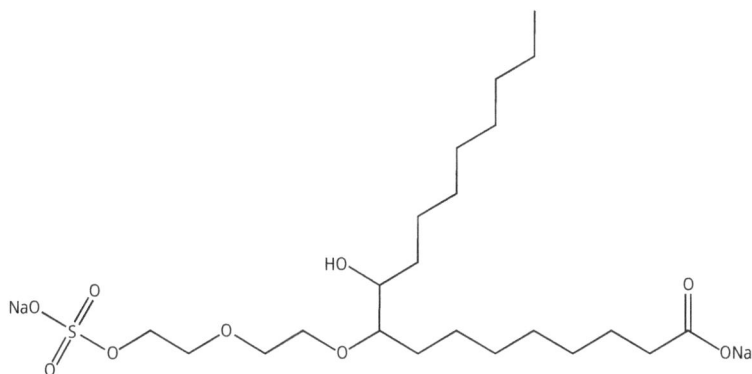

Figure 10.3: Candidate 16.

Other soy-based aromatic sulfonate structures were also developed; however, many were produced with low yields and were not evaluated further. Another candidate was produced through alternative chemistry versus epoxidation. Candidate 17 seen in Figure 10.4 was produced by reacting hydroformylated and hydrogenated methyl HOSO with chlorosulfonic acid followed by saponification.

Evaluations were performed for laundry applications using ASTM D4265 where results of ±10% are statistically equivalent. When substituting Candidate 16 for SDBS in a standard formulation (Laundry Formulation 1), it was found that performance was acceptable when replacing 50% of the SDBS. When replacing 100% of the SDBS,

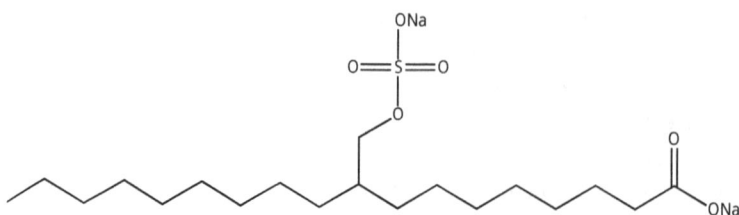

Figure 10.4: Candidate 17.

we were no longer in the 10% range. The results can be seen in Table 10.1. The standard was held at 100% while the other mixtures were seen performing below 100%. A direct comparison to SLS was also recorded. It was found during this research that sulfonates and sulfates were the greatest factor in attacking grass stains. However, we were surprised at the improvement in grass removal when substituting at 50% with SDBS versus the 100% substitution. This result showed that while we could not replace the SDBS by 100%, there may be a blend levels that are advantageous to overall performance.

Table 10.1: Stain removal results when replacing SDBS.

Soil	Substrate	Standard	Candidate 16C 50% SDBS replacement	Candidate 16C 100% SDBS replacement	SLS bio-standard (normalized)	Sig Dif
		120 mL	120 mL	120 mL	120 mL	
Clay	Cotton	58.8	53.8	58.6	60.8	6.9
Coffee	Cotton	51.2	51.5	44.8	42.2	8.6
Dust sebum	Cotton	56.3	44.1	40.4	44.7	5.6
EMPA 101 (olive oil)	Cotton	8.4	8.6	8.4	8.9	2.4
EMPA 112 (cocoa)	Cotton	7.1	4.4	11.1	9.3	5.9
EMPA 116 (blood, milk, ink)	Cotton	22.1	18.8	18.4	32.4	2.2
Grass	Cotton	59.1	72.8	48.3	33.9	6.2
Make up	Cotton	34.9	25.4	18.1	20.8	7.8
Red wine	Cotton	34.9	38.7	36.1	46.1	5.6
Spaghetti	Cotton	75.6	72	74.5	85.2	7.8
Overall soil Removal totals		**418.4**	**390.1**	**358.7**	**384.2**	**59.0**
% of Best		Best	93%	86%	92%	

10.3 Nonionic surfactants

Nonionic surfactants are the second largest group by volume at about 35% of the world's surfactant manufacture [4]. Fatty alcohol ethoxylates, alkyl phenol ethoxylates, fatty acid esters, and alklylpolyglucosides fall into this category. The hydrophilic groups are covalently bonded to the aliphatic tail allowing for controlled miscibility and dispersibility. This class of surfactants tends to be less sensitive to water hardness and pH changes but do suffer in applications with temperature changes.

The main nonionic surfactants in laundry applications are based on ethoxylates of dodecanol (lauryl alcohol). Palm kernel and coconut fatty acids are rich in lauric acid (C12) which is converted to alcohol through hydrogenation of the carboxylic acid. That hydroxyl is then alkoxylated at various chain lengths depending on the application needs. Similar reaction processes have been used with stearyl (C18) and cetyl alcohol (C16). These longer chain alcohols require greater amounts of ethoxylate groups to achieve water miscibility versus the lauric derivatives.

For nonionic surfactant replacement, we utilized laundry formulation 1 which contained a nonionic surfactant based on C12 alcohol with approximately 15 ethoxylate groups. For this reason, Tergitol® 15-s-15 was used. As mentioned above, our anionics were found to be good replacements for the nonionic portions of the formulation. We also found in other formulation evaluations that the stain correlating to ethoxylate groups was dust sebum removal; also known as ring around the collar. Our initial candidates for replacing nonionics were the epoxidized HOSO that was ring-opened with triethylene glycol, PEG 200, or PEG 400 followed by saponification of the triglyceride. Candidate 4 (the result of ring opening HOSO with triethylene glycol followed by saponification) can be seen in Figure 10.5. The benefits to producing through this reaction scheme are that no alkoxylation is being performed therefore removing the concern for dioxane formation. What was found in laundry testing was that there wasn't significant improvement when going from three ethoxylates to approximately eight ethoxylates. The three ethoxylates soy product is water miscible so no increase in the amount of ethoxylate groups is needed to achieve solubility.

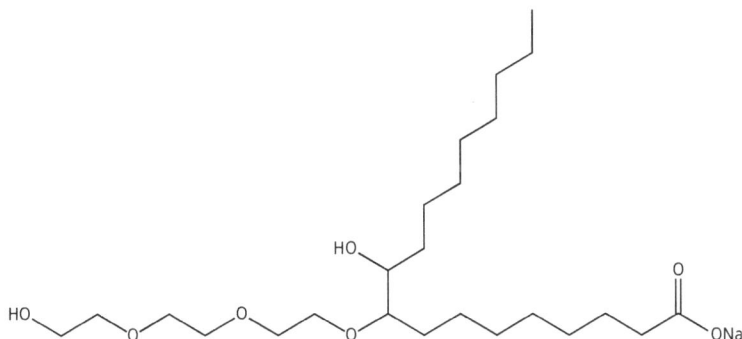

Figure 10.5: Candidate 4.

Later, other derivatives were also produced where alcohol amines were reacted with the intermediate to amidify versus the saponification route. An example of this can be found in Figure 10.6. The benefit is that the material will be less affected by formulation pH changes. However, it was not evaluated in laundry applications. A limited number of glucosides based on this structure were produced but tested in other applications such as fracking fluid additive.

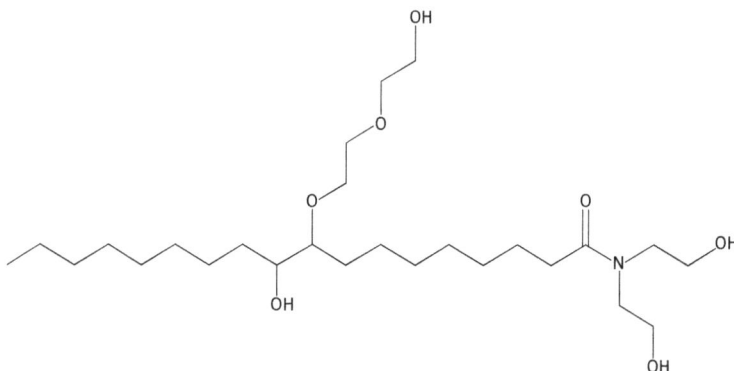

Figure 10.6: Candidate 52.

A similar candidate (Candidate 22E) was found to perform well in hard surface cleaning formulations. Candidate 22E is produced by ring opening the epoxidized HOSO with ethanol followed by amidifying with diethanolamine (DEA). The results are presented and discussed in "Hard surface" section. The benefit was that the candidate achieved statistical equivalence in stain removal as a drop-in replacement for cocamide DEA. Cocamide DEA is the result of amidification of coconut fatty acids with DEA which is rich in C12 fatty acid.

10.4 Cationic surfactants

Cationic surfactants typically have excellent antibacterial properties, provide good corrosion protection, and can be good demulsifiers [5]. Quaternary ammonium compounds and pyridines fall into this category and the surfactants tend to be pH-dependent. They also consist of primary, secondary, or tertiary amines. Some examples of these types of surfactants are cetrimonium bromide and cetylpyridinium chloride. A soy-based example of this is our Candidate 12 seen in Figure 10.7. This compound is produced by reacting epoxidized HOSO in the presence of excess ethylene diamine in an autoclave. The product can be achieved with or without catalyst. Candidate 12 was found to be not biodegradable due to the antimicrobial nature of the structure killing the microbes of ready biodegradation testing.

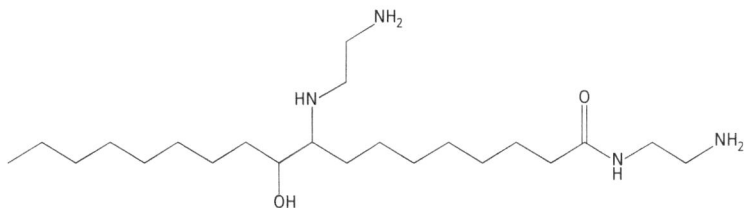

Figure 10.7: Candidate 12.

Work at Battelle has found that this class of surfactants works well in oilfield applications. Testing results can be found later in this chapter. Another soy cationic structure that performed well in oilfield testing can be seen in Figure 10.8. This compound started with either stearyl alcohol treated with epichlorohydrin (as seen below), with epoxidized SBO, or epoxidized methyl soyate. The soy-based products improve oil recovery while, in some cases, require less surfactant than that is currently used in the market.

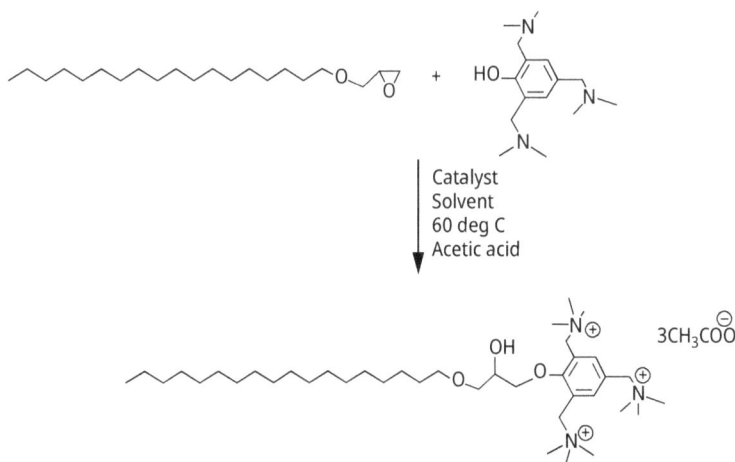

Figure 10.8: Reaction scheme for cationic soy surfactant [6].

10.5 Amphoteric surfactants

Finally, the last type of surfactants are amphoteric surfactants; also known as Zwitterionic surfactants. These surfactants can behave as a cation or anion depending on pH. They consist of structures such as phosphatides and lauroamphoacetates, have low reactivity with other materials, and are increasingly used in personal care products such as baby shampoo and facial cleanser. The cationic portion can be based on primary, secondary, or tertiary amines or quaternary ammonium cations. The anionic part can vary and include sulfonates such as hydroxysultaines and cocamidopropyl betaine.

As mentioned earlier, utilizing soy as the substrate to replace coconut fatty acid derivatives requires epoxidation and ring opening or hydroformylation and hydrogenation. The reaction scheme in Figure 10.9 is an example of that used to create a soy-based betaine (Candidate 43).

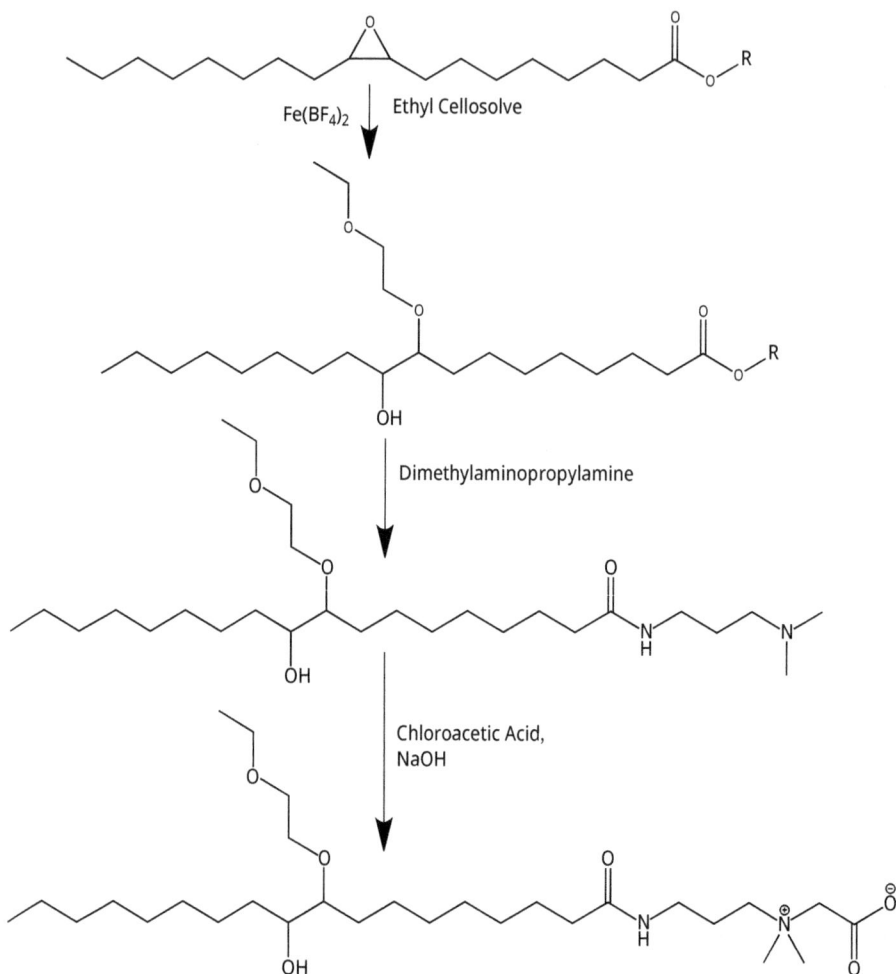

Figure 10.9: Reaction scheme for the production of Candidate 43.

Two benefits were found in the soy betaines. The first benefit was that we were able to directly replace SDBS in standard laundry formulation 1 with statistically equivalent performance. The stain removal results can be seen in Table 10.2.

Table 10.2: Replacement of SDBS with soy-betaine.

Soil	Substrate	Standard	Candidate 25 100% SDBS	Sig Dif
		120 mL	120 mL	
Clay	Cotton	56	50.5	3.1
Coffee	Cotton	42.5	43.7	2.6
Dust sebum	Cotton	64.3	62.1	6.2
EMPA 101 (olive oil)	Cotton	8.7	11.1	1.7
EMPA 112 (cocoa)	Cotton	1.1	2	5.2
EMPA 116 (blood, milk, ink)	Cotton	12.8	10.3	1.7
Grass	Cotton	47	42.6	1.6
Make up	Cotton	59.2	63.7	6.5
Red wine	Cotton	31.5	29.8	0.9
Spaghetti	Cotton	85	79.2	3.5
Overall soil Removal totals		**408.1**	**395**	
% of best		Best	97%	

The second benefit was found in our laundry standard formulation 2. In this formulation, both commodity and HOSO-based betaines performed well as drop-in replacements for cocamidopropyl betaine. The results are shown in Table 10.3.

Candidates 43 and 48 were HOSO and commodity soy-based, respectively. Both are produced by first reacting the epoxides with ethyl cellosolve followed by the same chemistry used to produce cocamidopropyl betaine and used as a weight-to-weight drop-in replacement. The Soy Mix 1 was based on laundry formulation 1 and consisted of a blend of soy surfactants including a soy-based builder.

10.6 Laundry formulations

Two formulations were used when evaluating the soy surfactants in laundry applications. The first was a very simple formulation seen in Table 10.4. The formulation consists of sodium carbonate as builder, Tergitol™ 15-s-15 as the nonionic surfactant, SDBS as the anionic surfactant, and sodium xylene sulfonate (SXS) as the hydrotrope to keep everything compatible. This formulation was 39% solids.

The second formulation, seen in Table 10.5, was derived from patent WO2007/068390. This formulation was 18.7% solids and utilized SLS, potassium cocoate, Tergitol™ 15-s-7 and 15-s-3, coco glucoside, and cocamidopropyl betaine.

Table 10.3: Standards and soy formulations versus off-the-shelf standards.

Soils on cotton	Off-shelf tide	Off-shelf 7th generation	Off-shelf purex	Standard 1	Standard 2	Soy mix 1	Candidate 43 (30 gal scale)	Candidate 48	Sig. Dif.
Clay	63.8	60.0	53.7	61.6	68	57.6	71.8	69.0	3.2
Coffee	54.7	51.4	46.0	52.8	48.3	64.1	48.1	49.5	3.9
Dust sebum	63.9	60.1	53.8	61.7	59.4	60.6	69.2	65.5	5.3
EMPA 101 (olive oil)	8.7	8.2	7.3	8.4	9.7	9.9	8.7	7.8	0.9
EMPA 112 (cocoa)	2.6	2.4	2.2	2.5	3.8	11	5.1	6.0	3.2
EMPA 116 (blood, milk, ink)	8.8	8.3	7.4	8.5	7.7	15.6	10.4	10.6	0.9
Grass	55.8	52.5	47.0	53.9	47.2	45.4	59.9	53.6	4
Make up	59.6	56.0	50.1	57.5	71.3	54.4	61.1	55.2	9.8
Red wine	30.8	28.9	25.9	29.7	26.9	39.1	27.5	25.5	2.4
Spaghetti	86.0	80.9	72.3	83	82.9	84.7	80.6	78.7	2
Overall soil Removal totals	**434.7**	**408.8**	**365.7**	**419.6**	**425.2**	**442.4**	**442.4**	**421.4**	
% of best	102%	96%	86%	99%	100%	104%	104%	99%	

Table 10.4: Laundry standard formulation 1.

	Function	Standard formulation 1
		Mass (g)
Water	Diluent	305
Sodium xylene sulfonate	Hydrotrope	50
Sodium dodecylbenzene sulfonate	Anionic surfactant	50
Tergitol™ 15-S-15	Nonionic surfactant	90
Sodium carbonate	Builder	5
	Total	500

Table 10.5: Laundry standard formulation 2.

	Function	Standard formulation 2
		Mass (g)
Water	Diluent	243.8
Sodium lauryl sulfate	Anionic surfactant	15.3
Potassium cocoate	Anionic surfactant	6
Tergitol™ 15-S-7	Nonionic surfactant	11.7
Coco glucoside	Nonionic surfactant	9
Cocamidopropylbetaine	Amphoteric surfactant	13.2
Tergitol™ 15-S-3	Nonionic surfactant	1
	Total	300

10.7 Other applications

10.7.1 Hard surface cleaning

A wide variety of functionalities are used in hard surface cleaning surfactants. The cleaners are used to remove dust, dirt, and stains. They typically consist of nonionic surfactants but can also include anionic and amphoteric surfactants. One hard surface surfactant is that of cocamide DEA. The concern with cocamide DEA is that it is listed in California Prop 65 as a known cancer-causing agent.

We initially produced candidates for the main purpose of replacing cocamide DEA in a standard formulation. The soy candidate 22, mentioned earlier, was created to contain similar chemistry to that of cocamide DEA. The formulations 1 and 2 evaluated are shown in Table 10.6. Formulation 1 was a high builder formulation that contained 30% solids while Formulation 2 was only 5% solids. In formulation 2, a builder based on soybean flour was utilized.

Table 10.6: Standard formulations for hard surface cleaning.

Formulation 1

Ingredient	%wt	Function
Tetrapotassium pyrophosphate	20	Builder
Cocamide DEA	8	Surfactant
Sodium xylene sulfonate	2	Hydrotrope
Water	70	Diluent

Formulation 2

Ingredient	%wt	Function
Tergitol™ 15-s-15	2	Surfactant
Sodium Xylene sulfonate	2	Hydrotrope
Soybuilder 18	1	Builder
Water	95	Diluent

The evaluations can be seen in Table 10.7. For Formulation 1, Candidate 22 (ethanol ring-opened diethanolamide) and Candidate 1 (soy-citrate) were considered statistically equivalent to the cocamide DEA standard. Candidate 4 (soy-triethylene glycol) and Candidate 13 (soy-EDTA) were just outside the 10% window. While Candidate 22 performed well, the other soy candidates are interesting as they don't contain the diethanolamide associated with the Prop 65 concern.

For the low solids' formulation, Candidates 25 (soy-betaine) and 1 both performed exceptionally well. However, more testing is required as less ethoxylate groups are typically used for the nonionic surfactant component in hard surface cleaners.

10.7.2 Oilfield application testing

An imbibition test is a comparison of the imbibition potential of water and oil into a rock. The wettability of the rock is determined by which phase imbibes more [7]. Many factors lead to good performance in oil removal from rock deposits. However, we focused on the overall imbibition testing and contact angles. Imbibition results can be seen in Figure 10.10. Candidate 12 was the best performer reaching 31% original oil in place (OOIP) which correlates to 72% improvement in oil recovery.

Contact angle with two oil basins were evaluated: Turner and Niobrara. The Turner results can be seen in Figure 10.11. The unique property was the diminished contact angle at the 1 gpt (gallon per trillion) level.

Table 10.7: Stain removal results for hard surface cleaning.

Soil	Type	Substrate	Cocamide DEA	Can. 22	Can. 1	Can. 4	Can. 13	Sig. Dif.
ASTM D4488-A3	Grime	Vinyl tile	76.5	74.2	66.7	63.9	72.5	7.2
ASTM D4488-A5	General purpose	Vinyl tile	89.6	89.3	83.6	85.1	83.7	4.9
ASTM D4488-A2	Kitchen grease	Wallboard	35.6	31.5	31.2	28.3	19.8	3.4
Overall soil Removal totals			**201.7**	**195**	**181.5**	**177.3**	**176**	
% of best			100%	96.70%	90.00%	87.90%	87.30%	

Soil	Type	Substrate	Tergitol™ 15-s-15	Can. 25	Can. 1	Can. 4	Sig. Dif.
ASTM D4488-A3	Grime	Vinyl tile	33.7	39.9	22.5	6.5	3.2
ASTM D4488-A5	General purpose	Vinyl tile	23.6	22.1	21	23	7.3
ASTM D4488-A2	Kitchen grease	Wallboard	6.7	3.5	20.1	3.6	8.6
Overall soil Removal totals			**64.0**	**65.5**	**63.6**	**33.1**	
% of best			100%	102.30%	99.40%	51.70%	

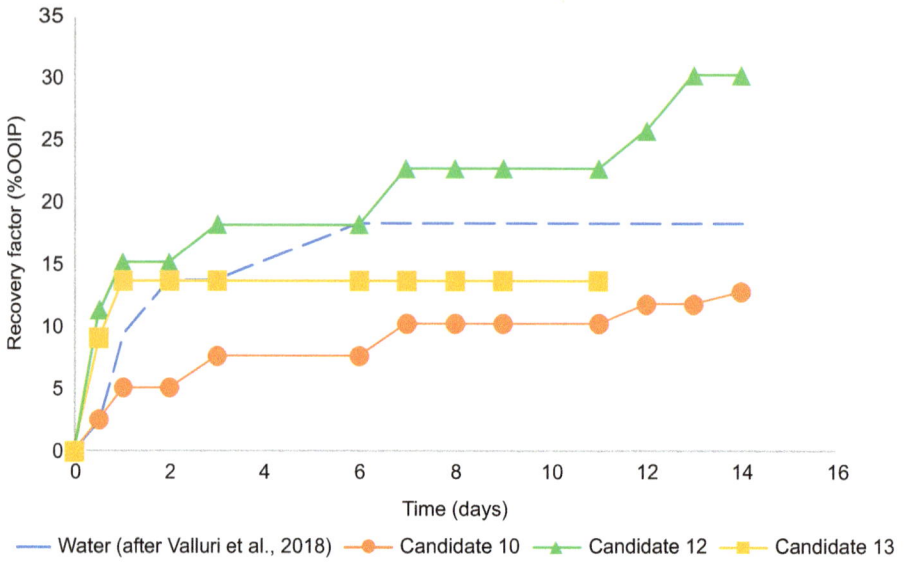

Figure 10.10: Imbibition testing.

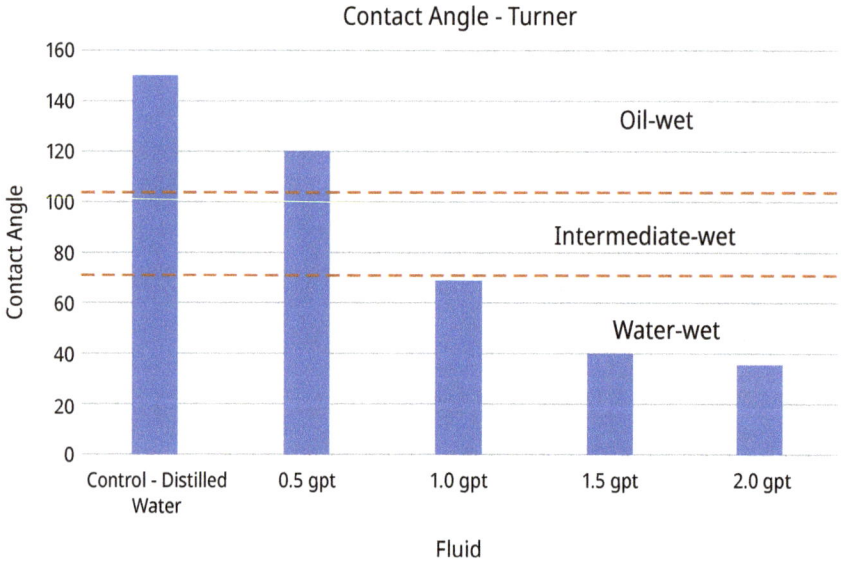

Figure 10.11: Contact angle with Turner basin oil.

10.8 Biodegradation testing

OECD 310 is an aerobic biodegradation test that predominantly measures ready biodegradability by the evaluation of CO_2 in sealed vessels. In the OECD 310 test method, the CO_2 evolution resulting from the ultimate aerobic biodegradation of the test substance is determined by measuring the inorganic carbon (IC) produced in sealed test bottles [8]. Four soy-based candidates were evaluated for ready biodegradation testing. Candidate 1 (soy-citrate) results can be seen in Figure 10.12. The candidate is very close to the 10-day window quickly achieves the 60% threshold for testing. However, Candidate 2 (soy-lactate) and Candidate 25 (soy-diethylene glycol betaine) reached levels of about 50% biodegradation. It was suggested that these be run at lower concentrations and will likely achieve the 60% threshold. The only candidate that didn't perform well was Candidate 12 (soy-ethylene diamine). This candidate only achieved 10% biodegradation in the time window observed. The amine groups tend to kill microbes needed for breaking down the surfactant. While this opens up the possibility of preservatives, it does not perform well in this testing.

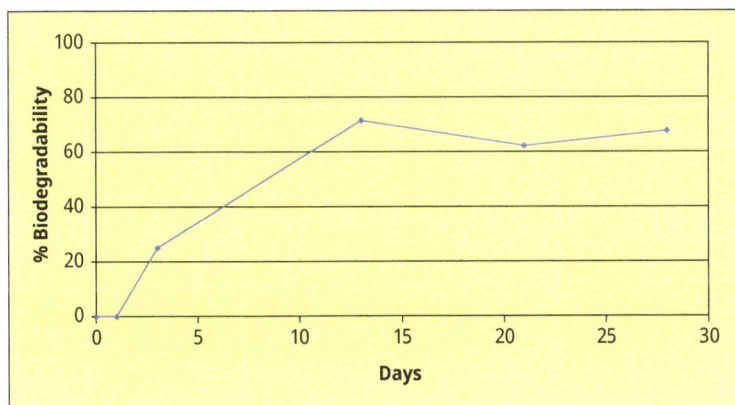

Figure 10.12: OECD 310 test results of Candidate 1.

10.9 Surface tension testing

Surface tension is the tendency of liquid surfaces at rest to shrink into the minimum surface area possible. Surface tension (dynes/centimeter) is what allows objects with a higher density than water such as razor blades and insects (e.g., water striders) to float on a water surface without becoming even partly submerged [9]. Surface tension is a good gauge for surfactant performance in cleaning applications as lower surface tensions typically lead to better rinsing of stains on a substrate. Table 10.8 shows that many of the soy surfactants are in the same range as those used and accepted in

Table 10.8: Comparison of soy surfactants versus standards.

Sample	ST	Standard	ST
Water	71.2	SDBS	34
Candidate 1	32.5	Pluronic® L64	37
Candidate 2	36.6	Tergitol™ 15-S-3	
Candidate 4	34.5	Tergitol™ 15-S-7	
Candidate 5	33.6	Tergitol™ 15-S-15	36
Candidate 10	32.0	Tergitol™ 15-S-40	45
Candidate 11	36.5		
Candidate 12	32.4		
Candidate 13	35.2		
Candidate 14	32.0		
Candidate 15	37.3		
Candidate 16	39.8		
Candidate 25	28.2		
Candidate 29	26.8		
Candidate 32	27.1		
Candidate 39	27.1		
Candidate 43	25.5		

the market. When switching to soy-betaine-based candidates, we were able to drop the surface tension to below the level of 30 °C.

10.10 Critical micelle concentration

In colloidal and surface chemistry, the critical micelle concentration (CMC) is defined as the concentration of surfactants above which micelles form and all additional surfactants added to the system will form micelles [10]. The micelles can pack in various ways but are typically spherical. We determined the CMC values (Figure 10.13) of select candidates and found that many were in the expected 0.01–0.5 range. For CMC testing, the lower the value the better.

10.11 Ross-Miles foam height testing

The Ross-Miles method is used for measuring the foamability of surfactant solutions and the stability of the foam produced, which is based on height measurement. It is named after the authors John Ross and Gilbert D. Miles [11]. The foam test determines a foam profile over time for various functionalities of surfactants. Some select results can be seen in Figure 10.14. Many of our anionic surfactant candidates followed a similar pattern to the standard nonionic Tergitol™ surfactant. Candidate 16C is interesting

from plots	candidate1	candidate2	candidate3	candidate4	candidate5	candidate10	candidate11	candidate12	candidate13	candidate14	candidate15	candidate16c
CMC approximate	wt%	wt%	wt%	wt%	wt%	wt%	wt%	wt%	wt%	wt%	wt%	wt%
concentration	0.1	0.1	0.01	0.5	Weak Transition	0.05	0.1	0.01	0.1	0.05	0.1	0.01
1% Turbidity *NTU	1234.8	184.5	187.3	200.2	65.0	26.1	138.2	466.1	153.0	105.8	351.2	731.0
*NTU approximation by UV-vis												

Figure 10.13: CMC results of select candidates.

Ross Miles Foam Height Test Results

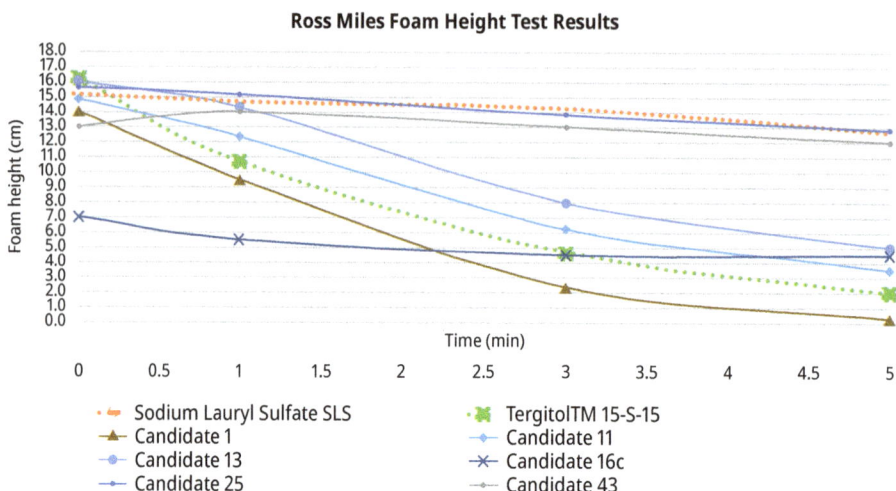

Figure 10.14: Ross-Miles foam height test results.

as it is a sulfonate which holds its foam height over extended time but starts at half the height of SLS at the same concentration. This potentially has implications where low foam is needed such as high efficiency laundry applications. The soy-betaine candidates contained similar foam patterns to SLS.

10.12 Summary

HOSO and commodity SBO can be utilized as platform chemicals to create a variety of surfactants with a multitude of functionalities. C18-based fatty acids have not traditionally been used as surfactants as they typically have lower water miscibilities, versus shorter chains, which can only be overcome by blending with shorter chain surfactants. Insertion of functional groups at the olefin site not only alters the molecules miscibility but also can affect the overall performance of the surfactant. This affect is shown through structure to stain relationships identified through evaluation in cleaning applications and further by properties such as CMC value, Ross-Miles foam height testing, and surface tension lowering determination. Examples of structure-to-stain relationships are lactate to make-up, sulfonates/sulfates to grass, PEGs to dust sebum, and citrate to coffee, cocoa, blood, milk, ink, and red wine. A fully soy-based laundry formulation has been shown to perform equivalently to off the shelf fully formulated standards. The properties, such as foam height and surface tension, of many of the soy surfactants are in-line with standards currently in the marketplace made from predominately C12 chains.

The ability to perform weight for weight replacement of coconut fatty acid-based surfactants has become increasingly important as supply line concerns increase. The United States has no domestic source of C12 fatty acids other than petroleum derivatives. These soy surfactants are now ready for formulation trials showing promise for the future of sustainable green materials.

This platform chemistry was developed through funding from the United Soybean Board.

References

[1]   Surfactant. Merriam-Webster.com Dictionary, Merriam-Webster, https://www.merriam-webster. com/dictionary/surfactant. Accessed 27 Jun. 2022.
[2]   Garbark, et al. Laundry builders and surfactants derived from bio-based hydroxyacids and epoxides. US Patent 11168284.
[3]   Garbark, et al. Bio-based surfactants, US Patent Application 20220089983.
[4]   Kumar A. Global Markets for Surfactant Chemicals and Materials, BCC Research, 2018.
[5]   Carmen Prieto-Blanco M, Fernández-Amado M, López-Mahía P, Muniategui-Lorenzo S, Prada-Rodríguez D. Chapter 11 – Surfactants In Cosmetics: Regulatory Aspects and Analytical Methods. In: Salvador A, Chisvert A, editors. Analysis of Cosmetic Products, 2nd ed., Elsevier; 2018, pp. 249–87. ISBN 9780444635082. https://doi.org/10.1016/B978-0-444-63508-2.00011-4.
[6]   Lalgudi, et al. Bio-based surfactants, cationic surfactants comprising an ether link, US Patent Application 20210139409.
[7]   Imbibition. Oilfield Glossary, Schlumberger, https://glossary.oilfield.slb.com/Terms/i/imbibition. aspx#:~:text=An%20imbibition%20test%20is%20a%20comparison%20of%20the,which%20phase% 20imbibes%20more.%20See%3A%20drainage%2C%20oil-wet%2C%20water-wet Accessed 27Jun. 2022.
[8]   OECD 310 – ready biodegradation – CO2. Situ Biosciences, https://www.situbiosciences.com/biodeg radation/oecd-310-ready-biodegradation-co2/#:~:text=OECD%20310%20is%20an%20aerobic%20bio degradation%20test%20that,inorganic%20carbon%20%28IC%29%20produced%20in%20sealed% 20test%20bottles.
[9]   Surface tension – an overview | sciencedirect topics. www.sciencedirect.com. Retrieved 2021-12-30.
[10]   IUPAC. Compendium of chemical terminology, 2nd ed. (the "Gold Book") (1997). Online corrected version: (2006–) "critical micelle concentration". doi: 10.1351/goldbook.C01395.
[11]   Ross J, Miles GD. An apparatus for comparison of foaming properties of soaps and detergents. Oil Soap 1941;99–102.

Michelle Young*, Rachel Yoho

Chapter 11
Bioconversion products of microbial electrochemical systems

11.1 Introduction

There are few things in life more valuable than creating something from nothing. Creating something of value from invaluable materials is a rare and special human endeavor. Human history has a rich tradition of seeking to create value-added products from raw materials with little to no monetary value or usefulness. Medieval alchemy sought to turn less valuable metals into gold. Creating wine from water is a popular story in one major world religion. These examples illustrate the desire to do something more than other types of creation that we may encounter every day. Cooking, sculpting, painting, and many forms of creation shape outcomes of beauty and wonder (and valuable end products) from raw materials.

The idea of creating something valuable from nearly useless materials is not new. Yet, there are few activities as remarkable and versatile as microbial systems to create a wide range of products or facilitate such a vast diversity of processes. Microbial systems clean up wastewater [1], remediate metals [2], remove hazardous organic compounds [3], synthesize many products from nanomaterials to hydrocarbons [4, 5], recover phosphorous [6], and so much more. The possibilities and applications of a number of designs and systems seem nearly limitless.

When considering the challenges facing humanity, the intertwined climate and energy crises come to mind as the pressing issue, as well as the drivers of several other global pressing needs. The climate crisis elicits a number of potential emotional responses and shapes actions and decision-making [7]. These emotional and intellectual responses have led to numerous innovations in the research sector over the last couple of decades. Barriers to progress stem largely from the perceptions of [8] and implementation of [9] renewable energy technologies rather than the development of the technologies themselves.

*Corresponding author: Michelle Young, Carollo Engineers, Inc., 390 Interlocken Crescent, Suite 800, Broomfield, CO 80021, USA, e-mail: MYoung@carollo.com
Rachel Yoho, George Mason University, 4400 University Dr., Fairfax, VA 22030, USA, e-mail: ryoho@gmu.edu

https://doi.org/10.1515/9783110791228-011

11.1.1 Some recent trends

The advancement of renewable energy technologies, alongside the development of methods for creating valuable products from wastes, bridges the gap in terms of addressing needs, adding clear value, driving sustainability, and demonstrating practical application. Microbial electrochemical systems are one such area of versatile application.

The last two decades have seen the development and advancement of microbial electrochemistry, particularly with practical applications. These practical applications [10] range from energy production [11], wastewater treatment and resource recovery [12], niche energy availability for remote sensing [13], and removal of environmental contaminants [14]. The microbial research community grew in prominence particularly after the potential engineering applications were realized, specifically, the waste-to-energy processes [15]. In these early examples, microorganisms oxidized organic compounds in wastewater to generate energy. This work had important connections not only to decreasing the costs of wastewater treatment but also more broadly to climate needs and the energy crisis [15].

The technologies, broadly, demonstrate a large versatility in terms of application and product output. The designs, materials, and microorganisms then also vary widely. In this work, we work backward, highlighting from the products generated back to specific details about the design.

11.1.2 Microbial electrochemical cells

Broadly, the microbial electrochemical cells (MXCs) represent a unique class of technologies relying on microorganisms and spanning wastewater treatment, bioremediation, and energy generation through multiple bioprocesses [16]. Important distinctions in this space are the acronyms occasionally representing different time periods in the development of the field, advancements, and applications. Here, we will use the general term of MXC as the overarching term for a microorganism-based system that generates many different potential products. Other terms that appear in this space are microbial fuel cell (MFC), typically focused on energy output [15] and microbial electrolysis cell (MEC), typically require energy input to produce valuable products including hydrogen [17]. Certain applications have their own acronyms as well including the microbial desalination cell (MDC) [18] and plant applications (PMFCs) [19].

Beyond the importance of the related acronyms, other key identifying factors are whether the reaction of interest occurs on the bioanode or the biocathode, the biological processes, and any other applicable living organisms. For example, these include biocathode bioremediation [20, 21], product synthesis [22, 23], and fermentation [24], among others [10].

11.2 Bioanodes

Generally understood as the growth of a film of some sort of living microorganism on the surface of an electrode, the bioanode presents a number of important benefits and challenges to MXCs. Overall, microbial electrochemical systems necessitate the microbial colonization of an anode in order to generate the electrical current [25]. These are the best-known biofilms in the field, yet many unknowns remain. Importantly, MXCs are taking advantage of the "enzymatic machineries" in the varied microorganisms to facilitate reactions not possible in nonbiological engineered systems [15]. The factors that define and shape the formation of bioanodes are still poorly understood [25], even considering their deep history and importance in the field.

The performance of the bioanode shapes the overall performance of the MXC [26]. The many factors affecting the performance of the bioanode are well outside of the scope of this chapter; however, the bioanode itself shows promise in adding opportunity in the field. For example, capacitive electrodes in the anode compartment of an MXC create a situation in which a capacitive bioanode can form. This type of integrated system allows the system to both generate and store renewable energy [27].

The poor understanding of biofilm formation on anodes presents an important challenge in the scale-up of systems [25]. Enrichment processes typically occur in batch systems over multiple cycles until current generation is stable [25, 28, 29].

Early challenges to MXC systems were the use of various mediators in the system. Early iterations of these systems relied on soluble mediator compounds to facilitate electron transfer [30]. These mediators were generally toxic and impossible to use in open systems [30], including potassium ferricyanide [31], anthraquinone 2, 6-disulfonic acid, cobalt sepulchrate [32], thionine [33], neutral red [34], and azure A [35].

11.3 Anodophilic microorganisms

Microorganisms with the ability to use a solid material external to the cell as a final electron acceptor in the respiratory process generally are considered to be anodophilic microorganisms. These organisms typically oxidize some sort of organic material in a liquid medium or liquid waste stream and transfer the electrons to the final, external electron acceptor [36].

The mechanisms through which these organisms transfer the electrons outside of their cells vary among microbial species and are also highly contested in the field. Cytochromes, particularly the outer membrane c-type cytochromes, are considered as key components of these organisms' ability to respire a final extracellular electron acceptor [37]. Others in the field assert the "metal-like conductivity" of the microbial nanowires as pili [38]. This group further espouses the metal-like conductivity as due to the overlapping π–π orbitals of aromatic amino acids within the nanowires themselves [38].

More widely studied are the specific cytochrome mechanisms, for example, with multiple pathways of electron transport observed depending on the anode potential for *Geobacter sulfurreducens* [39] and *Geoalkalibacter ferrihydriticus* [40]. Specific cytochromes appear essential to potential-dependent extracellular electron transport. For example, *imcH* in *G. sulfurreducens* is an inner membrane cytochrome identified to be required for high redox potential extracellular electron acceptors [41]. Other work indicates "electron conduits" each with several key components: a periplasmic multiheme *c*-type cytochrome, an integral outer membrane anchor, and out membrane redox lipoproteins [42]. This further supports the necessity of cytochrome-dependent pathways for extracellular electron transport, particularly depending on the applied potential of the anode [39].

Importantly, real-life biofilms are unlikely to be pure cultures. The reality of the extracellular electron transport is much more likely to be some mix among the two proposed mechanisms. Naturally, real life is wildly complex. Mixed culture extracellular electron transport in larger systems and in the environment likely have these complexities and many more.

Here, we highlight several important organisms to this type of work, notably *Geobacter* and *Shewanella* as well as a few related extremophiles with particularly compelling and niche applications.

11.4 Geobacter

Among mixed culture systems, members of the *Geobacter* genus are essential in both generating current and decreasing charge transfer resistance in the anode biofilms [25, 43, 44].

Extracellular electron transport is a complex process, with notable advancements in the field occurring only in the last 10 years. Essential to these advancements were the study of pure culture biofilms rather than mixed cultures dominated by *Geobacter* or another exoelectrogenic species.

The family, *Geobacteraceae*, and genus *Geobacter* are deltaproteobacteria considered to be ubiquitous in many environmental settings including soil, sediments, and similar subsurface environments. In these environments, metal reduction is a typical practice [45]. This family, identified just over 30 years ago, provides researchers and practitioners with essential services such as degrading hydrocarbon contaminants, reduction of radionucleotides, toxic metals, and nitrogen. Therefore, these organisms appear as frequently studied and applied components of several bioapplications: bioremediation, bioenergy, and bioelectronics [45].

Among the *Geobacter* species, the most well-known and well-studied is *G. sulfurreducens*. The first identified, however, was *G. metallireducens*, following investigations of sediments from the Potomac River in the eastern portion of the United States [46]. In the over three decades since, *G. sulfurreducens* emerged as the more studied organism, particularly as the first to have a sequenced genome and more understood genetic system [47]. For *Geobacter*, the outer membrane OmcE and OmcS outer membrane *c*-type cytochromes are understood to mediate the extracellular electron transfer mechanisms [37].

11.5 Shewanella

Aside from the rich history of *Geobacter*, the other most well-known organism in this field is *Shewanella*. Overall, the genus has more than 50 known species across different types of marine environments and ecosystems [48]. These are aquatic microbes found worldwide [49], much like *Geobacter*. What sets these organisms apart are their diversity in respiratory pathways and ability to survive, and thrive, at low temperatures [45].

The biotechnology applications of *Shewanella* are diverse. From energy generation through biocatalysis to remediation of chlorinated compounds, radionucleotides, and pollutants to the environment, *Shewanella* presents a complex physiology [45]. Interestingly, beyond the heavy metal remediation and MXC applications, members of this same genus are becoming more well known as human pathogens, particularly from occupational or marine exposure [48]. These human illnesses typically are skin infections, middle ear inflammation and infections, and blood infections (bacteremia) [48]. Research indicates that the human infections are limited to *Shewanella algae* [48], so risk in biotechnology research likely is negligible.

Shewanella oneidensis stands out among the genus for research and biotechnology applications. For *S. oneidensis* MR-1, the MtrC and OmcA are understood to be the *c*-type cytochromes involved in extracellular electron transport [37]. Overall, multiheme cytochromes exist in a number of phylogenetically diverse microorganisms and enable long-range electron transfer (tens of nanometers) [50]. These abilities have important global-scale impacts in the biogeochemical cycling of N, S, and Fe [50]. More specifically, *S. oneidensis* is well known to be able to use ferric iron [Fe(III)], a popular research application choice, as an external final electron acceptor [51, 52].

Interestingly, flavins are also found to have a role in extracellular electron transport for *Shewanella*. In a study of *S. oneidensis* MR-1 and MR-4 grown on poised anodes, the intact biofilms revealed redox-active molecules inside the biofilms. These were riboflavin and riboflavin-5′-phosphate, with riboflavin the dominant molecule for established biofilms. Removal of the biofilm from the electrode also indicated that riboflavins remained on the surface of the electrode as well as could adhere to other materials of interest in these systems like Fe(III) and Mn(IV) oxy(hydro)oxides [53]. Other molecules of interest are water-soluble molecules secreted by the organism. For example, *S. oneidensis* secretes water-soluble molecules acting as electron shuttles or Fe(III) complexing ligands, quinones and quinone moieties are observed to accelerate Fe(III) reduction by functioning as mediators, and extracellular appendages ("nanowires") have also been observed [51]. By comparison, *Geobacter* are not known to produce any electron shuttles [22, 30].

Shewanella as a genus overall is unique and varied, particularly within this context of interest to biotechnology applications and bioproduct development. Much research has been done and is ongoing on key species, particularly *S. oneidensis*.

11.6 Other species

Other relatives of *Geobacter* have gained prominence in pure culture studies of anode biofilms. In particular, these have been the "extremophiles" and are all members of the δ-proteobacteria [10] such as *G. ferrihydriticus* [54, 55] and *Geoalkalibacter subterraneus* [54, 56]. Even outside of the δ-proteobacteria, other organisms are able to respire anodes. *Thermincola ferriacetica* is an example of this from the class *Firmicutes* [10, 57–59]. As the name implies, *Thermincola ferriacetica* is best known for its thermophilic existence and ability to respire an anode while under high temperatures.

11.7 Biocathodes

As compared to early, and ongoing, work in the field that grows biofilms on anodes, the biocathode has emerged as an interesting area of research. As the name implies, the biocathode has a living biofilm growing on the cathode of the system. The systems then are microbial-cathode MFCs, or generally, biocathodes [60], or can broadly be called biocathode MXCs.

The "traditional" MFC or MXC have the biological anode and abiotic cathode and occasionally require the use of a catalyst or electron mediator. This increases costs and limits application possibilities [61]. Biocathodes specifically are seen as a way to overcome these limitations [61].

Biocathodes are versatile, with oxygen as an electron acceptor, or the use of contaminants [60]. This allows for several different applied processes for nutrient removal or bioremediation and can be co-occurring with biological production of electricity. These systems can be applied to convert electricity into reduced products [60]. Overall, the classification of the biocathode depends on which final electron acceptor is used in the reaction, including aerobic biocathodes (oxygen) and anaerobic biocathodes (nitrate or sulfate, e.g.) [61]. The general reactions for aerobic have the electron mediators (e.g., iron and manganese) reduced by the cathode in an abiotic process then reoxidized by the microbes [61]. The anaerobic biocathodes reduce the final electron acceptors through a process of accepting electrons from the cathode via microbial metabolism [61]. In a fascinating application of both engineering and microbiology, it is possible to have an MXC with both a bioanode and biocathode, including without requiring the use of chemical catalysts [62].

11.8 MXCs for products

Living systems are typically more complex to understand and control than other systems; however, this has been addressed in major areas such as wastewater treatment. The living fuel-cell systems have important advantages over other methods of energy generation. Fuel cells, whether biological or not, avoid the inefficiencies of combustion engines, as oxidation and reduction can occur on the electrode surfaces [30]. The inclusion of the biological aspects of the highly efficient fuel-cell system incorporates the advantages of microorganisms to break down organic compounds and, generally, produce electricity [30, 63]. Of the dozen or so processes identified for the microbial electrochemical technologies, the "X" in MXC dilutes out the meaning into many different technical goals [15].

11.8.1 Production and recovery of valuable products

In this work, we highlight several notable products or activities of microbial systems: electricity, hydrogen, hydrogen peroxide (H_2O_2), ammonium (NH_4^+), volatile fatty acids (VFAs), and heavy metals. We chose these specific processes, as they represent examples of widely studied or niche applications, great value-adds to specific systems, or opportunities specifically available with microbial systems.

11.8.1.1 Electricity

The most well-known product of the MXC system is electricity. From early days in the field, production of electricity in niche applications and more broadly appears to have sustained the general interest in the technology as a whole. For electricity generation from biological systems, many details are shared in earlier sections, including on *Geobacter*, *Shewanella*, and bioanodes.

Multiple fields of study are involved in bioproduct generation from MXC systems. Starting from the microbiological areas of interest for a new type of biological process to engineering applications, research areas are abundant. The realization of the ability of microorganisms to perform redox reactions directly on electrodes, specifically without the use of chemical mediators, provides unique research areas in both scientific and engineering fields [15]. Anecdotally, high current densities (~11 A/m^2) and ~70% Coulombic efficiencies were observed in *Geobacter*-enriched butyrate (an important fermentation product) oxidizing communities [64]. Importantly, this study indicated that mixing complementary cultures outperformed simply enriching cultures. Basically, the butyrate-oxidizing community combined with a *Geobacter* rich mixed culture created a new mixed community outperforming other research studies [61].

While this may be seen to further complicate the research field, the applications are profound. Carefully creating the physical systems and development of the biological communities may continue to enhance the field, including increased current densities and high Coulombic efficiencies.

11.8.1.2 Biogreen hydrogen

Although vehicle technology has moved on from past interest in hydrogen fuel cells, this particular potential bioproduct is still of interest from the MXC system. The microbial bioelectrochemical system research areas generally progressed from MFC to MXCs. The MFC systems are highlighted by the possible application in wastewater treatment and energy generation but have more recently expanded in one way through the use of biocathodes [65]. The systems and processes are so varied [61] that it is difficult to summarize well in this context.

Several technologies that can be used to produce hydrogen from a biological approach: biophotolysis, photo fermentation, dark fermentation, and hybrid biohydrogen production via electrochemical methods [66]. Several different types of organisms can be used to facilitate the reactions in these different systems from cyanobacteria and green microalgae to different types of bacteria [65].

Research applications to produce hydrogen vary widely. In the biocathode application, hydrogen can be produced by the microorganisms. A study with biocathode and bioanode application revealed approximately 20% hydrogen recovery at the cathode, but experienced losses after 1,600 h of operation due to calcium phosphate buildup on the cathode [62]. In an MEC application, a mixed culture of anodic syntrophic fermentative and anode respiring bacteria produced hydrogen from a milk synthetic wastewater source with the best current density (150 A/m^3) as compared to glycerol and starch synthetic wastewater sources [67]. Excellent hydrogen recoveries (~90%) and production (0.94 m^3/m^3/d) observed experimentally [67] provide great promise for the field.

11.8.1.3 Hydrogen peroxide

H_2O_2 is a valuable industrial chemical and oxidizing agent with a wide range of uses ranging from a bleaching agent for textiles and paper, a propellant, and an oxidant in chemical manufacturing [68]. Approximately 5% of the U.S.'s annual H_2O_2 usage is in the water and wastewater treatment industry [69]. This typically is in low concentrations for advanced oxidation coupled with either Fenton reactions or UV treatment for the removal of complex organic chemicals and emerging contaminants [68, 70, 71] and for disinfection [72–76]. H_2O_2 is also used in advanced oxidation processes to remove biological products or contaminants that contribute to taste and odor issues [74, 77, 78]. This makes it an important value-added bioproduct, particularly due to the importance of addressing taste and odor for the consumer. H_2O_2 is utilized at wastewater treatment plants (WWTPs) for hydrogen sulfide removal in scrubbing towers to reduce odor emissions [79, 80].

Electrolysis of H_2O_2 has been commercially available for more than 100 years but is economically disadvantageous due to large energy requirements [68]. During electrolysis, H_2O_2 is produced through oxygen reduction at the cathode:

$$O_2 + 2H_2O + 2e^- \rightarrow H_2O_2 + 2OH^- \quad \left(E^{o'} = +0.424\ V_{Ag/AgCl}\right)$$

One potentially energy-neutral application of MXC technology is cathodic electrosynthesis of H_2O_2 using electricity produced by anodophilic microorganisms using microbial peroxide-producing cells (MPPCs). One advantage of H_2O_2 production using MPPCs is that it requires simple catalysts like carbon rather than Pt or Au for electricity and H_2 production.

Researchers have produced H_2O_2 in MPPCs at concentrations suitable for tertiary water and wastewater treatment with little to low energy input (30–50 mg H_2O_2/L for disinfection and 55–20 mg H_2O_2/L for micropollutant removal) when fed with acetate- or lactate-based synthetic wastewater at the laboratory scale [71, 81]. Rozendal et al. [82] produced 3.9 g H_2O_2/L-d and concentrations as high as 1.3 g/L with energy input of 0.93 W-h/g H_2O_2. To maximize H_2O_2 production, Modin and Fukushi [83] produced 9 g/L H_2O_2 at a rate of 11.1 g/L-d while applying a high amount of energy at 3.0 W-h/g H_2O_2. Griffin et al. [84] produced 4.6 g H_2O_2/L at a production rate of 4.6 g/L-d with 1.6 Wh/g H_2O_2 energy input.

Young et al. [85, 86] demonstrated that H_2O_2 production and concentration are a trade-off on most MPPC: reactors produced as high as 3.1 g H_2O_2/L and a rate of 0.76 g H_2O_2/L at a 4 h hydraulic retention time (HRT), but the net production rate of H_2O_2 was higher 1.02 g H_2O_2/L-d at a 1 h HRT with an effluent concentration of 1 g/L. The decrease in concentration is attributed to degradation of H_2O_2 in the reactor through reaction with the cathode catalyst.

A few pilot-scale MPPC studies are available. Sim et al. [87] operated at 110 L MPPC with acetate as the anode media and tap water as the catholyte. The reactor achieved 98 mg H_2O_2/L production with low conversion efficiency (7.2%). Conversion efficiency was likely impacted by the low conductivity of tap water, which did not provide adequate anions or cations for transfer to the anode chamber to maintain charge neutrality.

11.8.1.4 Ammonia

Ammonia (NH_3) is an important part of the food-energy-water nexus. The NH_3 used in fertilizers is produced via the energy-intensive Haber-Bosch process, which requires 28–30 GJ of energy per tonne of NH_3 produced [88] and 949 m^3 of natural gas as a substrate [89]. At the same time, urine and domestic wastewater are high in nitrogen that, when naturally hydrolyzed, becomes inorganic NH_3. It is estimated that total Kjeldahl nitrogen (TKN), which is the sum of organic N and NH_3, is about 50 mg/L in domestic wastewater, and humans produce about 12 g TKN per person per day [80, 90]. Thus, there is ample opportunity to use wastewater and urine treatment as potential sources of NH_3.

One potential method of separation from waste streams are MXCs, where the focus at the anode is used for COD removal and the cathode for NH_3 recovery using a MFC, MEC, or MDC. In the MXC, COD is utilized by anodophilic microorganisms to produce electricity, releasing NH_4^+ during substrate utilization and current that is transferred to the cathode [91]. To maintain charge neutrality within the cell, NH_4^+ can transfer to the cathode chamber when a cation exchange membrane (CEM) is used to separate the anode and cathode chambers. When internal resistances are low and the current high, a recovery system can operate at an MFC producing electricity, while recovery systems with low current production operate at an MEC.

Many researchers have focused on NH_3 removal using MECs and MFCs through promotion of nitrification and denitrification processes in the anode chamber and using a biocathode [92–95]. Other researchers have focused on maximizing NH_3 transport to and off-gassing at the cathode chamber. Villano et al. [96] operated an MEC with a methane (CH_4)-producing biocathode captured a modest 33 mg/day of NH_3 (50% of daily input at the cathode) in the cathode chamber at a reasonable 8.3 h HRT in the cathode chamber. Off-gassing of NH_3 was not observed mostly due to the pH being 8.2, well below the pK_a of 9.35 [97]. Kuntke et al. [98] used MFCs with up to 3 g/L NH_4^+ in the influent to isolate 124 mg NH_4^+/day of at the cathode while producing 0.23 W/m^2 of energy. Qin and He [99] used an MFC coupled with forward osmosis to capture and concentrate 81% of NH_4^+ supplied to the anode chamber. Aeration of the cathode chamber encouraged stripping of NH_3 from the catholyte, with two-thirds of NH_3 off-gassing from the catholyte.

Significant research has been dedicated to NH_3 from urine. Using real urine at 4 g NH_4^+/L, Kuntke et al. [100] recovered 3.29 $g/m^2/d$ with 40% off-gassing as NH_3 while producing 0.05 W/m^2 of energy. Kuntke et al. [101, 102] used different concentration methods to recover NH_3 at rates as high as 162 $g/m^2/day$ with 2.3 Wh/gN recovered. In the only scaled-up reactor, Zamora et al. [103] operated an MEC with 2.5 L anode and cathode volumes using real urine to recover NH_4^+. Ninety-two percent of influent NH_4^+ migrated from the anode to the cathode chamber, of which 31% was recovered.

In MDCs, the anode and cathode chambers are separated by a center chamber that acts as a desalination chamber. The anode and desalination chambers are separated by anion exchange membrane (AEM), and the desalination and cathode chambers are separated by a CEM. Zhang and Angelidaki [104, 105] applied a lab-scale modified submerged MDC for *in situ* NH_3 recovery in an anaerobic digester to help prevent NH_3 inhibition of methanogenesis. The MDC recovered an average of 80 $gN/m^2/day$ and achieved a maximum power density of 0.71 W/m^2, making the cell a net energy producer.

Other NH_4^+-rich waste streams have also been explored for MFC and MEC applications including pig slurry [106], landfill leachate [107], and livestock water [108, 109]. Most notably, Sotres et al. [106] used an MEC to recover up to 25.5 $g/m^2/day$ from piggery waste with 0.4 V applied voltage. In the same study, an MFC was used to recover 7.2 $g/m^2/day$ while producing 0.46 W/m^2. Sotres et al. [106] also saw increased NH_4^+ recovery rates (2.3–7.2 $g/m^2/day$) with increasing N and organic loading rates that correlated with increased current production.

11.8.1.5 Volatile fatty acids

VFAs are valuable commodities used in a variety of industries including chemical manufacturing, food manufacturing, consumer and personal care products, and bioplastics [110]. Short-to-medium chain VFAs are classified as carboxylic acids with two to six linearly chained carbons with a carboxyl group at one end. During typical anaerobic digestion, carbohydrates, proteins, and lipids are hydrolyzed to amino acids, monosaccharides, and larger fatty acids [80, 111]. These hydrolysis products can be fermented to VFAs by acetogenic and acidogenic fermenters. In a typical anaerobic digester, the presence of methanogenic *Archaea* would convert the VFAs to CH_4; however, VFAs can accumulate to high concentrations when methanogenesis is inhibited [111]. One major focus is VFA recovery from fermentation of organic-rich waste streams like municipal sludge, food waste, fats, oils, and grease (FOG), and other municipal solid wastes [110, 112, 113].

MXCs have demonstrated CO_2 valorization to VFAs using two different methods: as a separation process for high-strength VFA waste streams in systems similar to desalination processes and as a production source and recovery system using microbial electrosynthesis (MES) cells and biocathodes [114–116]. These MXC technologies lend themselves well to VFA separation since membranes and cathode operating potential can potentially be tailored to select for specific VFAs [117]. For example, production of acetate from CO_2 requires a theoretical potential of −0.28 V.

11.8.1.5.1 Source separation via desalination cells
Similar to NH_4^+ recovery, MDCs are effective at isolating VFAs in a middle chamber. Zhang and Angelidaki [114] used a system similar to a MDC to isolate VFAs from a fermentation broth consisting of acetate, propionate, and butyrate. The microbial bipolar electrodialysis cell was fed to fermentation broth to dialysis chambers while the anode and cathode were operated using domestic wastewater and NaCl solution,

respectively. The system achieved 98% VFA recovery from the fermentation broth with 1.2 V applied potential during a 94 h retention time.

11.8.1.5.2 Production via microbial electrolysis and biocathodes

Biocathode systems that produce VFAs are based on conversion of CO_2 valorization to carboxylic acids [116]. A variety of pure and mixed acetogenic biocathode cultures have demonstrated the ability to reduce CO_2 to VFAs with the addition of electrons from either external electricity sources or anodophilic bacteria in the anode chamber [118].

Acetate has been a major focus of MES technologies. Acetate recoveries have ranged 0.02–1,330 g/m^2/day at applied current densities ranging from 0.01 to 200 A/m^2 [119–122]. Generally, increased supplied current to the cathode resulted in higher acetate production rates. Most notably, Jourdin et al. [120, 121] used abiotic anodes to supply electricity to several layers of macroporous carbon felt cathode stacked together and inoculated with a mixed culture. They achieved 1,330 g/m^2/day acetate production at neutral pH (6.7) and with 200 A/m^2 of applied current. More importantly, Jourdin et al. reported >99% current recovery to acetate. Yu et al. [119] used the thermophilic bacteria *Moorella thermoautotrophica* to produce acetate and formate up to 2.9 and 3.5 g/m^2/day, respectively, with about 1 A/m^2 of added current density. Roy et al. [123] utilized unpurified CO_2 from a brewery as a substrate source for MES performance. The CO_2 was 97.9% pure and supplied to a mixed-culture biocathode producing up to 66 g/m^2/day of acetate while producing small amounts of formate and butyrate and 84% conversion of electrons to VFAs.

Other VFAs have also been recovered via MES, primarily caproate [124–126], butyrate [124, 125, 127–129], and formate [119, 123]. These VFAs were generally produced simultaneously with acetate [119, 123, 125, 128, 130]. Production rates for different VFAs ranged from 0.04 to 5.7 g/L/day for butyrate, 0.95–2.4 g/L/day for caproate, and 0.36 g/L/day for formate [130, 131]. Jourdin et al. [124] manipulated cathode HRT and CO_2 loadings to favor butyrate and caproate production. At a 14-day HRT and 173 L/day CO_2 loading, Jourdin et al. [124] observed chain elongation of ethanol with acetate to produce 106 and 64 g/m^2/day of *n*-butyrate to *n*-caproate, respectively. Batlle-Vilanova et al. [129] selectively produced butyrate over other products by maintaining a pH around 5 and H_2 partial pressures >1 atm.

11.8.1.6 Heavy metals

Microbial electrochemical cells have significant promise for the remediation and recovery of different heavy metals. As shown in Figure 11.1, many of the heavy metals that are common contaminants in water and acid mine drainage have positive redox potentials, making them ideal for recovery at cathode, although some anode processes exist [132]. Table 11.1 summarizes select works on heavy metal recovery from MXCs. No pilot-scale studies have been performed using these technologies, largely due to limitations in membrane materials and mass transfer limitation due to electrode spacing and reactor volume. Unless otherwise noted in the figure, all MXCs were dual chamber in design, with either a CEM or AEM separating the anode and cathode chambers and performed as batch experiments.

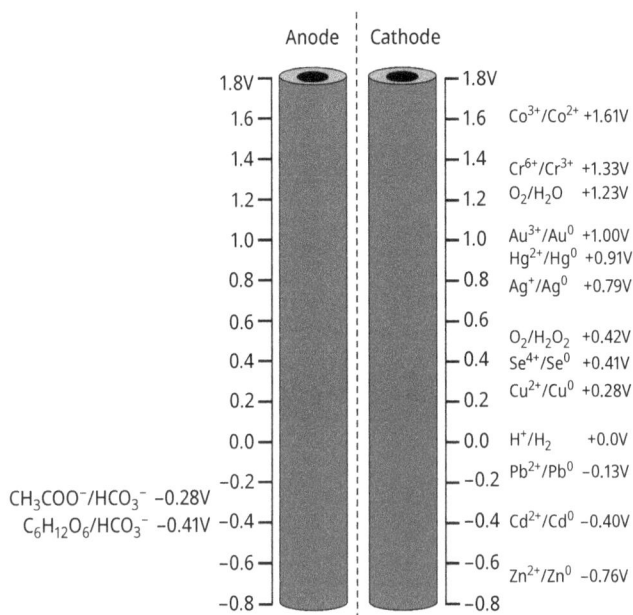

Anode | Cathode

1.8V	1.8V
1.6	1.6 Co^{3+}/Co^{2+} +1.61V
1.4	1.4 Cr^{6+}/Cr^{3+} +1.33V
1.2	1.2 O_2/H_2O +1.23V
1.0	1.0 Au^{3+}/Au^0 +1.00V Hg^{2+}/Hg^0 +0.91V
0.8	0.8 Ag^+/Ag^0 +0.79V
0.6	0.6
	O_2/H_2O_2 +0.42V
0.4	0.4 Se^{4+}/Se^0 +0.41V
0.2	0.2 Cu^{2+}/Cu^0 +0.28V
0.0	0.0 H^+/H_2 +0.0V
−0.2	−0.2 Pb^{2+}/Pb^0 −0.13V
CH_3COO^-/HCO_3^- −0.28V	
$C_6H_{12}O_6/HCO_3^-$ −0.41V −0.4	−0.4 Cd^{2+}/Cd^0 −0.40V
−0.6	−0.6 Zn^{2+}/Zn^0 −0.76V
−0.8	−0.8

Figure 11.1: Redox potentials for common products recovered in microbial electrochemical cells (adapted from Nancharaiah et al. [132]).

11.8.1.6.1 Chromium (VI)

Chromium is a common groundwater contaminant and common industrial chemical in the metallurgy, electroplating, leathering, dyes, and corrosion control industries [132]. The two most common forms are hexavalent (Cr(VI)) , which is highly soluble at all pHs and toxic, mutagenic, and carcinogenic, and trivalent (Cr(III)) , which is significantly less toxic than Cr(VI) and tends to precipitate to chromium hydroxide (Cr(OH)$_3$) under moderately acidic to alkaline conditions. The reduction reaction of aqueous Cr (VI) has a highly favorable redox potential at the cathode that is even more favorable than O$_2$ reduction ($E^{o\prime}$ = +1.23 V_{SHE}), making Cr(VI) reduction an electricity-producing process [133]:

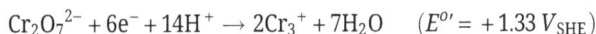

$$Cr_2O_7^{2-} + 6e^- + 14H^+ \rightarrow 2Cr_3^+ + 7H_2O \quad (E^{o\prime} = +1.33\ V_{SHE})$$

Complete Cr$_2$O$_7$$^{2-}$ was generally achievable at concentrations less than 150 mg/L while using simple carbon-based catalysts. Wang et al. [133] were the first to demonstrate Cr (VI) reduction from contaminated catholyte in a batch cathode chamber using a graphite

Table 11.1: Select metals recovery works using MXCs.

Authors	Cathode setup	Membrane type	Cathode HRT (h)	Anolyte/buffer	Catholyte	Maximum % contaminant remediation and reduction rate	pH of cathode	Current density/power density at maximum removal	Columbic efficiency
Chromium (VI)									
Wang et al. [133]	Graphite plate	PEM	150	20 mM acetate	100 mg/L $Cr_2O_7^{2-}$	100% 0.67 g/m^3/h	2.0	0.34 A/m^2 and $R = 1,000\ \Omega$ 155 mW/m^2	59%
Li et al. [134]	Rutile	PEM	26	27 mM acetate	26 mg/L $K_2\ Cr_2O_7$	97%	2.0	0.3 A/m^2	
Yi et al. [135]	Carbon cloth	PEM	72	13 mM acetate	80 mg/L $Cr_2O_7^{2-}$	100%	2.0	1.6 A/m^2 and 700 mW/m^2	
Liu et al. [136]	Carbon cloth with polypyrrole MnO_2 catalyst	PEM	32	9 mM acetate	50 mg/L $Cr_2O_7^{2-}$	100%	2.0	3.7 A/m^2 and 1,429 mW/m^2	
Gupta et al. [137]	Carbon nanofibers with Al-Ni catalyst	PEM		12 mM acetate and 100 mM sucrose	200 ppm $Cr_2O_7^{2-}$	85%	2.0	7.86 A/m^2 and 1530 mW/m^2	93%
Zhang et al. [138]	Carbon rod	CEM	4	12 mM acetate	50 mg/L Cr(VI)	68%	2.0	0.55 A/m^2 and 3.1 W/m^2	69%

Cobalt

Huang et al. [139]	MFC: Graphite felt MEC: Carbon rod	CEM	MFC: 6 MEC: 6	17 mM acetate	MFC: 10 mM HCl MEC: 250 mg/L Co(II)		MFC: 1.0 MEC: 6.0	13.7 A/m² and 0.2 V applied voltage	MFC: 41% MEC: 75%
Jiang et al. [140]	Graphite felt	CEM	6.2	17 mM acetate	847 µm Co(II)	92%			
Huang et al. [141]	GAC and Pt-coated carbon felt	AEM	12 d	18 mM acetate	40–60 mg/L CoSO$_4$	95%	8.4	8.2 W/m³	22%
Saad et al. [142]	Stainless steel screens	CEM	6	12 mM acetate	50 ppm	99%	7.4–8.4	3.2 mA and 1.8 V applied voltage	
Huang et al. [161]	Graphite felt	CEM	6	17 mM acetate	0.5 mM Co(II)	4.7 g/L/h	5.6	13.3 A/m³ and 1.5 W/m³	
Shen et al. [143]	Graphite felt	CEM		--	40 mg/L Co(II) and 50 mg/L Cu(II)	5.3 g/L/h	5.5	4.8 W/m³	54%

Copper

Shen et al. [143]	Graphite felt	CEM		--	40 mg/L Co(II) and 1,000 mg/L Cu(II)	83 g/L/h	2.9	26 W/m³	55%
ter Heijne et al. [144]	Graphite foil	Bipolar	Anaerobic: 7d Aerobic: 6d	20 mM acetate	1 g/L Cu(II)	>99.9%	3	Anaerobic: 0.43 W/m² Aerobic: 0.80 W/m²	
Tao et al. [145]	Graphite felt	PEM	264	28 mM glucose	200 mg/L Cu(II)	96%	4.7	0.11 W/m²	

(continued)

Table 11.1 (continued)

Authors	Cathode setup	Membrane type	Cathode HRT (h)	Anolyte/buffer	Catholyte	Maximum % contaminant remediation and reduction rate	pH of cathode	Current density/power density at maximum removal	Columbic efficiency
Wu et al. [146]	Carbon cloth with Pt catalyst	*N/A	28	16 mM acetate	12.5 mg/L Cu(II)	98.3%	6	10.2 W/m^3	30%
Modin et al. [147]	Titanium wire	AEM	2	20 mM acetate	0.8 g/L Cu(II), 0.4 g/L Pb(II), 0.8 g/L Cd(II), and 0.3 g/L Zn(II)	84.3%	< 2	0.1–7.0 A/m^2 and no applied	77%
Zhang et al. [138]	Carbon rod	CEM	4	12 mM acetate	50 mg/L Cu(II)	60%	2.0	0.55 A/m^2 and 3.2 W/m^2	54%
Ai et al. [148]	Carbon cloth	*N/A	24	20 mM acetate				–	–
Cadmium (II)									
Modin et al. [143]	Titanium wire	AEM	20–24	20 mM acetate	0.8 g/L Cu(II), 0.4 g/L Pb(II), 0.8 g/L Cd(II), and 0.3 g/L Zn(II)	62%	< 2	0.1–6.0 A/m^2 and 0.51 V applied	5.3%
Choi et al. [149]	Carbon cloth	AEM	60	12.2 mM acetate	50–200 mg/L Cd(II)	90–94%	6	–	
Abourached et al. [150]	Carbon cloth	*N/A	30	60 mM acetate	200 µM Cd(II) and 400 µM Zn(II)	90%	–	3.6 W/m^2	34%

Colantonio et al. [151]	Stainless steel mesh	*N/A	48	8.3 mM acetate	10 mg/L Cd(II)	95%		0.42–0.45 mA and 0.6 V applied	34–37%
Ai et al. [148]	Carbon cloth	*N/A	24	20 mM acetate	132 mg/L Cd(II)	>99.9%	7.1	--	--
Zhang et al. [138]	Titanium sheet	CEM	4	12 mM acetate	50 mg/L Cr(VI), 50 mg/L Cu(II), and 50 mg/L Cd(II)	29%	2.0	0.4 A/m^2 and N/A**	28%
Silver									
Choi and Cui [152]	Carbon cloth	AEM	8	12 mM acetate	200 mg/L Ag(I)	>99%	2.0	3.8 A/m^2 and 2.6 W/m^2	
Tao et al. [153]	Graphite plate	PEM	5.5	21 mM acetate	1 mM AgNO$_3$	>89%	2.0	--	
Tao et al. [148]	Graphite plate	PEM	12	21 mM acetate	1 mM [AgS$_2$O$_3$]$^-$	84%	6.3	--	
Wang et al. [154]	Graphite	Bipolar	21	27 mM acetate	1.6 g/L Ag$^+$	>99%	9.2	--	
Lim et al. [155]	Graphite felt	AEM		10 mM acetate	4,000 ppm Ag$^+$	92%	7	4.25 A/m^2 and 1.9 W/m^2	
Kamperidis et al. [156]	Graphite cloth with Pt	PEM	24	8.3 mM glucose	50 mg/L	>93%	2 and 7	248 W/m^2	
Gold									
Choi and Hu [157]	Carbon cloth	CEM	12	12 mM acetate	200 mg/L Au(III)	99.8%	2.0	2.1 A/m^2 and 1.9 W/m^2	80%

(continued)

Table 11.1 (continued)

Authors	Cathode setup	Membrane type	Cathode HRT (h)	Anolyte/ buffer	Catholyte	Maximum % contaminant remediation and reduction rate	pH of cathode	Current density/ power density at maximum removal	Columbic efficiency
Lead (II)									
Modin et al. [143]	Titanium wire	AEM	20–24	20 mM acetate	0.8 g/L Cu(II), 0.4 g/L Pb(II), 0.8 g/L Cd(II), and 0.3 g/L Zn(II)	48%	<2	0.1–5.1 A/m^2 and 0.34 V applied	5.3%
Selenite									
Catal et al. [158]	Carbon cloth with Pt	*N/A	72	1 mM glucose	200 mg/L Se^{2-}	99%	--	1,000 W/m^2	38%
Zinc (II)									
Modin et al. [143]	Titanium wire	AEM	20–24	20 mM acetate	0.8 g/L Cu(II), 0.4 g/L Pb(II), 0.8 g/L Cd(II), and 0.3 g/L Zn(II)	44%	<2	16.8 24.7 A/m^2 and 1.5 V applied	1.2%
Abourached et al. [150]	Carbon cloth	*N/A	50	60 mM acetate	200 µM Cd(II) and 400 µM Zn(II)	97%	--	--	34%
Mercury									
Wang et al. [159]	Carbon paper	AEM	8	17 mM acetate	100 mg/L Hg(II)	99.5%	2	1.44 A/m^2 and 0.4 W/m^2	4%

*Single-chamber MXC.

**Powered by MFCs with no direct energy production.

rod, achieving 100% remediation of 100 mg/L of $Cr_2O_7^{2-}$ at pH 2 in 150 h. Li et al. [134] coupled Cr(VI) reduction at the cathode with light irritation to achieve 97% Cr(VI) reduction. Li et al. [135] achieved 100% conversion of 80 mg/L $Cr_2O_7^{2-}$ to Cr(III) while producing 700 mW/m². In the same study, Li et al. [135] performed SEM on the carbon cloth cathode and showed that the chromium was precipitating on the cathode.

Other researchers have explored catalysts attached to carbon electrodes to improve Cr(VI) reduction speeds and energy production. Alumina-nickel nanoparticles were dispersed on a carbon nanofiber cathode, increasing the Cr (VI) reduction rate to 2.13 $g/m^3/h$ [137]. Deposition of the nanoparticles improved the general electroconductivity of the cathode, improving power density to 1,530 mW/m². However, the system was only about to remove 85% of the 200 ppm present as Cr (VI) . Liu et al. [136] achieved 100% remediation of 50 mg/L $Cr_2O_7^{2-}$ in less than 32 h using a carbon cloth coated with a polypyrrole-MnO_2 catalyst. Results were repeatable over more than six cycles, with intermediate removal of Cr(III) from the cathode using a 100 mM NaOH wash. This is the first study to indicate that the Cr(III) could easily be recovered from the cathode surface and the cathode reused.

11.8.1.6.2 Cobalt (II)

Recovery of Co(II) has become an increasingly important sustainability process due to its presence in Li-ion batteries in the form of $LiCoO_2$. The $LiCoO_2$ present in batteries must first be isolated from the cathode materials and then the Co(III) leached into aqueous solution for recovery using MECs. Co(III) can be reduced to Co(II)

$$LiCoO_2 + 4H + e^- \rightarrow Co^{2+}(aq) + Li(aq) + 2H_2O \quad (E^{o\prime} = +1.61\ V_{SHE})$$

and Co (II) reduced to Co(0) through

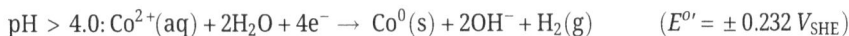

$$pH < 4.0: Co^{2+}(aq) + 2e^- \rightarrow Co^0(s) \qquad\qquad (E^{o\prime} = \pm 0.232\ V_{SHE})$$

$$pH > 4.0: Co^{2+}(aq) + 2H_2O + 4e^- \rightarrow Co^0(s) + 2OH^- + H_2(g) \qquad (E^{o\prime} = \pm 0.232\ V_{SHE})$$

Since the energy produced during Co(III) is more than sufficient to power Co(II) reduction to Co(0), Huang et al. [139] developed a combined MFC–MEC system that first releases Co (II) from $LiCoO_2$ at an MFC cathode and then reduces Co(II) at an MEC cathode. The systems were relatively small, ranging up to 25 mL in anode and cathode chamber volumes. In the MFC, cobalt was leached from 1 g/L of $LiCoO_2$ loaded on the cathode at a rate of 46 mg/L/h. In the MEC, Co(II) was reduced to Co(0) at 7 mg/L/h. More importantly, the entire system was run without external energy input. Jiang et al. [140] recovered Co(0) while producing H_2 gas using an MEC. Peak Co(II) recovery of 92% occurred at 0.5 V applied voltage and 13.2 A/m², after which H_2 production increased with the increased power being added to the system. SEM and XRD confirmed the presence of crystalline Co (0) structures on the cathode surface. Huang et al. [141] increased surface area at the cathode by utilizing GAC touching the cathode surface, Co(0) deposition to the GAC surface.

Co(0) was effectively removed in that form from the cathode and GAC surfaces using a stripping solution. Saad et al. [160] utilized multiple-stacked stainless steel screens as the cathode to increase surface area and conductivity. With an applied voltage of 1.8 V, the system remediated >99% of 50 ppm Co(II) . The porous nature of the screen improved mass transport by reducing the diffusion layer between the bacteria and bulk liquid.

Researchers have used biocathode systems to make Co(II) remediation energy positive [143, 161]. Huang et al. [139] recovered 4.6 mg Co(II) /L/h while producing 1.5 W/m^3 of power. Shen et al. [143] performed concomitant Co(II) and Cu(II) recovery 5.3 mg Co(II) /L/h using a biocathode. For this important contaminant, these biobased systems provide a promising step to address environmental issues.

11.8.1.6.3 Copper (II)

Copper is a metal that is essential for cell metabolism but is a broad concern, with long-term exposure causing toxicity and organ failure. Copper waste sites are associated with mining, metals and electronic manufacturing, and some pesticides. MFC remediation of Cu(II) is an emerging recovery technique due to copper's positive redox potential making it an excellent candidate for recovery on the cathode:

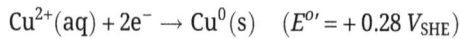

$$Cu^{2+}(aq) + 2e^- \rightarrow Cu^0(s) \quad (E^{o\prime} = +0.28\ V_{SHE})$$

There have been limited uses of MFCs for Cu recovery, but all researchers reported producing pure Cu(0) on the cathode surface [143–145]. All researchers reported Cu(II) was reduced to pure Cu(0), not CuO or CuO$_2$, and was deposited on the cathode surface.

ter Heijne et al. [144] performed one of the first studies on Cu(II) remediation in a cathode chamber, finding little difference in results under aerobic and anaerobic conditions. More than 99.9% removal was achieved from 1 g/L Cu(II) initial concentration while producing 0.43 and 0.8 W/m^2 under anaerobic and aerobic conditions, respectively.

Lower power densities were as effective at removing Cu(II): Tao et al. [145] was able to achieve >96% Cu(II) recovery as Cu(0) with 0.11 W/m^2 power density. Shen et al. [143] used an MFC to produce electricity for simultaneous Cu(II) and Co(II) on a MEC biocathode consisting primarily of *Proteobacteria* and *Firmicutes*. There were increased removal rates of Cu(II) (83 mg/L/h) and Co(II) (6.4 mg/L/h) with increased influent Cu(II) concentration up to 1,000 mg Cu(II)/L.

Single-chamber MFCs have also been used to recover Cu(II) [146, 148]. The MFC reduced low concentrations <12.5 mg/L Cu(II) with >98% efficacy [146]. Consistent with the microbial community data from Shen et al. [143], Cu was consumed by a *Proteobacteria* and *Firmicutes* biocathode and reduced to Cu$_2$O or elemental copper. Another study performed simultaneous removal of Cu(II) and Cd(II) in simulated acid mine drainage, achieving >99.9% recovery of Cu mostly as Cu(0) with trace amounts of Cu$_{gl}$O [148]. The advantage of a single chamber system is the absence of a membrane, which is more costly and increases Ohmic resistance in the system.

11.8.1.6.4 Cadmium (II)

Cadmium (II) is a carcinogenic and toxic metal highly regulated in waterways. Cd(II) is generally used in electroplating, pigments, and battery manufacturing. Traditional activated sludge processes are ineffective at removing it from wastewater. Cd(II) reduction to Cd(0) has a low redox potential:

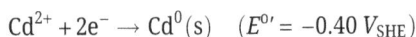

$$Cd^{2+} + 2e^- \rightarrow Cd^0(s) \quad (E^{0\prime} = -0.40\ V_{SHE})$$

Because of the low redox potential, Cd(II) is often recovered at a bioanode of a MEC using Cd(II) as a substrate for biomass growth like acetate or at the cathode with power input from an outside source. Reduction to solid Cd is dominant at a low pH ($\leq$2). At higher pH, Cd can be precipitated as cadmium hydroxide (Cd $(OH)_2$) or cadmium carbonate ($CdCO_3$).

Cd(II) recovery research has focused on both single- and dual-chamber MECs, with many researchers coupling MEC Cu(II) recovery at a bioanode with an MFC to power the system. Choi et al. [149] used two dual-chamber MXCs to compensate for the energy required to remediate Cd(II): they used a Cr(VI) -reducing MFC to power cathodic Cd(II) removal from a separate MEC. Choi et al. [149] obtained 90–94% Cd(II) recovery as Cd(0) while entirely powered by the Cr(VI) MFC. Zhang et al. [138] used Cr (IV) and Cu(II)-recovery MFCs in series and parallel to power a Cd(II)-recovery MEC. Cd(II) recovery was highest (29%) when powered in parallel connection by both MFCs. Colantonio and Kim [151] used a single-chamber MEC to achieve 95% Cd (II) removal in 48 h when applying 0.6 V potential. Another single-chamber MXC used a novel anodophilic bacteria, *Pseudomonas* sp. E8, to achieve >99.9% cathodic removal of Cd(II) and Cu(II) in simulated acid mine drainage [148]. SEM, XRD, and EDXS confirmed that the Cd(II) was reduced to Cd(0) at the cathode surface.

Other researchers explored concomitant removal of Cd with other metal contaminants. Modin et al. [147] performed cathodic Cd recovery from fly ash contaminated with copper, lead, cadmium, and zinc in a dual-chamber MEC. The applied potential of the system was varied to remove each contaminant independently. For Cd(II), 62% of Cd was recovered as Cd(0) with 0.51 V of applied potential. Abourached et al. [150] used a single-chamber MEC to recover Cd(II). The MEC obtained 90% removal at 200 μM Cd(II) but exhibited signs of decreased voltage likely due to microorganisms toxicity at 300 μM Cd(II), indicating the potential need for a membrane separator to prevent inhibition of anode microorganisms.

11.8.1.6.5 Silver (I) and gold (III)

Precious metals like silver and gold are used in a variety of applications including jewelry, acid mine drainage, electronics manufacturing, and dental applications. Silver and gold (III) are also common in wastewater streams [162]. Both metals have highly favorable redox potentials for capture at a cathode. Gold is generally present in solution as $AuCl_4^-$ and can be converted to solid gold via:

$$AuCl_4^- + 3e^- \rightarrow Au(s) + 4Cl^- \quad (E^{o\prime} = +1.00\ V_{SHE})$$

Silver can be present as multiple soluble compounds including Ag^+, $Ag\ (NH_3)^{2+}$, and $[AgS_2O_3]^-$. While these compounds have slightly less favorable redox potentials versus gold, they are generally more omnipresent, providing more opportunities for capture.

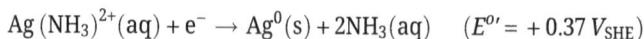

$$Ag^+(aq) + e^- \rightarrow Ag^0(s) \qquad\qquad (E^{o\prime} = +0.80\ V_{SHE})$$

$$Ag\ (NH_3)^{2+}(aq) + e^- \rightarrow Ag^0(s) + 2NH_3(aq) \quad (E^{o\prime} = +0.37\ V_{SHE})$$

$$[AgS_2O_3]^-\ (aq) + e^- \rightarrow Ag^0(s) + S_2O_3^{2-}(aq) \quad (E^{o\prime} = +0.25\ V_{SHE})$$

Limited research has been performed with gold, with 99.8% removal in 12 h with an initial solution concentration of 200 mg Au(III) /L and limited power production (1.9 W/m^2) [157]. SEM confirmed that Au(0) deposited on the cathode as clusters of crystals.

Conversely, significant research has pursued silver recovery using MFCs. Silver removal in MFCs has consistently exhibited high removal efficiencies of high initial concentration Ag(I) waste streams [152–156]. Initial research by Tao et al. [153] and Choi and Cui [152] demonstrated >95% removal of 1.6 g/L Ag $(NH_3)^{2+}$ and 200 mg/L AgNO$_3$, respectively, in less than a day through electrochemical deposition of Ag(0) crystals on the cathodes. Interestingly, removal of the Ag ions resulted in the cathode chamber pH increasing from 2.0 to about 7.0, making the system more sustainable for general water treatment [152, 153].

Other researchers have demonstrated >99% removal of Ag in slightly alkaline cathodic conditions. Yun-Hai et al. [163] achieved >99% removal of Ag at high initial concentrations (1.6 g Ag $(NH_3)_2^+$/L).

11.8.1.6.6 Other metals

As summarized in Table 1, limited research has been performed in the removal of other metals, including several divalent metals including zinc (II) [147], lead (II) [147], and mercury (II) [159] as well as selenite [158] with relatively positive results. Since lead and zinc have negative redox potentials, they require an MEC configuration for remediation. Conversely, the positive redox potentials of mercury and selenite means their remediation has the potential of being energy-generating. These metals were usually investigated as part of an amalgam of heavy metals, making each specific metal difficult to remove.

11.9 Challenges to MXC implementation

MXCs provide several opportunities for valuable product recovery with power generation, including electricity production, H$_2$O$_2$ production, and heavy metals recovery. Regardless of the product produced, implementation of valuable product recovery in

MXCs has faced many challenges. Scaling up MXCs for practical application in wastewater treatment and heavy metal recovery requires building robust systems at large capacities. To obtain power or produce H_2 at levels sufficient to power a water resource recovery facility, multiple MXCs will be required to be operated either in parallel or series depending upon the desired outcome [164]. Several groups have explored stacking systems to increase overall voltage and current output but have produced energy far below expectations [165–167]. Stacking systems will require complex control systems to prevent voltage reversal to less active cells [168].

Materials provide a particular challenge for MXC implementation. The separator membrane in a dual-chambered MXC must be able to withstand different head pressures, pHs, and chemicals which can deteriorate the membrane's integrity [85, 86]. Aerobic cathodes can require expensive catalysts like Pt to effectively produce electricity [61, 62] and a gas diffusion layer that is hydrophobic to retain the catholyte while allowing for sufficient transfer of O_2 to the cathode chamber. For metals recovery, the membrane and cathode require materials resistant to the acids often required to recover the metals.

In addition, H_2O_2, NH_4^+, and heavy metals recovery research have not discerned if the bacteria on anodes or biocathodes will be affected or inhibited by transfer of ions across a membrane. For H_2O_2 produced at pH > 11.35, the chemical deprotonates to HO_2^- which can transport from the cathode to the anode chamber across a membrane and potentially destroy the anode biofilm. Ki et al. [169] determined that fungi grow at the anode and consume destructive HO_2^- before it can contact the anode biofilm. It is unclear if the transport of chemicals across the membrane can cause antimicrobial-resistant gene expression at the anode chamber. Work performed by Chen et al. [170] found that *Geobacter* can biodegrade oxytetracycline without gene expression.

However, possibly the most inhibitory part of MXC manufacturing is cost. One of the most popular AEM and CEM manufacturers charges $200/m^2 of membrane [171]. Ge and He [172] reported that over 60% of MXC costs were associated with the membrane. Precious metal catalysts for energy generation are extremely expensive.

Overall, biocathodes are seen as one of the "low-hanging fruit" in the field. It is an area ripe for continued development. Bioprospecting in this is abundant and remains well open to new and important developments. Also, biocathodes have some of the easier challenges in these fields. The focus areas for the MXC, MEC, and similar acronym systems certainly have shifted in the decades since the early identification of *Geobacter* in sediments [46]. As the focus areas continue to evolve, signs point to greater focus on specific bioproduct and bioremediation applications.

11.10 Summary

Microbial electrochemical systems demonstrate incredible versatility. From unique niche applications to product generation to materials recovery, these systems have a short but rich research history and promising future. These systems can perform value-added services from recovering rapidly depleting materials from waste streams to electricity generation to bioremediation. The future of both research areas and practical applications for microbial electrochemical systems and microbial electrochemical technologies is bright. As the climate crisis continues to escalate, advancements are needed across multiple sectors. The bioproducts produced from these microbial systems may not power the world, but they will have important, niche applications. The demand for harnessing microbial abilities to either recover or create valuable products seems to be a promising direction for a more sustainable future.

References

[1]   Daims H, Taylor MW, Wagner M. Wastewater treatment: a model system for microbial ecology. Trends Biotechnol 2006; 24(11):483–89.

[2]   Roane TM, Pepper IL, Miller RM. Microbial remediation of metals. Biotechnol Res Ser 1996; 6:312–40.

[3]   Kobayashi H, Rittmann BE. Microbial removal of hazardous organic compounds. Environ Sci Technol 1982; 16(3):170A–183A.

[4]   Gericke M, Pinches A. Microbial production of gold nanoparticles. Gold Bull 2006; 39(1):22–28.

[5]   Ladygina N, Dedyukhina EG, Vainshtein MB. A review on microbial synthesis of hydrocarbons. Process Biochem 2006; 41(5):1001–14.

[6]   Rittmann BE, Mayer B, Westerhoff P, Edwards M. Capturing the lost phosphorus. Chemosphere Internet 2011; 84(6):846–53. Available from: http://dx.doi.org/10.1016/j.chemosphere.2011.02.001

[7]   Davidson DJ, Kecinski M. Emotional pathways to climate change responses. Wiley Interdiscip Rev: Clim Change 2022; 13(2):1–19.

[8]   Sovacool BK. The cultural barriers to renewable energy and energy efficiency in the United States. Technol Soc Internet 2009; 31(4):365–73. Available from: http://dx.doi.org/10.1016/j.techsoc.2009.10.009

[9]   Susskind L, Chun J, Gant A, Hodgkins C, Cohen J, Lohmar S. Sources of opposition to renewable energy projects in the United States. Energy Policy Internet 2022; 165 February:112922. Available from: https://doi.org/10.1016/j.enpol.2022.112922

[10]   Yoho R Energy and the environment: Electrochemistry of electron transport pathways in anode-respiring bacteria and energy technology and climate change in science textbooks. Arizona State University; 2016.

[11]   Rabaey K, Verstraete W. Microbial fuel cells: Novel biotechnology for energy generation. Trends Biotechnol 2005; 23(6):291–98.

[12]   Munoz-Cupa C, Hu Y, Xu C, Bassi A. An overview of microbial fuel cell usage in wastewater treatment, resource recovery and energy production. Sci Total Environ [Internet] 2021; 754:142429. Available from: https://doi.org/10.1016/j.scitotenv.2020.142429

[13]   Ewing T, Ha PT, Babauta JT, Tang NT, Heo D, Beyenal H. Scale-up of sediment microbial fuel cells. J Power Sources [Internet] 2014; 272:311–19. Available from: http://dx.doi.org/10.1016/j.jpowsour. 2014.08.070

[14]   Vijay A, Khandelwal A, Chhabra M, Vincent T. Microbial fuel cell for simultaneous removal of uranium (VI) and nitrate. Chem Eng J [Internet] 2020; 388:January 124157. Available from: https://doi.org/10.1016/j.cej.2020.124157

[15]   Torres CI. On the importance of identifying, characterizing, and predicting fundamental phenomena towards microbial electrochemistry applications. Curr Opin Biotechnol 2014; 27:107–14.

[16]   Yoho R, Popat S, Fabregat-Santiago F, Giménez S, Heijne A, Torres C. Electrochemical Impedance Spectroscopy as a Powerful Analytical Tool for the Study of Microbial Electrochemical Cells. In: Biofilms in Bioelectrochemical Systems. 2015. p. 249–80.

[17]   Hasany M, Mardanpour MM, Yaghmaei S. Biocatalysts in microbial electrolysis cells: A review. Int J Hydrogen Energy [Internet] 2016; 41(3):1477–93. Available from: http://dx.doi.org/10.1016/j.ijhy dene.2015.10.097

[18]   Cao X, Huang X, Liang P, Xiao K, Zhou Y, Zhang X, et al. A new method for water desalination using microbial desalination cells. Environ Sci Technol 2009; 43(18):7148–52.

[19]   Deng H, Chen Z, Zhao F. Energy from plants and microorganisms: Progress in plant-microbial fuel cells. Chem Sus Chem 2012; 5(6):1006–11.

[20]   Aulenta F, Reale P, Canosa A, Rossetti S, Panero S, Majone M. Characterization of an electro-active biocathode capable of dechlorinating trichloroethene and cis-dichloroethene to ethene. Biosens Bioelectron [Internet] 2010; 25(7):1796–802. Available from: http://dx.doi.org/10.1016/j.bios.2009.12.033

[21]   Tandukar M, Pavlostathis SG. Co-digestion of municipal sludge and external organic wastes for enhanced biogas production under realistic plant constraints. Water Res [Internet] 2015; 87:432–45. Available from: http://dx.doi.org/10.1016/j.watres.2015.04.031

[22]   Nevin KP, Lovley DR. Lack of production of electron-shuttling compounds or solubilization of Fe(III) during reduction of insoluble Fe(III) oxide by geobacter metallireducens. Appl Environ Microbiol 2000; 66(5):2248–51.

[23]   Rabaey K, Rozendal RA. Microbial electrosynthesis – revisiting the electrical route for microbial production. Nat Rev Microbiol 2010; 8(10):706–16.

[24]   Dennis PG, Harnisch F, Yeoh K, Tyson GW. Dynamics of cathode-associated microbial communities and metabolite profiles in a glycerol-fed bioelectrochemical system. Appl Environ Microbiol 2013; 79 (13):4008–14.

[25]   Ortiz-Medina JF, Call DF. Electrochemical and microbiological characterization of bioanode communities exhibiting variable levels of startup activity. Front Energy Res 2019; 7(September):1–11.

[26]   Pham TH, Aelterman P, Verstraete W. Bioanode performance in bioelectrochemical systems: recent improvements and prospects. Trends Biotechnol 2009; 27(3):168–78.

[27]   Deeke A, THJA S, Hamelers HVM, Buisman CJN. Capacitive bioanodes enable renewable energy storage in microbial fuel cells. Environ Sci Technol 2012; 46(6):3554–60.

[28]   Hutchinson AJ, Tokash JC, Logan BE. Analysis of carbon fiber brush loading in anodes on startup and performance of microbial fuel cells. J Power Sources 2011; 196(22):9213–19.

[29]   Paitier A, Godain A, Lyon D, Haddour N, Vogel TM, Monier JM. Microbial fuel cell anodic microbial population dynamics during MFC start-up. Biosens Bioelectron [Internet] 2017; 92 October 2016:357–63. Available from: http://dx.doi.org/10.1016/j.bios.2016.10.096

[30]   Bond DR, Lovley DR. Electricity production by *Geobacter sulfurreducens* attached to electrodes. Appl Environ Microbiol 2003; 69(3):1548–55.

[31]   Emde R, Swain A, Schink B. Anaerobic oxidation of glycerol by *Escherichia coli* in an amperometric poised-potential culture system. Appl Microbiol Biotechnol 1989; 32(2):170–75.

[32] Emde R, Schink B. Enhanced propionate formation by propionibacterium freudenreichii subsp. freudenreichii in a three-electrode amperometric culture system. Appl Environ Microbiol 1990; 56(9):2771–76.

[33] Kim N, Choi Y, Jung S, Kim S. Effect of initial carbon sources on the performance of microbial fuel cells containing proteus vulgaris. Biotechnol Bioeng 2000; 70(1):109–14.

[34] Park DH, Zeikus JG. Electricity generation in microbial fuel cells using neutral red as an electronophore. Appl Environ Microbiol 2000; 66(4):1292–97.

[35] Choi Y, Jooyoung S, Seunho J, Sunghyun K. Optimization of the performance of microbial fuel cells containing alkalophilic bacillus sp. J Microbiolo Biotechnol 2001; 11(5):863–69.

[36] Pinto RP, Srinivasan B, Manuel MF, Tartakovsky B.A two-population bio-electrochemical model of a microbial fuel cell. Bioresour Technol [Internet] 2010; 101(14):5256–65. Available from: http://dx.doi.org/10.1016/j.biortech.2010.01.122

[37] Shi L, Richardson DJ, Wang Z, Kerisit SN, Rosso KM, Zachara JM, et al. The roles of outer membrane cytochromes of *Shewanella* and *Geobacter* in extracellular electron transfer. Environ Microbiol Rep 2009; 1(4):220–27.

[38] Malvankar NS, Lovley DR. Microbial nanowires for bioenergy applications. Curr Opin Biotechnol 2014; 27:88–95.

[39] Yoho RA, Popat SC, Torres CI. Dynamic potential-dependent electron transport pathway shifts in anode biofilms of *Geobacter sulfurreducens*. Chem Sus Chem 2014; 7(12):3413–19.

[40] Yoho RA, Popat SC, Rago L, Guisasola A, Torres CI. Anode biofilms of *Geoalkalibacter ferrihydriticus* exhibit electrochemical signatures of multiple electron transport pathways. Langmuir 2015; 31(45):12552–59.

[41] Levar CE, Chan CH, Mehta-Kolte MG, Bond DR. An inner membrane cytochrome required only for reduction of high redox potential extracellular electron acceptors. Mbio 2014; 5(6).

[42] Jimenez Otero et al 2018–Identification of different putative outer membrane electron conduits necessary for Fe (III) citrate, Fe (III) oxide, Mn (IV) oxide, or electrode reduction by *Geobacter sulfurreducens*.pdf.

[43] Marsili E, Sun J, Bond DR. Voltammetry and growth physiology of *Geobacter sulfurreducens* biofilms as a function of growth stage and imposed electrode potential. Electroanalysis 2010; 22(7–8):865–74.

[44] ter Heijne A, Schaetzle O, Gimenez S, Navarro L, Hamelers B, Fabregat-Santiago F. Analysis of bio-anode performance through electrochemical impedance spectroscopy. Bioelectrochemistry [Internet] 2015; 106:64–72. Available from: http://dx.doi.org/10.1016/j.bioelechem.2015.04.002

[45] Reguera G, Kashefi K. The electrifying physiology of geobacter bacteria, 30 years on. Adv Microb Physiol 2019; 74:1–96.

[46] Lovley DR, Phillips EJP. Rapid assay for microbially reducible ferric iron in aquatic sediments. Appl Environ Microbiol 1987; 53(7):1536–40.

[47] Tabares M, Dulay H, Reguera G. *Geobacter sulfurreducens*. Trends Microbiol [Internet] 2020; 28(4):327–28. Available from: https://doi.org/10.1016/j.tim.2019.11.004

[48] Janda JM, Abbott SL. The genus *Shewanella*: From the briny depths below to human pathogen. Crit Rev Microbiol 2014; 40(4):293–312.

[49] Hau HH, Gralnick JA. Ecology and biotechnology of the genus *Shewanella*. Annu Rev Microbiol 2007; 61:237–58.

[50] Breuer M, Rosso KM, Blumberger J, Butt JN. Multi-haem cytochromes in *Shewanella* oneidensis MR-1: Structures, functions and opportunities. J R Soc Interface 2015; 12(102).

[51] Shi L, Rosso KM, Clarke TA, Richardson DJ, Zachara JM, Fredrickson JK. Molecular underpinnings of Fe(III) oxide reduction by *Shewanella oneidensis* MR-1. Front Microbiol 2012; 3(Feb):1–10.

[52] Myers CR, Nealson KH. Respiration-linked proton translocation couples to anaerobic reduction of manganese(iv) and iron(iii) in *Shewanella putrefaciens* MR-1. J Bacteriol 1990; 172(11):6232–38.

[53] Marsili E, Baron DB, Shikhare ID, Coursolle D, Gralnick JA, Bond DR. *Shewanella* secretes flavins that mediate extracellular electron transfer. Proc Natl Acad Sci USA 2008; 105(10):3968–73.

[54] Badalamenti JP, Krajmalnik-Brown R, Torres CI. Generation of high current densities by pure cultures of anode-respiring Geoalkalibacter spp. under alkaline and saline conditions in microbial electrochemical cells. Mbio 2013; 4(3).

[55] Zavarzina DG, Kolganova TV, Boulygina ES, Kostrikina NA, Tourova TP, Zavarzin GA. *Geoalkalibacter ferrihydriticus* gen. nov. sp. nov., the first alkaliphilic representative of the family geobacteraceae, isolated from a soda lake. Microbiol (N Y) 2006; 75(6):673–82.

[56] Greene A, Patel B, Yacob S. Geoalkalibacter subterraneus sp. nov., an anaerobic fe (iii)-and mn (iv)-reducing bacterium from a petroleum reservoir, and emended descriptions of the family desulfuromonadaceae and the genus geoalkalibacter. Int J Syst Evol Microbiol 2009; 59(4):781–85.

[57] Parameswaran P, Bry T, Popat SC, Lusk BG, Rittmann BE, Torres CI. Kinetic, electrochemical, and microscopic characterization of the thermophilic, anode-respiring bacterium thermincola ferriacetica. Environ Sci Technol 2013; 47(9):4934–40.

[58] Marshall CW, May HD. Electrochemical evidence of direct electrode reduction by a thermophilic Gram-positive bacterium, thermincola ferriacetica. Energy Environ Sci 2009; 2(6):699–705.

[59] Zavarzina DG, Sokolova TG, Tourova TP, Chernyh NA, Kostrikina NA, Bonch-Osmolovskaya EA. Thermincola ferriacetica sp. nov., a new anaerobic, thermophilic, facultatively chemolithoautotrophic bacterium capable of dissimilatory Fe(iii) reduction. Extremophiles 2007; 11(1):1–7.

[60] Huang L, Regan JM, Quan X. Electron transfer mechanisms, new applications, and performance of biocathode microbial fuel cells. Bioresour Technol 2011; 102(1):316–23.

[61] He Z, Angenent LT. Application of bacterial biocathodes in microbial fuel cells. Electroanalysis 2006; 18(19–20):2009–15.

[62] Jeremiasse AW, Hamelers HVM, Buisman CJN. Microbial electrolysis cell with a microbial biocathode. Bioelectrochemistry 2010; 78(1):39–43.

[63] Wingard LB, Shaw CH, Castner JF. Bioelectrochemical fuel cells. Enzyme Microb Technol 1982; 4(3):137–42.

[64] Miceli JF, Garcia-Peña I, Parameswaran P, Torres CI, Krajmalnik-Brown R. Combining microbial cultures for efficient production of electricity from butyrate in a microbial electrochemical cell. Bioresour Technol 2014; 169:169–74.

[65] Harnisch F, Schröder U. From MFC to MXC: Chemical and biological cathodes and their potential for microbial bioelectrochemical systems. Chem Soc Rev 2010; 39(11):4433–48.

[66] Azwar MY, Hussain MA, Abdul-Wahab AK. Development of biohydrogen production by photobiological, fermentation and electrochemical processes: a review. Renew Sustain Energy Rev [Internet] 2014; 31:158–73.Available from: http://dx.doi.org/10.1016/j.rser.2013.11.022.

[67] Montpart N, Rago L, Baeza JA, Guisasola A. Hydrogen production in single chamber microbial electrolysis cells with different complex substrates. Water Res 2015; 68:601–15.

[68] Campos-Martin JM, Blanco-Brieva G, Fierro JLG.Hydrogen peroxide synthesis: an outlook beyond the anthraquinone process. Angew Chem Int Ed [Internet] 2006; 45(42):6962–84. Available from: http://dx.doi.org/10.1002/anie.200503779.

[69] FMC Corporation. Vision. 2012.

[70] de la Cruz N, Giménez J, Esplugas S, Grandjean D, de Alencastro LF, Pulgarin C. Degradation of 32 emergent contaminants by UV and neutral photo-fenton in domestic wastewater effluent previously treated by activated sludge. Water Res 2012; 46(6):1947–57.

[71] Yang W, Zhou H, Cicek N. Treatment of organic micropollutants in water and wastewater by uv-based processes: A literature review. Crit Rev Environ Sci Technol 2014; 44(13):1443–76.

[72] Glaze WH, Kang JW, Chapin DH. The chemistry of water treatment processes involving ozone, hydrogen peroxide and ultraviolet radiation. Ozone Sci Eng [Internet] 1987; 9(4):335–52. Available from: http://dx.doi.org.ezproxy1.lib.asu.edu/10.1080/01919518708552148.

[73] Wagner M, Brumelis D, Gehr R. Disinfection of wastewater by hydrogen peroxide or peracetic acid: development of procedures for measurement of residual disinfectant and application to a physicochemically treated municipal effluent. Water Environ Res 2002; 33–50.

[74] Ksibi M. Chemical oxidation with hydrogen peroxide for domestic wastewater treatment. Chem Eng J [Internet] 2006 Autumn; 119(2–3):161–65. Available from: http://www.sciencedirect.com.ezproxy1.lib.asu.edu/science/article/pii/S1385894706001331.

[75] Kruithof JC, Kamp PC, Martijn BJ. UV/H$_2$O$_2$ treatment: a practical solution for organic contaminant control and primary disinfection. Ozone Sci Eng [Internet] 2007; 29(4):273–80. Available from: http://search.ebscohost.com/login.aspx?direct=true&db=eih&AN=26099550&site=ehost-live.

[76] Snyder S, Lei H, Wert E, Westerhoff P, Yoon Y. Removal of edcs and pharmaceuticals in drinking water. Water Environ Res Foundation 2008.

[77] Acero JL, von Gunten U. Influence of carbonate on the ozone/hydrogen peroxide based advanced oxidation process for drinking water treatment. Ozone Sci Eng [Internet] 2000; 22(3):305–28. Available from: http://dx.doi.org.ezproxy1.lib.asu.edu/10.1080/01919510008547213.

[78] Drogui P, Elmaleh S, Rumeau M, Bernard C, Rambaud A. Oxidising and disinfecting by hydrogen peroxide produced in a two-electrode cell. Water Res 2001; 35(13):3235–41.

[79] Charron I, Feliers C, Couvert A, Laplanche A, Patria L, Requieme B. Use of hydrogen peroxide in scrubbing towers for odor removal in wastewater treatment plants. Water Sci Technol 2004; 50(4):267–74.

[80] Metcalf & Eddy Inc. Wastewater Engineering: Treatment and Reuse. 5th edition. New York: McGraw-Hill Education; 2014.

[81] Rajala-Mustonen RL, Heinonen-Tanski H. Effect of advanced oxidation processes on inactivation of coliphages. Water Sci Technol 1995; 31(5):131–34.

[82] Rozendal RA, Leone E, Keller J, Rabaey K. Efficient hydrogen peroxide generation from organic matter in a bioelectrochemical system. Electrochem Commun 2009 Sep; 11(9):1752–55.

[83] Modin O, Fukushi K. Production of high concentrations of H$_2$O$_2$ in a bioelectrochemical reactor fed with real municipal wastewater. Environ Technol 2013; 34(19):2737–42.

[84] Griffin J, Taw E, Gosavi A, Thornburg NE, Pramanda I, Lee HS, et al. Hybrid approach for selective sulfoxidation via bioelectrochemically derived hydrogen peroxide over a niobium(v)-silica catalyst. ACS Sustain Chem Eng 2018; 6(6):7880–89.

[85] Young MN, Links MJ, Popat SC, Rittmann BE, Torres CI. Tailoring microbial electrochemical cells for production of hydrogen peroxide at high concentrations and efficiencies. ChemSusChem 2016; 9(23).

[86] Young MN, Chowdhury N, Garver E, Evans PJ, Popat SC, Rittmann BE, et al. Understanding the impact of operational conditions on performance of microbial peroxide producing cells. J Power Sources 2017; 356.

[87] Sim J, Reid R, Hussain A, An J, Lee HS. Hydrogen peroxide production in a pilot-scale microbial electrolysis cell. Biotechnol Rep [Internet] 2018; 19:e00276. Available from: https://doi.org/10.1016/j.btre.2018.e00276

[88] Kirova-Yordanova Z. Exergy analysis of industrial ammonia synthesis. Energy 2004; 2912-15 SPEC. ISS.:2373–84.

[89] Fowler D, Coyle M, Skiba U, Sutton MA, Cape JN, Reis S, et al. The global nitrogen cycle in the twenty-first century. Philos Trans R Soc B Biol Sci 2013; 368:1621.

[90] Randall DG, Naidoo V. Urine: the liquid gold of wastewater. J Environ Chem Eng 2018; 6(2):2627–35.

[91] Kelly PT, He Z. Nutrients removal and recovery in bioelectrochemical systems: A review. Bioresour Technol [Internet] 2014; 153:351–60. Available from: http://dx.doi.org/10.1016/j.biortech.2013.12.046

[92]   Clauwaert P, Rabaey K, Aelterman P, de Schamphelaire L, Pham TH, Boeckx P, et al. Biological denitrification in microbial fuel cells. Environ Sci Technol 2007; 41(9):3354–60.

[93]   You SJ, Ren NQ, Zhao QL, Kiely PD, Wang JY, Yang FL, et al. Improving phosphate buffer-free cathode performance of microbial fuel cell based on biological nitrification. Biosens Bioelectron 2009; 24(12):3698–701.

[94]   Virdis B, Rabaey K, Yuan Z, Keller J. Microbial fuel cells for simultaneous carbon and nitrogen removal. Water Res 2008; 42(12):3013–24.

[95]   Virdis B, Rabaey K, Rozendal RA, Yuan Z, Keller J. Simultaneous nitrification, denitrification and carbon removal in microbial fuel cells. Water Res 2010; 44(9):2970–80.

[96]   Villano M, Scardala S, Aulenta F, Majone M. Carbon and nitrogen removal and enhanced methane production in a microbial electrolysis cell. Bioresour Technol 2013; 130:366–71.

[97]   Snoeyink VL, Jenkins D. Water Chemistry. New York: John Wiley & Sons, Inc.; 1980.

[98]   Kuntke P, Geleji M, Bruning H, Zeeman G, Hamelers HVM, Buisman CJN. Effects of ammonium concentration and charge exchange on ammonium recovery from high strength wastewater using a microbial fuel cell. Bioresour Technol [Internet] 2011; 102(6):4376–82. Available from: http://dx. doi.org/10.1016/j.biortech.2010.12.085

[99]   Qin M, He Z. Self-supplied ammonium bicarbonate draw solute for achieving wastewater treatment and recovery in a microbial electrolysis cell-forward osmosis-coupled system. Environ Sci Technol Lett 2014; 1(10):437–41.

[100]   Kuntke P, Śmiech KM, Bruning H, Zeeman G, Saakes M, Sleutels THJA, et al. Ammonium recovery and energy production from urine by a microbial fuel cell. Water Res 2012; 46(8):2627–36.

[101]   Kuntke P, Zamora P, Saakes M, Buisman CJN, Hamelers HVM. Gas-permeable hydrophobic tubular membranes for ammonia recovery in bio-electrochemical systems. Environ Sci: Water Res Technol 2016; 2(2):261–65.

[102]   Kuntke P, THJA S, Saakes M, Buisman CJN. Hydrogen production and ammonium recovery from urine by a microbial electrolysis cell. Int J Hydrogen Energy [Internet] 2014; 39(10):4771–78. Available from: http://dx.doi.org/10.1016/j.ijhydene.2013.10.089

[103]   Zamora P, Georgieva T, ter Heijne A, Sleutels THJA, Jeremiasse AW, Saakes M, et al. Ammonia recovery from urine in a scaled-up microbial electrolysis cell. J Power Sources [Internet] 2017; 356:491–99. Available from: http://dx.doi.org/10.1016/j.jpowsour.2017.02.089

[104]   Zhang Y, Angelidaki I. Submersible microbial desalination cell for simultaneous ammonia recovery and electricity production from anaerobic reactors containing high levels of ammonia. Bioresour Technol 2015; 177:233–39.

[105]   Zhang Y, Angelidaki I. Counteracting ammonia inhibition during anaerobic digestion by recovery using submersible microbial desalination cell. Biotechnol Bioeng 2015; 112(7):1478–82.

[106]   Sotres A, Cerrillo M, Viñas M, Bonmatí A. Nitrogen recovery from pig slurry in a two-chambered bioelectrochemical system. Bioresour Technol [Internet] 2015; 194:373–82. Available from: http://dx.doi.org/10.1016/j.biortech.2015.07.036

[107]   Qin M, Molitor H, Brazil B, Novak JT, He Z. Recovery of nitrogen and water from landfill leachate by a microbial electrolysis cell-forward osmosis system. Bioresour Technol 2016; 200:485–92.

[108]   Qin M, Hynes EA, Abu-Reesh IM, He Z. Ammonium removal from synthetic wastewater promoted by current generation and water flux in an osmotic microbial fuel cell. J Cleaner Prod [Internet] 2017; 149:856–62. Available from: http://dx.doi.org/10.1016/j.jclepro.2017.02.169

[109]   Zou S, Qin M, Moreau Y, He Z. Nutrient-energy-water recovery from synthetic sidestream centrate using a microbial electrolysis cell – forward osmosis hybrid system. J Cleaner Prod [Internet] 2017; 154:16–25. Available from: http://dx.doi.org/10.1016/j.jclepro.2017.03.199

[110]   Veluswamy GK, Shah K, Ball AS, Guwy AJ, Dinsdale RM. A techno-economic case for volatile fatty acid production for increased sustainability in the wastewater treatment industry. Environ Sci: Water Res Technol 2021; 7(5):927–41.

[111] Rittmann BE, McCarty PL. Environmental Biotechnology: Principles and Applications. 2nd edition. New York: McGraw-Hill; 2020.

[112] Dahiya S, Sarkar O, Swamy YV, Venkata Mohan S. Acidogenic fermentation of food waste for volatile fatty acid production with co-generation of biohydrogen. Bioresour Technol [Internet] 2015; 182:103–13. Available from: http://dx.doi.org/10.1016/j.biortech.2015.01.007

[113] Puyol D, Batstone DJ, Hülsen T, Astals S, Peces M, Krömer JO. Resource recovery from wastewater by biological technologies: opportunities, challenges, and prospects. Front Microbiol 2017; 7(Jan):1–23.

[114] Zhang Y, Angelidaki I. Bioelectrochemical recovery of waste-derived volatile fatty acids and production of hydrogen and alkali. Water Res [Internet] 2015; 81:188–95. Available from: http://dx.doi.org/10.1016/j.watres.2015.05.058

[115] Liu B, Kleinsteuber S, Centler F, Harms H, Sträuber H. Competition between butyrate fermenters and chain-elongating bacteria limits the efficiency of medium-chain carboxylate production. Front Microbiol 2020; 11(March):1–13.

[116] Jiang Y, May HD, Lu L, Liang P, Huang X, Ren ZJ. Carbon dioxide and organic waste valorization by microbial electrosynthesis and electro-fermentation. Water Res 2019; 149:42–55.

[117] Aktij SA, Zirehpour A, Mollahosseini A, Taherzadeh MJ, Tiraferri A, Rahimpour A. Feasibility of membrane processes for the recovery and purification of bio-based volatile fatty acids: a comprehensive review. J Ind Eng Chem [Internet] 2020; 81:24–40. Available from: https://doi.org/10.1016/j.jiec.2019.09.009

[118] May HD, Evans PJ, LaBelle EV. The bioelectrosynthesis of acetate. Curr Opin Biotechnol [Internet] 2016; 42:225–33. Available from: http://dx.doi.org/10.1016/j.copbio.2016.09.004

[119] Yu L, Yuan Y, Tang J, Zhou S. Thermophilic moorella thermoautotrophica-immobilized cathode enhanced microbial electrosynthesis of acetate and formate from CO_2. Bioelectrochemistry [Internet] 2017; 117:23–28. Available from: http://dx.doi.org/10.1016/j.bioelechem.2017.05.001

[120] Jourdin L, Grieger T, Monetti J, Flexer V, Freguia S, Lu Y, et al. High acetic acid production rate obtained by microbial electrosynthesis from carbon dioxide. Environ Sci Technol 2015; 49(22):13566–74.

[121] Jourdin L, Freguia S, Flexer V, Keller J. Bringing high-rate, Co_2-based microbial electrosynthesis closer to practical implementation through improved electrode design and operating conditions. Environ Sci Technol 2016; 50(4):1982–89.

[122] LaBelle EV, May HD. Energy efficiency and productivity enhancement of microbial electrosynthesis of acetate. Front Microbiol 2017; 8(May):1–9.

[123] Roy M, Yadav R, Chiranjeevi P, Patil SA. Direct utilization of industrial carbon dioxide with low impurities for acetate production via microbial electrosynthesis. Bioresour Technol [Internet] 2021; 320(PA):124289. Available from: https://doi.org/10.1016/j.biortech.2020.124289

[124] Jourdin et al 2019–enhanced selectivity to butyrate and caproate above acetate in continuous bioelectrochemical chain elongation from CO_2-steering with CO_2 loading rate and hydraulic retention time.pdf.

[125] Vassilev I, Hernandez PA, Batlle-Vilanova P, Freguia S, Krömer JO, Keller J, et al. Microbial electrosynthesis of isobutyric, butyric, caproic acids, and corresponding alcohols from carbon dioxide. ACS Sustain Chem Eng 2018; 6(7):8485–93.

[126] Jiang Y, Chu N, Qian DK, Jianxiong Zeng R. Microbial electrochemical stimulation of caproate production from ethanol and carbon dioxide. Bioresour Technol [Internet] 2020; 295 (September 2019):122266. Available from: https://doi.org/10.1016/j.biortech.2019.122266.

[127] Ganigué R, Puig S, Batlle-Vilanova P, Balaguer MD, Colprim J. Microbial electrosynthesis of butyrate from carbon dioxide. Chem Commun 2015; 51(15):3235–38.

[128] Raes SMT, Jourdin L, Buisman CJN, DPBTB S. Continuous long-term bioelectrochemical chain elongation to butyrate. Chemelectrochem 2017; 4(2):386–95.

[129] Batlle-Vilanova P, Ganigué R, Ramió-Pujol S, Bañeras L, Jiménez G, Hidalgo M, et al. Microbial electrosynthesis of butyrate from carbon dioxide: production and extraction. Bioelectrochemistry 2017; 117:57–64.

[130] Jourdin L, Raes SMT, Buisman CJN, DPBTB S. Critical biofilm growth throughout unmodified carbon felts allows continuous bioelectrochemical chain elongation from CO_2 up to caproate at high current density. Front Energy Res 2018; 6(Mar):1–15.

[131] Dessì P, Rovira-Alsina L, Sánchez C, Dinesh GK, Tong W, Chatterjee P, et al. Microbial electrosynthesis: towards sustainable biorefineries for production of green chemicals from CO2 emissions. Biotechnol Adv 2021; 46(December l):2020.

[132] Nancharaiah YV, Venkata Mohan S, Lens PNL. Metals removal and recovery in bioelectrochemical systems: a review. Bioresour Technol 2015; 195:102–14.

[133] Wang G, Huang L, Zhang Y. Cathodic reduction of hexavalent chromium [Cr(vi)] coupled with electricity generation in microbial fuel cells. Biotechnol Lett 2008; 30(11):1959–66.

[134] Li Y, Lu A, Ding H, Jin S, Yan Y, Wang C, et al. Cr(vi) reduction at rutile-catalyzed cathode in microbial fuel cells. Electrochem Commun [Internet] 2009; 11(7):1496–99. Available from: http://dx.doi.org/10.1016/j.elecom.2009.05.039.

[135] Li M, Zhou S, Xu Y, Liu Z, Ma F, Zhi L, et al. Simultaneous Cr(vi) reduction and bioelectricity generation in a dual chamber microbial fuel cell. Chem Eng J [Internet] 2018; 334 (November 2017):1621–29. Available from:: https://doi.org/10.1016/j.cej.2017.11.144.

[136] Liu X, Yin W, Liu X, Zhao X. Enhanced Cr reduction and bioelectricity production in microbial fuel cells using polypyrrole-coated MnO_2 on carbon cloth. Environ Chem Lett 2020; 18(2):517–25.

[137] Gupta S, Yadav A, Verma N. Simultaneous Cr(vi) reduction and bioelectricity generation using microbial fuel cell based on alumina-nickel nanoparticles-dispersed carbon nanofiber electrode. Chem Eng J [Internet] 2017; 307:729–38. Available from: http://dx.doi.org/10.1016/j.cej.2016.08.130

[138] Zhang Y, Yu L, Wu D, Huang L, Zhou P, Quan X, et al. Dependency of simultaneous Cr(vi), Cu(ii) and Cd(ii) reduction on the cathodes of microbial electrolysis cells self-driven by microbial fuel cells. J Power Sources 2015; 273:1103–13.

[139] Huang L, Yao B, Wu D, Quan X. Complete cobalt recovery from lithium cobalt oxide in self-driven microbial fuel cell – microbial electrolysis cell systems. J Power Sources Internet 2014; 259:54–64. Available from: http://dx.doi.org/10.1016/j.jpowsour.2014.02.061

[140] Jiang L, Huang L, Sun Y. Recovery of flakey cobalt from aqueous Co(ii) with simultaneous hydrogen production in microbial electrolysis cells. Int J Hydrogen Energy 2014; 39(2):654–63.

[141] Huang T, Song D, Liu L, Zhang S. Cobalt recovery from the stripping solution of spent lithium-ion battery by a three-dimensional microbial fuel cell. Sep Purif Technol [Internet] 2019; 215 (January):51–61. Available from: https://doi.org/10.1016/j.seppur.2019.01.002

[142] Saad DR, Alismaeel ZT, Abbar AH. Cobalt removal from simulated wastewaters using a novel flow-by fixed bed bio-electrochemical reactor. Chem Eng Process – Process Intensif [Internet] 2020; 156 (May):108097. Available from: https://doi.org/10.1016/j.cep.2020.108097

[143] Shen J, Sun Y, Huang L, Yang J. Microbial electrolysis cells with biocathodes and driven by microbial fuel cells for simultaneous enhanced Co(ii) and Cu(ii) removal. Front Environ Sci Eng 2015; 9(6):1084–95.

[144] Heijne AT, Liu F, R van der W, Weijma J, Buisman CJN, Hamelers HVM. Copper recovery combined with electricity production in a microbial fuel cell. Environ Sci Technol 2010; 44(11):4376–81.

[145] Tao HC, Liang M, Li W, Zhang LJ, Ni JR, Wu WM. Removal of copper from aqueous solution by electrodeposition in cathode chamber of microbial fuel cell. J Hazard Mater [Internet] 2011; 189 (1–2):186–92. Available from: http://dx.doi.org/10.1016/j.jhazmat.2011.02.018.

[146] Wu Y, Zhao X, Jin M, Li Y, Li S, Kong F, et al. Copper removal and microbial community analysis in single-chamber microbial fuel cell. Bioresour Technol [Internet] 2018; 253(November 2017):372–77. Available from: https://doi.org/10.1016/j.biortech.2018.01.046.

[147] Modin O, Wang X, Wu X, Rauch S, Fedje KK. Bioelectrochemical recovery of Cu, Pb, Cd, and Zn from dilute solutions. J Hazard Mater 2012; 235–236:291–97.

[148] Ai C, Hou S, Yan Z, Zheng X, Amanze C, Chai L, et al. Recovery of metals from acid mine drainage by bioelectrochemical system inoculated with a novel exoelectrogen, pseudomonas sp. E8. Microorganisms 2020; 8(1).

[149] Choi C, Hu N, Lim B. Cadmium recovery by coupling double microbial fuel cells. Bioresour Technol 2014; 170:361–69.

[150] Abourached C, Catal T, Liu H. Efficacy of single-chamber microbial fuel cells for removal of cadmium and zinc with simultaneous electricity production. Water Res 2014; 51:228–33.

[151] Colantonio N, Kim Y. Cadmium (II) removal mechanisms in microbial electrolysis cells. J Hazard Mater [Internet] 2016; 311:134–41. Available from: http://dx.doi.org/10.1016/j.jhazmat.2016.02.062

[152] Choi C, Cui Y. Recovery of silver from wastewater coupled with power generation using a microbial fuel cell. Bioresour Technol 2012; 107:522–25.

[153] Tao HC, Gao ZY, Ding H, Xu N, Wu WM. Recovery of silver from silver(i)-containing solutions in bioelectrochemical reactors. Bioresour Technol 2012; 111:92–97.

[154] Yun-Hai W, Bai-Shi W, Bin P, Qing-Yun C, Wei Y. Electricity production from a bio-electrochemical cell for silver recovery in alkaline media. Appl Energy [Internet] 2013; 112:1337–41. Available from: http://dx.doi.org/10.1016/j.apenergy.2013.01.012

[155] Lim BS, Lu H, Choi C, Liu ZX. Recovery of silver metal and electric power generation using a microbial fuel cell. Desalin water treat 2015; 54(13):3675–81.

[156] Kamperidis T, Tremouli A, Remoundaki E, Lyberatos G. Silver recovery from wastewater, simulating the chemical extract originating from a PV panel using microbial fuel cell technology. Waste Biomass Valorization [Internet] 2022; 7(0123456789). Available from: https://doi.org/10.1007/s12649-022-01793-y

[157] Choi C, Hu N. The modeling of gold recovery from tetrachloroaurate wastewater using a microbial fuel cell. Bioresour Technol 2013; 133:589–98.

[158] Catal T, Bermek H, Liu H. Removal of selenite from wastewater using microbial fuel cells. Biotechnol Lett 2009; 31(8):1211–16.

[159] Wang Z, Lim B, Choi C. Removal of Hg^{2+} as an electron acceptor coupled with power generation using a microbial fuel cell. Bioresour Technol [Internet] 2011; 102(10):6304–07. Available from: http://dx.doi.org/10.1016/j.biortech.2011.02.027

[160] Saad DR, Alismaeel ZT, Abbar AH. Cobalt removal from simulated wastewaters using a novel flow-by fixed bed bio-electrochemical reactor. Chem Eng Process – Process Intensif [Internet] 2020; 156 (August):108097. Available from: https://doi.org/10.1016/j.cep.2020.108097

[161] Huang L, Liu Y, Yu L, Quan X, Chen G. A new clean approach for production of cobalt dihydroxide from aqueous Co(ii) using oxygen-reducing biocathode microbial fuel cells. J Cleaner Prod 2015; 86:441–46.

[162] Westerhoff P, Lee S, Yang Y, Gordon GW, Hristovski K, Halden RU, et al. Characterization, recovery opportunities, and valuation of metals in municipal sludges from U.S. wastewater treatment plants nationwide. Environ Sci Technol 2015; 49(16):9479–88.

[163] Yun-Hai W, Bai-Shi W, Bin P, Qing-Yun C, Wei Y. Electricity production from a bio-electrochemical cell for silver recovery in alkaline media. Appl Energy 2013; 112:1337–41.

[164] Gajda I, Greenman J, Ieropoulos IA. Recent advancements in real-world microbial fuel cell applications. Curr Opin Electrochem [Internet] 2018; 11:78–83. Available from: https://doi.org/10.1016/j.coelec.2018.09.006

[165] Kim T, Kang S, Kim HW, Paek Y, Sung JH, Kim YH, et al. Assessment of organic removal in series- and parallel-connected microbial fuel cell stacks. Biotechnol Bioprocess Eng 2017; 22(6):739–47.

[166] Yang W, Li J, Ye D, Zhang L, Zhu X, Liao Q. A hybrid microbial fuel cell stack based on single and double chamber microbial fuel cells for self-sustaining ph control. J Power Sources [Internet] 2016; 306:685–91. Available from: http://dx.doi.org/10.1016/j.jpowsour.2015.12.073.

[167] Oh SE, Logan BE. Voltage reversal during microbial fuel cell stack operation. J Power Sources 2007; 167(1):11–17.

[168] Kim B, Mohan SV, Fapyane D, Chang IS. Controlling voltage reversal in microbial fuel cells. Trends Biotechnol [Internet] 2020; 38(6):667–78. Available from: https://doi.org/10.1016/j. tibtech.2019.12.007.

[169] Ki DW. Anaerobic Conversion of Primary Sludge to Resources in Microbial Electrochemical Cells. Arizona State University; 2016.

[170] Chen J, Yang Y, Liu Y, Tang M, Wang R, Tian Y, et al. Bacterial community shift and antibiotics resistant genes analysis in response to biodegradation of oxytetracycline in dual graphene modified bioelectrode microbial fuel cell. Bioresour Technol [Internet] 2019; 276(January):236–43. Available from: https://doi.org/10.1016/j.biortech.2019.01.006.

[171] Inc. MI. Price List [Internet]. 2022 [cited 2022 Jun 4]. Available from: https://ionexchangemembranes.com/price-list/.

[172] Ge Z, He Z. Long-term performance of a 200 liter modularized microbial fuel cell system treating municipal wastewater: Treatment, energy, and cost. Environ Sci: Water Res Technol 2016; 2(2):274–81.

Section III: **Synthetic biology**

Beth Bannerman, Sunil Chandran

Chapter 12
Clean chemistry: powered by biology

12.1 Introduction

"By far, nature is the best chemist of all time." This is a statement by Nobel Laureate Frances Arnold, who conducted the first directed evolution of enzymes – proteins that catalyze chemical reactions. The uses of her results include more environmentally friendly manufacturing of chemical substances, such as pharmaceuticals, and the production of renewable fuels.

Nature as we know has been catalyzing chemical reactions for 4.5 billion years. It is indisputable that organisms have adapted and evolved during the history of life on Earth.

Natural chemical reactions and biological evolution, the process by which species of life (plants and animals) adapt over time in response to their changing environment, have created enormous diversity of life. There is a variety of evidence for evolution – one is molecular biology. DNA and the genetic code reflect the shared ancestry of life.

We see evidence in fossils too. From the shape of a human skeleton and skull, we can see how humans have evolved over time and differently by geography. What is not as easy to see, but no less clear, are the imprints of microscopic organisms left in rocks about 3.7 billion years old [1].

Although the biotechnology developed by Amyris Inc. is the most advanced in the world, it is simply science powered by nature. Amyris leverages nature's own catalytic conversion process (fermentation) to produce molecules that already exist in nature.

That is how we define clean manufacturing. In this chapter, we will describe the science in more detail, the near limitless opportunity of its application and benefits to nature as well as the economy. We will also describe our own evolution as a company through mid-2022 (the time of this writing) and share the lessons we have learned along the way.

Beth Bannerman, Amyris, Inc., San Francisco, CA, USA, e-mail: bannerman@amyris.com
Sunil Chandran, Amyris, Inc., San Francisco, CA, USA, e-mail: chandran@amyris.com

https://doi.org/10.1515/9783110791228-012

12.2 The founders

The year was 2002. In a laboratory at the University of California, Berkeley, four post-doctoral fellows along with the professor of Chemical and Biomolecular Engineering were discovering new ways to apply engineering principles to biological systems.

Working with single-cell organisms, such as bacteria or yeast, known for their catalytic capabilities to ferment beer, bread, and wine, they discovered how to transform yeast into a "living factory" that could produce almost any compound that exists in the natural world through fermentation. It was a major scientific breakthrough with enormous potential to revolutionize medicine and every other commercial field with a dependence on natural compounds. Especially for natural compounds that were rare or difficult and expensive to manufacture by conventional chemical routes, this clean manufacturing process was significantly faster [2] and less costly. In addition, the natural compounds produced using this new method were pure and bioidentical to those that existed in nature. The process was sustainable and less destructive to the environment.

Professor Jay Keasling had already been widely known as a pioneer in synthetic biology. Together with Vince Martin, Jack Newman, Kinkead Reiling, and Neil Renninger [3], these scientists recognized the potential of their discovery and in 2003, started Amyris Biotechnologies to change the world.

GALLEY SIDEBAR:
The name Amyris (plant name *Amyris balsamifera*) was chosen by Jack Newman's wife. An avid gardener, she had been taken by the plant's use as a substitute for real sandalwood. The promise of substituting nature-for-nature fit well with the researchers' work to use biology to develop an alternative source for Artemisinin. The scientists' passion for their work was also neatly captured in the word Amyris, a homophone for Amorous.

With a company name chosen and the realization that their technology could be applied to almost any chemical compound to address a range of global problems, their first question was where to begin?

12.3 Purpose-driven from the start

That same year, the Africa Malaria report[1] was published by the World Health Organization (WHO) and UNICEF, which said that malaria remained "the single biggest cause of death of young children in Africa and one of the most important threats to the health of the pregnant women and their newborns."

---

1 http://apps.who.int/iris/bitstream/handle/10665/67869/WHO_CDS_MAL_2003.1093.pdf;jsessionid=63282432317F59B29A64066179B73771?sequence=1.

According to Renninger, "We saw this as a great application of the technology. If you can make one compound, you've built most of the technology required to make any one of those compounds. This was a great way to launch the company."

The team published a paper in *Nature Biotechnology* describing how they created a pathway by inserting genes into *Escherichia coli*, a common bacteria. The findings demonstrated how biotechnology could help advance the United Nations' Millennium Development Goals [4], aligned to reversing global poverty, hunger, and disease affecting billions of people. The UN called for reducing the deaths from malaria, a preventable mosquito-borne infectious disease and WHO was recommending artemisinin-based combination therapies (ACTs) as an effective malaria treatment.

The challenge was the availability of the key ingredient, artemisinin, a natural compound traditionally found in the extract of *Artemisia annua*, or the sweet wormwood plant. While ACTs are the most effective treatment, *A. annua* crop yields can be unpredictable from year to year, causing price volatility ranging from $400 to $1,200 per kg [5] and impacting overall availability of this life-saving drug.

The next year, in 2004, the Bill and Melinda Gates Foundation [6] granted $42.6 million to the Institute for OneWorld Health (a nonprofit pharmaceutical company) to create new approaches to developing a more affordable cure for malaria. Amyris partnered with OneWorld Health and UC Berkeley to develop a scalable, stable, and cost-effective production process for artemisinin. This funding set Amyris on the path to engineer its first molecule, proving its capability as a biotech pioneer on a global stage, and paving the way to a dependable and secure supply of antimalarial treatments produced in a sustainable and scalable way.

Amyris successfully engineered a microbe to produce artemisinic acid, a molecule that can be converted to artemisinin allowing for on-demand, clean manufacturing to meet the global need for artemisinin and ACTs. Amyris demonstrated that its technology not only demonstrated how modern manufacturing could stabilize costs and supply but also save lives. Amyris went on to license its artemisinic acid production strain and process to Sanofi via OneWorld Health and later, Amyris was recognized with a United Nations Global Citizen Award [7] based on its contributions toward several aspects of the UN's Millennium Development Goals, including the production of artemisinic acid through clean manufacturing powered by biology.

By mid-2005, after receiving the Gates grant, Amyris was exploring new applications and chemical compounds with a connection to their earlier work. Also at this time, the team was seeking new investment from Silicon Valley venture capitalists, who happened to be particularly interested in global warming and renewable fuels. Amyris had identified a few interesting fuel candidates belonging to the isoprenoid class of molecules (also called terpenoids), a source of complex hydrocarbons, which are the dominant component of crude oil, gasoline, and diesel fuel.

This would prove to be an important turning point for the company.

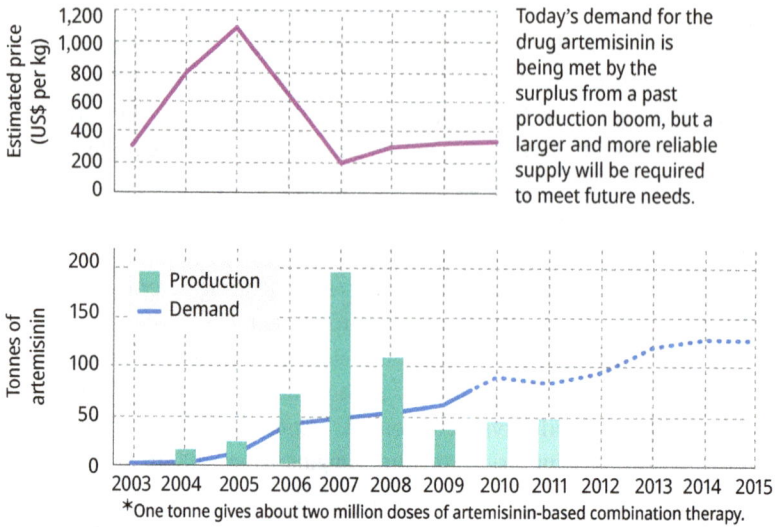

Today's demand for the drug artemisinin is being met by the surplus from a past production boom, but a larger and more reliable supply will be required to meet future needs.

*One tonne gives about two million doses of artemisinin-based combination therapy.

Figure 12.1: Boom and bust for artemisia farmers.
Source: *iGem Academy.*

12.4 Purpose-driven: clean air

Fossil fuel use is one of the biggest sources of greenhouse gas emissions that contribute to global warming. The decade ahead (2010s) would become known for its climate change consequences, human-driven extreme weather events including heatwaves, floods, droughts, and wildfires impacting millions of people around the world from Russia to Pakistan and East Africa. Global temperatures soared to new records and by 2019; global levels of carbon dioxide in the atmosphere reached 415 parts per million for the first time in human history.

Amyris was positioned well with technology to help address this global problem. With demonstrated capability in the pharmaceutical sector, Amyris had also secured $20 million [8] in Series A funding from Khosla Ventures, Kleiner Perkins Caufield & Byers, and Texas Pacific Group Ventures and, at the same time, in 2007, brought on a new Chief Executive Officer, John Melo, formerly President of US fuel operations at the oil firm, BP.

According to Amyris Board Director Geoff Duyk, Amyris was realistic in terms of what it would take to fully realize the value of its technology. "Unless you are vertically integrated, too much of the rent flows away from you," he said. "Amyris recognized the need to be responsible for both building and deploying." Driven by the belief that being vertically integrated would capture the most value over time, Amyris prepared itself for what would be a major change for the company in the years to come.

In the next few years, Amyris would partner with one of the largest sugar and ethanol distributors in Brazil and form a joint venture with an ethanol-producing mill to build a production plant for renewable fuels. By early 2009, against the backdrop of a global financial collapse, Amyris' renewable diesel biology was technically viable, the product was of the highest quality but production costs were not yet competitive. The company had grown to 200 employees in Emeryville, CA with a team of 40 people working in Brazil. Manufacturing operations had been established in partnership with Brazilian producers of sugarcane – the feedstock for Amyris' fermentation and the essential element to its clean manufacturing.

Innovation in the global biofuels market came in several waves: the first from corn which then raised concerns about food availability and questions about the amount of energy corn produces being lower than the amount of energy it takes to produce; the second wave of biofuel production was from biomass such as food crop waste, eliminating the main issues of the first generation; and the third generation used algae, more efficient than previous generations. Amyris participated in the fourth wave of biofuels, which used biomass that captured CO^2 even more efficiently [9]. Global production of biofuels was up, but the cost of petroleum had swung wildly between 2000 and 2010; from $30/barrel in 2000 to $133/barrel in July 2008 and back down to $39/barrel in early 2009 [10]. Ultimately, the economics of production could not compete with the variability of the oil market and so Amyris began exploring extensions of its science and technology platform, beyond biofuels.

In 2010, Amyris became a public company, selling its stock in an initial public offering [11]. By the end of 2012, Amyris was producing its first commercial molecule, farnesene, for specialty chemicals and renewable fuels. Branded by Amyris as Biofene, this terpenoid, which is found in many plants including green apples, can also be converted to other compounds, including ingredients used for flavors, fragrances, and cosmetics. In 2014, Amyris earned the distinction of the Presidential Green Chemistry Award from the Environmental Protection Agency [12] for its clean manufacturing approach to farnesane, a breakthrough renewable hydrocarbon for use as diesel and jet fuel: Initial studies showed that farnesane can reduce greenhouse gas emissions by as much as 80% when compared to fossil jet fuels [13].

By this time, Amyris had reached another crossroads. Its science had so much optionality that the challenge once again was to evaluate and choose the best next step. The company had increasingly become interested in supplying ingredients to consumer product companies because their products used many petroleum-based ingredients that could be replaced by farnesene alternatives.

The company's growth plan would rely on partnerships, which take time to develop. Said two Amyris executives at the time, "We don't plan to market the consumer products ourselves, so we are working with partners; however, it might take these companies some time to research and launch potential applications. As a result, it is hard to predict the volume and pricing of the chemical products at this point. It's more than convincing our customers. Our [business] customers convince their consumers, that the new formulation is better."

Because the production pipeline at the time was singularly focused on farnesene, a precursor to other chemical compounds, there was minimal impact on their manufacturing strategy and a wide array of opportunity for end uses. To determine the best end-use for farnesene, Amyris looked at three primary criteria: technical fit, financial impact, and regulatory approval and timelines. Amyris recognized that the first company with commercially viable products would lead the market and the perception of the biotech industry. So Amyris invested further in R&D with new robotics, machine learning, and artificial intelligence. The company developed its own programming code, Genotype Sequencing Language. Amyris soon began optimizing the speed and costs of developing new yeast strains, efficient living factories, to get the most conversion from sugars from the fermentation process to the final product. By now, there were more chemical compounds for specialty ingredients in active development and testing.

Three major insights influenced the direction Amyris took next with its science: (1) Amyris was leading the biotech sector with its ability to commercialize products by converting sugarcane syrup into specific chemical compounds, providing the company with optionality to disrupt virtually any industry, particularly consumer goods; (2) by moving up the value chain from ingredient supplier to consumer goods producer, Amyris could realize a greater percentage of the end products' profitability, and (3) Amyris could potentially gain 10–30 times better capital efficiency [14] by producing specialty ingredients (lower volume, higher margin) compared to biofuel or other commodity ingredients (high volume, low margin).

Galley sidebar:
Commodity chemicals and renewable fuels generated less than $1 of revenue for every kilogram of product and less than $0.2 per kilogram in gross margins, while specialty chemicals generated $39 per kilogram of revenue and $25 per kilogram of gross margins [15].

12.5 The business of lab-to-market™

These insights marked an important pivot in Amyris' evolution. Amyris could now begin positioning itself with broader value: Amyris is disrupting conventional production systems that rely on destructive and nonsustainable practices and replace them with highly scalable, clean manufacturing powered by biology. Its science and technology platform enables a scientist to precisely engineer microbes like *Saccharomyces cerevisiae* (brewer's yeast) to ferment sugar and produce a variety of bioidentical molecules which serve as natural ingredients that consumers use on a day-to-day basis. Using its technology platform, Amyris accesses better performing natural ingredients at lower cost and then produces them at scale in a way that is far better for the environment by not depleting natural resources.

A microbe like *Saccharomyces cerevisiae*, which has evolved for billions of years, can be further modified to consume sugar to produce a target molecule. These microbes can be seen as hardware and the genetic code as software to turn the hardware into tiny living factories that go on to make ingredients used in skincare, fragrances, flavors, health, and nutritional supplements in addition to biofuels. Coding is complex enough in computer science, let alone in living organisms. Amyris accelerates the cycle of design, test, learn, and iterate with every microbe it programs.

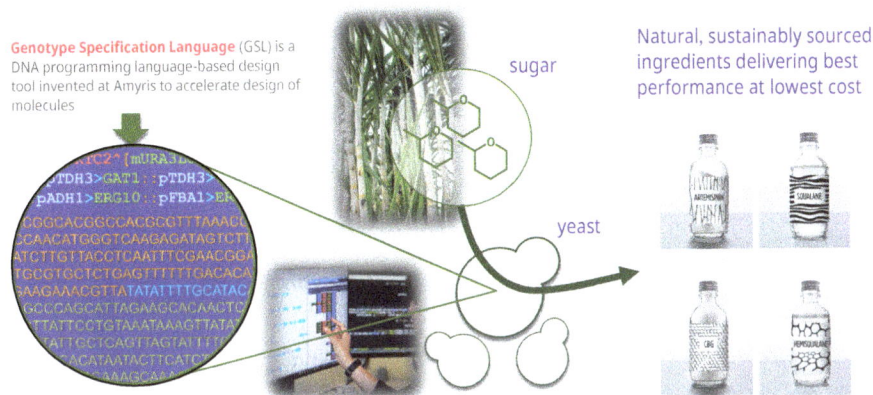

Genotype Specification Language (GSL) is a DNA programming language-based design tool invented at Amyris to accelerate design of molecules

sugar

Natural, sustainably sourced ingredients delivering best performance at lowest cost

yeast

Figure 12.2: We program cells to create high performing sustainable ingredients.
Source: *Amyris.*

Amyris improves the predictability and repeatability of its success using its proprietary Lab-to-Market™ technology platform. This platform has evolved to continuously learn from previous iterations via highly optimized and automated molecular biology, analytical and process development tools, combined with machine learning algorithms and statistical models, enabling the collection of larger amounts of quality data. A fully integrated, end-to-end, R&D, process development, and manufacturing solution can disrupt how companies traditionally approach this problem. This integration means that Amyris starts every project with the end in mind. Before design on the first strain begins, the team already has a good understanding of its manufacturing requirements, cost targets, and project timeline.

Since its start, Amyris has grown its business designing, developing, and producing clean chemistry. By shifting to sustainably sourced, clean manufacturing, Amyris is advancing its mission to accelerate the world's transition to sustainable consumption. A key to its success is the prioritization of opportunities. Amyris has applied its technology to solve supply chain volatility and make rare molecules from nature abundant while also being sustainably sourced. This aligns with a great consumer demand trend of our time and the need for companies to lower costs and access reliable supply. The flexibility of its technology platform has enabled Amyris to build an expansive product portfolio of molecules.

What Amyris can make with this platform is almost unbounded in diversity. Already, Amyris has successfully engineered strains that have produced greater than 250 molecules through fermentation from over 20 different biological pathways. These 250 molecules unlock thousands more: not only can Amyris access upstream biochemical intermediates, but every time the company optimizes a pathway, it can also produce hundred more related compounds. Moreover, Amyris can use simple chemistry to further diversify every scaffold. All told, Amyris can access hundreds of thousands of potential targets, or over half of all small molecule diversity found in nature.

This technology is at the core of Amyris' value – the ability to make molecules and the ability to scale molecules –13 to date – to the metric ton or even kiloton scale, at a speed unparalleled in the sector to date. With each successful scale-up, Amyris leverages its know-how to optimize its next success in pathway engineering, analytics to measure the molecules, developing processes for fermentation and purification, and running unit operations at scale.

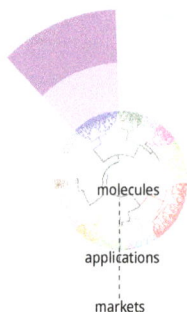

Ingredient	Application	Market
Farnesene	Polymers	High Performance Materials
Vitamins	Nutrition Ingredient	Human and Animal Health
Squalane	Emollient	Clean Beauty
Hemisqualane	Silicone Replacement	Clean Beauty/Personal Care
Squalene	Immunostimulant	Pharma

Figure 12.3: A single class of molecules creates significant commercial value.
Source: *Amyris.*

Technology on its own is not as impactful as its application. In the realm of personal care, Amyris has commercialized effective formulations, which are on the market today – as borne out by extensive clinical study and benchmarking. Like Amyris' core technology, this know-how builds on itself and becomes exponentially stronger over time as Amyris enhances its knowledge base. In applications where Amyris is not an expert, it forms partnerships with market leaders to stay focused on only the most relevant and urgent applications and to drive adoption for accelerated scale and impact.

The real value and impact are realized once Amyris' ingredients reach consumers at a massive scale. Amyris has leveraged economies of scale across its formulation, manufacturing, distribution, digital infrastructure, marketing, and consumer experience capabilities to grow brands that will reach new audiences. And this again is why Amyris selects top partners in their respective markets.

One example of how value shows up through Amyris' capabilities of scaling ingredients, identifying applications, and getting them to market is a single molecule known as farnesene. Farnesene has given rise to billions of dollars of value to the company through a variety of applications that can access a wide range of end markets. Derivatives of farnesene include vitamin E, squalane, hemisqualane, and squalene, proving how combining fermentation and chemistry amplifies value. As with all its molecules, Amyris provides a No Compromise™ solution: best performance in its category, for the best cost, using a sustainable process.

When choosing the next target molecule to make, Amyris applies three filters: market, technology, and efficacy.

Ideally, there is a large total addressable market for a target molecule made from animals or plants within a challenged supply chain that is subject to variable availability, price, and quality. There may be further value driven for our brands and partnerships through offering a distinctive advantage in cost and performance.

Amyris also considers the technology investment that would be required. How much existing experience in R&D, process development, and manufacturing can be leveraged? What strains, processes, and capital expenditures are already in place? If little, does the market opportunity justify new investment?

Efficacy is critical to Amyris. First and foremost, the molecule must be safe for its application. Second, the target molecule must have best-in-class activity in its respective applications. Finally, there must be complementary and synergistic functions to ingredients in Amyris' existing portfolio to capture added value from the portfolio of ingredients available for formulation.

Over the years, Amyris has been continuously innovating and investing in its technology platform. Its first commercial product took nearly 40 months to develop from strain to pilot plant run. And today, Amyris averages less than 12 months. The cost of product development has dropped by 90%; time-to-market has reduced by 80%, and Amyris has been able to do this by increasing the bandwidth of its R&D pipeline by 500% since 2012, while increasing operating expenditures by only 20%.

12.6 Purpose-driven: clean beauty

One notable chemical compound that Amyris developed from farnesene was squalane. For decades, squalene had been harvested from sharks' liver and then converted via a hydrogenation process into squalane. Livers of an estimated 3,000 sharks are required to produce just under 1 ton of squalane. Up to 2.7 million deep-sea sharks [16] a year were killed to meet the global demand for squalane in the cosmetic industry alone.

After environmentalists pushed for the cosmetic industry to stop using shark liver oil, scientists turned to olives, a plant-based source of squalene.

The undisputable skin health benefits [17] of squalane are the reason for its long-standing global appeal:

– Provides superior moisturization to the skin for longer compared to other oils
  – Moisture is vital for healthy skin. Your skin produces oil naturally to support moisture, strength, and elasticity. When applied to the skin squalane mimics the natural oils your skin produces and because of these biomimetic qualities, squalane has been proven to have exceptional moisturizing properties. Light-weight and nongreasy, squalane is efficiently absorbed into the skin, without leaving an oily residue and works quickly to replenish moisture in the skin, evenly, helping to balance the skin's natural oil production.
– Lowers transepidermal water loss (TEWL) by 13.9%
  – Moisturized skin also looks and feels firmer and reduces the appearances of fine lines and wrinkles. Studies have shown that squalane has better absorption than other moisturizers like Jojoba and Argan oils with greater short-term moisturization, a 51% immediate improvement in moisturization and continued moisturization for 24 h. Squalane has also been proven to reduce TEWL by 13.9%, keeping that moisture locked in.

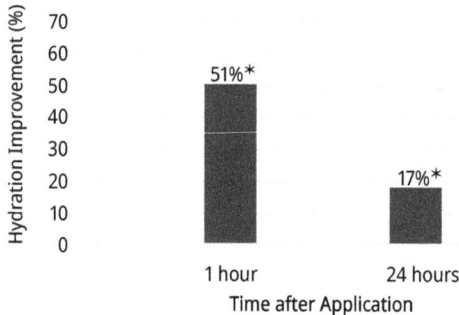

Figure 12.4: Moisturizes skin.
Source: *Amyris.*

– Reduces skin roughness by 28%
  – A deficiency in essential skin lipids can result in the skin feeling tight and itchy with a dull appearance and rough or even cracked texture. A compromised skin barrier further exacerbates dryness as moisture easily escapes from the skin. Regular use of squalane replenishes the skin's lipid content and prevents moisture loss due to its outstanding occlusive emollient properties.

– Promotes healthier skin with diverse microbiome
  – The microbiome is an ecosystem of bacteria living on the skin's surface that keeps it healthy. The microbiome aids in wound healing, limits exposure to allergens, minimizes oxidative damage, keeps the skin plump and moist, and even protects it from harmful UV rays. Studies have shown that younger look-ing skin can be achieved by a healthy and well-balanced skin microbiome. A microbiome that's out of balance can leave skin vulnerable to concerns such as rosacea, atopic dermattitis, psoriasis, acne, eczema, and inflammation. Clinical studies have shown that squalane not only supports a healthy skin microbiome but also increases microbiome diversity.

– Promotes cell turnover by 34%
  – Skin cell turnover is the continuous process of shedding dead skin cells and replacing them with younger cells. This process is essential to healthy skin as dead, dull skin cells are shed to reveal a fresh, radiant complexion. The lon-ger the cell turnover cycle takes, the more dead cells are left to build up on the surface of the skin. This proliferation of cellular debris can make skin ap-pear dry, and dull, clog the pores leading to breakouts, make lines and wrin-kles appear more pronounced, and make age spots and hyperpigmentation appear darker. Healthy cell turnover is key to smoother, more even toned, healthy, vibrant skin. Regular use of squalane has been shown to promote cell turnover by 34% which in turn improves the texture and tone of the skin.

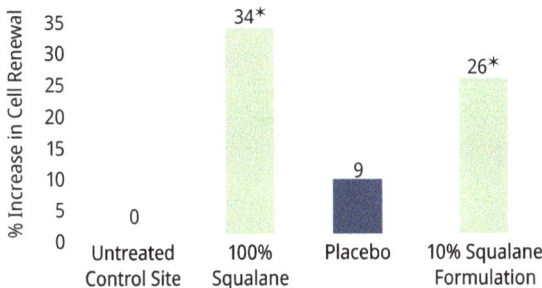

Figure 12.5: Improves cell turnover.
Source: *Amyris.*

– Repairs and reinforces the skin barrier
  – A healthy barrier is critical to normal skin function but everything from environ-mental changes and stress, to the use of harsh soaps and overexfoliation can compromise it. When your barrier is compromised, your skin is not able to per-form its two primary functions: inhibit TEWL and protect itself from environ-mental pollutants. Squalane's emollient properties support and enhance the skin barrier. Your skin barrier is also slightly acidic. This acidity (the acid mantle) is

crucial to a healthy skin barrier. With a pH range at 5.5–6.5, squalane protects and maintains the skin's acid mantle and promotes optimal barrier function.

– Suitable for all skin types
  – We have all heard that oils can cause breakouts but not all oils are created equal. Squalane is an oil that mimics squalene, an essential component of SSL. When skin is dry as a result of a lack of lipids, environmental factors, harsh cleansers, stress, or even a hormonal imbalance, this can lead to an overproduction of sebum to compensate, both congesting the skin and leaving it feeling excessively oily. Squalane is ideal for all skin types, even oily skin and sensitive skin, and all ages from birth up. Squalane is a lightweight, colorless, odorless, anti-inflammatory oil that is noncomedogenic, meaning it will not clog your pores or leave a greasy residue while giving skin the moisture it needs. Squalane is nonirritating up to 100% concentration and is safe for use on all skin from birth onward.

– Supercharges the skincare routine
  – Because squalane is so readily absorbed in the skin, it can also enhance absorption of other active ingredients in your skin care products. Squalane is the most effective emollient for delivering CBD compared to other traditional emollients.

GALLEY sidebar:

What is Amyris squalane? Amyris squalane, branded as Neossance™ is a plant-derived, sustainable, ethical, and renewable version of nature's most effective emollient, squalene. In humans, squalene is one of the main, naturally occurring components of our skin surface lipids (SSL), a complex mixture of lipids that moisturize the skin and protect it from the environment. Newborn babies have the greatest concentration of squalene in their skin, but the reserve begins to drop suddenly between 30 and 40 years. Because of squalene's superior function as an occlusive emollient, both moisturizing and preventing TEWL, it has long been used in skincare products as a moisturizer.

In 1910, Japanese scientist Mitsumaro Tsjuiimoto discovered squalene in shark liver oil – specifically, sharks from the family *Squalidae* – and so, for many years, sharks were killed for their livers and the squalene extracted for use in cosmetic products. This unethical practice contributed to many shark species becoming endangered and had a devastating environmental impact. Thanks to advances in the field of biotechnology, Amyris has pioneered a process through which to create a clean, sustainable, and ethical squalene alternative known as squalane: a biomimetic lipid that uses the power of science to mimic nature with improved purity, delivery, and efficacy.

In an alternative process reliant on 600× more land and the harvesting of olive trees, the squalene is extracted from olive pulp, skin, and pits and then converted via a hydrogenation process into squalane. Olive-derived squalane however is priced higher than shark-derived squalane and has a lower purity (squalane from olive oil wastes can be contaminated with by-products from processing – including plant waxes, free fatty acids, phytosterols, and neutralization by-products) with a characteristic odor and light yellow color.

Furthermore, both natural sources – shark and olive – presented a structural mismatch between the organization and the resource needs of the market, leading to cycles of shortages and higher prices. Shark liver oil and the unsaponifiable fraction of olive oil are by-products of the food industry;

hence the amounts available varied depending on many environmental factors over which squalane manufacturers had no control.

Seeing the need for a truly renewable, sustainable, and innovative solution, Amyris pioneered the technology to produce squalane from sustainable sugarcane.

Amyris squalane comes from 100% sustainably and ethically grown and harvested, renewable sugarcane which is then biofermented to create the purest squalane available. Lightweight, odorless, colorless, antibacterial, noncomedogenic, and suitable for all skin types, squalane is nature's most powerful emollient.

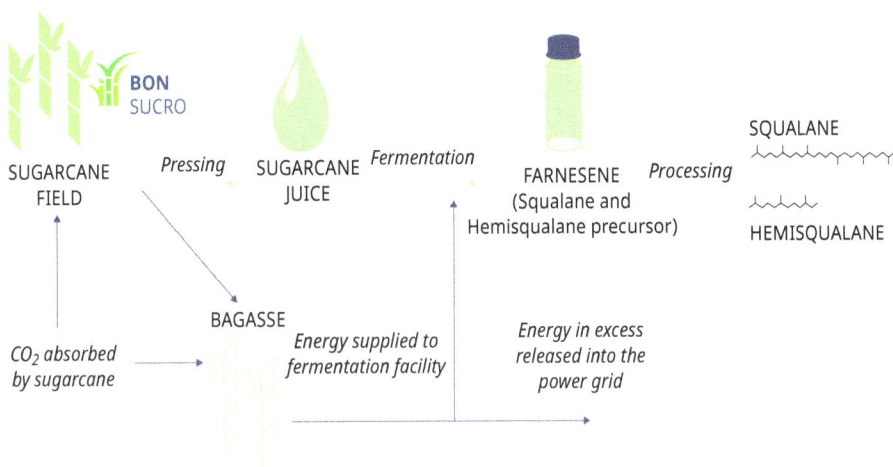

BON SUCRO

SUGARCANE FIELD   *Pressing*   SUGARCANE JUICE   *Fermentation*   FARNESENE (Squalane and Hemisqualane precursor)   *Processing*   SQUALANE

HEMISQUALANE

BAGASSE

Energy supplied to fermentation facility

Energy in excess released into the power grid

CO₂ absorbed by sugarcane

Figure 12.6:
Source: *Amyris.*

The Amyris squalane clean manufacturing process starts in the sugarcane fields of Brazil. The sugarcane is certified by Bonsucro, a nonprofit that promotes sustainable sugarcane production. Amyris believes that independent certification is important to ensure sustainability at our ingredient's source.

Amyris' sustainably and ethically grown and harvested non-GMO sugarcane crops are located well outside the threatened Amazon region and do not contribute to deforestation of the rainforest nor compete with food agriculture. The sugarcane sourced by Amyris is a fast growing, rapidly renewable crop that requires minimal irrigation due to abundant rainfall in Brazil. The sugarcane is harvested and crushed to extract the juice from the plant that contains that sugar. This sugary juice is then boiled until thickened and spun through a centrifuge to remove any impurities. This process is done utilizing sustainable production practices, including the use of cogeneration systems to supply energy. Through cogeneration, the by-products of converting sugarcane into sugar are used to produce energy to power Amyris' plant.

The cane juice is then fermented with Amyris' engineered yeast to produce β-farnesene, the chemical precursor to squalane. The resulting β-farnesene undergoes a distillation and hydrogenation process to become pure squalane. Amyris' process is more efficient than the olive oil squalane process from seed to market, requiring four less manufacturing steps. Neossance™ squalane has 20 times less impurities than olive oil, ensuring safe, high-quality products.

Neossance Squalane Process
– 30% more efficient
– Clean, simple, circular manufacturing
– Pure, consistent Squalane

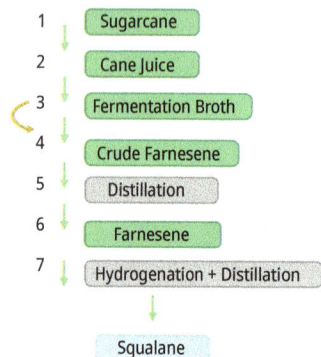

1 Sugarcane
2 Cane Juice
3 Fermentation Broth
4 Crude Farnesene
5 Distillation
6 Farnesene
7 Hydrogenation + Distillation

  Squalane

Olive Oil Squalane Process
– Dirty and intensive chemical processing
– Lower purity, inconsisten Squalane

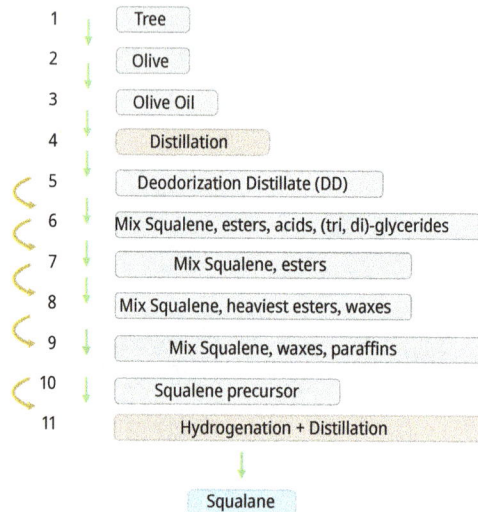

1 Tree
2 Olive
3 Olive Oil
4 Distillation
5 Deodorization Distillate (DD)
6 Mix Squalene, esters, acids, (tri, di)-glycerides
7 Mix Squalene, esters
8 Mix Squalene, heaviest esters, waxes
9 Mix Squalene, waxes, paraffins
10 Squalene precursor
11 Hydrogenation + Distillation

  Squalane

Figure 12.7:
Source: *Amyris.*

Sugarcane-derived Amyris squalane is the most chemically pure variety of squalane available with 94–97% purity. Squalane has the same chemical structure regardless of its origin. However, there are issues with sustainability and purity depending on the source. For example, olive oil derived squalane can be contaminated with by-products from processing, including plant waxes, free fatty acids, phytosterols, and neutralization by-products, and has a characteristic odor and light yellow color.

These impurities can range anywhere from 6 to 18% depending on the manufacturing process including:

– Isosqualane (3–5%)
– Monocyclosqualane (1–3%)
– Hemisqualane (C15; 0–1%)
– Sesquisqualane (C45; 0–1%) Control

PROPERTIES	FIRST GENERATION	SECOND GENERATION	THIRD GENERATION
	SHARK **Squalane**	OLIVE **Squalane**	NEOSSANCE **Squalane (sugar)**
C30 Purity	Pure	Less Pure	Less Pure
Initial Odor Intensity	80% In-spec	45% In-spec	100% In-spec
Saturation	Medium	Low	Medium
Volatile Impurities	Some Impurities	More Impurities	Fewest Impurities
Sustainability	Not Sustainable	Not Sustainable	Most Sustainable
Consistency	Inconsistent	Inconsistent	Always Available

Figure 12.8:
Source: *Amyris.*

Galley sidebar:
What role does squalane play in skin health? Our skin is our largest organ and the stratum cor-
neum (SC) is the skin's outermost layer, acting both as the barrier against external aggressors like
bacteria, pollution, and UV radiation, while simultaneously preventing water loss (TEWL) to keep skin
hydrated. The SC layer of the skin is also where SSL, a film that is part of the epidermal barrier, is
produced. SSL is a layer of lipids, ceramides, and fatty acids with squalene accounting for around 13%
of these lipids. Studies have shown that as we age, the SC displays a 30% reduction in total lipid content. The importance of skin lipids in the maintenance of skin homeostasis (healthy, balanced skin) is
well-documented, and the disturbance of these lipids is linked to conditions including dry skin, rough
surface texture, tightness, dullness, and volume loss; skin diseases such as acne, atopic dermatitis, and
psoriasis; and the development of chronic, low-grade inflammation. Fortunately, the topical application
of lipids can combat these conditions, improve barrier repair, and reduce TEWL. In other words, re-
plenishing our lipid levels with topical applications of squalane has multiple benefits for skin health.
  While squalene is one of the most important human skin cell lipids, due to its multiple double
bonds, it is unstable, very sensitive to air, and is considered a fragile compound for most practical
uses. In squalane, these double bonds have been eliminated making it more stable with a total absence of toxicity, odorless and colorless, with excellent sensorial properties and a longer efficacy. If we
use the analogy that skin is made up of "bricks" (protein from skin cells) and "mortar" (lipids), the lipid
layer is what holds the walls together and keeps the skin strong, hydrated, and soft. The composition,
size, and structure of squalane makes for the perfect "mortar" diffusing into the spaces between cells
and providing moisture and filling in any breaches in the skin barrier.

Because Amyris controls custody of its ingredients – from Lab-to-Market™ – the biotech
company can confidently claim that its squalane is 100% plant-based and sustainable.

  In February 2010, Amyris entered a partnership with Soliance, a provider of ingredients to the French cosmetic industry, to sell large quantities of its squalane to
the cosmetic industry. In 2011, Amyris also entered into a multiyear agreement to

provide large quantities of squalane to Nikko Chemicals Co. Ltd. for distribution in the Japanese market.

In Amyris' 2014 Annual Report, Melo said, "Our business model has two key sources of revenue: sales of our renewable products and inflows from product development collaborations. Our business model is simple – we partner with leading companies to solve their supply, performance, cost, and sustainability problems. Our partners provide market insight and market access, which helps us to prioritize molecules for development as well as an effective channel access to the market."

Said Melo at the time, "In order to effectively scale our business, we need to be in a production environment. Only then will we learn what isn't working at larger volumes. We will learn from being in the market with real products." This statement would play out later when Amyris expanded its business into new markets in the consumer sector, flavors and fragrances, beauty, and health and wellness.

In 2015, Melo shared his plan to begin selling specialty, formulated products directly to industrial and consumer markets by establishing Amyris' own consumer brands in markets where the company already had the best performing molecule and a significant cost advantage, "Our objective for adding these new points of access is to accelerate our market reach and deliver positive impact on the planet while generating as much as 10 times the revenue and three to five times the margin dollars for each liter of product we produce."

The same year, Amyris announced a partnership with Squalane Natural Health, a privately held personal care products company based in the Netherlands, to enter in agreements for the production and marketing of squalane. Amyris branded its squalane ingredient Neossance™ and sold through global distributors to large beauty companies around the world and formulated in a range of products from antiaging skincare to sun care and deodorant.

Galley sidebar:
Amyris squalane is:
- 100% naturally derived
- USDA Certified Biobased Product
- Approved by Ecocert
- Microbiome friendly
- My Green Lab Platinum Certified
- EWG Skin Deep Rated
- Bonsucro Certified

In 2016, starting with Biossance™, Amyris launched its own in-house consumer brands championing their squalane. Clean beauty skincare brand, Biossance™ started as an experiment with a soft internal launch before officially launching on the Home Shopping Network. The following year, the brand launched in Sephora and then expanded beyond the US to Canada and Brazil. Since its launch, Biossance has grown 300% year over year, with global reach across five continents, earning over 50 global awards and attracting millions of consumers to its direct-to-consumer site, Biossance.com, which now offers more than 20 clean beauty products. The success of Biossance is a testament to the quality performance of the squalane ingredient and demonstrated an impressive consumer response, which was not lost on major consumer products companies looking for a sustainable alternative to their shark and olive squalane.

Biossance was launched as a clean beauty brand just as the term "natural" began to lose its label luster due to consumer confusion around its definition and the greenwashing by many brands. Clean beauty focuses on ingredients that are safe and nontoxic as well as manufacturing processes that are sustainable and socially responsible. Amyris' clean manufacturing and commitment to safe, effective, and sustainable ingredients propelled Biossance to lead the clean beauty trend. Positioned perfectly for the megatrend of healthy living from the outside in and inside out, Biossance launched the Clean Academy, a beauty industry first, connecting directly with consumers to educate about the ingredients, formulations, and offering guidance on healthy skin care routines.

At the time of this writing, the Amyris family is made up of nine consumer brands, with three using squalane formulations: Biossance, Pipette™, and Rose Inc.™ with more consumer brand launches planned in 2022 and beyond.

As with Biossance, squalane is the hero ingredient powering the Pipette line of baby and mother care products. Amyris launched Pipette in 2019, earning certifications by EWG Verified, National Eczema Association, Leaping Bunny and approvals by pediatricians, Good Housekeeping, Parents magazine, Allure, and WWD. Distributed internationally, Pipette products have seen a 900% growth in points in distribution and are now sold at Target, Ulta, Walmart, and Walgreens. Free of sulfates, silicone, mineral oil and petrolatum, the clean formulation and manufacturing have a special appeal to new parents.

In August 2021, Amyris launched Rose Inc., a color cosmetics brand with founder Rosie Huntington-Whiteley, entrepreneur and super model. Applying Amyris' technology and unique, biodesigned, and fermented ingredients, including squalane to color was a logical next step. Today, the brand is sold globally including at Sephora and direct-to-consumer, and offers 14 products including color, skincare, and skincare tools.

At the same time, Amyris also announced the launch of JVN™, a brand of clean hair products created for all hair types, with founder Jonathan Van Ness, hair stylist, TV personality, and *New York Times* best-selling author. Free of silicone and sulfates, each of the 10 products addresses the top hair health and styling concerns; they are color-safe, cruelty-free, and vegan. Already an award winner including from Vogue, the JVN line is sold internationally, including at Sephora and direct-to-consumer. The retail launch orders exceeded the first full year of Biossance revenues, an indication that clean manufacturing was gaining traction with consumers.

Figure 12.9:
Source: *Amyris.*

12.6.1 Amyris family of award-winning consumer brands

The JVN product line is formulated with a signature ingredient, hemisqualane, a derivative of farnesene created by Amyris through its fermentation, distillation, and hydrogenation process. Amyris leveraged its Lab-to-Market™ technology platform and clean manufacturing to create hemisqualane to provide a clean, sustainable alternative to silicones and existing silicone replacements. Amyris' process is more efficient and more sustainable than leading alternatives to silicone which are either petrochemically derived or derived from nonsustainable plant sources like palm oil, a major driver of deforestation of some of the world's most biodiverse forests, destroying the habitat of already endangered species. Hemisqualane is also biodegradable and does not harm aquatic life. Hemisqualane is the best performing, natural alternative to silicones in haircare. Unlike silicones used in rinse-off hair products, hemisqualane penetrates hair, nourishing it, strengthening it, and preventing color degradation.

The performance results of Amyris hemisqualane and its comparative benefits to nonsustainable hair ingredients is quickly winning loyal customers.
– Protects and repairs damaged hair
    – Hemisqualane improves your hair over time by creating healthier, stronger strands. It penetrates the hair's core better to deeply repair, protect, and smooth all types and textures.
    – 40% reduction in hair cortex damage after one treatment
    – 15% damage reduction from bleaching after one treatment, seals cuticles
– Makes hair easier to comb, wet and dry
    – 30% reduction in combing energy (WET)
    – 27% reduction in combing energy (DRY)
– Improves hair elasticity
    – 7% increase in elongation Improves elasticity
    – strengthens the hair, prevents breakage
– Maintains color
    – 30% increase in color maintenance
– Reduces frizz
    – In clinical studies, hemisqualane has shown a significant, instant reduction in frizz. Hemisqualane keeps frizz at bay long after applying and improves hair health over time.
    – 29% reduction in frizz after application

Galley sidebar:

What is hemisqualane?

Hemisqualane is a plant-derived, clean, sustainable, biodegradable silicone alternative. Amyris developed hemisqualane as a safe, sustainable solution to growing consumer concerns about silicone toxicity and their impact on the environment. Cyclomethicones (D4 and D5) have been essential components of personal care and beauty for decades but have been classified as a toxic "Substance of Very High Concern" by the European Union (EU).

Silicones are also not biodegradable, leading to concerns about their environmental impact on waterways, oceans, and accumulation inside of wildlife. Amyris leverages its Lab-to-Market™ technology to create a clean, sustainable, and biodegradable hemisqualane, the most safe, sustainable silicone alternative available. Lightweight, odorless, transparent, nongreasy noncomedogenic, and suitable for all skin and hair types, hemisqualane is the best-performing natural silicone replacement available.

Silicones have been essential components of personal care and beauty for decades as they deliver even coverage, spreadability and a light, dry feel in antiperspirant, sun care, color cosmetics, and makeup removal applications. In haircare, silicones support combability, thermal and color protection, antifrizz, and elongation features. In recent years, however, there has been a growing concern about silicone toxicity and its impact on the environment. Environment Canada assessments concluded that cyclotetrasiloxane and cylcopentasiloxane – also known as D4 and D5 – are toxic, persistent, and have the potential to bioaccumulate in aquatic organisms as they are not biodegradable.

The EU classifies D4 as an endocrine disruptor, based on evidence that it interferes with human hormone function, and a possible reproductive toxicant that may impair human fertility. In laboratory experiments, exposure to high doses of D5 has been shown to cause uterine tumors and harm to the reproductive and immune systems. D5 can also influence neurotransmitters in the nervous system. Structurally similar to D4 and D5, cyclohexasiloxane (or D6) is persistent and has the potential to bioaccumulate.

Environment Canada's assessment of D6 concluded that this third siloxane is not entering the environment in a quantity or concentration that endangers human health or the environment but noted significant data gaps concerning its toxicity. Common silicones in cosmetics are: dimethicone, cyclomethicone, cyclohexasiloxane, cetearyl methicone, and cyclopentasiloxane. As a result, cyclomethicones are restricted by the EU and have been banned by a host of clean beauty brands and retailers.

Hemisqualane is sustainably sourced from renewable sugarcane. Amyris' clean manufacturing process reduces CO_2 emission compared to traditional petrochemical processes and respects biodiversity. Hemisqualane is also biodegradable and does not harm aquatic life. Hemisqualane is superior to silicones in skincare in haircare. In addition to concerns about the impact of silicone on health and the environment, there are additional drawbacks to the use of silicones in skincare and haircare.

While D5 is well known for its ability to remove long-wear and waterproof makeup, its occlusive properties can act as a barrier trapping oil, dirt, dead skin cells, and other comedogenic substances, thereby increasing the chance of clogged pores and potentially leading to acne. Hemisqualane is noncomedogenic.

In hair products, silicones create a seal that keeps hair hydrated from within, but this protective layer can block other nourishing ingredients from penetrating into the hair follicle. Over time, silicone can also build up on hair, resulting in a dry feel and dull appearance. Unlike silicones, hemisqualane penetrates the hair shaft for maximum nourishment and protection.

Both consumer and expert evaluations demonstrate that hemisqualane provides excellent sensorial and spreading properties and is ideal for replacing silicones in color cosmetics, suncare, haircare, and more. It also has makeup removal benefits and is a top replacement for isohexadecane, amodimethicone, and D5.

12.7 Purpose-driven: health and wellness

From its earliest days, Amyris was focused on health and wellness. Years after its success with artemisinin, Amyris partnered with a pharmaceutical intermediate Nenter & Co. to convert farnesene to vitamin E for distribution to the global nutraceuticals market [18]. This fat-soluble antioxidant vitamin is important for a strong immune system and healthy skin and eyes [19]. Human consumption of vitamin E supplements makes up a fraction of the market. Nearly 90% of the vitamin E produced goes into animal feed to promote health and immunity. Amyris' clean manufacturing lowered the traditional manufacturing cost by as much as 35% by simplifying the process and reducing the number of required process steps. In 2019, Amyris sold its vitamin E royalties to DSM for $57 million.

Just as Amyris was exiting the vitamin E business, it was entering the sugar substitute market, which today is projected to reach $20.6 billion globally by 2025, recording a CAGR of 4.5% during the forecast period [20].

In 2018, Amyris launched its newest compound, Rebaudioside M (Reb M), which is the active ingredient in the stevia leaf and found only at such extremely low concentrations (less than 0.1%), that it is very difficult to isolate the molecule using traditional extraction and harvesting processes. Using its fermentation and clean manufacturing process, Amyris successfully produces pure Reb M at scale, without reliance on leaf extraction or the associated challenges. Amyris Reb M sweetener, which received FDA GRAS (Generally Regarded as Safe) designation, is 350×–500× sweeter than sugar depending on the application from baking to beverages.

Requiring only miniscule amounts of Reb M in formulations, this pure, high potency sweetener tastes like sugar without the lingering and bitter aftertaste normally associated with stevia. It also has a zero glycemic index making it suitable for people living with diabetes. One in three Americans live with prediabetes with minority and low income communities most impacted. Seven in ten deaths are caused by illness are related to sugar intake. In the US alone, $3 trillion is spent annually on healthcare costs related to diabetes, obesity, and heart disease.

In 2021, Amyris signed a partnership valued at $100 million with Ingredion, a market leader in the global distribution of sweeteners. Its global customer reach, formulation capabilities, and commercial teams accelerate the availability and adoption of the fermentation-based, zero-calorie sweetener. Ingredion is also a minority owner of Amyris' Barra Bonita manufacturing facility that comes fully online in 2022. The partnership underscores Amyris' recognized leadership in synthetic biology in successfully developing, scaling, and commercializing pure ingredients that are sustainably sourced and produced by fermentation.

"We are excited by the addition of this breakthrough ingredient to our sugar reduction portfolio, which complements our PureCircle stevia product line and will allow us to provide our customers with the broadest selection of nature-based, high-intensity sweeteners on the market," said Jim Zallie, Ingredion's president and chief executive officer. "Our partnership with Amyris positions Ingredion well to meet our customers' increasing demand for quality ingredients that will drive transformational change in the food industry."

According to Ingredion, which offers Reb M made from all three production methods, Reb M from fermentation and bioconversion outperformed Reb M extracted from the leaf across all four key sustainability metrics (GHG emissions, land use, water scarcity, and culmulative energy demand), with the fermentation-based approach (which doesn't involve stevia plants at all) generating the lowest environmental impact by some degree. Ingredion went on to say that fermentation-based Reb M is the cheapest option, in addition to having the lowest environmental footprint [21].

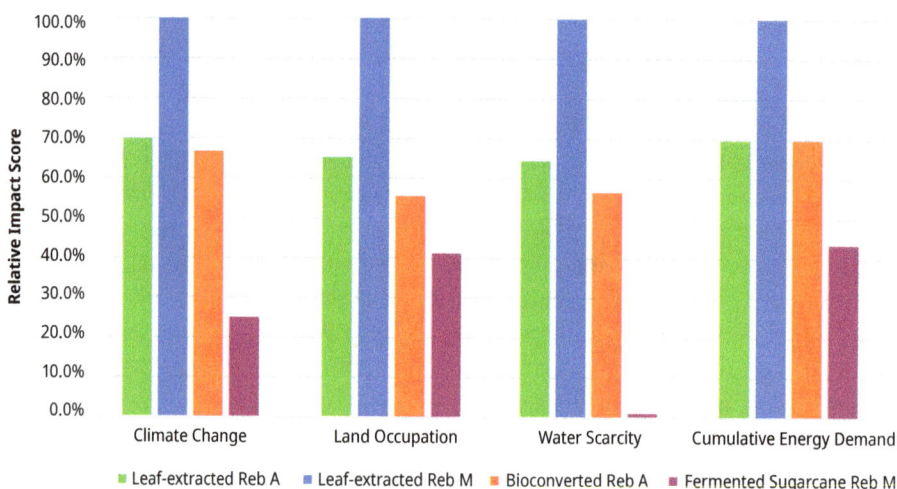

Figure 12.10:
Source: *Ingredion.*

Amyris also launched Purecane™, its own consumer brand of no-calorie sweetener powered by the hero ingredient, Reb M. The brand was named the No. New Release on Amazon.com, consistently earning 4+ star reviews for its products including zero-calorie packets and variety of zero-calorie, culinary, and baking sweeteners.

One of the best known, clean manufactured ingredients by Amyris is squalane, as discussed earlier. However, Amyris also produces squalene. While squalane is a stable, noncomedogenic lipid compound that is better suited for cosmetics squalene is an active ingredient in an important class of adjuvants (including influenza vaccines), which are used in vaccines to increase the immune response and help the vaccine to work more effectively. Squalene is traditionally harvested from deep-sea shark livers, which has the potential to devastate shark populations and contribute to long-term environmental issues. With funding from the National Institutes of Health (NIH), Amyris developed a pharmaceutical-grade squalene adjuvant that is a sustainable and scalable alternative to shark-derived adjuvants (vaccine boosters). Amyris is in conversation with leading pharma partners to commercialize the production and use of its squalene vaccine adjuvant. In 2022, Amyris entered into a joint venture with ImmunityBio to accelerate the commercialization of a next-generation Covid-19 vaccine.

Amyris and ImmunityBio combine important vaccine technology and manufacturing capabilities in the joint venture. Upon completion of successful human trials and regulatory approval, the joint venture's goal is to start delivering the second-generation vaccine as soon as is practically possible with a goal of delivering immunity for Covid-19 and access to underserved parts of the world where current vaccine technology is challenged due to cost and supply chain limitations.

"We are pleased to combine our expertise in human trials, T-Cell technology and our access to RNA manufacturing capacity with the Amyris and Infectious Disease Research Institute (IDRI [22]) RNA technology platform and Amyris' adjuvant technology," said Patrick Soon-Shiong, M.D., Executive Chairman and Global Chief Scientific and Medical Officer at ImmunityBio.

12.8 Partnerships are key

Partnering is key to Amyris' strategy. As a recognized leader in the biotechnology sector, Amyris has made deliberate choices to partner with industry leaders to maximize value from its technology and to demonstrate its ability to scale and commercialize ingredients faster, at a better cost point and with greater purity. Amyris has earned its leadership position in biotechnology from its repeated success at leveraging its proprietary Lab-to-Market™ technology platform to commercialize fermentation-derived ingredients.

> **Galley sidebar:**
> "We will meet the growing demand for clean ingredients not by depleting nature, but through clean chemistry, working with nature. Clean chemistry will continue to become a bigger component in applications for everyday consumer products as consumers continue to demand more natural and sustainable solutions. Our future growth is about more molecules, into more products and used at higher rates in each application. We're able to offer a reliable, high-quality supply of molecules that are lower cost and sustainable. It's the better choice in every way." Amyris President & CEO John Melo

Amyris first partnered with global leader Firmenich in 2011 to produce three, biobased flavor and fragrance ingredients and then later added new partnerships with Givaudan and International Flavors & Fragrances – two of the largest and best known flavor and fragrances houses in world. Ten years later, in a $50 million deal, Amyris licensed rights to DSM to assume the supply of farnesene to Givaudan for the production and sale of a single specialty ingredient. Later the same year, Amyris sold its flavors and fragrances (F&F) biobased intermediates business to DSM for $150 million. DSM has been an important strategic partner to Amyris over the years. In 2017, Amyris sold its manufacturing facility (Brotas 1) to DSM and accelerated its own plans for the construction of a new, state-of-the-art manufacturing plant for specialty products.

In April 2022, Amyris announced the commissioning of its newest, state-of-the-art fermentation plant in Barra Bonita, Brazil, with abundant availability of sustainable raw materials (sugarcane), securing production capacity for its rich pipeline of sustainable and biobased compounds. The new manufacturing site at Barra Bonita is large: 185,000 m^2 land area; 1,500 pieces of process equipment; 50,000 m of process piping; with a 4,000-ton steel structure.

Brazil has always been important to Amyris and will continue to play an important role in its future. Its new manufacturing plant is located next to the Raizen Barra Bonita sugar mill, which is the world's second largest sugar mill, at the heartland of Brazilian sugar cane production, in São Paulo State. Amyris has access to Barra Bonita's economic and sustainable sugar cane, the feedstock for Amyris' specialty fermentation process.

Amyris' feedstock supply partner, Raizen is the largest sugar cane producer in the world. They have led the way in ensuring that their Barra Bonita sugar cane is Bonsucro-certified. This certification ensures that our feedstock is ethically and sustainably produced and harvested under the strictest guidelines to ensure societal contributions and benefits. Amyris is the only biotechnology company that also has received this certification, an important distinction for Amyris' commercial customers and distribution partners, because it provides traceability for sustainable supply chains.

The Barra Bonita plant aids Amyris in its efforts to achieve a zero-carbon footprint in fermentation production. The plant's electric substation is colocated with Raizen's cogeneration plant. As of the time of this writing, Amyris' electrical plant converts the fiber waste stream from the sugarcane production into energy. And Amyris is exploring future innovations to reduce emissions including the use of sustainable biogas, processed from the vinasse waste stream. There are already plans underway to use biogas as sustainable fuel for its steam.

Amyris did not lose sight of the opportunity to positively impact other social aspects of Barra Bonita's community. The company prioritized local labor since the start of construction with nearly 80% of the workforce on the construction site from the local Barra Bonita region. Amyris has also partnered with technical and professional schools for the training of skilled operators at its highly automated plant.

The new manufacturing plant has a proprietary design with a process tower two times taller than Amyris' former Brotas plant which allows Amyris to reduce the amount of energy required in production and to take full advantage of gravity with the vertical fermentation process.

The clean manufacturing process starts at the top where the sugarcane syrup enters. Sterilizing and enriching the sugarcane into several variations of fermentation feedstock happen at the next level. Each feedstock is optimized to precise specifications for Amyris' different compounds. The materials go through three stages of fermentation: inoculation, seed growth, and then full-scale fermentation. The controls needed to achieve precise fermentation conditions for feeding, cooling, and production are located at the center of the tower. At the base, Amyris controls the harvesting of its fermentation broth before moving it to the purification stage.

The tower clusters five fermentation "mini factories," capable of concurrently producing any combination of Amyris' 13 different ingredients. Each line, which consists of two fermentation tanks, operates independently of the other. In total, these lines can produce around 6,000 tons per year of fermented ingredients sourced sustainably.

Amyris continues to invest in transformative technologies that benefit the environment and that are biology-based. As one of the earliest synthetic biology companies to successfully reach full-scale manufacturing with fermentation technology, Amyris has a unique perspective on what it takes to scale up fermentation and downstream processing technology. As a fundamental principle, Amyris prioritizes scaling up as quickly as possible to gain competitive business advantages. By reaching industrially relevant scales faster, Amyris can produce larger samples quickly, sample customers early and often for feedback, initiate any regulatory processes early on, and quickly assess how the technology is performing in manufacturing facilities.

Individually, any of those factors is compelling enough to rapidly push to industrial scale, and when leveraged holistically, Amyris has an unmatched view into manufacturing operations and market potential. From strain engineering to process development to full-scale deployment, Amyris has achieved impressive manufacturing milestones with 13 sustainable products in market and almost two decades of experience in 17 full-scale production facilities across seven countries, in addition to its new specialty precision fermentation plant in Barra Bonita.

With every successful fermentation run at industrial scale, Amyris gains another opportunity to closely examine sustainability metrics – like CO_2 footprint and water usage – to identify areas for optimization. As a result, Amyris continuously refines its life cycle analysis (LCA), receiving more precise feedback at every stage from inception to final product shipping to waste disposal. This feedback informs where the team should direct technology development efforts to improve the sustainability of Amyris products.

Through this understanding, Amyris can also better estimate the sustainability impact for new product opportunities being explored and thus make better choices for new technology development and deployment.

Clean manufacturing at full-scale capacity has allowed Amyris to deliver on its commitment to people and the planet – to Make Good. No Compromise™. As the company continues to expand its reach to even more consumers around the world, the Amyris LCA engine will only grow stronger, enabling the highest level of quality and performance for the world's sustainable ingredients.

Amyris is delivering clean products that today's consumers demand, while building a robust supply chain that allows the company to increase its clean manufacturing capacity. Amyris has invested in advanced, specialty fermentation capability. With its own manufacturing capability, Amyris gains more control over its ability to grow sustainably as a company. Amyris has learned that it is impossible to have real impact on sustainability and the health of people and the planet, without clean manufacturing at scale. Because Amyris believes that fermentation is the future of chemistry, we have a responsibility to clean up the world's chemistry in the fastest way possible.

Galley sidebar:
Amyris has commercialized 13 biobased ingredients
Artemisinin is a key ingredient used in antimalarial drugs, and ACTs are the first-line treatment recommended by the WHO. Artemisinin is traditionally found in the extract of Artemisia annua or the sweet wormwood plant. While ACTs are the most effective treatment, Artemisia crop yields can be unpredictable from year to year, causing price volatility of the plant-derived molecule and impacting overall availability of this life-saving drug. Amyris successfully engineered yeast to produce artemisinic acid, a molecule that can be converted to artemisinin at a significantly lower cost and without the risks of climate related crop variability. Using synthetic biology, this approach allows for "on demand" clean manufacturing to meet the global need for artemisinin and ACTs.

    Biofene, or trans-β-Farnesene, is a sustainable and versatile ingredient, produced via fermentation and synthetic biology. Biofene is a precursor molecule that with a bit of clean chemistry can be converted to other natural ingredients that Amyris produces. Applications of Amyris' Biofene include performance materials, adhesives, fragrances, surfactants, stabilizers, oligomers and polymers, resins, emulsifiers, vitamin precursors, crop protection, and foams, coatings, and sealants.

    Biosilica™. Silica is used widely in personal care and cosmetics and is traditionally sourced from nonrenewable sand dredging, which requires significant energy consumption and emits large amounts of CO_2. Amyris' sustainable and renewable Biosilica™ is derived from discarded sugarcane ashes and has similar or better performance characteristics compared to industry benchmarks. Applications suited for Biosilica™ are foundations, creams, lotions, and powders, as a sustainable and better performing alternative to extractive silica and microplastics. Biosilica™ has unique sphericity, high oil absorption, particle size, and oil/water absorption ratio. These characteristics offer premium benefits such as elegant texture, sebum control, mattification, and antiaging effects without drying the skin. Biosilica™ is also 100% biogenic (produced by nature), homogenous, soft to the skin, and has a silky application without blurring characteristics.

    Bisabolol is used in personal care products such as lotions, sunscreens, and face/eye creams. It has been a popular personal care ingredient for hundreds of years because of its skin healing antimicrobial, anti-inflammatory, and antiirritant properties. Bisabolol is traditionally sourced from either the German chamomile (*Matricaria recutita*) or more commonly, the Brazilian Candeia tree. The challenge with botanical sources comes down to availability, purity, and sustainability. When extracted from German Chamomile, the amount of bisabolol produced can vary greatly. It also requires a significant amount of plant material, making it cost prohibitive. The more common botanical source is from the endangered Brazilian Candeia tree. On average, it takes 12 years to grow a new tree, and a metric ton of tree material yields only about 7 kg of bisabolol. The limited availability of the Candeia tree also leads to significant price instability and fluctuations in product supply. Growing efforts to preserve Brazilian biodiversity have also led to innovation to identify alternative sources for bisabolol.

    CBG (Cannabigerol) is the precursor from which all other cannabinoids are synthesized and has exciting therapeutic potential to address consumer skincare concerns. To date, cannabinoids have been extracted directly from hemp plants, which require significant land usage, are difficult to scale, and often have regulatory risks related to THC. Using biotechnology and sugarcane fermentation, Amyris produces clean CBG, without the hemp plant. This process guarantees no THC in the final product and, therefore, carries no regulatory risk or challenges related to THC. Additionally, the entire process ensures environmental and social responsibility, starting with its Bonsucro-certified sugarcane feedstock in the fermentation process. Studies demonstrate CBG provides benefits to address common skincare conditions like dry skin, irritation, and redness. This fermented ingredient gives consumer goods companies a new option to formulate a completely new variety of oils, creams, and lotions with clean CBG.

Neossance™ **Hemisqualane** is the preferred sustainable and naturally derived alternative to silicone by top global brands. As a sustainable ingredient in virtually unlimited supply, it provides an equivalent sensory and functional performance to unhealthy, banned silicones like isohexadecane and cyclomethicone in skincare, haircare, suncare, makeup removal, and color cosmetics. Certain types of silicones can build up on hair that block out moisture and subsequently and prevent the hair from receiving moisture. Over time, silicones can make the hair appear dull and can also result in weaker hair due to lack of moisture. Hemisqualane is ideal for conditioners, treatments, and styling products as a sustainable and high-performing silicone replacement for hair care. Hemisqualane moisturizes and repairs the hair shaft at the root and from the inside out. It delivers color and thermal production, improved manageability, reduced frizz, and reduced greasiness without drying the scalp.

Manool is a key ingredient used to make woody, amber notes in the fragrance industry. Traditionally, manool is sourced from fallen Manoao pine trees, a native New Zealand species. Manoao pine trees are an endangered species and currently, the harvesting of fallen Manoao pines is restricted to one company to preserve New Zealand's natural biodiversity. As a result, manool is only available in small quantities from the traditional agricultural source and long-term availability can be difficult to predict. Amyris has developed a renewable solution to help protect the Manoao pine species as well as ensure a reliable supply of manool for the future.

Patchouli oil, native to Southeast Asia, has been traded for centuries along the Silk Road. It is extracted from the leaves of *Pogostemon cablin*, an exotic mint, and is one of the largest fragrance crops globally. Patchouli oil is used around the world, most commonly in fragrances and personal care products, with production estimated at 1,200 tons per year. Currently, the patchouli plant is one of the largest, but most variable fragrance crops in the world. In its botanical state, the fragrance is inconsistent and can have burnt or rubbery undertones, due to extraction methods and variability in region and environmental conditions. Amyris' fermentation-produced patchouli-type oil provides a cleaner, more consistent fragrance profile and is not subject to fluctuations in cost or availability. Because of the clean manufacturing, Amyris' patchouli-type oil doesn't have the harsh notes typical of patchouli extracted from agricultural sources, offering a unique, elevated woody note. For one hectare of land, traditional plant-based sources for patchouli yield less than 0.15 metric tons, whereas Amyris' fermentation-based approach yields 0.73 metric tons. Further, the land that patchouli grows on can be continuously cultivated on a single plot of land for only 3–4 years before soils are depleted and unsuitable for patchouli production; in comparison sugarcane – Amyris' feedstock – can be cultivated on the same plot of land for up to 200 years. Amyris' patchouli-type oil helps meet the growing demand from formulators interested in this fragrance and offers higher purity, stabilized pricing, and a reliable supply chain source. Due to the elevated, clean scent of Amyris' patchouli-type oil, it has also led to innovative product applications and can be found in fine fragrances, haircare, body care, soap, candles, laundry detergents, fabric softeners, room sprays, and more.

Purecane™ **Reb M** is a zero-calorie sweetener – pure, sustainable with a zero glycemic index, and is certified non-GMO. Reb M offers formulators a high-purity, high-potency sweetener with a taste profile that mimics sucrose with no lingering aftertaste. Reb M is bioidentical to stevia's sweetest molecule, present in the leaves at only very low concentrations (less than 0.1%), making it difficult to isolate using traditional extraction and harvesting processes. But Amyris creates the pure Reb M molecule through fermentation, by engineering yeast and fermenting sugarcane, so none of the challenges that exist with leaf purification exist with Amyris' clean chemistry.

Santalols are the active components of sandalwood oil that gives it its signature woody aroma. Sandalwood has been deeply tied to Indian culture over generations and has traditionally been harvested with great respect and a mind toward sustainability – a sandalwood tree must grow for 30 years to reach full maturity before its essential oil can be harvested. But over the course of decades, between population growth and globalization, the trees have been overharvested and the supply has dwindled. As a result, pricing of sandalwood has skyrocketed, it has become inaccessible, and today, it is a threatened species.

Sclareol is a key component in plants ranging from clary sage to key limes that makes aromas linger longer in products. Sclareol has come to replace a fragrance material called ambergris, which is derived from sperm whale secretion and is difficult to find and cost prohibitive – its key chemical constituent is ambroxide. As a result of the limited supply available from sperm whales, the industry now uses an alternative to ambroxide: sclareol, commonly derived from clary sage plants, a flowering herb native to Europe.

Squalane. Amyris' Neossance™ squalane is a high-quality and versatile emollient used in a wide variety of applications such as skincare, sun care, color cosmetics, makeup removal, and deodorants. Squalane has proven skin therapy benefits in formulations, including an increase in skin firmness and the reduction in the appearance of fine lines and wrinkles. Squalane boosts skin health with moisturizing, antiaging, and brightening skin improvements from a natural ingredient that is ethically and sustainably sourced. Squalane is the more stable, hydrogenated version of squalene, a compound we are born with as babies but naturally diminishes as we age. Traditionally, beauty companies have relied on shark livers for this ingredient, threatening shark populations with long-lasting consequences to our ocean's biodiversity. An alternate traditional source, olive oil, carries dependencies on weather, crop yields, and environmental supply chain risks. Neossance™ squalane offers purity and quality, helping brands formulate consistent and high-performing products, without causing environmental harm.

Squalene is a natural oil that is the active component in an important class of vaccine adjuvants. Adjuvants are a key ingredient used in vaccines to increase the immune response and ultimately help the vaccine work more effectively. Adjuvants thus allow less of the immunogen (the microbe or protein that the vaccine is targeting) to be used when manufacturing vaccines. Squalene has known properties as an effective carrier system for immunogens. Traditionally, squalene is harvested from deep-sea shark livers, which has the potential to devastate shark populations and have long-lasting ecological consequences. Amyris' pharmaceutical grade squalene is molecularly identical to the traditional shark source and can also provide a sustainable, traceable, and reliable source of a critical vaccine ingredient.

References

[1]   Dodd MS, et al. Evidence for early life in Earth's oldest hydrothermal vent precipitates. Nature 2017;543:60–64.

[2]   Interlandi J. Newsweek 2008.

[3]   Magat C. Stanford University Graduate School of Business Case Study on Amyris, 2009.

[4]   Amyris News Release, Globe newswire, (2016). Specter M. The New Yorker, 2009.

[5]   Cohen JM, Singh I, O'Brien ME. Predicting global fund grant disbursements for procurement of artemisinin-based combination therapies. Malar J 2008;7:200. 10.1186/1475-2875-7-200. PMID: 18831742; PMCID: PMC2570684.

[6]   Sanders R. UC Berkeley News 2004.

[7]   Amyris News Release, Globe newswire, 2015.

[8]   Gormley B. *Wall Street Journal* 2016.

[9]   Villas-Boas JM. Berkeley Haas Case Series, Amyris, 2019.
[10]   Villas-Boas JM Berkeley Haas Case Series, Amyris, 2019.
[11]   Nasdaq.com, 2010.
[12]   Environmental Protection Agency, 2014.
[13]   Amyris Press Release, Globenewswire, 2014.
[14]   Villas-Boas JM. Berkeley Haas Case Series, Amyris, 2019.
[15]   Amyris Inc. Q2 2018 Earnings Call, 2018.
[16]   Ducos L, et al. Bloom Association, 2015.
[17]   McPhee D, et al, Cosmetics & Toiletries magazine. 2014.
[18]   Amyris News Release, Globe newswire, 2017.
[19]   National Institutes of Health, Office of Diary Supplements, 2021.
[20]   Sugar Substitutes Market Report, MarketsandMarkets, 2020.
[21]   Food Navigator. Which Reb M production method is best for the environment? Ingredion LCA probes stevia sustainability metrics. August 2022.
[22]   AAHI Press release. IDRI now AAHI, Businesswire, 2022.

Crispinus Omumasaba, Alain A. Vertés, F. Blaine Metting*,
Hideaki Yukawa**

Chapter 13
Hydrogen bacteria: a developing platform for the sustainable carbon dioxide-based bioindustry

13.1 Introduction

Autotrophic CO_2 fixation generated the fossil coal, gas, and petroleum reservoirs that currently satisfy more than 80% of global energy demand [1]. Humankind has depended on these fossil fuels since the Industrial Revolution. Fossil fuel-derived energy is used to supply a wide range of products, fuel global transportation systems, and industries and to heat homes and cook food. The unbridled exploitation of fossil fuels has led to serious concerns about long-term availability and sustainability, encompassed in the concept of "peak oil," on top of its global environmental impact. The consensus that climate change manifested by global warming can no longer be disputed or ignored has spurred research and application of alternative, renewable energy and product sources; that is, reduced carbon feedstocks at global industrial scale. Terrestrial ecosystems and oceans are increasingly burdened with industrial pollutants, wastes, and plastics even as the atmosphere accumulates greenhouse gases (primarily CO_2 and methane) that drive climate change. Farmlands and forests are increasingly impacted by changing climate as reflected by excessive drought, flooding, and wildfire. Also the "food vs. fuel" debate has intensified in view of the need to meet the caloric and nutritional requirements of a human population projected to approach or exceed 10 billion within three decades. In response, biofuel and bioproducts industries have grown rapidly as have demands for raw materials [2, 3].

Meanwhile, international debate on how to address climate change through improved basic understanding, mitigation, and adaptation has been ongoing since the 1980s. Intergovernmental negotiations have produced several important accords, notably the 1987 Montreal Protocol [4], the 1992 UN Framework Convention on Climate Change [5],

---

Note: Pacific Northwest National Laboratory, Retired.

**Corresponding author: Hideaki Yukawa, Utilization of Carbon Dioxide Institute Co. Ltd., 2-4-32 Aomi, Koto, Tokyo, 135-0064, Japan, e-mail: hyukawa@co2.co.jp
*Corresponding author: F. Blaine Metting, Utilization of Carbon Dioxide Institute Co. Ltd., 2-4-32 Aomi, Koto, Tokyo, 135-0064, Japan
Crispinus Omumasaba, Alain A. Vertés, Utilization of Carbon Dioxide Institute Co. Ltd., 2-4-32 Aomi, Koto, Tokyo, 135-0064, Japan

https://doi.org/10.1515/9783110791228-013

the 2005 Kyoto Protocol [6], the 2015 Paris Agreement [7], and the 2021 Glasgow Climate Pact [8]. There is general consensus that anthropogenic CO_2 release is a key driver of climate change and that immediate measures be taken to prevent the global average temperature from rising beyond 1.5–2 °C. Limited global warming necessitates major transitions in the energy sector with not only a substantial shift away from fossil fuels but also dramatically improved energy efficiency. To avoid catastrophic climate change impacts and minimize global warming, the Intergovernmental Panel on Climate Change proposed 116 action plans, all but 15 of which involve carbon capture and storage [9].

Atmospheric CO_2 capture is very challenging as is determining how to efficiently and economically utilize the collected carbon. Outside of processes that capture solar energy to fix CO_2, biological processes that tap energy stored in low value carbon sources share the capacity for self-renewal and chemical energy storage largely compatible with existing industrial and commercial infrastructure. Nature is replete with fermentative microorganisms that grow on $CO_2/H_2/O_2$ feed streams via mechanisms that are not yet entirely elucidated. Regardless, understanding and consequently exploiting biological CO_2 fixation mechanisms, premised on the fact that autotrophic fixation is responsible for the production of most global organic matter, will eventually unlock unrealized potential for sustainable energy and value-added chemicals. As dwindling fossil fuel reserves move global society toward strategies to control CO_2 emissions, future industrial practices will become increasingly CO_2 emission-sensitive in the pursuit of a carbon neutral economy.

Sound industrial strategies for attaining carbon neutrality lie in implementation of the so-called circular economy [10], with no net CO_2 release to the atmosphere. Reaching carbon neutrality is an ongoing technical and economic challenge which few industries can claim total compliance with the full slate of existing legislative and policy goals [11].

It long has been known that some bacteria use atmospheric trace gases, including H_2, for energy and carbon fixation. These HOB, alternatively called hydrogen bacteria or "knallgas bacteria," have been isolated from diverse ecological niches ranging from desert soils [12, 13] to deep sea [14, 15] and geothermal environments [16, 17]. Growing autotrophically, HOB utilize external H_2 as an electron donor with oxygen (O_2) as electron acceptor to reduce CO_2. Among diverse taxonomic units, HOB are distinct from hydrogen-oxidizing acetogens [18, 19] that similarly oxidize H_2 but without autotrophic CO_2 fixation. They also differ from chemolithoautotrophs that utilize H_2 under anaerobic conditions with sulfate (SO_4) or CO_2 as electron acceptors [20] and electrolithoautotrophs which derive electrons from electric current [21, 22].

Ever since a report of "knallgas bacteria" as sources of single-cell protein [23] attracted interest from NASA [24], the ability of these bacteria to fix CO_2 in the absence of light has intrigued the scientific community. So long as photorespiration remains a principal drawback to agriculture and other biotechnologies that rely on primary producers such as plants and algae, HOB represent the best option to realize high yield bioprocessing [25]. Importantly, HOB typically accumulate protein at levels (50–83% dry cell weight) well beyond those seen in most plants and algae [26], with essential amino acid scoring patterns

(Table 13.1) that remarkably match the FAO/WHO/UNU-recommended standard pattern for human nutrition [27]. With ample evidence that HOB-derived biomass qualifies as an excellent source of nutritive protein, they are an important alternative source that can help meet the needs of an increasingly affluent yet constrained global population [3]).

More than half a century since the seminal report on HOB as a source of single-cell protein [23], only a few strains have since received attention for industrial application [28]. By far the most prolific of these is *Cupriavidus necator* H16 (formerly *Ralstonia eutropha* H16), a Gram-negative, facultative chemolithoautotroph able to assimilate CO_2 via the Calvin cycle [29]. Others with industrial potential that have received significant research attention include *Rhodococcus opacus*, a Gram-positive bacterium with exciting potential for lignin and industrial waste valorization [30] and *Xanthobacter autotrophicus*, a Gram-negative rod-shaped bacterium which produces considerable amounts of the food dye and antioxidant zeaxanthin [30], and has shown potential remediation of groundwater [31, 32]. Current research is focused on finding new strains, substrates, or cultivation conditions to economically produce microbial protein and value-added chemicals.

13.2 HOB in the CO_2-based bioindustry

A CO_2-based bioindustry is evolving toward established production methods in the chemical industry [19], with characteristic distinctions in which biomass itself is the desired end product. Consequently, the CO_2-based bioindustry can be said to be organized in value-added chains from raw materials to desired end-products via specific intermediates. Currently, CO_2-containing feed stream gases are mostly of fossil fuel origin. Driven by competitiveness and the pursuit of carbon neutrality, however, use of gas streams from renewable materials is increasing. Viable industrial-scale practices based on gas streams from renewable materials cannot be at the expense of end-product cost-effectiveness. Also, products resulting from these processes must minimally meet equivalent quality and performance as from fossil resources. As a result, irrespective of the raw material, each bioprocess is subject to the following constraints:

(i)   Processes for drop-in fuels and chemicals require inherent potential to match or exceed the efficiency and economic performance of mature production processes;

(ii) Product and process specifications must be sensitive to variability among raw materials and the consequences of biosynthesis;

(iii) Methods, techniques, and equipment to economically handle large volumes of aqueous systems and purify molecules of interest in an aqueous environment are indispensable;

(iv) Over the long term, factoring in economic reality and depreciation of existing assets, value-added chains will be altered to take full advantage of bioprocesses which might lead to more new products because biomass feedstocks are typically in a more oxidized redox state.

The gaseous $CO_2/H_2/O_2$ feed stream as a substrate has several advantages including global availability, noncompetitiveness with food or feed, predictable pricing through well-proven market mechanisms, and full metabolic utility. Moreover, HOB use results in few impurities derived from nonreactive carbon, good stability, reduced risk of microbial contamination, and accessibility from waste streams. Last but not least, the gaseous $CO_2/H_2/O_2$ feed stream as a substrate is an important enabler for circular economy considerations.

13.3 Native and heterologous pathways for chemicals

The Calvin–Benson–Bassham (CBB) cycle upon which HOB depend to fix CO_2 is the most prevalent CO_2 assimilation mechanism on Earth having evolved for optimal synthesis of C3 metabolites, but not for the production of the C2 building block acetyl-CoA [33]. Carbon lost in producing acetyl-CoA from step-wise decarboxylation of C3 sugar limits the maximum carbon yield of the CO_2 fixation process but avails two crucial starting points for anaplerotic CO_2 assimilation pathways catalyzed by PEP carboxylase (ppc)

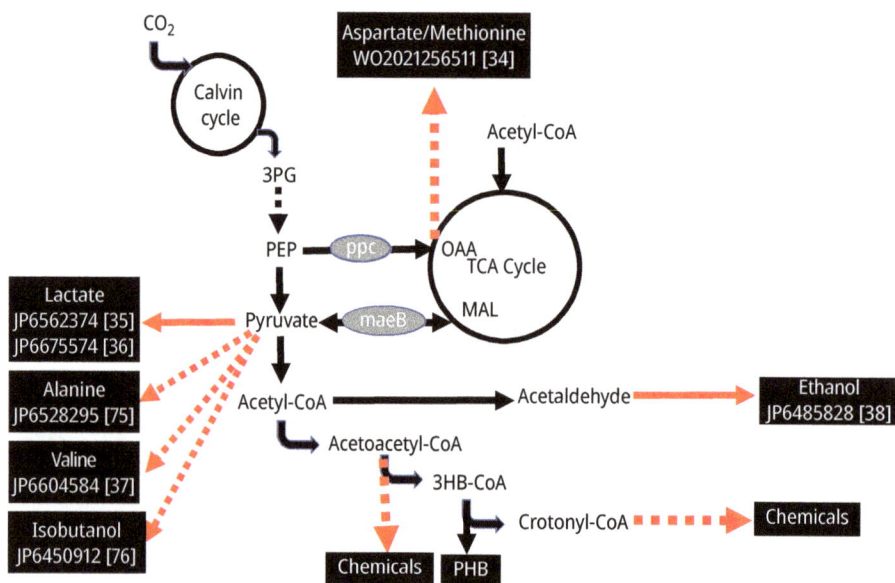

Figure 13.1: Schematic diagram of CO_2 fixation pathway in HOB showing bioproducts and patents (where applicable) assigned to Utilization of Carbon Dioxide Institute (UCDI). Black arrows indicate native reaction steps. Red arrows represent pathways where at least one step is not native. Dotted arrows signify multiple steps. Abbreviations: 3 PG, 3-phosphoglycerate; PEP, phosphoenolpyruvate; OAA, oxaloacetate; MAL, malate; 3HB-CoA, 3-hydroxybutyryl-CoA; ppc, PEP carboxylase; maeB, malic enzyme.

and malic enzyme (maeB) as well as heterologous pathways to valuable chemicals (Figure 13.1).

Early success has been achieved in constructing heterologous pathways starting from pyruvate and a number of products starting from PEP are in the pipeline. Interest in leveraging these off-ramps parallels those in cyanobacteria in which the carbon flux in two prolific strains has been redirected to produce a number of chemicals via a heterologous pathways centered on the CBB cycle [39]. The rising number of patents on HOB-based bioproduct formation from CO_2 is testament to the improving feasibility of HOB as industrial strains.

Atmospheric CO_2 in photosynthesis or bicarbonate (HCO_3^-) ions in nonphotosynthetic CO_2 fixation is reduced to 3-phosphoglycerate (3 PG), the C3 metabolite which serves as a precursor for all cellular constituents and most of the reduced carbon on Earth [40]. This cycle exhibits a number of limitations, including O_2 sensitivity to that previous research addressed by modifying relevant enzymes [41–45] or by augmenting the cycle with synthetic pathways that are insensitive to O_2 [40]. Given that CO_2 fixation by HOB is relatively resilient to O_2 [46] arguably due to the presence of multiple hydrogenases [47] that not only catalyze the oxidation of H_2 to form $2e^-$ and $2[H^+]$ [42] but also support cell growth and survival [48, 49]. It is noteworthy that the genomes of HOB reveal a carbonic anhydrase enzyme which should function to generate HCO_3^- ions from CO_2 gas bubbled into reaction medium. Not much has been reported about the impact of this enzyme on the observed efficacy of CO_2 fixation by HOB. Regardless, the conversion efficiency of CO_2 to acetyl-CoA is no doubt improved by the aforementioned anaplerotic CO_2 assimilation pathways and may contribute to the fast growth of HOB cells. The essential amino acid content of the biomass compares favorably with the best alternative proteins emerging on the market (Table 13.1).

Table 13.1: Amino acid content (% of total, dry weight basis) of candidate alternative protein replacements.

Amino acid	Hydrogenophilus thermoluteolus	Fusarium venenatum[a]	Tofu[a] (Glycine max)	Green peas[a] (Pisum sativa)
Isoleucine	2.62	2.32	2.82	0.92
Leucine	4.92	3.83	4.61	1.53
Histidine	1.46	1.17	1.43	0.51
Lysine	3.15	4.04	2.93	1.50
Methionine	1.33	0.88	0.70	0.39
Phenylalanine	2.64	2.13	2.77	0.95
Threonine	2.77	2.57	2.60	0.96
Tryptophan	1.17	0.85	0.78	0.18
Valine	3.80	5.12	2.89	1.11
Tyrosine	2.34	1.64	2.32	0.54
Serine	2.02	2.36	3.36	0.86

Table 13.1 (continued)

Amino acid	Hydrogenophilus thermoluteolus	Fusarium venenatum[a]	Tofu[a] (Glycine max)	Green peas[a] (Pisum sativa)
Proline	2.53	2.22	3.59	0.82
Glycine	3.00	2.30	2.43	0.87
Alanine	4.56	2.99	2.56	1.14
Glutamic acid	6.4	5.51	10.90	3.51
Aspartic acid	5.08	4.67	2.35	2.35
Arginine	4.09	2.96	4.54	2.02
Cysteine	0.43	0.40	0.19	0.15

[a]GRAS Notice (GRN) No. 904 https://www.fda.gov/food/generally-recognized-safe-gras/gras-notice-inventory.

13.4 HOB strains and products of interest

At the turn of the millennium, a number of companies exploiting fermentative micro-organisms to utilize CO_2 as a feedstock began to emerge around the world. It is difficult to ascertain exactly what strains are favored and by which companies, but strains belonging to the genera *Cupriavidus, Azohydromonas, Herbaspirillum, Pseudomonas, Paracoccus* (formerly *Flavobacterium*), *Sulfuricurvum, Azonexus*, and *Xanthobacter* are commonly used. *Cupriavidus necator* H16 is hardy and produces biomass containing 50–60% single-cell protein (SCP) even in the presence of carbon monoxide (CO), a bacterial growth inhibitor [50]. It is worth noting that the production of acetate from CO_2 by *Sporomusa ovata* and polyhydroxybutyrate by closely related *Cupriavidus* strains from acetate was reported [51, 52]. *Rhodococcus opacus* is attractive for lipid production for nutrition, biofuel, and commodity chemicals [53]. *Xanthobacter autotrophicus* produces copious amounts of zeaxanthin, a carotenoid used as a food dye. Conventionally produced from plants via a costly, labor-intensive operation with downstream purification requiring multiple processing steps involving harsh solvent extractions [54], the dye can be produced by some *Paracoccus* [55]. However, a keenly anticipated application of *Paracoccus* is in the degradation of chlorinated hydrocarbons [32].

The Tokyo-based UCDI is pioneering the use of a fast-growing (doubling time of 1 h is one-half that of *C. necator*), moderately thermophilic HOB, *Hydrogenophilus thermoluteolus* [56]. The first thermophilic HOB to be used in SCP production, this opens doors to high-temperature bioprocessing not possible with mesophilic HOB strains. Moreover, *H. thermoluteolus* accumulates a much higher protein content than typical bacteria or other HOBs. Christened UCDI® Protein, the protein mix produced by this organism is not only sustainable but also rich in RuBisCO (ribulose-1,5-bisphosphate carboxylase/oxygenase). RuBisCO is the predominant enzymatic mechanism in the biosphere by which autotrophic bacteria, algae, and terrestrial plants fix CO_2 into organic biomass via the CBB reductive pentose phosphate pathway. Universal, but catalytically modest, RuBisCO is the subject of intense interest by researchers aiming to enhance carbon fixation in plants and prokaryotes [57, 58]. RuBisCO is known to enhance the secretion of serotonin (happiness hormone) and melatonin (sleep regulation hormone) and elevate tryptophan levels in the human brain [59]. UCDI® Protein is the basis of food ingredients expected to contribute to mental and physical health. UCDI® Protein therefore holds the promise of an appealing animal protein alternative with excellent taste characteristics and good health outcomes. In parallel, UCDI is advancing technology for fossil fuel-free plastics. UCDI® PLASTIC contains polylactic acid and polyethylene using proprietary technologies for producing lactic acid from CO_2. Final materials are derived without resorting to petroleum and offer a wide range of potential applications including shopping bags, food trays, agricultural materials, tableware, stationery, and construction materials.

While SCP is the most common product from companies employing HOBs, value-added products such as oils, biochemicals, bioplastics, and biofuels can be produced from a variety of raw materials. For example, Finland-based Solar Foods manufactures an SCP for which Novel Food license approval is pending in the European Union. Similarly, Belgium-based Avecom has developed several SCPs from low-value substrates. Likewise, UK-based Deep Branch Biotechnology, which produces SCP with a tailored amino acid profile, pilots its SCP production using CO_2 from flue gas. As HOBs begin to permeate contemporary food and chemicals industries, they contribute to value creation and provide new jobs and livelihoods based on the nascent CO_2 industry. The combined market for meat, eggs, dairy, and seafood products is projected to grow seven-fold from $35 billion to reach at least $290 billion by 2035 as growing consumer sentiment drives unparalleled growth in plant-, microorganism-, and animal cell-based alternatives [60].

13.5 HOBs for bioremediation

The Japanese Ministry of Economy, Trade and Industry (METI) awarded US-based waste management specialist Kurion a $10 million grant to demonstrate technology to remove tritium from contaminated water for possible deployment at Fukushima following the great Tohoku earthquake and tsunami of 2011. Remarkably, HOBs have been shown capable of reducing orthophosphate to ultra-low concentrations in a fed-batch reactor [61]. Perchlorate (ClO_4^-) is a strong oxidizing agent of interest due to its reactivity, occurrence, persistence in surface water, groundwater, soil, and food [46]. Perchlorate-contaminated water is a widespread problem as more sites are identified worldwide [62]. Perchlorate contamination can be a health concern due to its ability to disrupt the use of iodine by the thyroid gland and the production of metabolic hormones and has been linked to brain damage in infants [63]. The chemical is a component of rocket fuel, ammunition, fireworks, and explosives and airbag initiators for vehicles, matches, and signal flares. Exposure can also damage the development of fetuses and children. The US Environment Protection Agency has twice opted not to regulate perchlorate since 2020.

Biological perchlorate reduction is a promising alternative to conventional physical/chemical treatment processes and has the advantage of converting perchlorate to the benign products chloride and oxygen. A number of bacteria are capable of reducing perchlorate using a variety of electron donors including organic carbon compounds, hydrogen, iron, and reduced sulfur compounds. Treatment technologies that can remove perchlorate from drinking water without introducing organic chemicals that stimulate bacterial growth in water distribution systems are ideal. Hydrogen is an ideal energy source for bacterial degradation of perchlorate as it leaves no organic residue and is sparingly soluble [64].

13.6 Future perspectives

Several fundamentally different CO_2 fixation pathways have evolved that vary in their reaction sequences and types of carboxylases, cofactors, and electron donors used to fix inorganic carbon into biomass [65, 66]. The conversion of CO_2 into value-added products attracts great attention to autotrophic metabolism, rendering it a popular topic for fundamental and applied research. The conventional manufacturing approach linking two such processes is to purify the acetate intermediate between synthesis steps; the feasibility of avoiding this otherwise costly step has recently been demonstrated [67]. Major strides in basic understanding of HOB continue to expand the potential range of applications in the production of industrial chemicals [68] and cultivation of protein-rich biomass for human and animal nutrition, resource recovery, and bioremediation [69, 70].

Plastics are ubiquitous in contemporary society with the ever-increasing use of these fossil-based materials responsible for emissions of 1.7 $GtCO_2eq/y$, corresponding to roughly 3.8% of global greenhouse gas emissions [71]. Moreover, their disposal exerts undue stress on the environment with up to 8 million tons of plastic waste collecting annually in the world's oceans with particularly detrimental effects of microplastics on aquatic life and, by extension, human health [72].

Concerns about availability, sustainability, and cost of fossil fuel reserves have invited warnings that 60% of Earth's remaining oil and natural gas and 90% of the remaining coal must stay locked underground if potentially catastrophic rise in global temperatures are to be avoided [73]. In order to reduce global reliance on fossil fuels, renewable energy must be brought on line as a matter of urgency. And, innovative ideas for engineering novel autotrophic, CO_2-fixation pathways into industrially tractable organisms must be pursued.

Beyond the resetting of the energy mix of every nation, especially those that cover much of their needs from coal-fired power stations, countering global warming necessitates a total rethinking of future food and energy security. The cost, volume, and energy density of alternative fuels remain concerns that must be addressed by better technologies to improve cost-efficiencies even as costs of climate disasters are not factored into the cost of goods sold. Concerted efforts to reverse the adverse effects of fossil fuels while not sacrificing energy security drove interest in biofuels, with Brazil managing to replace conventional fuels for transportation with ethanol. Nevertheless, current biofuels remain susceptible to fluctuations in oil prices, with the resulting extreme variability of biorefinery profit margins a major hurdle to large-scale deployment of biomass as an industrial feedstock. Biodiesel and bioethanol, the two primary biofuels, suffer from their dependence on a narrow range of agricultural feedstocks (notably grains and sugarcane) as well as a limited variety of production processes. Carbon-based synthetic biofuels which can be produced from any type of biomass could address these limitations with appropriate biotechnologies. HOB can enter this domain of synthetic biofuels where fuels are manufactured via biochemical conversion processes from "defossilised" CO_2 sources such as point source capture from industrial processes and direct capture from air.

With the HOB ability of 1-aminocyclopropane-1-carboxylate conversion and phosphate solubilization, the bacteria can address the functions of biofertilizers. The enriched HOB can recover nitrate from wastewater without any secondary nitrogen pollution, extending HOB application for resource recovery from wastewater [74]. The fundamental attribute of HOB being that these autotrophic organisms are able to use H_2 as an electron donor, making them particularly interesting in co-cultures as a means to extend the potential range of useful biotechnological processes. Anaerobic bacteria are particularly interesting because anaerobiosis is typically less challenging to achieve at large industrial scale. What is more, reducing microbial metabolism enables HOB-driven biotechnological processes to be incorporated into the pathways of conventional petro chemistries. This is very much worth exploring to achieve economies of scale and scope that industrial biotechnology needs to achieve to compete with fossil-fuel-driven petrochemical processes.

References

[1] Ducat DC, Silver PA. Improving carbon fixation pathways. Curr Opin Chem Biol 2012;3-4:337–44. doi: 10.1016/j.cbpa.2012.05.002.

[2] Ezeh, AC, Bongaarts, J, Mberu, B. Global population trends and policy options. *Lancet* 2012;**380** (9837): 142–148 doi: 10.1016/s0140-6736(12) 60696–5

[3] World Population Prospects. 2022. https://www.un.org/development/desa/pd/sites/www.un.org. development.desa.pd/files/wpp2022_summary_of_results.pdf (Accessed 10 September 2022).

[4] Montreal Protocol. (https://treaties.un.org/doc/publication/unts/volume%201522/volume-1522-i-26369-english.pdf).

[5] UN Framework Convention on Climate Change. https://unfccc.int/files/essential_background/background_publications_htmlpdf/application/pdf/conveng.pdf.

[6] Kyoto Protocol. https://unfccc.int/resource/docs/convkp/kpeng.pdf.

[7] Paris Agreement. https://sustainabledevelopment.un.org/frameworks/parisagreement.

[8] Glasgow Climate Pact. (https://unfccc.int/sites/default/files/resource/cma2021_10_add1_adv.pdf).

[9] IPCC Climate Change 2022: Mitigation of Climate Change, https://www.ipcc.ch/report/sixth-assessment-report-working-group-3/.

[10] Liguori R, Faraco V. Biological processes for advancing lignocellulosic waste biorefinery by advocating circular economy. Bioresour Technol 2016;215:13–20.

[11] Nachmany M, Fankhauser S, Setzer J, Averchenkova A. Global trends in climate change legislation and litigation – 2017 update, 1997. (http://www.lse.ac.uk/GranthamInstitute/wp-content/uploads/2017/04/Global-trends-in-climate-change-legislation-and-litigation-WEB.pdf).

[12] Ehsani E, Dumolin C, Arends JBA, Kerckhof FM, Hu X, Vandamme P, Boon N. Enriched hydrogen-oxidizing microbiomes show a high diversity of co-existing hydrogen-oxidizing bacteria. Appl Microbiol Biotechnol 2019;103:8241–53. doi: 10.1007/s00253-019-10082-z.

[13] Jordaan K, Lappan R, Dong X, Aitkenhead IJ, Bay SK, Chiri E, Wieler N, Meredith LK, Cowan DA, Chown SL, Greening C. Hydrogen-oxidizing bacteria are abundant in desert soils and strongly stimulated by hydration. mSystems 2020;5:e01131–20. doi: 10.1128/mSystems.01131-20.

[14] Nakagawa S, Takai K. Deep-sea vent chemoautotrophs: diversity, biochemistry and ecological significance. FEMS Microbiol Ecol 2008;65:1–14. doi: 10.1111/j.1574-6941.2008.00502.x.

[15] Sass K, Güllert S, Streit WR, Perner M. A hydrogen-oxidizing bacterium enriched from the open ocean resembling a symbiont. Environ Microbiol Rep 2020;12:396–405.

[16] Goto E, Kodama T, Minoda Y. Isolation and culture conditions of thermophilic hydrogen bacteria. Agric Biol Chem 1977;41:685–90.

[17] Vésteinsdóttir H, Reynisdóttir DB, Örlygsson J. *Hydrogenophilus islandicus* sp. nov., a thermophilic hydrogen-oxidizing bacterium isolated from an icelandic hot spring. Int J Sys Evol Microbiol 2011;61:290–94.

[18] Leigh JA, Mayer F, Wolfe RS. *Acetogenium kivui*, a new thermophilic hydrogen-oxidizing, acetogenic bacterium. Arch Microbiol 1981;129:275–80.

[19] Takors R, Kopf M, Mampel J, Bluemke W, Blombach B, Eikmanns B, Bengelsdorf FR, Weuster-Botz D, Durre P. Using gas mixtures of CO, CO2 and H2 as microbial substrates: The do's and don'ts of successful technology transfer from laboratory to production scale. Microbial Biotechnol 2018;11 (4):606–25. doi: 10.1111/1751-7915.13270.

[20] Aragno M, Schlegel HG. The Hydrogen-Oxidizing Bacteria. In: Starr MP, Stolp H, Trüper HG, Balows A, Schlegel HG, editors, The Prokaryotes, Berlin, Heidelberg: Springer; doi: 10.1007/978-3-662-13187-9_70.

[21] Ishii T, Kawaichi S, Nakagawa H, Hashimoto K, Nakamura R. From chemolithoautotrophs to electrolithoautotrophs: CO_2 fixation by fe(ii)-oxidizing bacteria coupled with direct uptake of electrons from solid electron sources. Front Microbiol 2015;6:994. doi: 10.3389/fmicb.2015.00994.

[22] Yamada S, Takamatsu Y, Ikeda S, Kouzuma A, Watanabe K. Towards application of electro-fermentation for the production of value-added chemicals from biomass feedstocks. Front Chem 2022;9:805597. doi: 10.3389/fchem.2021.805597.

[23] Schlegel HG, Lafferty R. Growth of 'knallgas' bacteria (Hydrogenomonas) using direct electrolysis of the culture medium. Nature 1965;205:308–09. doi: 10.1038/205308b0.

[24] Mateles RI, Baruah JN, Tannenbaum SR. Growth of a thermophilic bacterium on hydrocarbons: A new source of single-cell protein. Science 1967;157:1322–23.

[25] Hasegawa S, Suda M, Uematsu K, Natsuma Y, Hiraga K, Jojima T, Inui M, Yukawa H. Engineering of *Corynebacterium glutamicum* for high-yield l-valine production under oxygen deprivation conditions. Appl Environ Microbiol 2012;79:1250–57.

[26] Matassa S, Boon N, Pikaar I, Verstraete W. Microbial protein: future sustainable food supply route with low environmental footprint. Microbial Biotechnol 2016;9:568–75. doi: 10.1111/175-7915.2369.

[27] FAO/WHO/UNU Amino acid scoring patterns, https://www.fao.org/3/M3013E00.htm.

[28] Kerckhof F-M, Sakarika M, Van Giel M, Muys M, Vermeir P, De Vrieze J, Vlaeminck SE, Rabaey K, Boon N. From biogas and hydrogen to microbial protein through cocultivation of methane and hydrogen oxidizing bacteria. Front Bioeng Biotechnol 2021;9:733753. doi: 10.3389/fbioe.2021.733753.

[29] Claassens NJ, Scarincia G, Fischer A, Flamholz AI, Newell W, Frielingsdorf S, Lenz O, Bar-Even A. Phosphoglycolate salvage in a chemolithoautotroph using the calvin cycle. PNAS 2020;117:22452–61. doi: 10.1073/pnas.2012288117.

[30] Chatterjee A, DeLorenzo DM, Carr R, Moon TS. Bioconversion of renewable feedstocks by *Rhodococcus opacus*. Curr Opin Biotechnol 2020;64:10–16.

[31] Gómez MA, Rodelas B, Sáez F, Pozo C, Martínez-Toledo MV, Hontoria E, González-López J. Denitrifying activity of *Xanthobacter autotrophicus* strains isolated from a submerged fixed-film reactor. *Appl Microbiol Biotechnol.* 2005;**68**(5):680–5. doi: 10.1007/s00253-005-1937-y.

[32] Meusel M, Rehm H. Biodegradation of dichloroacetic acid by freely suspended and adsorptive immobilized *Xanthobacter autotrophicus* GJ10 in soil. Appl Microbiol Biotechnol 1993. doi: 10.1007/BF00170446.

[33] Tang K-H, Tang Y, Blankenship R. Carbon metabolic pathways in phototrophic bacteria and their broader evolutionary implications. Front Microbiol 2011;2. doi: 10.3389/fmicb.2011.00165.

[34] Kasai N, Sueoka H, Yukawa H, Ohtani N. Transformant of *Hydrogenophilus* Bacterium Capable of Producing Aspartic acid and Methionine. Patent No. WO2021256511A1. 2021.

[35] Yukawa H, Ohtani N. Lactic Acid-Producing *Hydrogenophilus* Bacterium Transformant. Japanese Patent No. JP6562374. 2021.

[36] Yukawa H, Ohtani N. A Recombinant of *Hydrogenophilus* Bacterium Producing Lactic Acid. Japanese Patent No. JP6675574B1, 2019.

[37] Yukawa H, Ohtani N. A Recombinant of *Hydrogenophilus* Bacterium with Enhanced Ability to Produce Valine. Japanese Patent No. JP6604584. 2018.

[38] Yukawa H, Ohtani N, Ishii M. Genus *Hydrogenophilus* bacterium transformant Japanese Patent No. JP6485828. 2018.

[39] Zhang A, Carroll AL, Atsumi S. Carbon recycling by cyanobacteria: Improving CO_2 fixation through chemical production. FEMS Microbiol Lett 2017;364. doi: 10.1093/femsle/fnx165.

[40] Yu H, Li X, Duchoud F, Chuang DS, Liao JC. Augmenting the calvin–benson–bassham cycle by a synthetic malyl-coa-glycerate carbon fixation pathway. Nature Communications 2018;9:2008. doi: 10.1038/s41467-018-04417-z.

[41] Feng L, Wang K, Li Y, Tan Y, Kong J, Li H, Li Y, Zhu Y. Overexpression of sbpase enhances photosynthesis against high temperature stress in transgenic rice plants. Plant Cell Rep 2007;26:1635–46.

[42] Li Z, Xin X, Xiong B, Zhao D, Zhang X, Bi C. Engineering the Calvin-Benson-Bassham cycle and hydrogen utilization pathway of *Ralstonia eutropha* for improved autotrophic growth and polyhydroxybutyrate production. Microb Cell Fact 2020;19:228. doi: 10.1186/s12934-020-01494-y.

[43] Lin MT, Occhialini A, Andralojc PJ, Parry MA, Hanson MR. A faster rubisco with potential to increase photosynthesis in crops. Nature 2014;513:547–50.

[44] Ruan CJ, Shao HB, Teixeira da Silva JA. A critical review on the improvement of photosynthetic carbon assimilation in C3 plants using genetic engineering. Crit Rev Biotechnol 2012;32:1–21.

[45] Sharwood RE, Ghannoum O, Whitney SM. Prospects for improving CO_2 fixation in C3-crops through understanding C4-rubisco biogenesis and catalytic diversity. Curr Opin Plant Biol 2016;31:135–42.

[46] Wilde E, Schlegel HG. Oxygen tolerance of strictly aerobic hydrogen-oxidizing bacteria. Antonie Van Leeuwenhoek 1982;48(2):131–43. doi: 10.1007/BF00405198.

[47]   Cramm R. Genomic view of energy metabolism in *Ralstonia eutropha* H16. J Mol Microbiol Biotechnol 2009;16:38–52.

[48]   Greening C, Biswas A, Carere CR, Jackson CJ, Taylor MC, Stott MB, Cook GM, Morales SE. Genomic and metagenomic surveys of hydrogenase distribution indicate H_2 is a widely utilised energy source for microbial growth and survival. ISME J 2016;10:761–77.

[49]   Islam ZF, Welsh C, Bayly K, Grinter R, Southam G, Gagen EJ, Greening C. A widely distributed hydrogenase oxidises atmospheric H_2 during bacterial growth. ISME J 2020;14:2649–58. doi: 10.1038/s41396-020-0713-.

[50]   Jiang Y, Yang X, Zeng D, Su Y, Zhang Y. Microbial conversion of syngas to single cell protein: The role of carbon monoxide. Chem Eng J 2022;450:138041. ISSN 1385-8947. doi: 10.1016/j.cej.2022.138041.

[51]   Aryal N, Tremblay PL, Lizak DM, Zhang T. Performance of different *sporomusa* species for the microbial electrosynthesis of acetate from carbon dioxide. Bioresoure Technol 2017;233:184–90.

[52]   Kedia G, Passanha P, Dinsdale RM, Guwy AJ, Esteves SR. Evaluation of feeding regimes to enhance PHA production using acetic and butyric acids by a pure culture of *cupriavidus necator*. Biotechnol Bioprocess Eng 2014;19:989–95. doi: 10.1007/s12257-014-0144-z.

[53]   Holder JW, Ulrich JC, DeBono AC, Godfrey PA, Desjardins CA, Zucker J, Zeng Q, Leach ALB, Ghiviriga I, Dancel C, Abeel T, Gevers D, Kodira CD, Desany B, Affourtit JP, Birren BW, Sinskey AJ. Comparative and functional genomics of *Rhodococcus opacus* PD630 for biofuels development. Richardson PM, editor. PLOS Genetics 2011;**7**(9):e1002219.

[54]   Li J, Engelberth AS. Quantification and purification of lutein and zeaxanthin recovered from distillers dried grains with solubles (DDGS). Bioresour Bioprocess 2018;5:32. doi: https://doi.org/10.1186/s40643-018-0219-3.

[55]   Ram S, Mitra M, Shah F, Tirkey SR, Mishra S. Bacteria as an alternate biofactory for carotenoid production: a review of its applications, opportunities and challenges. J Funct Foods 2020;67 (103867):ISSN 1756–4646. doi: 10.1016/j.jff.2020.103867.

[56]   Goto E, Kodama T, Minoda Y. Growth and taxonomy of thermophilic hydrogen bacteria. Agric. Biol. Chem. 1978;**42**(1):1305–1308

[57]   Banda DM, Pereira JH, Liu AK, Orr DJ, Hammel M, He C, Parry MAJ, Carmo-Silva E, Adams PD, Banfield JF, Shih PM. Novel bacterial clade reveals origin of form i rubisco. Nat Plants 2020;6:1158–66. doi: 10.1038/s41477-020-00762-4.

[58]   Liu D, Ramya RCS, Mueller-Cajar O. Surveying the expanding prokaryotic rubisco multiverse. FEMS Microbiol Lett 2017;1:364. doi: 10.1093/femsle/fnx156.

[59]   Young SN. The effect of raising and lowering tryptophan levels on human mood and social behaviour. Philos Trans R Soc Lond B Biol Sci 2013;368(1615):20110375. doi: 10.1098/rstb.2011.0375.

[60]   Witte B, Obloj P, Koktenturk S, Morach B, Brigl M, Rogg J, Schulze U, Walker D, Von Koeller E, Dehnert N, Grosse-Holz F. Food for Thought: The protein Transformation. Boston Consulting Group & Blue Horizon Coporation, 2021.

[61]   Barbosa RG, Sleutels T, Verstraete W, Boon N. Hydrogen oxidizing bacteria are capable of removing orthophosphate to ultra-low concentrations in a fed batch reactor configuration. Biores Technol 2020;311:123494. doi: 10.1016/j.biortech.2020.123494.

[62]   Cao F, Jaunat J, Sturchio N, Cancès B, Morvan X, Devos A, Barbin V, Ollivier P. Worldwide occurrence and origin of perchlorate ion in waters: A review. Sci Total Environ 2019;661:737–49. ISSN 0048-9697. doi: 10.1016/j.scitotenv.2019.01.107.

[63]   Valentin-Blasini L, Blount BC, Otero-Santos S, Cao Y, Bernbaum JC, Rogan WJ. Perchlorate exposure and dose estimates in infants. Environ Sci Technol 2011;45(9):4127–32.

[64]   Zhang H, Bruns MA, Logan BE. Perchlorate reduction by a novel chemolithoautotrophic, hydrogen-oxidizing bacterium. 2002. doi: 10.1046/j.1462-2920.2002.00338.x.

[65]   Berg A. Ecological aspects of the distribution of different autotrophic CO_2 fixation pathways. Appl Environ Microbiol 2011;77:1925–36. doi: 10.1128/AEM.02473-10.

[66]   Berg S, Wagner B, Cremer H, Leng MJ, Melles M. Diatom abundance in sediment profile Co1011. Pangea 2010. doi: doi.org/10.1594/PANGAEA.744755.

[67]   Cestellos-Blanco S, Friedline S, Sander KB, Abel AJ, Kim JM, Clark DS, Arkin AP, Yang P. Production of PHB from CO_2-derived acetate with minimal processing assessed for space biomanufacturing. Front Microbiol 2021;12:700010. doi: 10.3389/fmicb.2021.700010.

[68]   Parodi A, Jorea A, Fagnoni M, Ravelli D, Samorì C, Torri C, Gallettia P. Bio-based crotonic acid from polyhydroxybutyrate: synthesis and photocatalyzed hydroacylation. Green Chem 2021;23:3420–27.

[69]   Barbosa RG, van Veelen HPJ, Pinheiro V, Sleutels T, Verstraete W, Boon N. Enrichment of hydrogen-oxidizing bacteria from high-temperature and high-salinity environments. Appl Environ Microbiol 2021;87:e02439–20. doi: 10.1128/AEM.02439-20.

[70]   Lin L, Huang H, Zhang X, Dong L, Chen Y. Hydrogen-oxidizing bacteria and their application in resource recovery and pollutant removal. Sci Total Environ 2022;835:155559. doi: 10.1016/j.scitotenv.2022.155559.

[71]   Zheng JJ, Suh S. Strategies to reduce the global carbon footprint of plastics. Nat Clim Change 2019;9:374–78. doi: 10.1038/s41558-019-0459-z.

[72]   Campanale C, Massarelli C, Savino I, Locaputo V, Uricchio VF. A detailed review study on potential effects of microplastics and additives of concern on human health. Int J Environ Res Public Health 2020;17(4):1212. doi: 10.3390/ijerph17041212.

[73]   Welsby D, Price J, Pye S, Ekins P. Unextractable fossil fuels in a 1.5 °C world. Nature 2021;597:230–34. doi: 10.1038/s41586-021-03821-8.

[74]   Zhang W, Niu Y, Li Y-X, Zhang F, Zeng RJ. Enrichment of hydrogen-oxidizing bacteria with nitrate recovery as biofertilizers in the mixed culture. Bioresour Technol 2020. doi: 10.1016/j.biortech.2020.123645.

[75]   Yukawa H, Ohtani N, Ishii, M. *Hydrogenophilus* genus transformant. Japanese Patent No. JP6528295.2019

[76]   Yukawa H, Ohtani N, Ishii, M. *Hydrogenophilus* genus transformant. Japanese Patent No. JP6450912.2018

Vikas Kumar*, Anisa Mitra

Chapter 14
Renewable waste feedstocks for a sustainable aquaculture industry

14.1 Introduction

Aquaculture holds great promise for the global economies, environments, and food security. The projected rise in global population has led to an increase in aquaculture operations [1]. However, commercial aquaculture development faces several obstacles and is alleged to be extremely wasteful, harmful to the environment, and unable to produce healthy outcomes [2]. Intensive fisheries depend heavily on artificial feed requirements, which raise production costs [3]. "Aquafeed 2.0" refers to the recent evolution of aquafeed, which was primarily fish meal (FM) and fish oil (FO)-based before becoming primarily terrestrial-based [4]. In today's food production system, resources are limited, and the cost of energy, particularly from fossil fuels, continues to rise, putting additional strain on the achievement of the sustainable development goals of the aquaculture industry. As a result, in recent years, the concept of "regenerative" aquaculture has emphasized "precision aquaculture," "sustainable feeds," and "circular economy" [5]. In recent years, the circular bioeconomy has been used to produce ingredients for "Aquafeed 3.0" to improve aquaculture's sustainability in terms of reducing its environmental footprint. To address this critical bottleneck, the aquaculture industry must embrace circular economy (CE) to reduce environmental stress [6] rather than the traditional, linear economy approach of "take-make-dispose" [7]. CE principles encourage valorization or the value of biomass cascading down the value chain, by reducing waste, increasing efficiency, and supporting more sustainable systems. The concept of CE involves greater diversification and resilience by using renewable natural resources, with the goal of achieving a sustainability ecological conservation and food production. By reusing by products or co products?? in aquaculture, waste is reduced, and the value extracted from natural resources is increased. A reduction in feed waste due to improved fish nutrition can increase financial sustainability [8]. The modern-day aquafeed 3.0 is in consonance with the circular bioeconomy approach to the use of more nutrient-dense ingredients and makes our domesticated carnivorous species much more omnivorous. Aquaculture needs to create novel techniques where production and

*Corresponding author: Vikas Kumar, Aquaculture Research Institute, Department of Animal, Veterinary and Food Sciences, University of Idaho, Moscow, ID 83844-2330, USA,
e-mail: vikaskumar@uidho.edu
Anisa Mitra, Department of Zoology, Sundarban Hazi Desarat College, Pathankhali, South 24 Parganas, 743611, India

https://doi.org/10.1515/9783110791228-014

profitability are at par with reduced environmental impact and carbon footprint [4, 9]. Amidst a changing climate, consistently tumultuous economies, and rapidly changing social dynamics to decrease reliance on pricey imported feeds, there is a threat to the availability, affordability, environmental viability, and social acceptability of raw materials for aquafeed which are still being tested to find the best formulation. The promotion of alternate ingredients as aquafeed can help to achieve sustainable development by profit maximization and livelihood opportunities [10]. In this chapter, we focus on the alternate ingredients mainly the by-products/coproducts from nonconventional sources are to illustrate their importance and efficiency as aquafeed.

14.2 Waste coproducts

14.2.1 Distiller's dried grains with solubles

In recent decades, growing demand for ethanol as a fuel additive and decreasing reliance on fossil fuels have resulted in a dramatic increase in ethanol production from variety of cereal grains are produced [11]. Following ethanol removal from cereal grains via a series of particulate coproducts known as distiller's dried grains with solubles (DDGS) (Figure 14.1) [12, 13]. In comparison to other traditional alternative protein sources, DDGS are widely available and reasonably priced. DDGS has a sufficient amount of protein (usually between 250 and 450 g/kg, depending on the type) (Table 14.1) [14, 15] and can act as an alternate protein source as aquafeed. However, a few essential amino acids (EAA) are also lacking in DDGS; However, by including these compounds in the feed formulation, these deficiencies can be corrected. The incorporation of DDGS into trout diets is further complicated by the fact that the quality and nutritional composition of DDGS grain sources and processing plants can differ significantly [14, 16]. But the primary factor limiting higher levels of DDGS incorporation in fish feed is high-structural fiber content. However, study has proved that fractionated DDGS products can be tolerated to a higher degree than nonfractionated DDGS and have been used successfully up to 300 g/kg in rainbow trout diets without EAA supplementation or adverse effects on growth performance. Previous researchers demonstrated the importance of DDGS in rainbow trout [17], channel catfish [Lim et al., 2009; 18], Tilapia [19, 20], and sunshine bass [21]. Zhou et al. [22] proposed that growth in channel catfish juveniles fed diets containing 35–40% DDGS was comparable to fish fed a standard commercial manufactured formulation. Dietary inclusion of 30% DDGS in Nile tilapia improved feed consumption, feed efficiency, and growth [23] and incorporation of DDGS up to 30% in Mrigal [24] and Channel catfish [18] resulted in higher growth and feed conversion. A mixture of DDGS and soybean meal can be incorporated up to 49.5% in yellow perch (*Perca flavescens*) to increase growth rate [25]. DDGS can be incorporated in the diet of *Oreochromis niloticus* at 20% level without affecting growth performance

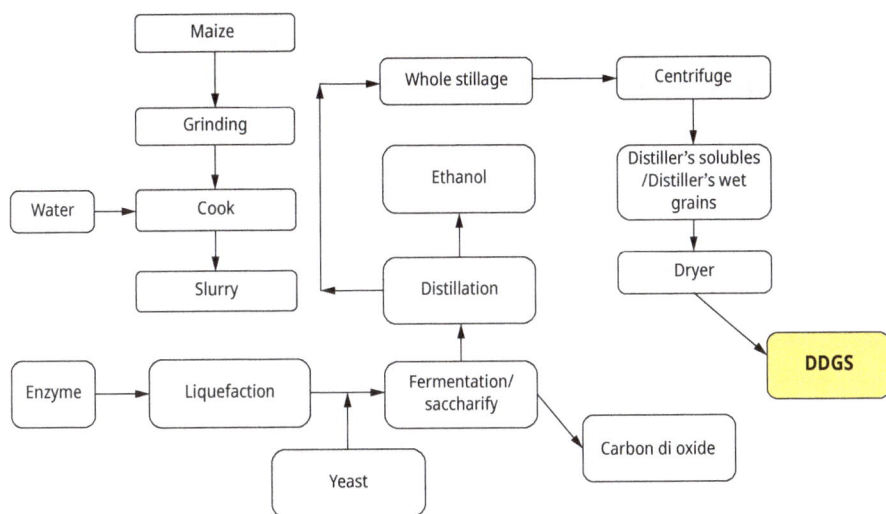

Figure 14.1: Process to produce maize distillers dried grains with soluble (DDGS) using dry-grind technology (modified after [33, 34]).

Table 14.1: Comparison of dietary component of DDGS from different sources.

Dietary component	DDGS source		
	Maize[a,b] (mg/kg)	Wheat[c] (mg/kg)	Sorghum grain[d] (mg/kg)
Dry matter	910 (870–920)	930 (920–940)	870–890
Crude protein	280 (260–310)	510 (390–590)	89–92
Fat	100 (90–117)	37 (29–50)	27–30
Crude fiber	110 (102–130)	77 (62–92)	21–24
Ash	52 (45–67)	50 (47–53)	14

[a][35],[b][36],[c][37], [d][129].

and body composition [26]. In rainbow trout (*Oncorhynchus mykiss*) DDGS and high protein DDGs were used to determine the efficiency of these two coproducts. DDGS, substituting 50% (DDGS50 diet) of the plant protein, resulted in higher feed intake and weight gain (WG) and lower feed conversion ratio (FCR) in trout fed, while feeding the DDGS100 diet resulted in a lower FCR compared with the control and the DDGS50 diets. DDGS100-containing diets tended to increase energy compared to fish fed the control and DDGS50 diets, hence high energy diet led to lower fish performance. However, the energy digestibility of rainbow trout fed the HPDDG100 and HPDDG50 diets was higher than control. The relative weight of the distal intestine, intestinal enzyme activity, or plasma metabolites was unaffected by either the DDGS or the HPDDG diets. When rainbow trout diets were substituted for the typical plant ingredient-based diets, it was demonstrated that both DGGS and HPDDG were suitable for energy and protein sources up to the level tested [27]. Significantly improved per-hectare profit and benefit-cost ratios were used to illustrate the economic benefit of high DDGS inclusion in common carp feeds. It was found that traditional cereal-based semi-intensive carp farming performed worse in terms of production, nutrient use, and economic performance than when DDGS-based compound feeds were combined with the most effective use of pond food [28]. Another experiment using high protein distillers dried grains (HP-DDG) with 25% of FM in *Pangasius hypopthalmus* is acceptable for positive growth and feed utilization [29] (Table 14.2).

Table 14.2: Effects of feeding DDGS to Pacific whiteleg shrimp and fish.

Pacific White leg shrimp

Body weight (initial – final, g)	DDGS %	Ingredients replaced	Trial duration (days)	Fish meal (%)	Supplemental lysine (%)	Optimum DDGS%	Reference
0.49–7.2	0–40	Soyabean meal, corn starch	56	6	0–0.13	40	Rhodes et al.
0.99–6.1	0–30	Fish meal, soybean meal, wheat flour	56	0		Partial replacement of soybean meal with DDGS in diets containing no fish meal reduced growth performance	[38]
0.04–16.3	10	Sorghum	126	0	None	10	[39]
0.45–25	0–10	Sorghum, fish meal	53	0	None	Up to 10	[40]

Fish

Body weight (initial–final, g)	DDGS (%)	Ingredients replaced	Trial duration (days)	Optimum DDGS%	Reference
Cyprinus carpio Juvenile 45.4 ± 0.6 to 497.9 ± 25.7	40	Plant ingredients	155 days	40	[41]
Pangasius hypophthalmus	0%, 25%, 50%, and 75% HP-DDG	Fish meal	14 weeks	25% of fish meal by HP-DDG	[6]

To evaluate the suitability of corn DDGS as a protein source for European catfish (*Silurus glanis*) an *in vivo* digestibility assessment was done to determine apparent digestibility coefficients (ADCs) for protein, lipid, phosphorus, and amino acids available in DDGS and growth, nutrient utilization, and metabolism in juveniles. ADC of dry matter, crude protein, crude fat, and phosphorus of DDGS ingredient for European catfish juveniles were found to be 49.42%, 73.39%, 77.38%, and 87.98%, respectively. The study concluded that 30% DDGS can be included in the diet of European catfish without compromising the growth performance and nutrient utilization [30]. Thus it is evident that the use of protein-enriched DDGS or high protein distiller's grains shows promise but the lack of data and familiarity with the feed ingredient, misconceptions about DDGS [31], and large-scale production is still the main constraint [32].

14.2.2 Beer industry coproduct

Brewer's spent grains (BSG) are the main by-coproducts (80–85% of total) of beer manufacturing industries [42] and widely available throughout the world [43]. Brewer's yeast is rich in protein and has 30% EAAs of total protein [44], vitamins, and minerals [45], and other bioactive compounds such as β-glucans, mannan oligosaccharides, vitamins, minerals, and nucleic acids [46]. After beer production, the different left-over forms of brewer's waste are the potential environmental hazard [47], and their safe disposal is a major environmental concern for the industry. But it is a good source of undegradable protein and water-soluble vitamins [48]. Proximate analysis suggests that it contains 43.51% crude protein, 1.05% lipids, and 8.33% total ash [49, 50]. Moreover, in recent past, several attempts were carried out to exploit its beneficial use in aquafeed [44, 51–53]. Several authors [54–56] cited this by-product as a potential alternative to FM in the feed for fish [17, 51, 57–60] and crustaceans [61] due to its high content in protein and fiber as well as lipids, minerals, and vitamins [62].

Jayant et al. [59] made an effort to develop an environment friendly feed formulation in *Pangasianodon hypophthalmus* fingerlings formulated with 0%, 25%, 50%, 75%, or 100% BSGs in replacement for soybean meal. The BSG contributed 0%, 24.26%, 47.10%, 69.51%, and 91.26% of the total protein in the diets. The study suggested that BSGs can serve as an alternate protein source with 50% level substitution of soybean meal without any adverse effect on growth, nutrient utilization, and feed conversion.

A trial was conducted to analyze the effect of the inclusion of yeast and spent grain obtained from breweries in feeds for rainbow trout (*Oncorhynchus mykiss*). The study showed that the combination of 20% yeast and 15% spent grain in the feed, formulated with only 15% inclusion of FM, showed 87–89% protein digestibility, and increased growth [63]. Another experiment showed that partial replacement of FM by inclusion of up to 30% brewers' spent yeast and 15% spent grain in isoproteic and isolipidic diets for gilthead sea bream (*Sparus aurata*) gave similar results in terms of growth, food conversion, and fillet final composition to a feed with FM as the main protein source and showed a protein digestibility of 89–95% [64]. Brewery waste (brewer's grains) was used at four different levels (10%, 20%, 30%, and 40% w/w) replacing rice bran in fish diet under a semi-intensive culture system and its impact on the growth of catla, *Catla catla*; rohu, *Labeo rohita*; and mrigal, *Cirrhina mrigala*, was studied. A better growth performance was attributed in *C. catla* and *L. rohita* fed on a diet containing 30% brewery waste in the feed, whereas *C. mrigala*, fed on a diet containing brewery waste at the above-mentioned levels, showed poorer growth than the control.

14.2.3 Soymeal coproduct-processed soybean meal (EnzoMeal™)

Soybean products and derivatives thereof are considered good alternatives to FM. There are many advantages in using soybean compared to FM such as competitive market price, stable supply, and available protein and balanced amino acids, among others. Despite the obvious advantages of using soybean, it still contains indigestible components such as nonstarch polysaccharides and other carbohydrates. Moreover, it has many antinutritional compounds and factors that restrict nutrient digestion and absorption. These compounds remain a hindrance to using the full potential of soybean products and derivatives as FM replacement in the diets of several aquatic animals especially fish [65–69]. Hence, through the efforts of soybean producers, conventional soybean meal has undergone several processes to improve exploit the potential of soybean meal. EnzoMeal™, an improved soybean meal developed by the Ohio Soybean Council in Battelle, Worthington, Ohio, USA, contains higher protein (20%) and lower antinutritional factors (50%) (Table 14.3). Table 14.4 shows the comparative amino acid profiles of EnzoMeal™ and conventional soybean meal.

The product EnzoMeal™ has been tested as an aquafeed in several aquatic animals. When used as a FM replacement for diets of yellow perch (*Perca flavescens*), a level of 50% replacement resulted in better growth and feed performance [70] compared to 100% FM replacement using conventional soybean meal and EnzoMeal™. The product used contained 580 g/kg crude protein and low levels of antinutritional compared to conventional soybean meal. However, like other plant-based proteins, EAA supplementation was implemented for EnzoMeal™-based diets. On the other hand, when EnzoMeal™ was used as a FM replacement for diets of Pacific white shrimp (*Litopenaeus vannamei*), the growth performance and product quality of the shrimp remained the same [71]. The study is notable as it effectively showed that shrimps could sustainably use diets with low levels of animal-based proteins when combined with alternative aquafeed ingredients such as EnzoMeal™. Last, when EnzoMeal™ was used as a long-term FM replacement for diets of rainbow trout (*Oncorhynchus mykiss*), growth and feed performance enhancement was achieved. In addition, the higher the inclusion levels of EnzoMeal™, the greater the performance reaching levels as seen in rainbow trout fed conventional soybean meal. This effect has been linked to lesser GH/IGF axis system disruptions which resulted in better

Table 14.3: Comparative nutritive value of soybean meal vs. modified soybean meal/EnzoMeal™.

	Commercial SBM	EnzoMeal™	EM benefits over SBM
Protein (%)	45.9	56	22% increase
Carbohydrate (%)	39.85	27.5	45% decrease
Oligosaccharide (%)	15	<0.05	~100% removal
Phytic acid (%)	1.52	0.88	42% decreased

Table 14.4: Amino acid profile (%) of commercial soybean meal and modified soybean meal.

	Soybean meal	EnzoMeal
Essential amino acids		
Lysine	2.07	2.4
Methionine	0.64	0.83
Histidine	1.22	1.41
Threonine	1.84	2.25
Tryptophan	0.71	0.84
Leucine	3.69	4.32
Isoleucine	2.19	2.65
Phenylalanine	2.41	2.81
Valine	2.34	2.77
Nonessential amino acids		
Tyrosine	1.56	1.89
Cysteine	0.62	0.72
Arginine	3.36	3.64
Alanine	2.03	2.48
Aspartic acid	5.3	6.39
Glutamic acid	8.66	9.94
Glycine	1.98	2.4
Ornithine	0.03	–
Proline	2.41	2.77
Serine	2.3	2.86

nutrient usage. The IGFBP-1b1 expression in the hepatic tissues was upregulated in rainbow trout fed high amounts of conventional soybean meal. This particularly resulted in growth inhibition. No adverse effects have been observed in rainbow trout fed EnzoMeal™ [72]. Hence, modification of conventional soybean meal, particularly to increase protein contents and to lower antinutritional factors, may pave the way to more effective alternative FM replacements.

14.2.4 Poultry coproduct meal

The most common poultry-based ingredients used in feed formulations consists of ground rendered clean parts of the carcasses of slaughtered poultry [73, 74]. Poultry and livestock by-products such as poultry meat flour, feather, and blood meal (BM) are important sources of protein for use in feeding fish. Poultry-based meal (PBM) emerged as one of the most promising alternative ingredients as relatively cheap source of protein compared with FM and available in large quantities throughout the year, especially in poultry-producing regions such as Asia. They contain essential fatty acids, vitamins, and minerals and are virtually free of antinutrients unlike plant proteins. [75, 76]. However, like other animal-based proteins, variation in compositional quality is common and the manufacturing characteristics also vary based on the products (Table 14.5) [77–81]. However, advancements in processing technologies can counteract these challenges [82]. In addition, laws and regulations in the selection of raw materials), sanitary transportation, and handling play a vital role in safeguarding the product quality to stimulate its use in aquaculture feeds [73, 79, 83]. Studies have proved that PBM can successfully replace FM in different dietary inclusion levels for numerous finfish and shellfish species. Although in a recent study using meta-analysis reported that no significant effect on growth was detected with PBM-supplemented diets compared with control diets with 100% FM across a wide range of studies, species, and environmental conditions. Similar trends were observed for freshwater fish, marine water fish, and crustaceans for final weight. Significantly higher FCR was detected in fish compared with crustaceans and in freshwater fish compared with marine fish. In both freshwater and marine fish, effects varied across different species. Certain factors, such as nutritional quality (specifically amino acids or fatty acids), palatability, and digestibility of PBM, could also mediate the influence of PBM supplements on the growth and FCR of fish [84].

14.2.5 Blood meal

Blood Meal (BM) is considered as a proteinaceous concentrate [35]. But the protein digestibility of BM manufactured using different techniques has been shown to differ significantly [83, 85]. Blood proteins appear to be especially sensitive to heat damage, and the drying technique used can have a very significant effect on digestibility of BM. Bureau et al. [83] observed that the digestibility of crude protein in spray-dried blood products was significantly higher than that of roto-plate dried, steam-tube dried, and ring-dried BMs. Study suggests that BM can be a very good source of bioavailable amino acids. Spray- and ring-dried BMs are widely used in salmonid feeds due to their very high digestibility and consistent quality (Table 14.5). Good performance of fish has been observed for fish fed diets containing approximately 8–20% BM in conjunction with high (more than 20%) FM levels [86, 87]. A study was conducted by [88] to investigate the feasibility of utilizing BM in diets for feeding Nile tilapia fingerlings and determine the optimum level for the BM. An average percentage Weight gain (WG) was 165% and 157% in fish which were fed the diets containing 10% and 15% BM. BM also improved food conversion ratios (FCR) which were 2.06 and 2.16 for the 10% and 15% BM treatments, respectively, as compared to 2.33 for the control diet. Moreover, addition of BM to the basic diet reduced the cost of the diet substantially. The study showed that the 10% BM treatment gave better performance than the 15% BM treatment. This probably relates to the known effect of BM on palatability of diets particularly at high inclusion levels. Hence, it is recommended to use only 10% of BM in fish diets. In another feeding trial diets were formulated to replace 0% (control), 9.80%, 19.61%, and 29.41% of FM with chicken hemoglobin powder in Largemouth bass. However, the fish-specific growth rate was significantly reduced when diet replacement level was up to 19.61%, which may be related to the feed intake and ADC of protein and amino acids. But the feed efficiency and protein efficiency ratio were not significantly decreased until the replacement level beyond 29.41% [89].

14.2.6 Feather meal

Feather meal (FeM) is produced by processing the feathers obtained after slaughtering poultry. FeM is a low-quality protein source as it lacks EAAs found in many livestock species, including lysine, methionine, histidine, and tryptophan [90]. But the keratin being the primary component of feather proteins (80–100%), it is poorly digestible when raw [91] Table 14.5. Hence it is hydrolyzed before it can be used as a valuable source of protein in feed [92; 93]. Bureau et al. [83] estimated the digestibility of crude protein of FeMs found to be varied between 77% and 86%. A comparable ADC value was reported by [94] for a FeM fed to rainbow trout. [95] estimated that the digestibility of crude protein of FeM was about 79% for rockfish (*Sebastes schlegeli*). A most recent study showed that the addition of chicken FeM fermented with *Bacillus subtilis* could increase the activity of proteolytic and lipolytic activity and nutrient digestibility of silver pompano [96]. Studies reported that 15–25% FeM could be included in the diet of rainbow trout [97, 98] and 15% FeM (90% crude protein) replacing FM [99] could be included in the diet of Chinook salmon without effect on growth and feed efficiency. Bishop et al. [100] reported that FeM can replace 66% of FM and meat and bone meal (MBM) (9.9% of the total diet) without any significant effect on the growth of *Oreochromis niloticus* fry. Although FeM as a sole source of animal protein in fishfeed appears to result in a decrease in WG and associated growth parameters. A 16 week feeding trial on red hybrid tilapia (*Oreochromis* sp.) showed that up to 15% of FeM can be included into the fish diet with good growth performance and FCR [101].

14.2.7 Meat and bone meal

The regulatory definition of Meat and Bone Meal (MBM) is the rendered product from mammalian tissues, including bone and must contain a minimum of 4% phosphorus with a calcium level not exceeding 2.2 times the actual phosphorus level. MBM can be used to design aquafeed to provide protein, amino acids, phosphorus, calcium, energy, and other nutrients (Table 14.5 and 14.6) although regulations and directives vary worldwide and from country to country [102]. Meat meal): the degree of protein digestibility in MBM appears to vary based on the source and processing. For MBM fed to red drum, [103] and [104] observed protein digestibility of approximately 74–79% (*Sciaenops ocellatus*). Allan et al. [105] reported lower digestibility values for MBM fed to Silver perch (*Bidyanus bidyanus*). Many freshwater and marine fish species were found to be highly digestible to meat meal [106, 107]. The maximum level of MBM in rainbow trout feeds could incorporate up to 24% MBM or about 25% of the total digestible protein and significant amounts of MBM could be consumed by Mozambique tilapia, and gilthead seabream without having an adverse effect on their performance [98]. In a trial, MBM (53% CP, 15% CL) was used to replace FM in the diets of juvenile gilthead seabream (*Sparus aurata*). Three extruded experimental diets (45% CP; 20% CL) were created with 0, 50, and 75% of the protein coming from MBM (diets MBM0; MBM50; MBM75). Overall, without affecting growth performance, feed utilization, or nutrient retention, MBM protein can replace up to 50% of FM protein in diets for young gilthead seabream. In contrast to feed intake, growth performance and feed efficiency were lower with diet MBM75 but similar with diets MBM0 and MBM50. Except for crude lipid and energy content, which were lower with the diet MBM75, total body composition was unaffected by diet composition. The composition of the diet had no impact on the retention of protein or EAAs, but the diet MBM75 had a lower retention of energy. Economic efficiency showed that diets containing MBM reduced production costs, with the MBM diets achieving the lowest economic conversion ratio and the diet MBM50 achieving the highest economic profit [108]. The study's findings showed that replacing 50% of the FM with MBM had no negative effects on the growth and survival rate of tilapia (*O. niloticus*) fry and made fry production more affordable [109]. In a 90-day feeding trial, the impact of switching from FM to MBM in diets for juvenile *Pseudobagrus ussuriensis* was assessed. Six diets were developed with MBM to replace FM at 0 (S0), 200 (S20), 400 (S40), 600 (S60), 800 (S80), and 1,000 g/kg (S100), respectively. The findings revealed no discernible difference in WG between fish fed S0, S20, and S40 diets. However, when MBM was used in place of 600, 800, and 1,000 g/kg FM protein, WG was significantly reduced. The diets S80 and S100 had significantly lower ADCs for protein and dry matter than the other diets. According to the study's findings, the ideal dietary MBM replacement level was 34.3% [110]. In juvenile *Macrobrachium nipponense*, with 15 or 50% of the FM protein replaced by either MBM or PBM [MBM(15), MBM(50)] demonstrated that adding MBM to diets in place of FM had no effect on *M. nipponense's* ability to grow, though PBM(15) had a significantly higher specific growth rate than the other groups. However, compared to other groups, shrimp fed the MBM(15) diet had significantly higher survival rates [111].

Table 14.5: Crude protein (CP) content and apparent digestibility coefficients (ADC) of CP and gross energy (GE) of rendered animal protein ingredients of various origins (modified after 66).

Ingredients along with their processing conditions	CP (%)	ADC of CP (%)	ADC of GE (%)
Feather meals			
Steam hydrolysis 30 min at 276 kPa, disc dryer	75	81	80
Steam hydrolysis 448 min at 276 kPa, disc dryer	82	81	78
Steam hydrolysis 40 min at 276 kPa, ring dryer	76	81	76
Steam hydrolysis 40 min at 276 kPa, steam tube dryer	75	87	80
Meat and bone meals			
125–135 °C, 20–30 min, 17–34 kPa	57	83	68
Same as above but air classification of final product to reduce ash content	55	87	73
133 °C, 30–40 min, 54 kPa	50	88	82
128, 20–30 min, 17–34 kPa	48	87	76
132–138 °C, 60 min	50	88	82
127–132 °C, 25 min	54	89	83
Poultry by-product meals			
138 °C, 30 min	65	87	77
127–132 °C, 30–40 min, 54 kPa	63	91	87
Blood meals			
Steam coagulated blood rotoplate dryer	83	82	82
Steam coagulated blood ring dryer	84	88	88
Whole blood spray-dryer	83	96	92
Blood cells, spray-dryer	86	96	93
Blood plasma, spray-dryer	71	99	99
Steam coagulated blood steam tube dryer	91	84	79
Whole blood, spray-dryer	82	97	94
Steam coagulated blood steam ring-dryer	86	85	86

14.2.8 Rendered animal fats

Rendered animal fats, which have been used in fish feeds for decades due to their low cost and widespread availability (Table 14.7), could be an interesting alternative to FO. In the last 10 years, the price of inedible animal fats has dropped by 40–50% to around $0.30/ kg, hence FO replaced by rendered fats, the cost of aquafeed could be reduced significantly. however, due to a variety of factors, including poor digestibility, variable quality, effects on growth, and product quality, and more recently, concern over disease transmission, the use of these products in aquafeeds has historically been restricted or even avoided [112]. These worries are less important today, according to a growing body of research, and rendered animal fats can be beneficial components of fish feed. Studies have shown that most fish species can effectively use rendered animal fats as long as the diet contains enough mono- or polyunsaturated fatty acids to ensure proper digestion of saturated fatty acids and the essential fatty acid requirements are met. The evidence suggests that rendered animal fats at levels corresponding to 30–40% of total lipids have no negative effects on growth performance, feed efficiency, or product quality in most fish species studied. Overall, using these low-cost alternatives to FO could result in significant immediate savings. With the increasing use of this FO alternate, researches are needed to support expanded applications of rendered fat from various potential sources in aquafeeds [120]. The only remaining barrier to the widespread use of animal fats in aquaculture feeds is the widespread fear of Bovine Spongiform Encephalopathy and other Transmissible Spongiform Encephalopathies, which is still not confirmed [121].

Table 14.6: Proximate analysis of meat and bone meals [113–118], amino acid composition [116, 119], and mineral content reported by [115, 116, 118].

Characteristic	Typical values
Dry matter (%)	92.9–100
Crude protein (%)	49.5–59.4
Ether extract (%)	8.9–16.0
Crude fiber (%)	1–5.13
Ash	20.7–52.9%
Moisture content	1.9–5.7%
pH	5.89–7.19
Protein solubility	2.2–7.2%
Particle size	25.6–800 µm
Color	Light tan brown to dark brown

Table 14.6 (continued)

Amino acids	%CP
Alanine	9.19
Arginine	6.98–7.06
Aspartic acid	4.25
Cystine	0.94–1.01
Glutamine	5.40
Glutamic acid	7.79
Glycine	16.67
Histidine	1.89–2.29
Isoleucine	2.76–3.69
Leucine	6.13–6.85
Lysine	5.18–6.08
Methionine	1.40–2.12
Phenylalanine	3.36–3.56
Proline	11.3
Serine	4.00
Threonine	3.27–4.65
Tryptophan	0.58–0.75
Tyrosine	2.79
Valine	4.20–4.29

Mineral content	
Ca	10.60–13.50%
P	4.73–6.50%
Mg	0.24–1.20%
K	1.02–1.56%
Na	0.71–0.78%
Cl	0.44–0.80%
S	0.39%
Cu	1.5–10 mg/kg
Fe	500–602 mg/kg
Mn	12.3–22 mg/kg
Zn	94 mg/kg
Mo	2.7 mg/kg

Table 14.7: Types and characteristics of rendered fats (modified after [121] and [120]).

Type of rendered fat	Characteristics
Feed-grade animal fat	Derived from by-products of different species, primarily beef and pork unsaturated to saturated (U to S) ratio between 1:1 and 1.6:1 depending on material used.
Poultry fat	Fat from poultry by-products. U to S ratio 2:1. Specific gravity, at 15–15.6 °C 0.91, color light brown liquid/pale brown solid, gross energy (kcal/kg) 9,000
Choice white grease	Derived mainly from the rendering of pork tissue. U to S ratio 1:1 certain blends of beef, pork, and poultry fat are sold as choice white grease. Specific gravity at 15–15.6 °C 0.84 color: yellow liquid/pale brown solid, gross energy (kcal/kg) 9,350
Tallow	Derived from rendered beef tissue, but may contain other animal fats. U to S ratio 0.9 to 1.6:1. specific gravity at 15–15.6 °C 0.89–0.91 color: <3 red (AOCS Wesson) gross energy (kcal/kg) 9,020
Yellow grease	Mainly restaurant grease but can contain dead stock fat and/or dark color, high FFA, and high MIU fat from any type of rendering operation. Specific gravity, 15–15.6 °C:0.92, color: med-brown liquid/brown solid, gross energy (kcal/kg) 9,372
Blended animal and vegetable fat	Blended animal fat, vegetable oil, acidulated vegetable oil, soapstock, and/or restaurant grease.

14.3 Single-cell protein

Single-cell protein (SCP) products can be prepared from different microbial sources, including microalgae, yeast and other fungi, and bacteria (Figure 14.2). They are the sources of almost all EAAs, which may not be available in plant derivatives. The nutrient varies based on the source of the SCPs. The source from microbial cells contains 70–80% amino-nitrogen of the total nitrogen. Algae contains a rich amount of fat and various vitamins like A, B, C, D, and E with β-carotene, tocopherols, and vitamin B. Bacterial SCP is high in protein (around 80 percent of total dry weight) and certain essential amino acids. *Bacillus* species have carotenoid pigments with antioxidant properties [122]. SCP meals are the promising alternate aquafeed, showing positive feeding trial data on salmon, trout, and shrimp with modulation of intestinal microbiota, enhancement of innate immunity, and strengthened resistance against stress [123] (Table 14.8). But the main constraint of SCP production in mass scale is the processing price of the product, finding new strains, and developing new technologies [124]. SCPs can be produced in a range of environments using various technologies such as aerobic, anaerobic, and gas bioreactors. Though SPC might have a comparable protein content to FM, it has some major drawbacks including presence of toxins, contain antinutritional compounds like phytic acid, unbalanced

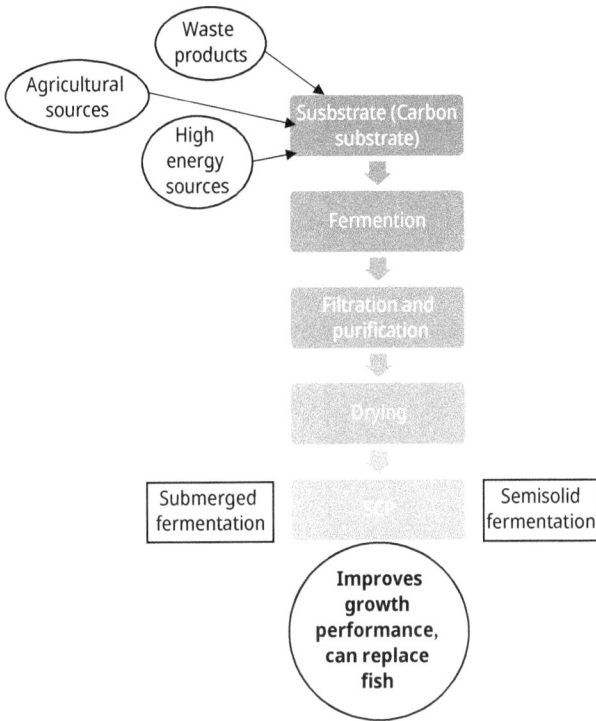

Figure 14.2: Production process of SCP.

Table 14.8: Advantages of SCP used as aquafeed.

– Recycling diverse types of wastes, such as those from agriculture, urban, and food, can provide a carbon source for SCP production
– Environmentally friendly and does not require large expanse of land
– Production is not seasonal nor climate-dependent
– High protein content with good amino acid profile, low fat content, and good carbohydrate content with high content of digestible proteins
– Large quantity of biomass can be produced in a comparatively shorter duration due to high rate of multiplication
– Modulate intestinal microbiota, enhance innate immunity, and strengthen resistance against stress

aminoacidic proportions, especially lysine and methionine, compared to FM, less palatable, difficulty in digestibility, or to scale-up the production process [123, 125].

14.4 Summary

Sustainable aquaculture requires aquafeeds that go beyond the utilization of fish meal (FM) and other standard protein replacements. The promotion of alternate ingredients as aquafeed can help to achieve sustainable development by profit maximization and livelihood opportunities. Various alternative ingredients mainly waste coproducts or by-products from different industries such as breweries, soymeal processing, poultry and meat processing, and even single-cell proteins are the promising alternates in partially and/or completely replacing FM from aquafeeds. They are proven as nutrient dense, digestible, and economical aquafeeds. Value-adding efforts and additional processing steps can improve protein content or reduce antinutrients seem to increase the viability of these alternatives. Though there is a threat to the availability, affordability, environmental viability, and social acceptability of raw materials for this renewable waste feedstocks which are still being tested to find the best formulation.

References

[1]   Bostock J, McAndrew B, Richards R, Jauncey K, Telfer T, Lorenzen K, Little D, et al. Aquaculture: Global status and trends. Philos Trans R Soc B Biol Sci 2010;365(1554):2897–912.
[2]   Macarthur, E. L. L. E. N., & Heading, H. E. A. D. I. N. G.(2019). How the circular economy tackles climate change. Ellen MacArthur Found, 1, 1–71.
[3]   Hertrampf JW, Piedad-Pascual F. Handbook on Ingredients for Aquaculture Feeds. Springer Dordrecht: Springer Science & Business Media; 2003.
[4]   Cottrell RS, Metian M, Froehlich HE, Blanchard JL, Jacobsen NS, McIntyre PB, Nash KL, et al. Time to rethink trophic levels in aquaculture policy. Revi Aquacult 2021;13(3):1583–93.
[5]   Centre for Tropical and subtropical aquaculture. Recycling feed byproducts shows promise for aquaculture nutrition, 2022.
[6]   Ellen MacArthur Foundation. Delivering the Circular Economy: A Toolkit for Policymakers. Ellen MacArthur Foundation; 2015.
[7]   Kusumowardani N, Tjahjono B. Circular economy adoption in the aquafeed manufacturing industry. Procedia CIRP 2020;90:43–48.
[8]   Ahmad AL, Chin JY, Harun MHZM, Low SC. Environmental impacts and imperative technologies towards sustainable treatment of aquaculture wastewater: A review. J Water Process Eng 2022;46:102553.
[9]   Colombo S, Turchini G. 'Aquafeed 3.0': Creating a more resilient aquaculture industry with a circular bioeconomy framework. Revi Aquacult 2021;13. 10.1111/raq.12567.
[10]   Mitra, A. (2020). Thought of Alternate Aquafeed: Conundrum in Aquaculture Sustainability? Proceedings of the Zoological Society, 74, 1–18.
[11]   Liu K, Han J. Changes in mineral concentrations and phosphorus profile during dry-grind processing of corn into ethanol. Bioresour Technol 2011;102(3):3110–18.
[12]   Bothast RJ, Schlicher MA. Biotechnological processes for conversion of corn into ethanol. Appl Microbiol Biotechnol 2005;67(1):19–25.
[13]   Welker, T. L., Lim, C., Barrows, F. T., & Liu, K. (2014). Use of distiller's dried grains with solubles (DDGS) in rainbow trout feeds. Animal Feed Science and Technology, 195, 47–57.
[14]   Ortín N, Waldo G, Yu P. Nutrient variation and availability of wheat DDGS, corn DDGS and blend DDGS from bioethanol plants. J Sci Food Agricult 2009;89(10):1754–61.

[15] Widyaratne GP, Zijlstra RT. Nutritional value of wheat and corn distiller's dried grain with solubles: Digestibility and digestible contents of energy, amino acids and phosphorus, nutrient excretion and growth performance of grower-finisher pigs. Can J Anim Sci 2007;87(1):103–14.

[16] Liu, S. X., Singh, M., & Inglett, G. (2011). Effect of incorporation of distillers' dried grain with solubles (DDGS) on quality of cornbread. LWT-Food Science and Technology, 44(3), 713–718.

[17] Cheng ZJ, Hardy RW. Effects of microbial phytase supplementation in corn distiller's dried grain with solubles on nutrient digestibility and growth performance of rainbow trout, oncorhynchus mykiss. J Appl Aquacult 2004;15(3–4):83–100.

[18] Li, M. H., Robinson, E. H., Oberle, D. F., & Lucas, P. M. (2010). Effects of various corn distillers by-products on growth, feed efficiency, and body composition of channel catfish, Ictalurus punctatus. Aquaculture Nutrition, 16(2), 188–193.

[19] Wu YV, Rosati R, Sessa DJ, Brown P. Utilization of protein-rich ethanol co-products from corn in tilapia feed. J Am Oil Chem Soc 1994;71(9):1041–43.

[20] Shelby, R. A., Lim, C., Yildrim-Aksoy, M., & Klesius, P. H. (2008). Effect of distillers dried grains with solubles-incorporated diets on growth, immune function and disease resistance in Nile tilapia (Oreochromis niloticus L.).

[21] Thompson, K. R., Rawles, S. D., Metts, L. S., Smith, R. G., Wimsatt, A., Gannam, A. L., . . . & Webster, C. D. (2008). Digestibility of dry matter, protein, lipid, and organic matter of two fish meals, two poultry by-product meals, soybean meal, and distiller's dried grains with solubles in practical diets for Sunshine bass, Morone chrysops× M. saxatilis. Journal of the World Aquaculture Society, 39(3), 352–363.

[22] Zhou P, Davis DA, Lim C, Yildirim-Aksoy M, Paz P, Luke AR. Pond demonstration of production diets using high levels of distiller's dried grains with solubles with or without lysine supplementation for channel catfish. North Am J Aquacult 2010;72(4):361–67.

[23] Lim C, Garcia JC, Yildirim-Aksoy M, Klesius PH, Shoemaker CA, Evans JJ. Growth response and resistance to streptococcus iniae of nile tilapia, oreochromis niloticus, fed diets containing distiller's dried grains with solubles. J World Aquacult Soc 2007;38(2):231–37.

[24] Paul, B. N., Das, S., Giri, S. S., Chattopadhay, D. N., Mandai, R. N., Pandey, B. K., & Chakraborti, P. P. (2012). Perfomance of Cirrhinus mrigala fingerlings on feeding dried distillers grain sowble. Indian Journal of Animal Research, 46(3), 272–275.

[25] Schaeffer TW, Brown ML, Rosentrater KA. Effects of dietary distillers dried grains with solubles and soybean meal on extruded pellet characteristics and growth responses of juvenile yellow perch. North Am J Aquacult 2011;73(3):270–78.

[26] Schaeffer, T. W., Brown, M. L., Rosentrater, K. A., & Muthukumarappan, K. (2010). Utilization of diets containing graded levels of ethanol production co-products by Nile tilapia. Journal of Animal Physiology and Animal Nutrition, 94(6), e348–e354.

[27] Øverland M, Krogdahl Å, Shurson G, Skrede A, Denstadli V. Evaluation of distiller's dried grains with solubles (DDGS) and high protein distiller's dried grains (HPDDG) in diets for rainbow trout (Oncorhynchus mykiss). Aquacult 2013;416:201–08.

[28] Sándor, Z. J., Révész, N., Varga, D., Tóth, F., Ardó, L., & Gyalog, G. (2021). Nutritional and economic benefits of using DDGS (distiller'dried grains soluble) as feed ingredient in common carp semi-intensive pond culture. Aquaculture Reports, 21, 100819.

[29] Allam BW, Khalil HS, Mansour AT, Srour TM, Omar EA, Nour AAM. Impact of substitution of fish meal by high protein distillers dried grains on growth performance, plasma protein and economic benefit of striped catfish (Pangasianodon hypophthalmus). Aquacult 2020;517:734792.

[30] Sándor, Z. J., Révész, N., Lefler, K. K., Čolović, R., Banjac, V., & Kumar, S. (2021). Potential of corn distiller's dried grains with solubles (DDGS) in the diet of European catfish (Silurus glanis). Aquaculture Reports, 20, 100653.

[31] U.S. Grains Council. DDGS – Production and Exports. 2020. https://grains.org/buying-selling/ddgs/.

[32] Welker TL, Lim C, Barrows FT, Liu K. Use of distiller' s dried grains with solubles (DDGS) in rainbow trout feeds. Anim Feed Sci Technol 2014;195:47–57. doi: 10.1016/j.anifeedsci.2014.05.011.

[33] Gyan, W. R., Yang, Q. H., Tan, B., Xiaohui, D., Chi, S., Liu, H., & Zhang, S. (2021). Effects of replacing fishmeal with dietary dried distillers grains with solubles on growth, serum biochemical indices, antioxidative functions, and disease resistance for Litopenaeus vannamei juveniles. Aquaculture Reports, 21, 100821.

[34] Pal, K., Maji, C., & Tudu, B. (2020). Journal of Food and Animal Sciences 1 (2020): 117–120. Journal of Food and Animal Sciences, 1, 117–120.

[35] Feedstuffs, 2011.

[36] Spiehs MJ, Whitney MH, Shurson GC. Nutrient database for distiller's dried grains with solubles produced from new ethanol plants in Minnesota and South Dakota. J Anim Sci 2002;80(10):2639–45.

[37] NRC. Nutrient Requirements of Swine. 10th edition, Washington DC: National Academy Press; 1998.

[38] Cummins, V. C., Webster, C. D., Thompson, K. R., & Velasquez, A. (2013). Replacement of fish meal with soybean meal, alone or in combination with distiller's dried grains with solubles in practical diets for pacific white shrimp, Litopenaeus vannamei, grown in a clear-water culture system. Journal of the world aquaculture society, 44(6), 775–785.

[39] Sookying, D., & Davis, D. A. (2011). Pond production of Pacific white shrimp (Litopenaeus vannamei) fed high levels of soybean meal in various combinations. Aquaculture, 319(1–2), 141–149.

[40] Roy, L. A., Bordinhon, A., Sookying, D., Davis, D. A., Brown, T. W., & Whitis, G. N. (2009). Demonstration of alternative feeds for the Pacific white shrimp, Litopenaeus vannamei, reared in low salinity waters of west Alabama. DoAquaculture research, 40(4), 496–503.

[41] Sándor ZJ, Révész N, Varga D, Tóth F, Ardó L, Gyalog G. Nutritional and economic benefits of using DDGS (distiller'dried grains soluble) as feed ingredient in common carp semi-intensive pond culture. Aquacult Rep 2021;21:100819.

[42] Tang D, Yin G, He Y, Hu S, Li B, Li L, Liang H, Borthakur D. Recovery of protein from brewer's spent grain by ultrafiltration. Biochem Eng J 2009;48:1–5.

[43] Essien J, Udotong I. Amino acid profile of biodegraded brewers spent grains (BSG). J Appl Sci Environ Manage 2010;12(1):12. 10.4314/jasem.v12i1.55582. ISSN: 1119–8362.

[44] Hassan MA, Aftabuddin M, Meena DK, Mishal P, Gupta SD. Effective utilization of distiller's grain soluble-an agro-industrial waste in the feed of cage-reared minor carp *labeo bata* in a tropical reservoir. India Environ Sci Pollut Res 2016;23:1e6.

[45] Ovie SO, Eze SS. Utilization of saccharomyces cerevisiae in the partial replacement of fishmeal in clarias gariepinus diets. Int J Adv Agricult Res 2014;2:83–88.

[46] Ferreira IMPLVO, Pinho O, Vieira E, Tavarela JG. Brewer's saccharomyces yeast biomass: characteristics and potential applications. Trends Food Sci Techno 2010;21(2):77–84.

[47] Hang YD, Splitstoesser DF, Woodams EE (1975) Utilization of brewery spent grain liquor by Aspergillus niger. Appl Microbiol 30(5):879–880.

[48] Taylor J, Taylor JRN. Some potential applications for Brewers Spent Grains the Protein-rich co-products, from sorghum lager beer brewing, 12th scientific and technical convention, 2009.

[49] Singh P, Paul BN, Rana GC, Mandal RN, Chakrabarti PP, Giri SS. Evaluation of breweries waste in the feed of catla catla fingerlings. Indian J Anim Nutri 2014;31(2):187–91.

[50] Singh P, Paul BN, Giri SS. Potentiality of new feed ingredients for aquaculture: A review. Agricult Rev 2018;39(4).

[51] Kaur VI, Saxena PK. Incorporation of brewery waste in supplementary feed and its impact on growth in some carps. Bioresour Technol 2004;91(1):101–04.

[52] Levic J, Djuragic O, Sredanovic S. Use of new feed from brewery by-products for breeding layers. Rom Biotechnol Lett 2010;15:5559–65.

[53] Zerai DB, Fitzsimmons KM, Collier RJ, Duff GC. Evaluation of brewer's waste as partial replacement of fish meal protein in nile tilapia, oreochromis niloticus, diets. J World Aquacult Soc 2008;39(4):556–64.

[54] Oliva-Teles A, Gonçalves P. Partial replacement of fishmeal by brewers yeast (*Saccaromyces cerevisae*) in diets for sea bass *(Dicentrarchus labrax)* juveniles. Aquacult 2001;202(3–4):269–78.

[55] Ozório ROA, Portz L, Borghesi R, Cyrino JEP. Effects of dietary yeast (*Saccharomyces cerevisia*) supplementation in practical diets of tilapia (*Oreochromis niloticus*). Animals 2012;2(1):16–24.

[56] Hauptman BS, Barrows FT, Block SS, Gibson Gaylord T, Paterson JA, Rawles SD, Sealey WM. Evaluation of grain distillers dried yeast as a fish meal substitute in practical-type diets of juvenile rainbow trout, *Oncorhynchus mykiss*. Aquacult 2014;432:7–14.

[57] Campos INÊ, Matos ELI, Aragão CLÁ, Pintado M, Valente LMP. Apparent digestibility coefficients of processed agro-food by-products in European seabass (*Dicentrarchus labrax*) juveniles. Aquacult Nutr 2018;24(4):1274–86.

[58] Huige NJ. Brewery Byproducts and Effluents. In: Priest FG, Stewart GG, Handbook of Brewing, 2nd edition, CRC Press; 2006. ISBN 9780824726577.

[59] Jayant M, Hassan MA, Srivastava PP, Meena DK, Kumar P, Kumar A, Wagde MS. Brewer's spent grains (BSGs) as feedstuff for striped catfish, *Pangasianodon hypophthalmus* fingerlings: An approach to transform waste into wealth. J Cleaner Prod 2018;199:716–22.

[60] Yamamoto T, Marcouli PA, Unuma T, Akiyama T. Utilization of malt protein flour in fingerling rainbow trout diets. Fish Sci 1994;60(4):455–60.

[61] Muzinic LA, Thompson KR, Morris A, Webster CD, Rouse DB, Manomaitis L. Partial and total replacement of fish meal with soybean meal and brewer's grains with yeast in practical diets for Australian red claw crayfish *Cherax quadricarinatus*. Aquacult 2004;230(1–4):359–76.

[62] Mussatto SI, Dragone G, Roberto IC. Brewers' spent grain: Generation, characteristics and potential applications. J Cereal Sci 2006;43(1):1–14.

[63] Estevez A, Padrell L, Iñarra B, Orive M, San Martin D. Brewery by-products (yeast and spent grain) as protein sources in rainbow trout (*Oncorhynchus mykiss*) feeds. Front Mar Sci 2022.

[64] Estevez A, Padrell L, Iñarra B, Orive M, San Martin D. Brewery by-products (yeast and spent grain) as protein sources in gilthead seabream (*Sparus aurata*) feeds. Aquacult 2021;543:736921.

[65] Hernández MD, Martinez FJ, Jover M, García B. Effects of partial replacement of fish meal by soybean meal in sharpsnout seabream (*Diplodus puntazzo*) diet. Aquacult 2007;263:159–67.

[66] Kumar V, Makkar HPS, Becker K. Detoxified *Jatropha curcas* kernel meal as a dietary protein source: Growth performance, nutrient utilization and digestive enzymes in common carp (*Cyprinus carpio* L.) fingerlings. Aquacult Nutr 2011a;17:313–26.

[67] Kumar V, Makkar HPS, Becker K. Nutritional, physiological and hematological responses in rainbow trout (*Oncorhynchus mykiss*) juveniles fed detoxified *Jatropha curcas* kernel meal. Aquacult Nutr 2011b;17:451–67.

[68] Kumar V, Makkar HPS, Becker K. Evaluations of the nutritional value of *Jatropha curcas* protein isolate in common carp (*Cyprinus carpio* L.). J Anim Feed Physiol Anim Nutr 2012a;96:1030–43.

[69] Kumar V, Sinha AK, Makkar HPS, De Boeck G, Becker K. Phytate and phytase in fish nutrition: A rev. J Animal Physiol Animal Nutr 2012b;96:335–64.

[70] Kumar V, Wang H-P, Lalgudi R, Cain R, McGraw B, Rosentrater KA. (2020). processed soybean meal as an alternative protein source for yellow perch (*Perca flavescens)* feed. Aquacult Nutr 25(4):917–31.

[71] Hulefeld R, Habte-Tsion H-M, Lalgudi R, Cain R, McGraw B, Tidwell JH, Kumar V. Nutritional evaluation of an improved soybean meal as a fishmeal replacer in the diet of pacific white shrimp, *Litopenaeus vannamei*. Aquacult Res 2018;49(4):1414–22.

[72] Kumar V, Lee S, Cleveland B, Romano N, Lalgudi R, Rubio M, McGraw B, Hardy R. Comparative evaluation of processed soybean meal (enzomealtm) vs. regular soybean meal as a fishmeal replacement in diets of rainbow trout (*Oncorhynchus mykiss*): effects on growth performance and growth-related genes. Aquacult 2019;516:734652.

[73] Cruz-Suárez LE, Nieto-López M, Guajardo-Barbosa C, Tapia-Salazar M, Scholz U, Ricque-Marie D. Replacement of fish meal with poultry by-product meal in practical diets for *Litopenaeus vannamei*, and digestibility of the tested ingredients and diets. Aquacult 2007;272(1–4):466–76.

[74] Dong FM, Hardy RW, Haard NF, Barrows FT, Rasco BA, Fairgrieve WT, Forster IP. Chemical composition and protein digestibility of poultry by-product meals for salmonid diets. Aquacult 1993;116(2–3):149–58.

[75] Abdul-Halim HH, Aliyu-Paiko M, Hashim R. Partial replacement of fish meal with poultry by-product meal in diets for snakehead, *Channa striata* (bloch, 1793), fingerlings. J World Aquacult Soc 2014;45(2):233–41.

[76] Gunben EM, Senoo S, Yong A, Shapawi R. High potential of poultry by-product meal as a main protein source in the formulated feeds for a commonly cultured grouper in Malaysia (*Epinephelus fuscoguttatus*). Sains Malays 2014;43(3):399–405.

[77] Davis DA, Arnold CR. Replacement of fish meal in practical diets for the pacific white shrimp, *Litopenaeus vannamei*. Aquacult 2000;185(3–4):291–98.

[78] Dawson MR, Shah Alam M, Watanabe WO, Carroll PM, Seaton PJ. Evaluation of poultry by-product meal as an alternative to fish meal in the diet of juvenile black sea bass reared in a recirculating aquaculture system. N Am J Aquacult 2018;80(1):74–87.

[79] Garza de Yta A, Davis DA, Rouse DB, Ghanawi J, Saoud IP. Evaluation of practical diets containing various terrestrial protein sources on survival and growth parameters of redclaw crayfish *Cherax quadricarinatus*. Aquacult Res 2012;43(1):84–90.

[80] Robinson EH, Li MH, Manning BB. Evaluation of corn gluten feed as a dietary ingredient for pond-raised channel catfish *Ictalurus punctatus*. J World Aquacult Soc 2001;32(1):68–71.

[81] Tacon AGJ, Hasan MR, Subasinghe RP FAO. Use of Fishery Resources as Feed Inputs to Aquaculture Development: Trends and Policy Implications. No. 1018. Rome: Food and Agriculture Organization of the United Nations; 2006.

[82] Najafabadi HJ, Moghaddam HN, Pourreza J, Shahroudi FE, Golian A. Determination of chemical composition, mineral contents, and protein quality of poultry by-product meal. Int J Poult Sci 2007;6(12):875–82.

[83] Bureau DP, Harris AM, Cho CY. Apparent digestibility of rendered animal protein ingredients for rainbow trout (*Oncorhynchus mykiss*). Aquacult 1999;180(3–4):345–58.

[84] Galkanda-Arachchige HSC, Wilson AE, Davis DA. Success of fishmeal replacement through poultry by-product meal in aquaculture feed formulations: A meta-analysis. Revi Aquacult 2020;12(3):1624–36.

[85] Cho CY, Slinger SJ, Bayley HS. Bioenergetics of salmonid fishes: energy intake, expenditure and productivity. Comp Biochem Physiol Part B 1982;73(1):25–41.

[86] Abery NW, Gunasekera RM, De Silva SS. Growth and nutrient utilization of murray cod *Maccullochella peelii* peelii (Mitchell) fingerlings fed diets with varying levels of soybean meal and blood meal. Aquacult Res 2002;33(4):279–89.

[87] Luzier JM, Summerfelt RC, Ketola HG. Partial replacement of fish meal with spray-dried blood powder to reduce phosphorus concentrations in diets for juvenile rainbow trout, *Oncorhynchus mykiss* (Walbaum) 1. Aquacult Res 1995;26(8):577–87.

[88] Obey, et al. 2019.

[89] Ding G, Li S, Wang A, Chen N. Effect of chicken haemoglobin powder on growth, feed utilization, immunity and haematological index of largemouth bass (*Micropterus salmoides*). Aquac Fish 2020;5(4):187–92.

[90] Crawshaw R. Co-product Feeds in Europe: Animal Feeds Derived from Industrial Processing. Lulu. comz; 2019.

[91] Moran Jr, E. T., Summers, J. D., & Slinger, S. J. (1967). Keratins as Sources of Protein for the Growing Chick: 2. Hog Hair, a Valuable Source of Protein with Appropriate Processing and Amino Acid Balance. Poultry Science, 46(2), 456–465.

[92] El Boushy, A. R., Van der Poel, A. F. B., & Walraven, O. E. D. (1990). Feather meal – A biological waste: Its processing and utilization as a feedstuff for poultry. Biological wastes, 32(1), 39–74.

[93] Papadopoulos MC. Processed chicken feathers as feedstuff for poultry and swine. A review. Agricult Wastes 1985;14(4):275–90.

[94] SUGIURA, S.H.; BABBITT, J.K.; DONG, F.M. et al. Utilization of fish and animal by-product meals in low-pollution feeds for rainbow trout Oncorhynchus mykiss (Waulbaum). Aquaculture Research, v.31, 585–593. 2000

[95] Lee, S. M. (2002). Apparent digestibility coefficients of various feed ingredients for juvenile and grower rockfish (Sebastes schlegeli). Aquaculture, 207(1–2), 79–95.

[96] Adelina A, Feliatra F, Siregar YI, Putra I, Suharman I Use of chicken feather meal fermented with *Bacillus Subtilis* in diets to increase the digestive enzymes activity and nutrient digestibility of silver pompano *Trachinotus Blochii* (Lacepede, 1801). F1000 Res. 2021;10:25. doi: 10.12688/ f1000research.26834.2. PMID: 33868644; PMCID: PMC8030120.

[97] Henrichfreise, B. 1989. Bewertung von aufgeschlossenem Getreide und hydrolysiertem Federmehl in der Ernahrung von Regenbogenforellenfuttern. Diss. Landwirtsch Fak. Rheinische Friedrich-Wilhelms-Univ. Bonn.

[98] Bureau DP, Harris AM, Bevan DJ, Simmons LA, Azevedo PA, Cho CY. Feather meals and meat and bone meals from different origins as protein sources in rainbow trout (*Oncorhynchus mykiss*) diets. Aquacult 2000;181(3–4):281–91.

[99] Fowler LG. Feather meal as a dietary protein source during parr-smolt transformation in fall chinook salmon. Aquacult 1990;89(3–4):301–14.

[100] Bishop, C. D., Angus, R. A., & Watts, S. A. (1995). The use of feather meal as a replacement for fish meal in the diet of Oreochromis niloticus fry. Bioresource Dnology, 54(3), 291–295.

[101] Yong ST, Mardhati M, Farahiyah IJ, Noraini S, Wong HK. Replacement of fishmeal in feather meal-based diet and its effects on tilapia growth performance and on water quality parameters. J Trop Agricult Food Sci 2018;46(1):47–55.

[102] Ockerman HW, Basu L. Production and consumption of fermented meat products. In: Toldrá F, Hui YH, Astiasaran I, Sebranek J, Talon R, editors. Handbook of Fermented Meat and Poultry. 2nd ed. 2014, pp.7–11.

[103] McGoogan BB, Reigh RC. Apparent digestibility of selected ingredients in red drum (*Sciaenops ocellatus*) diets. Aquacult 1996;141(3–4):233–44.

[104] Gaylord TG, Gatlin DM III. Determination of digestibility coefficients of various feedstuffs for red drum (*Sciaenops ocellatus*). Aquacult 1996;139(3–4):303–14.

[105] Allan GL, Parkinson S, Booth MA, Stone DAJ, Rowland SJ, Frances J, Warner-Smith R. Replacement of fish meal in diets for Australian silver perch, *Bidyanus bidyanus*: I. Digestibility of alternative ingredients. Aquacult 2000;186(3–4):293–310.

[106] Gomes EF, Rema P, Kaushik SJ. Replacement of fish meal by plant proteins in the diet of rainbow trout (*Oncorhynchus mykiss*): Digestibility and growth performance. Aquacult 1995;130(2–3):177–86.

[107] Watanabe T, Verakunpiriya V, Watanabe K, Viswanath K, Satoh S. Feeding of rainbow trout with non-fish meal diets. Fish sci 1998;63(2):258–66.

[108] Moutinho S, Peres H, Serra C, Martínez-Llorens S, Tomás-Vidal A, Jover-Cerdá M, Oliva-Teles A. Meat and bone meal as partial replacement of fishmeal in diets for gilthead sea bream (*Sparus aurata*) juveniles: Diets digestibility, digestive function, and microbiota modulation. Aquacult 2017;479:721–31.

[109] Hasan MT, Khalil I, Mohammed S, Kashem M, Hashem S, Mazumder S. Substitution of fish meal by meat and bone meal for the preparation of tilapia fry feed. Int J Anim Fish Sci 2012;5:464–69.

[110] Wang Y, Tao S, Liao Y, Lian X, Luo C, Zhang Y, Yang C, Cui C, Yang J, Yang Y. Partial fishmeal replacement by mussel meal or meat and bone meal in low-fishmeal diets for juvenile ussuri catfish (*Pseudobagrus ussuriensis*): Growth, digestibility, antioxidant capacity and IGF-I gene expression. Aquacult Nutr 2020;26(3):727–36.

[111] Yang Y, Xie S, Cui Y, Lei W, Zhu X, Yang Y, Yu Y. Effect of replacement of dietary fish meal by meat and bone meal and poultry by-product meal on growth and feed utilization of gibel carp, *Carassius auratus* gibelio. Aquacult Nutr 2004;10(5):289–94.

[112] Bureau, D. P. (2006). Rendered products in fish aquaculture feeds. Essential rendering. National Renderers Association, Alexandria, Virginia, USA, 179–184.

[113] Garcia R, Phillips J. Physical distribution and characteristics of meat & bone meal protein. J Sci Food Agricult 2009;89. 10.1002/jsfa.3453.

[114] Howie SA, Calsamiglia S, Stern MD. Variation in ruminal degradation and intestinal digestion of animal byproduct proteins. Anim Feed Sci Technol 1996;63(1–4):1–7.

[115] National Research Council and Canadian Department of Agriculture, 1971

[116] National Research Council, 2001

[117] Nengas I, Alexis MN, Davies SJ, Petichakis G. Investigation to determine digestibility coefficients of various raw materials in diets for gilthead sea bream, *Sparus auratus* l. Aquacult Res 1995;26(3):185–94.

[118] Preston RL. 2014 Feed Composition Table. Minneapolis, MN: Penton Media; 2014, pp. 18–26.

[119] Li YL, McAllister LTA, Beauchemin KA, He ML, McKinnon JJ, Yang WZ. Substitution of wheat dried distillers grains with solubles for barley grain or barley silage in feedlot cattle diets: Intake, digestibility, and ruminal fermentation. J Anim Sci 2011;89(8):2491–501.

[120] Trushenski JT, Lochmann RT. Potential, implications and solutions regarding the use of rendered animal fats in aquafeeds. Am J Anim Vet Sci 2009;4:4.

[121] Bureau, D. P., Gibson, J., & El-Mowafi, A. (2002). Use of animal fats in aquaculture feeds. Avances en Nutrición Acuícola.

[122] Bharti, V., Pandey, P. K., & Koushlesh, S. K. (2014). Single cell proteins: a novel approach in aquaculture systems. World Aquaculture, 45(4), 62–63.

[123] Pereira AG, Fraga-Corral M, Garcia-Oliveira P, Otero P, Soria-Lopez A, Cassani L, Cao H, Xiao J, Prieto MA, Simal-Gandara J. Single-cell proteins obtained by circular economy intended as a feed ingredient in aquaculture. Foods 2022;11:2831.

[124] Jones, S. W., Karpol, A., Friedman, S., Maru, B. T., & Tracy, B. P. (2020). Recent advances in single cell protein use as a feed ingredient in aquaculture. Current opinion in biotechnology, 61, 189–197.

[125] FAO. The Growth of Single-cell Protein in Aquafeed. Bangkok, 2022.

Section IV: **Some commercial bioproducts**

Dan Derr

Chapter 15
Microbially produced green surfactant bioproducts

15.1 Introduction

15.1.1 Surfactants

Surfactants [1] are chemicals that have an affinity for surfaces. Hence the name is a portmanteau for **surf**ace **act**ive **agent**. Surfactants are composed of a polar, hydrophilic head group, often depicted as a ball, and a nonpolar, hydrophobic, usually oleophilic tail, see Figure 15.1.

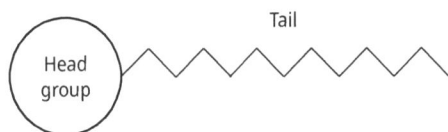

Figure 15.1: A commonly used cartoon model of a surfactant molecule.

Figure 15.1 is an appropriate model for a small molecule surfactant, but polymers can also be surfactants, though they will not be discussed in detail here. Nearly all well-characterized microbial biosurfactants are small molecules.

Briefly, surfactants' preference for residing at surfaces comes from the fact that their polar and nonpolar portions are connected, but spatially separated from each other. If a surfactant is put in water, it is driven to the surface where the head group can be just inside the water phase, and the tail sticks out into the air, or any other nonpolar phase that is in contact with the water.

Surfactants are an important commercial product in modern society but have also been used for at least 4,500 years by humanity. In fact, the saponification reaction using oil as a feedstock and producing soap is described as the first-recorded chemical reaction [2], at least the earliest found to date. Soap is the archetype of surfactants.

The current global market for surfactants is approximately 40 billion dollars per year [3]. Surfactants are used ubiquitously in commercial products and industrial processes. Applications include home cleaning, personal care, agriculture, energy production, paints and coatings, and manufacturing, among many others.

Dan Derr, Integrity BioChem, e-mail: dander99@msn.com

https://doi.org/10.1515/9783110791228-015

15.1.2 Microbial biosurfactants

Here biosurfactants are defined as surfactants that are what is known as "natural products" by chemistry professionals; in other words they are produced in natural systems. Specifically, the focus here will be on microbial biosurfactants. Microbial bio-surfactants are produced biosynthetically by microbes. Typically, microbes do not produce a pure compound biosurfactant; instead they produce a mixture of compounds that differ slightly in the length and unsaturation level of the tail, and in at least one case, the makeup of the head group. The exact structural details of each mixture will not be discussed at great length, here.

In addition to microbial biosurfactants, there are also plant-based biosurfactants, saponins being a well-studied example [4], but those will also not be discussed here.

There was a period in the twentieth century where surfactants, like many other commonly used chemicals, were dominated by synthetic products.[2] But over the last couple of decades, industry has shifted to using biological sources for synthetic surfactants, producing what are known as "bio-based" products. Natural products such as microbial bio-surfactants are an extension of this trend. Microbial biosurfactants currently have a miniscule share of the overall surfactants market, but the share is growing, and new products are now being introduced regularly, as will be discussed in more detail below.

It is notable that the first surfactant, soap, was – and is – a bio-based chemical: a natural oil is processed using synthetic chemistry – in this case simple alkaline hydrolysis chemistry, but synthetic nonetheless- to produce a useful product. Thus, the arc of surfactants history started with a bio-based solution several thousand years ago, traveled through a synthetic phase around the twentieth century, has reverted in the direction of bio-based, again, and is now moving toward natural products. That is not to say that synthetic surfactants will be totally displaced. Synthetic surfactants will likely remain a large slice of the market pie as there are many applications where synthetic surfactants are quantifiably the best choice when all three legs of the sustainability stool – economic, social, and environmental – are considered. But microbial biosurfactants will almost certainly continue to increase their market share for the foreseeable future.

15.1.3 Classes of surfactants

Without going into too much detail – whole books are written on this topic [5] – surfactants are often broken down into four classes:
1. Anionic – the head group has a negative charge
2. Nonionic – the entire molecule is neutral
3. Cationic – the head group has a positive charge
4. Zwitterionic or amphoteric – the head group carries both a positive and a negative charge

Microbial biosurfactants are either anionic or nonionic in nature. They tend to be good cleansers, emulsifiers, and sometimes have antimicrobial properties, making them attractive in personal and home care. Many biosurfactants are also being considered in applications with uncontrolled release, such as agriculture, wastewater treatment, remediation, and energy production because of their attractive environmental footprint.

Multiple biosurfactants will be nonionic in acidic conditions and anionic under less acidic conditions, though some are also nonionic at all pH they are stable at.

15.1.4 History

The scientific history of microbial biosurfactants begins around the middle of the twentieth century. The first use of the term biosurfactant appears to occur in 1960 [6], referring to rhamnolipids. Rhamnolipids had their structure elucidated in 1949 [7]. Another class of compounds that is now considered a biosurfactant, trehalose lipids, was first reported in 1933 [8]. In the 1960s, the term biosurfactant appears to be used in the academic literature a handful of times and becomes more common as the decades pass.

Fatty acids, the original soaps developed by human civilizations, can be produced microbially as what are known as free fatty acids (FFA). They would then be microbial biosurfactants. However, it is unlikely this is a viable production route, as soap is easily and cost effectively obtained from saponification of natural oils. As mentioned above, this makes them a bio-based surfactant. FFA will not be considered any further, here, as there is no incentive to make them as microbial biosurfactants.

Another surfactant that can be produced microbially is phospholipids. However, phospholipids are readily obtained as a biproduct during the production of vegetable oils. As they are produced today, phospholipids would be considered plant-derived biosurfactants, and as with FFA, there is no incentive to produce them via fermentation. Phopholipids will not be further considered here.

15.2 Microbial biosurfactants

Most microbial biosurfactants that are known to be commercially available today are glycolipids, in which Desai and Banat describe as "carbohydrates in combination with long-chain aliphatic acids or hydroxy aliphatic acids" in their seminal 1997 review [9]. Put more simply, glycolipids are a combination of sugar (glyco) and fat (lipid).

15.2.1 Sophorolipids

SL were the first biosurfactant commercialized in industrial quantities and they lead the microbial biosurfactant market, today. SL are glycolipids.

SL were first reported and described in 1961 by Gorin et al. [10]. The first mention that they are surfactants found was by Inoue and Ito in 1982 [11]. One of the reasons SL have been successfully commercialized is the productivities and titers of SL fermentations has been demonstrated at very high levels. For example, in 2004, Pekin et al. [12] showed 400 g/l titers with productivities approaching 1 g/l/h. The companies producing SL commercially have exceeded these productivity numbers, but it can be difficult to document commercial progress as companies often keep production metrics as trade secrets.

SL contain a sophorose sugar head group and a long-chain hydroxy fatty acid tail, as depicted in Figure 15.2. The carboxylic acid at the end of the tail can form a lactone with one of the sugar hydroxyl groups, as illustrated in the figure.

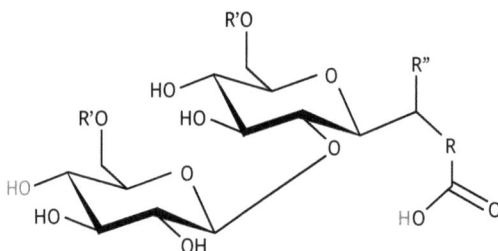

Figure 15.2: A general structure for SL. R = a hydrocarbon, often unsaturated, ~ 15 carbons in length, R' = (C = O)CH₃ or H, and R" = CH₃ or H. The gray OH on the terminal glucose and H on the carboxylic acid are commonly dehydrated to form a lactone ring in the SL produced biocatalytically.

As produced using modern production organisms, for example, *Candida bombicola*, SL tend to be richer in the lactone form than the ring-opened form, though this varies with fermentation conditions [13].

In the lactone form, SL are nonionic surfactants at all pH they are stable at and have low water solubility. Alkali conditions can hydrolyze the lactone, producing the ring-opened form, which is a nonionic surfactant at low pH, and an anionic surfactant at neutral and alkaline pH. Again, at low pH in the nonionic form, ring-opened SL has low water solubility. But the anionic surfactant at higher pH is highly water soluble.[13]

15.2.1.1 Commercial status

Evonik is almost certainly the largest volume producer of SL. They first began publicizing SL products around 2015 [14], and REWOFERM SL ONE [15] was commercialized in 2016 [16]. Evonik has indicated in the past that they produce SL at what they call their Fermas site in Slovakia. There has been conflicting information released on the fermentation capacity at this site over the years, but it is likely on the order of 1,000 m^3, and growing (see Rhamnolipids section), as of 2022.

Givaudan purchased Soliance in 2014 [17]. One of the products Soliance had developed was an SL product called Sopholiance S that continues to be marketed by Givaudan as of early 2022.

Also in 2014, DSM partnered with Synthezyme, which was a company commercializing sophorolipid technology at the time [18]. However, in this case no DSM or Synthezyme SL products could be found still marketed in 2022.

Holiferm is a biosurfactant company founded in 2018 that announced plans in 2021 to build a 1,100 tons/year SL plant [19] in Wallasey, UK [20]. At the time of this writing, they market a product line for personal care applications called HoliSurf and one for cleaning applications called HoneySurf. They use a proprietary continuous separation approach to improve productivities and titers compared to batch production.

BASF displayed an interest in SL technology in 2021 with an investment in Allied Carbon Solutions [21]. At the same time they announced an agreement with Holiferm to jointly develop glycolipids for a variety of end uses. Allied Carbon Solutions is a company headquartered in Japan that focuses on SL. The tradename for their SL product is ACS-Sophor.

Locus has been pursuing SL for several years. In 2022 their Performance Ingredients unit announced an exclusive agreement with Dow for SL as home and personal care ingredients [22]. At in-cosmetics in 2022, Dow announced EcoSense SL-60 HL and HA products for personal care applications [23].

It is difficult to capture all the products that contain SL, but a few highlights are worth noting. Saraya, a Japan-based global cleaning products company, had products as early as 2014 that contained SL [24]. Ecover has a number of homecare products they market in Europe that contain SL, and SL are now being used more and more in personal care products worldwide.

15.2.2 Rhamnolipids

RL, also a glycolipid, have recently entered the market as an ingredient in consumer care applications.

The structure of rhamnolipids was first elucidated in 1949 by Jarvis and Johnson.[7] Rhamnolipids typically have one or two rhamnose sugar units as the headgroup and two medium chain 3-hydroxy fatty acids as the tail, see Figure 15.3.

Jarvis and Johnson attributed the first isolation of rhamnolipids to Bergstrom, Theorell, and Davide in 1947, or possibly Birch-Hirschfeld in 1935. *Pseudomonas aeruginosa* was the originally produced organism. It then took more than a decade for anyone to report on their surfactant properties, which as mentioned in the introduction, appears to have occurred in 1960.[6] Productivities have now been improved to commercializable

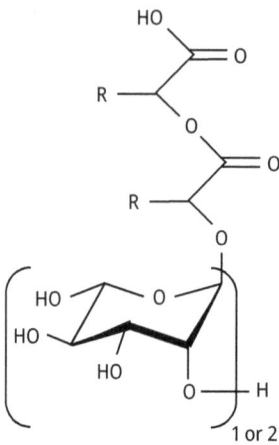

Figure 15.3: RL are made up of one or two rhamnose sugar units attached to two medium-chain length 3-hydroxy fatty acids.

levels [25]. While Pseudomonads are the most prolific natural producers, many genetically modified organisms (GMO) have also been developed that biosynthesize RL with commercializable metrics.[13]

15.2.2.1 Commercial status

Jeneil Biotech marketed an early commercial product that contained RL, a biofungicide called Zonix. They obtained EPA approval for this product in 2004 [26]. Stepan recently began distributing Zonix and it is still available today [27]. Jeneil continues to market RL for agriculture, bioremediation, consumer care, and antimicrobial applications on their website [28].

Evonik, with their market leading SL position, announced their first RL product, RHEANCE ONE, at in-cosmetics Global in April of 2018 for use in personal care and cosmetics [29]. Around the New Year of 2022, Evonik announced a "three digit million-euro" capital investment in a rhamnolipid production facility [30]. The new facility will be at their Fermas site in Slovakia, mentioned above in the SL section. That level of investment will likely result in thousands of tonnes, or even tens of thousands of tonnes, of production capacity for microbial biosurfactants. Evonik released REWO-FORM RL 100 in early 2022 [31].

In 2020, Stepan acquired NatSurFact from Logos Technologies [32]. (For full disclosure, note that this author worked on NatSurFact at Logos for several years.) Early in 2021 Stepan announced the acquisition of a commercial fermentation production plant in Lake Providence, LA, originally constructed by Myriant for succinic acid production. Stepan indicated they will use it for microbial biosurfactant production [33]. As noted above, they also distribute Zonix Biofungicide, an RL-based product.

The agreement between Holiferm and BASF mentioned above in the SL section also applies to RL [34].

Advanced BioCatalytics announced in early 2021 that they are marketing BioSS RL for oil & gas, cleaning, agriculture, and wastewater markets [35].

Without going into any further detail, a number of companies indicated an interest in RL sometime in the last few decades, but appear to have stopped pursuing them. Either there is no longer any evidence that they are working with RL, or their commercial momentum appears to have stalled.

Unlike for SL, above, RL are early enough in their commercial deployment that only a couple of products have been announced that contain them. In 2019, Unilever announced they were marketing a hand dishwashing soap in Chile called Quix that contained Evonik's RHEANCE ONE. Additionally, a niche personal care cleanser direct marketed by Booni Doon contains NatSurFact RL [36]. MarkNature, an online store that sells products suitable for farming, gardening, food and beverages, and laboratories, markets RL ingredients in small lot sizes, as well [37].

15.2.3 Mannosylerythritol lipids

It appears that the earliest elucidation of mannosylerythritol (MEL) structures was in 1983 by Kawashima, Nakahara, Oogaki, and Tabuchi [38]. The MEL portion of the molecule is the hydrophilic head group, and they have two ~ 6–10 carbon hydrophobic tails attached via ester linkages to the mannose, as shown in Figure 15.4. Kawashima et al. [38] used a *Candida sp.* to produce their MEL, but other yeasts such as *Pseudozyma* sp. and *Ustilago* sp. are also known as producers.[13]

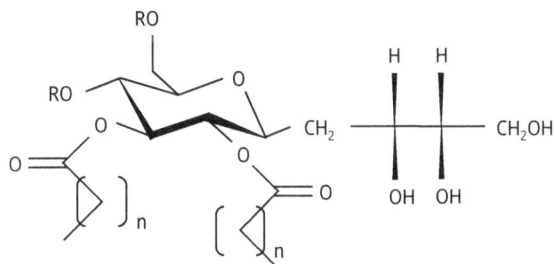

Figure 15.4: MEL are made up of a mannosylerythritol unit attached to two medium chain length fatty acids. R's are H or $(C = O)CH_3$ and $n = 6$–10.

15.2.3.1 Commercial status

MEL lipids were reported to be available from Toyobo in 2019.[15] However, Toyobo's website only lists their MEL product, Ceramela, as available in 1–5 kg lot sizes [39].

Advanced BioCatalytics announced in early 2022 the opening of a pilot plant that will be used to produce commercial quantities of MEL in the United States [40].

15.2.4 Surfactin

In contrast to most of the flycolipid microbial biosurfactants discussed here, surfactin is a lipopeptide or a combination of a hydrocarbon tail (lipo) with a cyclic, small protein head group (peptide). Surfactin was first isolated, characterized, and immediately classified as a surfactant in 1968 by Arima, Kakinuma, and Tamura [41]. As shown in Figure 15.5, the tail of the most common congener is 15 carbons long and attaches to the peptides through an α-hydroxy carboxylic acid. The head group of this congener has seven amino acid residues, including an aspartic acid. The aspartic acid provides a free carboxylic acid to allow the same pH behavior as described several times above: at highly acidic pH, it will be a nonionic surfactant, and at neutral and alkaline pH an anionic surfactant. This congener is produced using *Bacillus subtillis* as the producing organism.

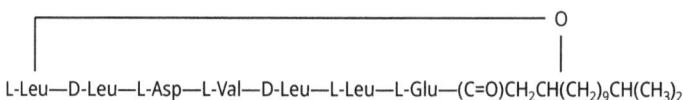

$$\text{L-Leu—D-Leu—L-Asp—L-Val—D-Leu—L-Leu—L-Glu—(C=O)}CH_2CH(CH_2)_9CH(CH_3)_2$$

Figure 15.5: A common congener of surfactin. The amino acids make up the head group, and the hydrocarbon chain is the tail.

15.2.4.1 Commercial status

In 2014, Kaneka announced they had succeeded in mass producing surfactin [42]. They currently market it as a cosmetics ingredient and to undisclosed other applications. Sabo is also involved in a partnership with Kaneka that includes surfactin [43].

15.2.4.2 Other lipopeptides

There are other lipopeptides that might operate as surfactants, for example, iturin, serrawettin (a.k.a. serratamolide), and lichenysin, though this has not been investigated to this author's knowledge. These materials appear to be of more interest as actives than surfactants.

15.2.5 Future options

There are a number of microbial biosurfactants that have been discussed in the academic literature or are used for applications other than as surfactants that will likely be commercialized in the near to not-too-distant future.

15.2.5.1 Cellobiose lipids

CL are a glycolipid that could be introduced in the next few years. The earliest mention that could be found of CL occurred in the mid-1980s and they were immediately classified as biosurfactants [44]. They have a cellobiose – a glucose dimer – headgroup with acyl esterification of some hydroxy groups and a ~16 carbon tail with a terminal α-hydroxy carboxylic acid. With a pK_a of ~3.5, CL will be nonionic at highly acidic pH and anionic at neutral and alkaline pH.

An interesting aspect of CL with regards to commercialization is that they are produced by the same organisms as MEL. Thus, the near-term commercial providers of MEL mentioned above will have institutional knowledge about their producing organism that may allow them to commercialize CL, as well.

15.2.5.2 Trehalose lipids

TL are a glycolipid that are likely further back in the pipeline, even though they discovered natural product earliest. They were first reported by Anderson and Newman in 1933.[7] They appear to be first classified as biosurfactants around 1980 [45]. The headgroup of TL is a trehalose moiety, which is another glucose dimer. The tail is composed of two comparatively long chain (>18 carbons) fatty acids. They have been found to be produced by a variety of organisms, including *Mycobacteria, Corynebacteria, Pseudomonads, Arthrobacteria, Rhodococcus,* and *Micrococcus*. One potential issue of trehalse lipids is they are a component of the cell wall in many producing species, which would make them more difficult to isolate in downstream processing than excreted products like all those listed above.[13] CL are not known to be commercially available, today.

15.2.5.3 Flavolipids

Even further back in the pipeline is flavolipids, which have a unique combination of head group and tails. Badour et al. [46] describe flavolipids as having a headgroup made up of a citric acid and two cadaverine moieties and two moderate length hydrocarbon tails. These are also referred to in the literature as siderolipids, but flavolipids appear to be the more common usage. They were first reported and classified as biosurfactants in 2004 by Bodour et al. [46]. Titers in the initial report are 0.1 g/l or less, but, as the authors note, could likely be improved with optimization. No production optimization studies have been found in the literature. In fact, the amount of interest in flavolipids appears to be low. What is of interest, here, is that as recently as 20 years ago, new microbial biosurfactants were still being discovered. The implication is that there are likely more out there, and one of them, including flavolipids, could be commercializable.

15.2.5.4 Polymer biosurfactants

Microbially produced polymers that have surfactant properties can be difficult to characterize. Examples include polysaccharide-protein-fatty acid complexes and lipo-polysaccharide-protein complexes. Because they are poorly characterized and are not marketed as surfactants, they are not discussed here in any more detail, with one exception, emulsan.

Emulsan is a well-characterized polymer that has repeating hydrocarbon chains pendant from a glycopolymer backbone. There are also carboxylic acid moieties every three units on the glycopolymer backbone. The pK_a of the carboxylic acid is reported to be 3 [47], meaning that the polymer chain will have some anionic character at pH's as low as 1.5. At pH's above ~ 5, it would be expected that the carboxylic acids are nearly 100% deprotonated, producing one negative charge for every three sugar units at pH > 5.

While emulsan is a well-characterized microbial biosurfactant, it is not known to be commercially available today. That could change in the future.

15.3 Perspective

It appears that in the last few years, after decades of expectation, microbial biosurfactants are finally becoming available in large commercial quantities. Specifically, SL have been available for the last several years, and RL are entering the market at the time of this writing, in 2022. There is also a healthy pipeline of other biosurfactants that can grow their current small market share or be introduced in the future.

Notably, the large capital investments by Evonik and Stepan – and partnerships announced by BASF and Dow – are a sign that global leaders in the surfactant market believe there is a future in microbial biosurfactants.

Because microbial biosurfactants are effectively new ingredients, they will likely open up new end uses or improve the functionality of current products that use surfactants. Additionally, because of their positive ecological profile, they can lower the environmental impact of the industry.

This author looks to the future of microbial biosurfactants with bubbly enthusiasm!

References

[1]   Myers D. Surfactant Science and Technology, John Wiley & Sons, Inc; 2006.
[2]   Verbeek H. Historical Review in Surfactants in Consumer Products: Theory, Technology and Application. Springer-Verlag; 1987, pp. 1–4.

[3]   Sakac N, Markovic D, Sarkanj B, Madunic-Cacic D, Hajdek K, Smoljan B, Jozanovic M. Direct
      potentiometric study of cationic and nonionic surfactants in disinfectants and personal care
      products by new surfactant sensor based on 1,3-dihexadecyl-1h-benzo[d]imidazole-3ium. Molecules
      2021;26:1366–81. doi: 10.3390/molecules26051366.

[4]   Moghimipour E, Handali S. Saponin: Properties, methods of evaluation and applications. Annu Res
      Rev Biol 2015;5(3):207–20. doi: 10.9734/ARRB/2015/11674.

[5]   Porter M. Handbook of Surfactants, 2nd edition, Blackie Academic & Professional; 1991.

[6]   Anderson K. Folding of outer membrane proteins in the anionic biosurfactant rhamnolipid. Febs
      Lett 1960;588(10):1955.

[7]   Jarvis F, Johnson M. A glyco-lipide produced by *Pseudomonas aeruginosa*. J Am Chem Soc 1949;71
      (12):4124–26.

[8]   Anderson R, Neman M. The chemistry of the lipids of tubercle bacilli: xxxiii. isolation of trehalose
      from the acetone-soluble fat of the human tubercle baillus. J Biol Chem 1933;101(2):499–504.

[9]   Desai J, Banat I. Microbial production of surfactants and their commercial potential. Microbiol Mol
      Biol Rev 1997;61(1):47–64.

[10] Gorin P, Spencer J, Tulloch A. Hydroxy fatty acid glycosides of sophorose from torulopsis magnoliae.
      Can J Chem 1961;39(4):846–55.

[11] Inoue S, Ito S. Sophorolipids from torulopsis bombicola as microbial surfactants in alkane
      fermentations. Biotechnol Lett 1982;4(1):3–8.

[12] Pekin G, Vardar-Sukan F, Kosaric N. Production of sophorolipids from *Candida bombicola* ATCC 22214
      using Turkish Corn oil and honey. Eng Life Sci 2005;5(4):357–62. doi: 10.1002/elsc.200520086.

[13] Soberon-Chavez G editor. Biosurfactants from Genes to Applications, Springer; 2011.

[14] McCoy M. American cleaning institute's 2015 conference. Chem Eng News 2015.

[15] Salek K, Euston S. Sustainable microbial biosurfactants and bioemulsifiers for commercial
      exploitation. Process Biochem 2019;85:143–55. doi: 101016/j.procbio.2019.06.027.

[16] Tullo A. Evonik industries is c&en's company of the year for 2016. Chem Eng News 2017.

[17] Reisch M. Givaudan will buy ingredients maker. Chem Eng News 2104.

[18] Bomgardner M. Biobased chemicals advance. Chem Eng News 2014.

[19] Culliney K. UK biotech startup holiferm eyes €8m for commercial biosurfactant upscale. Cosmet Des
      Eur 2021.

[20] Hatmaker J. Biosurfactant company signs for 90,000 sq ft wallasey industrial. *Place North West*,
      January 10, 2022.

[21] Bettenhausen C. BASF invests in biosurfactants. Chem Eng News 2021.

[22] Bettenhause C. Dow signs deal with locus performance ingredients for sophorolipids. Chem Eng
      News 2022.

[23] Happi https://www.happi.com/live-from-shows/in-cosmetics/2022-04-05/dow-introduces-high-
      performing-and-sustainable-technologies (accessed 2022–06–02).

[24] Kulakovskaya E, Kuakovskaya T. Prospects of Practical Application of Sophorolipids, Cellobios Lipids,
      and Mels. In: Kulakovskaya E, Kuakovskaya T, editors. Extracellular Glycolipids of Yeasts, Academic
      Press; 2014, pp. 75–83. doi: 10.1016/B978-0-12-420069-2.00006-6.

[25] Lohitharn N Enhanced Production of Rhmanolipids Using at Least Two Carbon Sources. US 11 142
      782 B2, 2021.

[26] US Environmental Protection Agency Office of Pesticide Programs. *Biopesticides registration action
      document hamnolipid biosurfactant*, March 23, 2004.

[27] Growing Produce. https://www.growingproduce.com/fruits/whats-new-for-2022-in-biological-crop-
      protection-products/ (accessed 2022–06–02).

[28] Jeneil Biotech, Inc. Product Offerings. https://www.jeneilbiotech.com/biosurfactants (accessed
      2022–06–02).

[29] Stuart E. Evonik to presetn innovations at in-cosmetics global 2018. Glob Cosmet Ind 2018.

[30] Bettenhausen C. Evonik invests in rhamnolipid biosurfactants. Chem Eng News 2022.

[31] Happi. https://www.happi.com/contents/view_breaking-news/2022-03-17/evonik-launches-rewoferm-rl-100 (accessed 2022–06–02).

[32] Bettenhausen C. Stepan buys biosurfactant business from logos. Chem Eng News 2020.

[33] Tullo A. Stepan makes a pair of acquisitions. Chem Eng News 2021.

[34] Bettenhausen C. Switching to sustainable surfactants. Chem Eng News 2022.

[35] PR Newswire. https://www.prnewswire.com/news-releases/advanced-biocatalytics-is-pleased-to-announce-the-commercial-release-of-bioss-rl-our-new-rhamnolipid-based-surfactant-and-part-of-our-fermactants-product-line-301293551.html (accessed 2022–06–02).

[36] Bettenhausen C. Rhamnolipids rise as a green surfactant. Chem Eng News 2020.

[37] MarkNature. https://www.marknature.com/products/rhamnolipid-biosurfactantandhttps://www.marknature.com/products/purify-rhamnolipid?pr_prod_strat=copurchase&pr_rec_id=8d88a6c91&pr_rec_pid=7216096149679&pr_ref_pid=7030679896239&pr_seq=uniform (accessed 2022–06–02).

[38] Kawashima H, Nakahara T, Oogaki M, Tabuchi T. Extracellular production of a mannosylerythritol lipid by a mutant of *Candida* sp. from *n*-alkanes and triacylglycerols. J Ferment Technol 1983;61 (2):143–49.

[39] Toyobo Cosmetic Ingredients Department. https://www.toyobo-global.com/products/cosme/category/ceramela/index.html) (accessed 2022–06–02).

[40] PR Newswire. https://www.prnewswire.com/news-releases/abc-opens-commercial-mel-biosurfactant-production-in-irvine-ca-301549977.html (accessed 2022–06–02).

[41] Arima K, Kakinuma A, Tanura G. Surfactin, a crystalline peptide lipid surfactant produced by *Bacillus subtilis*: Isolation, characterization and its inhibition of fibrin clot formation. Biochem Biophys Res Comm 1968;31(3):488–94.

[42] PHYS.ORG. https://phys.org/news/2014-10-drastic-reduction-synthetic-surfactants-power.html (accessed on 2022–06–02).

[43] COSSMA. https://www.cossma.com/business/article/sabo-partners-with-kaneka-34277.html (accessed 2022–06–02).

[44] Frautz B, Lang S, Wagner F. Formation of cellobiose lipids by growing and resting cells of *Ustilago maydis*. Biotechnol Lett 1986;8:757–62. doi: 10.1007/BF01020817.

[45] Cooper D, Zajic J. Surface-active compounds from microorganisms. Adv Appl Micrbiol 1980;26:229–53. doi: 10.1016/S0065-2164(08)70335-6.

[46] Bodour A, Guerrero-Barajas C, Jiorle B, Malcomson M, Paull A, Somogyi A, Trinh L, Bates R, Maier R. Structure and characterization of lavolipids, a novel class of biosurfactants produced by flavobacterium sp. Strain MTN11. Appl Environ Micrbiol 2004;70:114–20. doi: 10.1128/AEM.70.1.114-120.2004.

[47] Gutnick D. The emulsan polymer: Perspectives on a microbial capsule as an industrial product. Biopolymers 1987;26(S0):S223–S240. doi: 10.1002/bip.360260020.

Jacyr Quadros
Chapter 16
Epoxidation of vegetable oils and their use in plasticized polyvinyl chloride (PVC) applications

16.1 Chemistry of epoxidation

Epoxidation is a commonly used reaction to promote the modification of vegetable oils to generate chemicals that are useful in various applications, such as flexible PVC, polyurethanes, coatings, etc.

16.1.1 Epoxidation mechanism

Epoxidation of vegetable oils has been known for a long time. In the first experiments, the main reagent was peracetic acid [1], but formic acid and the *in situ* formation of performic acid [2] have been used industrially as the preferred method due to its better conversion and improved safety.

Epoxidation is described as a two-phase reaction [3], as follows:

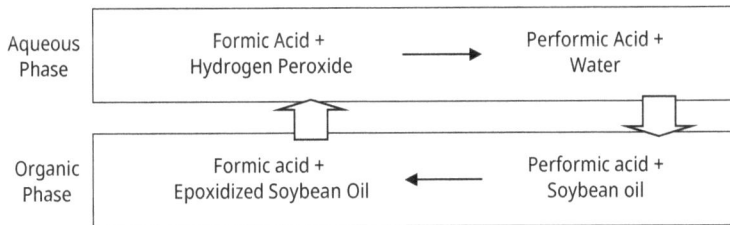

These reactions may be represented as follows. A single example of a molecule found in soybean oil (linolenic triglyceride) has been chosen for simplification.

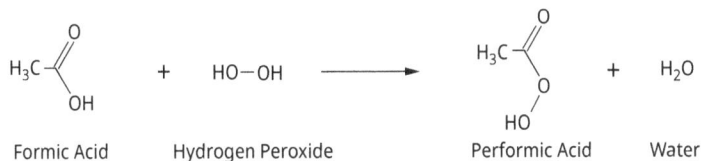

Jacyr Quadros, Innoleics Serv. e Cons. Ltda., São Paulo, Brazil, email:jquadros@innoleics.com

https://doi.org/10.1515/9783110791228-016

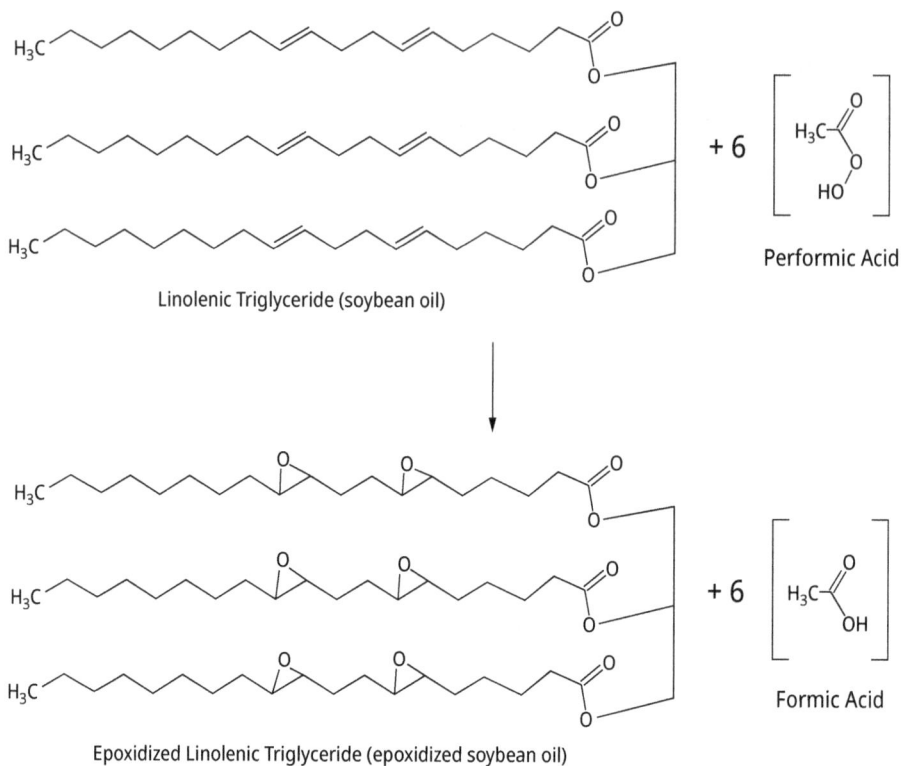

Linolenic Triglyceride (soybean oil)

+ 6 Performic Acid

Epoxidized Linolenic Triglyceride (epoxidized soybean oil)

+ 6 Formic Acid

Mass transfer between aqueous and organic phases is a limiting factor for the apparent kinetics of the reaction. Thus, a significant level of agitation and breakdown of droplets is critical.

At the same time, concurrent reactions that promote undesired ring-opening happen, as indicated below. These reactions are intensified by the presence of acids or excess hydrogen peroxide.

Epoxidized trigliceryde

+ H₂O

Hydroxyls formed from ring-opening

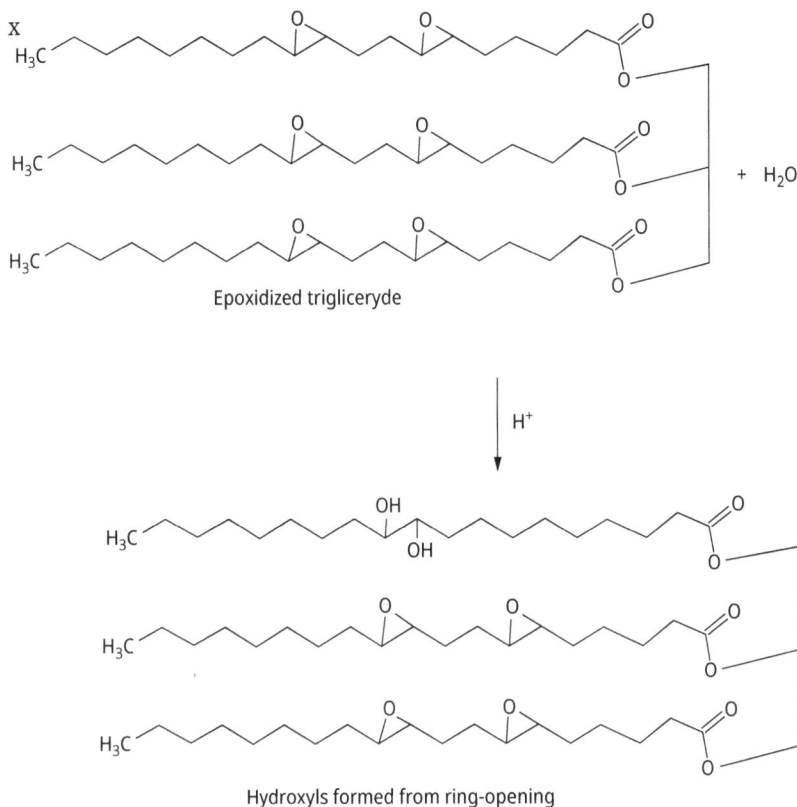

The main application for epoxidized soybean oil and derivatives is as plasticizers for PVC, and the oxirane index is directly related to the product's compatibility with the resin. The oxirane index, ASTM D 1652 method, measures the weight percent of oxirane ring oxygen atoms in a given molecule. It indicates the conversion of the epoxidation process. Double bonds are undesirable, as they are inversely related to the long-term stability of the epoxidized molecules [4]. A complete epoxidation of the double bonds while minimizing the ring-opening reaction is fundamental to maximizing plasticizer compatibility and permanence.

The epoxidation reaction is highly exothermic [5], and controlling the temperature is one of the key considerations in the commercial manufacture of epoxidized vegetable oils.

16.1.2 Industrial experience

The traditional industrial process relies on a controlled addition of hydrogen peroxide to the premixed formic and soybean oil. Intense agitation is fundamental to ensure good mass transfer between phases and maximize heat transfer between reaction media and cooling devices.

The ingredients are most commonly, refined soybean oil, formic acid at 85% purity, and hydrogen peroxide (ideally greater than 60% in water). Higher concentrations of hydrogen peroxide allow better conversion and higher reaction rates. Some producers preheat the mix before starting the addition of hydrogen peroxide, but that is not necessary, as the reaction is highly exothermic even at room temperature. The reaction times vary from 7 to 18 h, depending on reactor size and heat exchange capacity.

The intensification of the heat removal through increased agitation and greater exchange surface areas allows a faster addition of the peroxide, resulting in improved conversion and reduced reaction time [5].

Once the reaction has achieved its optimum conversion, where the maximum oxirane index is obtained, it is important to stop the reaction; otherwise, the ring-opening reactions will continue and decrease the oxirane index. Stopping the reaction can be done by adding water to dilute the hydrogen peroxide, cooling the reaction media, or ensuring quick phase separation.

After the reaction is complete, the next step is phase separation, which is typically done by decanting, normally requiring 12 to 24 h depending on vessel volume, or by centrifuging.

Once the phases are separated, the residual volatiles – water, peroxide, and formic acid – are removed by flash distillation followed by stripping in vacuum. Temperatures should ideally be maintained below 120 °C, a general rule to avoid additional ring-opening during these final processes.

16.1.3 Latest academic developments (state-of-the-art)

Many academic studies exploring different reaction models to improve conversion and reaction time have been published in recent years. Experiments involving new catalysts [6, 7], ultrasound intensification [8], and surfactants [9] have demonstrated that the reaction may have improved yields and selectivity, but the majority of recent developments have not yet been implemented in industrial settings. Enzymatic catalysis [10] has also been studied, demonstrating significant reduction of undesired ring-opening reactions, but with significant increases in reaction times and process complexity.

The most promising area seems to be process intensification, such as by using micro-reactors [11] or continuous processes [12], but additional research and cost improvements are required to make these options viable for industrial production sites.

16.2 Transesterification

Transesterification is one of the preferred methods for further modifying epoxidized vegetable oils to obtain useful chemicals by breaking down the triglycerides with alcohols. These materials can be used as primary plasticizers for PVC and offer improved performance as intermediates for other polymers, as lubricants, or even as additives for agricultural chemicals.

16.2.1 Other esterification technologies

Several technologies to split the triglycerides of vegetable oils are available, but transesterification has proven to be the most effective in terms of conversion, process simplicity, and cost.

16.2.2 Transesterification mechanism

Transesterification is a reaction where triglycerides react with a mono-alcohol to form fatty acid esters. The mono-alcohol normally ranges from one (methanol) to nine (isononanol) carbons.

The reaction can be carried out with base or acid catalysts, but base catalysts are preferred, such as sodium methoxide. In the case of the transesterification of epoxidized oils, acid catalysts could damage the oxirane rings, as discussed previously in this chapter.

For base catalysis, low humidity and low acidity of all reactants are fundamental to avoid poor conversion and undesirable soap formation.

The three-step reaction is represented below.

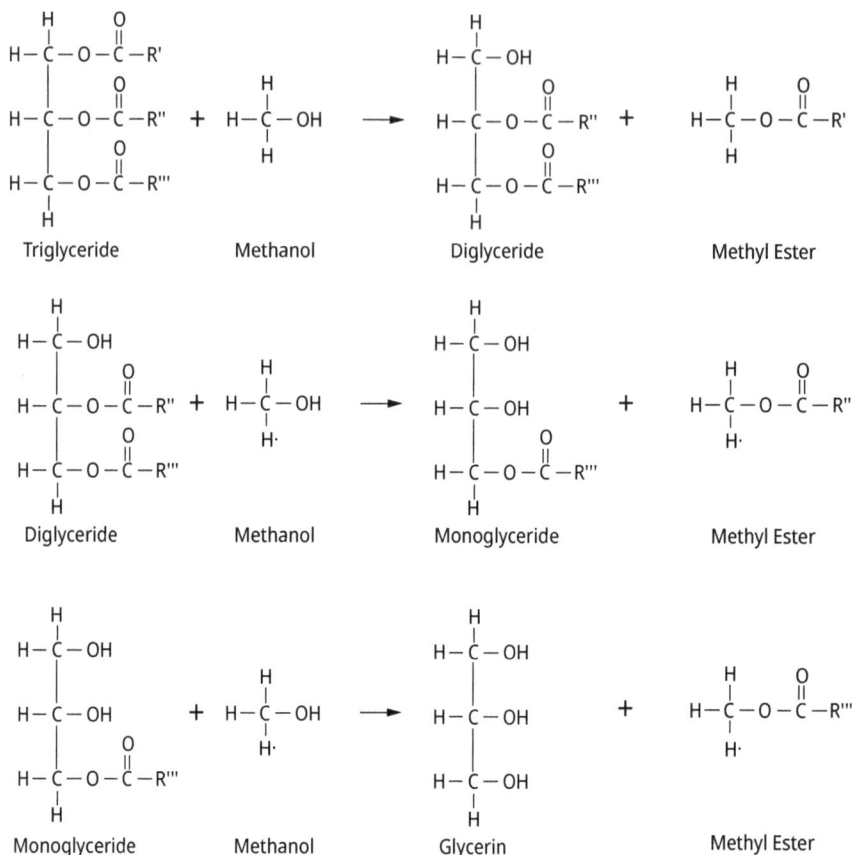

Triglyceride   Methanol   Diglyceride   Methyl Ester

Diglyceride   Methanol   Monoglyceride   Methyl Ester

Monoglyceride   Methanol   Glycerin   Methyl Ester

16.2.3 Equilibrium

Transesterification is a reversible equilibrium reaction. This equilibrium is dependent on reaction time and initial ratios of reactants.

The reaction is intensified by agitation, as methanol and glycerin have low solubility in the glycerides and esters. When carried out at 60 °C with excess methanol and intense agitation, such as a stirred reactor with high shear or ultrasound cavitation, the reaction achieves equilibrium in a few minutes, and the reaction media changes from a light yellow to a darker, reddish tone. After setting, the glycerin settles at the bottom and can be separated. From the top layer the excess alcohol is then removed by distillation, which can be reutilized. After the distillation, the final product is stored.

Should the reaction be interrupted earlier than at equilibrium or should there be a lower ratio of alcohol to triglycerides, the reaction will produce a combination of esters and tri-, di-, and monoglycerides. The proportions of these molecular species depend on the reaction conditions. The chart below shows a representation of the conversion over time for an excess ratio of methanol.

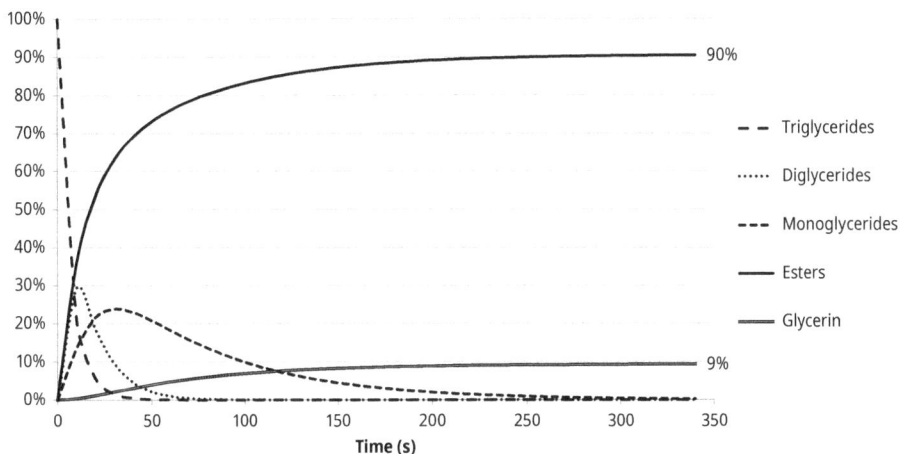

16.3 Acylation

Acylation or acetylation is a common reaction in the chemistry of vegetable oil and animal fat-based raw materials. The conversion of available hydroxyl groups into acetyl esters is instrumental in modifying molecules to achieve desired performances and behavior, as well as improving overall stability. In the case of PVC applications, hydroxyls are known to cause compatibility problems.

16.3.1 Acylation mechanism

The reaction is represented below, with the acylation of a diglyceride as an example:

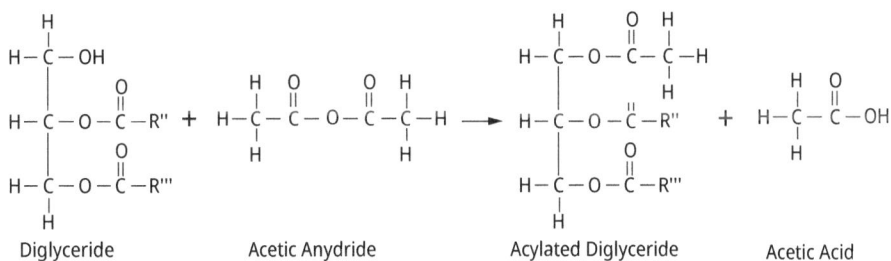

16.3.2 Industrial experience

The most commonly applied method of acylation that is widely adopted in the industry is the reaction of available hydroxyl groups with acetic anhydride carried out without the need for a catalyst.

After the reaction concludes, the removal of excess acetic anhydride and residual acetic acid is done by distillation. The removal of these acids is fundamental to ensuring the quality and long-term stability of the final material.

Once all higher volatility components are properly removed, the product is finished and stored.

16.4 Bio-based plasticizers

16.4.1 Bio-plasticizers gain momentum

The flexible PVC industry has an interesting opportunity to increase the contents of bio-based materials, promoting a reduction of the carbon footprint in the supply chain. Many segments, such as automotive, footwear, flooring, and synthetic leather, are introducing low carbon or even negative carbon footprint materials as part of their regular product lines.

This movement presents a great opportunity for PVC formulators, as bio-based plasticizers can be used to achieve these goals at a competitive cost.

16.4.2 Bio-based plasticizers generations

Epoxidized materials have been used as plasticizers for PVC since the early stages of flexible PVC technology. Plasticizers are normally high molecular weight esters that have solubility parameters within an ideal range, promoting plasticity and flexibility to the PVC resin, while reducing hardness.

The first vegetable-based PVC plasticizers used were simple epoxidized oils, such as soybean oil and linseed oil. These materials, considered the first generation of epoxy plasticizers, have presented an acceptable performance in some applications. Nevertheless, due to cost and performance limitations when compared to phthalate plasticizers, epoxidized plasticizers were relegated to a secondary role in the business. Epoxidized Soybean Oil (ESO) had been used typically as a co-stabilizer applied at low concentrations – below 2% – in PVC compounds, or at most, as a secondary plasticizer. Although presenting superior performance, Epoxidized Linseed Oil was commercially limited due to its higher cost. Despite these limitations, epoxy derivatives represented the largest volume of noncyclic plasticizers in the 1970s [4].

It was only in the 2000s that epoxy derivatives presented a significant penetration in the PVC plasticizers market, mainly due to increased interest in sustainability and increased petroleum-based materials costs. At the same time, ESO producers significantly improved product quality, enabling its use at higher concentrations.

As a consequence of these favorable market conditions other epoxy derivatives also began to gain momentum, especially epoxidized esters of soybean oil, due to a good combination of cost and performance. With improved compatibility, epoxidized esters of lower alcohols (methyl, ethyl, butyl, amyl, and 2-ethyl-hexyl) [13] entered the market as primary plasticizers, establishing what can be considered the second generation of epoxy derivatives. There were still a few limitations when compared to traditional phthalate plasticizers.

More recently, additional developments have introduced materials, such as epoxidized acylated mono- and diglycerides of soybean oil [14] that perform even better. These third-generation materials are replacing existing petroleum-based plasticizers with little performance deficits.

As a separate group, specialty bio-based materials such as acylated castor oil esters [15] present superior performance but their use is limited due to their higher cost.

16.4.3 Performance comparisons

Vegetable-based plasticizers can be considered primary plasticizers for flexible PVC applications. As these are molecules with different structures and functionalities when compared to traditional general-purpose plasticizers such as di-isononyl phthalate (DINP) or dioctyl terephthalate (DOTP), there are naturally a few performance differences. In most applications, these differences are either not significant or are adequately mitigated by formulation adjustments or special additives.

The charts below show some of the key parameters observed by formulators in the flexible PVC industry. A third-generation vegetable plasticizer with a 750 Da molecular weight was chosen for comparison with DOTP and Di(2-ethylhexyl) phthalate (DEHP or DOP).

Mass Loss

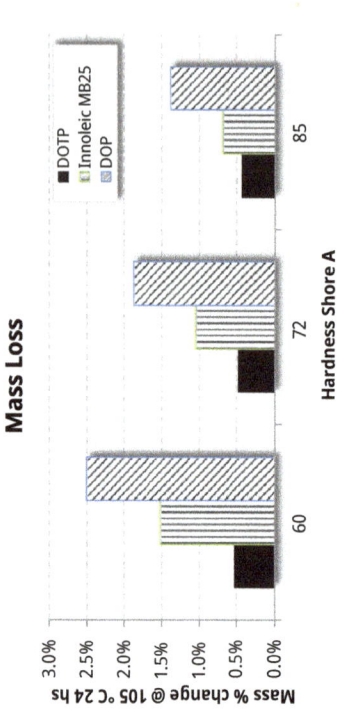

Elongation at Break

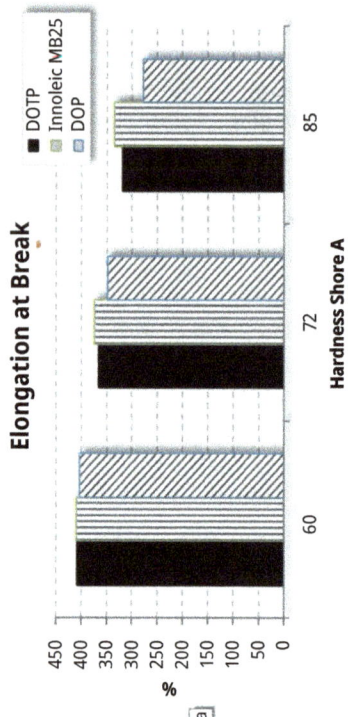

Efficiency

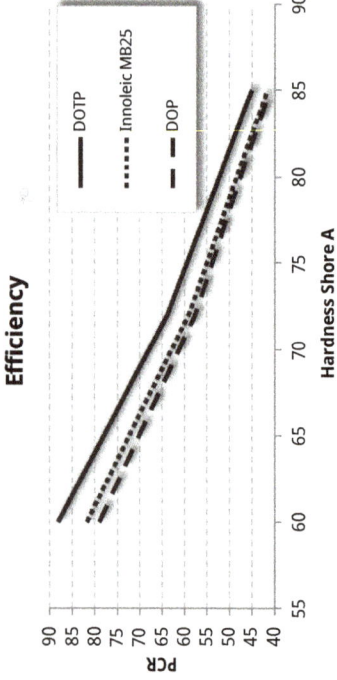

Tensile Strength

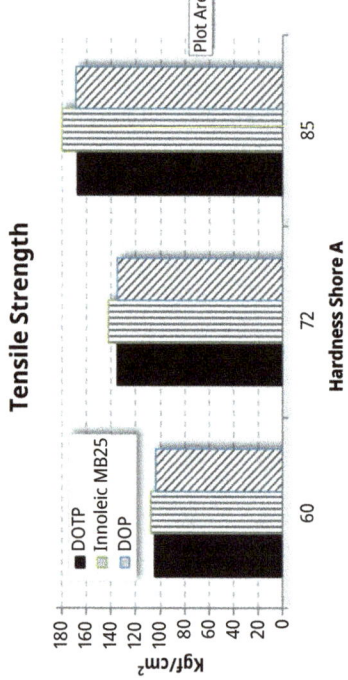

The high efficiency of the vegetable-based plasticizer is readily noticeable. The efficiency is similar to that of DOP, but with a considerably lower volatility. Also, a typical characteristic is noted in the mechanical properties: higher tensile strength, while also having a higher elongation at break, which can be observed in final applications where materials formulated with vegetable-based epoxy plasticizers present a higher resilience.

In addition to these characteristics, compounds made with epoxy plasticizers also present a considerably superior performance in thermal stability, as shown in the picture below, where methyl epoxy soyate is compared to DOP and di-isobutyl phthalate (DIBP). Due to this inherent thermal stability property, PVC compounded with vegetable oil plasticizers require less potentially harmful and toxic metal soap stabilizers that are used to ensure PVC heat stability. Such thermal stabilizers are also quite expensive, adding to the overall cost of compounded PVC formulations.

Another characteristic of epoxy plasticizers is viscosity increase in paste PVC applications that require adjustments or blends with other plasticizers. This is expected as epoxy materials are inherently more efficient and cause the resin particles to swell faster, increasing viscosity over time. The chart below shows this effect, comparing traditional plasticizers, including dioctyl adipate (DOA) with octyl epoxy soyate, methyl epoxy soyate, and a 470 Da molecular weight third-generation vegetable epoxy plasticizer.

Viscosity

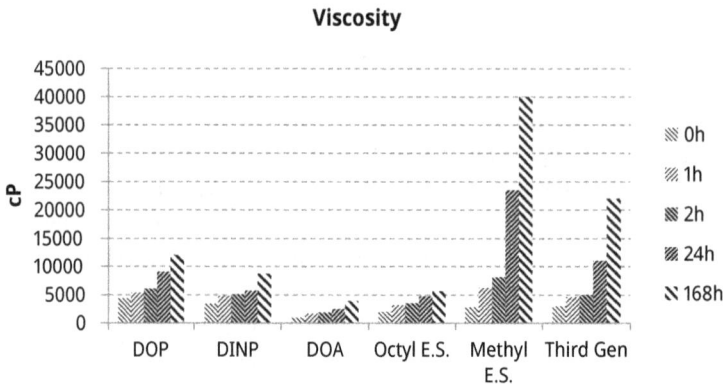

Adjustments can be made, depending on the formulation, with a combination of different PVC resins that have lower viscosity aging or by including lower viscosity plasticizers.

16.5 Challenges to growth and wider adoption

16.5.1 Key areas for technical improvements

Epoxy plasticizers present a few areas of improvement to enable higher growth and wider adoption.

One of the key sensitivities is higher water extraction when compared to petroleum-based materials. This is an important performance aspect, especially for outdoor and intense water-exposed applications, such as pool liners, hoses, roofing, and awnings. Since this is intrinsic to the polar nature of the epoxidized molecules, additives may play an important role in mitigating this effect. Hydrophobic components, such as silicone oils, have been considered and can be effective, but their use is limited due to the low compatibility with the PVC matrix. Another solution is to create physical barriers by layering different compositions to minimize the direct contact of water with the bio-based plasticizers.

Another area of concern is hydrolysis [16]. This is a chemical reaction where water breaks the oxirane rings forming hydroxyls, making the product incompatible with the PVC matrix. The consequence is an accelerated exudation of the plasticizer generating undesirable surface defects, such as tack or oil build. Although critical in accelerated weathering tests, this reaction is a concern, mostly in applications where articles are exposed to high humidity or moisture and high temperatures. In addition, proper processing of the epoxidized material to minimize acidity also significantly reduces the hydrolysis reaction rates.

Finally, a surface effect known as blooming [17] is also a sensitive area for epoxidized materials. This effect is noted in articles with darker colors and glossy finish. Occasionally, when submitted to low temperatures (−5–0 °C), a white bloom becomes visible on the surface. If heated to > 30 °C, the effect disappears completely. There is no conclusive study of the mechanism that generates this problem, but it could be associated with the presence of saturated fatty acid esters, which have a very high melting point. These molecules, present even at very low levels on the surface of the PVC, are not detectable visibly and might nucleate and crystallize, creating the visible whitish bloom. When heated, these molecules return to their liquid state and are not visible. This effect is not easy to replicate in laboratory conditions, as the rate of temperature drop and increase seems to be a key factor for its appearance. High oleic soybean oil, with reduced fatty acid saturates through genetic modification, could potentially mitigate this effect, but the cost of this emerging technology is still a limiting factor.

16.5.2 Marketing and regulatory opportunities

Governments and regulatory agencies are promoting the reduction of carbon emissions through programs and regulations, creating new opportunities for bio-based materials. Companies are now faced with sustainability and emission reduction goals, forcing them to seek process improvements and new materials.

A study [18] published in 2014 shows that the industry was not ready to adopt chemicals from renewable sources. The respondents indicated that the barriers to adopting bio-based materials were, in order of importance: high cost, little market demand, inferior performance, and sustainable claims credibility. In the current scenario for bio-based plasticizers, cost has significantly been reduced; the demand is increasing due to the environmental pressure; performance has improved significantly with the new generations of plasticizers; and sustainable claims can be certified through reliable ecolabels.

A study presented at PVC 2021 [19] concluded that companies with higher awareness about ecolabels have a higher perception of demand and premium value for renewable plasticizers. Also, there was a direct correlation between company size and its perception of demand for bio-based materials.

Younger consumers are also significantly more impacted by sustainability than older ones [20]. In response to this demand, companies are introducing new product lines directed at this group of consumers that is considerably more likely to change its purchasing behavior towards more sustainable products.

Bibliography

[1] Findley TW, Swern D, Scanlan JT. Epoxidation of unsaturated fatty materials with peracetic acid in glacial acetic acid solution. J Am Chem Soc 1945;3(6):412–14.

[2] Niederhauser WD, Koroly JE. Process for the epoxidation of esters of oleic and linoleic acids. US2485160 USA, 1949.

[3] Santacesaria E, et al. Biphasic model describing soybean oil epoxidation with H2O2 in continuous reactors. Ind Eng Chem 2011;51(26):8760–67.

[4] Sears JK, Darby JR. The Technology of Plasticizers. s.l.: Wiley; 1982.

[5] de Quadros JV, Jr, Giudici R. Epoxidation of soybean oil at maximum heat removal and single addition of all reactants. Chem Eng Process Intensifi 2016;100:87–93.

[6] Farias M, Martinelli M, Bottega DP. Epoxidation of soybean oil using a homogeneous catalytic system based on a molybdenum (VI) complex. Appl Catal 2010;1–2(384):213–19.

[7] Wai PT, Jiang P, Shen Y, Zhang P, Gu Q, Leng Y. Catalytic developments in the epoxidation of vegetable oils and the analysis methods of epoxidized products. RSC Adv 2019;65(9):38119–36.

[8] Han LJ, Li L, Liu GQ. Ultrasound-assisted acceleration of soybean oil epoxidation and the interfacial tension study. In Adv Mater Res 2011;201:2583–86

[9] Rethwisch D. et al. Epoxidation of Soybean Oil in a Microemulsion-Assisted Environment. In: The 2005 Annual Meeting. Cincinnati, OH, USA: s.n., 2005.

[10] Vlček T, Petrović ZS. Optimization of the chemoenzymatic epoxidation of soybean oil. J Am Oil Chem Soc 2006;83(3):247–52.

[11] Cortese B, de Croon MHJM, Hessel V. Novel Process Window for the Soybean Oil epoxidation-Simulation of High-p, T Processing Microreactors. In: 8th European Congress on Chemical Engineering (ECCE-8). Berlin, Germany: s.n.; 2011, p. 42.

[12] Olivieri GV, de Quadros JV Jr, Giudici R. Evaluation of potential designs for a continuous epoxidation reactor. Ind Eng Chem Res 2021;60(39):14099–112.

[13] Benecke HP, Vijayendran B, Elhard JD. Plasticizers derived from vegetable oils. US6797753B2 USA, January 31, 2002.

[14] Quadros J Jr., Carvalho JAD. Plasticized PVC composition. US 8,623,947 USA, 01 07, 2014.

[15] Galanakis CM editor. Biobased Products and Industries. s.l.: Elsevier; 2020, p. 390.

[16] Guo Y, Hardesty JH, Mannari VM, Massingill JL. Hydrolysis of epoxidized soybean oil in the presence of phosphoric acid. J Am Oil Chem Soc 2007;84(10):929–35.

[17] Lauridsen CB, Hansen LW, Brock-Nannestad T, Bendix J, Simonsen KP. A study of stearyl alcohol bloom on Dan Hill PVC dolls and the influence of temperature. Stud Conserv 2017;62(8):445–55.

[18] Special Chem. *Polymer-Additives*. [Online] February 1, 2014. [Cited: April 7, 2022.] https://polymer-additives.specialchem.com/tech-library/article/unquestionable-verdict-for-newer-bio-based-chemicals-or-additives-no-thanks-too-expensive.

[19] Quadros JQ, Garcia-Tousa D. *Eco-Labels, Applicability and Impact on the Flexible PVC Sector*. on-line: s. n., 2021.

[20] Nielsen. Global consumers seek companies that care about environment issues. *Nielsen*. [Online] 2018. https://www.nielsen.com/us/en/insights/article/2018/global-consumers-seek-companies-that-care-about-environmental-issues/.

Gunnar Lynum, Steven Lynum

Chapter 17
The positive impact bio-solutions have on the global construction industry

Addressing the COP26 Conference's preliminary opening on a Sunday afternoon, Abdulla Shahid, the President of the UN General Assembly called for an honest look at where things stand: "We have had decades to argue the facts about climate change, about the power of renewables, about the fine details of monitoring or cost-sharing. Yet, we have still failed to act with the conviction and determination required" [1]. In this chapter, SMD Products Company Inc. would like to share how bio-based solutions can have a positive impact on the global construction industry. As a small private enterprise, it is sometimes easier to impact change, thanks to public policy initiatives, as smaller enterprises are sometimes missing the large bureaucratic red tape that sometimes blocks innovation.

Note: (Soybean Field in USA, Autumn 2021, Courtesy of Gunnar Lynum).

Gunnar Lynum, Steven Lynum, SMD Products Company, Inc. & Strategic Market Development, LLC
P.O. Box 1634, Manchester, MO 63021, e-mail: customerservice@smdproductscompany.com,
www.smdproductscompany.com

https://doi.org/10.1515/9783110791228-017

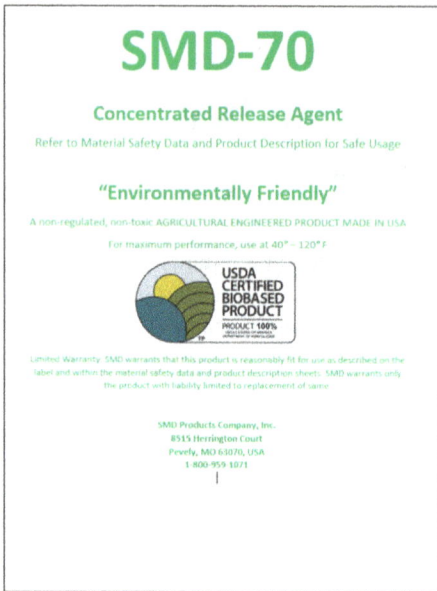

Note: (SMD-70 Product Label, courtesy of SMD Products Company, Inc.).

17.1 Introduction

The goal of this chapter is to walk the reader through one of the greatest environmental challenges we face as a civilization. Modern day teaching focuses on the environmental impact of several industries, which include the sectors of energy, transportation, mining, manufacturing, and so on. But one that is often overlooked is the one that impacts every person on this earth, namely the construction industry. Without buildings or modern-day structures, one can only fathom how the world would have evolved. It is believed that about 1.8 million years ago humans may have lived in what may have been the first "house" in Tanzania in Olduvai Gorge, and started heading down the path, which is now often referred to as environmental destruction, which in turn has been analyzed to exhaustion [1]. If one is interested in archeology, there are many sources that are highlighted at the end of this chapter. For avoidance of doubt, the authors of this chapter want to be clear to the reader that we are capitalists, realists, and not tenured environmentalists. We believe in free markets, the modernization of the world, and know this can be done in a sustainable and environmentally friendly manner, thanks to bio solutions that have evolved in the recent years. It is important to set the stage and understand the global impact that industrialization has had on the environment. It is also important to recognize the commonsense solutions that have occurred and continue to occur, thanks to the unified efforts of several industrialized nations that are trying their utmost to protect the world for future generations, as most recently observed at COP26 in Glasgow, Scotland [2]. We will explore, propose, and share the new way of doing things. It was Peter F. Drucker who stated, "If you want something new, you have to stop doing something old." Let us now proceed and look at something new, but first, by understanding how we got here and why we are proposing something new:

17.1.1 The history of construction (i.e. Concrete)

The first evidence of the use of a concrete-like material being used dates to roughly 3000 BC. The Egyptian pyramids are a perfect example of how a material, unlike modern-day concrete, was used to build solid structures. It can be argued that there must have been other structures built prior to the pyramids using a concrete-like substance, but for interest of getting to the end of our story, let us start with the Pyramids of Egypt as our jump-off point when trying to understand the impact that construction has had on all of us. It is believed that the Egyptians' definition of concrete was a unique mixture of gypsum, lime, mud, and straw that was used to bind the large blocks of stone that make up the pyramids. Nearly 2500 years later, it is clear that the Romans improved upon the work of the Egyptians and formed a structure like the Coliseum in Rome, which was built under the instruction of Emperor Vespasian, and completed during the time of Emperor Titus in 80 AD [3]. A fantastic marvel, the Amphitheatre, able to hold between 50,000 and 80,000 spectators, is sometimes called one of the Seven New Wonders of the World [4].

Between the Roman Empire and the Middle Ages, it gets a little murky as the formulation of concrete was believed to have been lost until it was found again in 1414 in old manuscripts [5]. The invention of early concrete is credited to John Smeaton. A perfect account of this historically significant fact goes to the Canadian authors at Giatec Scientific, who share the story on their website as follows:

> It wasn't until 1756 that the technology took a big leap forward when John Smeaton discovered a more modern method for producing hydraulic lime for cement. He used limestone containing clay that was fired until it turned into clinker, which was then ground into powder. He used this material in the historic rebuilding of the Eddystone Lighthouse in Cornwall, England in 1793.

Today's concrete is made from what is called Portland cement. Just over thirty years after Smeaton famously used his first modern day concrete in the Eddystone Lighthouse; Joseph Aspdin invented a mixture, which he made by burning grounded chalk and finely crushed clay in a lime kiln until carbon dioxide evaporated, producing a strong cement, later named "Portland Cement." The inventor – Joseph Aspidin – is credited with inventing Portland Cement in 1867 [3]. And, the idea of reinforcing concrete, the modern day use of rebar, was done by Joseph Monier through his 1867 patent of the concept that one can reinforce concrete using an iron mesh.

From the industrial revolution onward, the advancements and the use of concrete cannot be understated. It has become a part of every person's world – from streets to small structures to high-rise buildings to major dams. Without concrete, many things we take for granted would not have been possible. Can one imagine what large cities would look like if the world's first concrete high-rise in Cincinnati, Ohio had never been built in 1903? The Ingalls Building, a modest 16-story building is credited to be the world's first reinforced concrete skyscraper. This architectural first, listed in the National Register of Historic Places, has now been renovated, as of the timing of this publication, and is now operated as a Courtyard Marriott Hotel in downtown Cincinnati [6]. What if the Hoover Dam had not been built in 1936? At over 221 meters in height, it is still one of the largest hydroelectrical stations in the United States [7]. These achievements have helped society, but at what cost? Our next section shall discuss the environmental footprint the concrete industry has created.

17.1.2 The environmental footprint issue

Now that we understand the colorful and detailed history of the industry, let us step back and understand what environmental footprint has been created, and what opportunity there is for sustainable bio-based materials to be used to reduce this environmental impact on the future. It is interesting to note, as we just learned, the first high-rise was built in the Great State of Ohio, and one of the solutions we plan to introduce is also grown in the same great state of Ohio. But, before we find the solution, let us first understand the problem.

In 2019, the National Academy of Sciences brought attention to a paper that was published by the University of Rochester, which took a broad look at the problem [8]. It was cited that carbon-dioxide (CO2) from the cement industry accounts for up to 8% of the global CO_2 emissions. The industry has its challenges, giving off up to 2.8Gtons/year of environmentally unfriendly CO_2 gasses – CO_2 emissions due to the decomposition of $CaCO_3$ to CaO and the use of fossil fuels in that process. This is a problem, but one that will be solved in the coming years, as there are several well-known efforts in developing countries to produce a form of "green cement" that offers lower overall emissions during production, primarily due to the effort of taking industrial waste and using it as a raw material in the cement. So, it is safe to say that the core cement industry is already looking at various solutions in order to reduce its ecological footprint. One sub segment of the construction industry, which has not yet been addressed, is the often overlooked environmental impact of "form-release agent," or sometimes called "form oil." At a recent gathering of industry leaders in Washington, DC at the 2021 AFCC Biobased Economy Conference, we were able to hear the industry challenges and highlight what sustainable bio-based products can do to help the industry lower its environmental footprint [9].

The University of Cambridge dictionary defines environmental footprint as: the effect that a person, company, activity, etc. has on the environment [10]. For the purposes of this chapter, we will focus on the use of form oil and other release agents. It is said that form oil or form release touches nearly every concrete project being executed. So, what is a concrete form release? It is a chemical that is applied to prevent the adhesion of freshly placed concrete to the forming surface. The forming surface is usually plywood, steel, metal, silicon, or some combination of these materials. The form release market is segmented into various uses – PU molding, Rubber molding, Concrete molding, Die-Casting, plastics moldings, and others – and is dominated by industry powerhouses such as: Chem-Trend L.P. (US), Henkel AG & Co. KGaA (Germany), Croda International Plc (UK), LANXESS Group (Germany), Shin-Etsu Chemical Co., Ltd. (Japan), Daikin Industries Ltd. (Japan), Michelman, Inc. (US), Marbocote Limited (UK), McGee Industries, Inc. (US), Miller-Stephenson, Inc. (US), TAG Chemicals GmbH (Germany), and LORD Corporation (US) amongst others [11]. The concrete form release market is the third largest segment in the greater concrete release agent market, which has suppliers that are more from the petroleum industry, like Shell, Chevron, etc. Here the goal of a good concrete form release is to create a barrier between the fresh concrete and the form; thus, highly toxic and high VOC content oils are sometimes used from these suppliers as they are easy to apply in spray form. The environmental challenge is that this form release, which acts as a barrier runs off the form during the process and off the concrete after the process. One can only imagine what this concrete form release runoff does to the local eco-system when building a dam or bridge, which is close to local water sources. It in increasingly evident that demand for biodegradable form release agents will become even more evident.

The US Government attempted to limit the VOC (volatile organic compounds) emissions from architectural coatings under Section 183(e) of the Clean Air Act in 1999. The new regulation set form-release VOC content to 450 grams per liter or 3.8 pounds per gallon for newly manufactured products. In the act, form releases were classified as architectural coatings, but only applied to the manufacturers and not the users. These new regulations forced more processing steps in the creation of petroleum-based form-release agents, and for this reason, also increased the environmental footprint of these products. Though the intention of the regulation was noble, the large loophole continued to exist [12]. The usage of petroleum-based form releases cause strong smells due to the VOC content, which is still allowed, and skin irritations due to the petroleum content in the form release. These minor discomforts are easily avoidable by implementing a system whereby bio-based products are used, as they do not contain VOCs. The raw materials in biodegradable solutions have been shown to reduce skin-irritation, which continues to plague workers who handle form release daily.

17.2 The global initiative – UCOP26 / Paris accord / BIO-preferred

As noted at the beginning of this chapter, the most recent global initiatives for the reduction of environmental damage on a broad basis has been led by the Paris Accord and by, what was recently called, UCOP26 held in Scotland. Addressing the Conference's preliminary opening, Abdulla Shahid, the President of the UN General Assembly called for an honest look at where things stand: "We have had decades to argue the facts about climate change, about the power of renewables, about the fine details of monitoring or cost-sharing. Yet, we have still failed to act with the conviction and determination required." [13] For this reason, it became clear that certain governments and international agencies would need to step up and take action into their own hands. This movement has created the untiring demand for bio-based solutions that can not only contribute to reducing greenhouse gases, but also reduce the pollution of the waterways. One example of a roadmap for a sustainable future was presented by the Tokyo Organizing Committee of the Olympic and Paralympic Games [14]. In parallel, the USDA (United States Department of Agriculture) was also working for the promotion of Bio-based solutions to be used by federal contractors on infrastructure projects in the United States. This program run by the USDA is called the Bio-Preferred program, see the chapter 1 for more details, and involves third-party testing of products for their compliance to very rigid criteria, in which those that pass, are invited to attach the USDA Bio-Preferred label on their product [15]. Activities by entities such as the International Olympic Committee, United Nations, and the US Government help stimulate industry to come up with creative solutions to complex problems. But, as noted in the beginning of our chapter, the fine balance between regulation and capitalism should be maintained to stimulate innovation.

> Bio-based solutions can produce better concrete products, be sustainable, easier to handle, and cost effective.

Global trends of sustainability

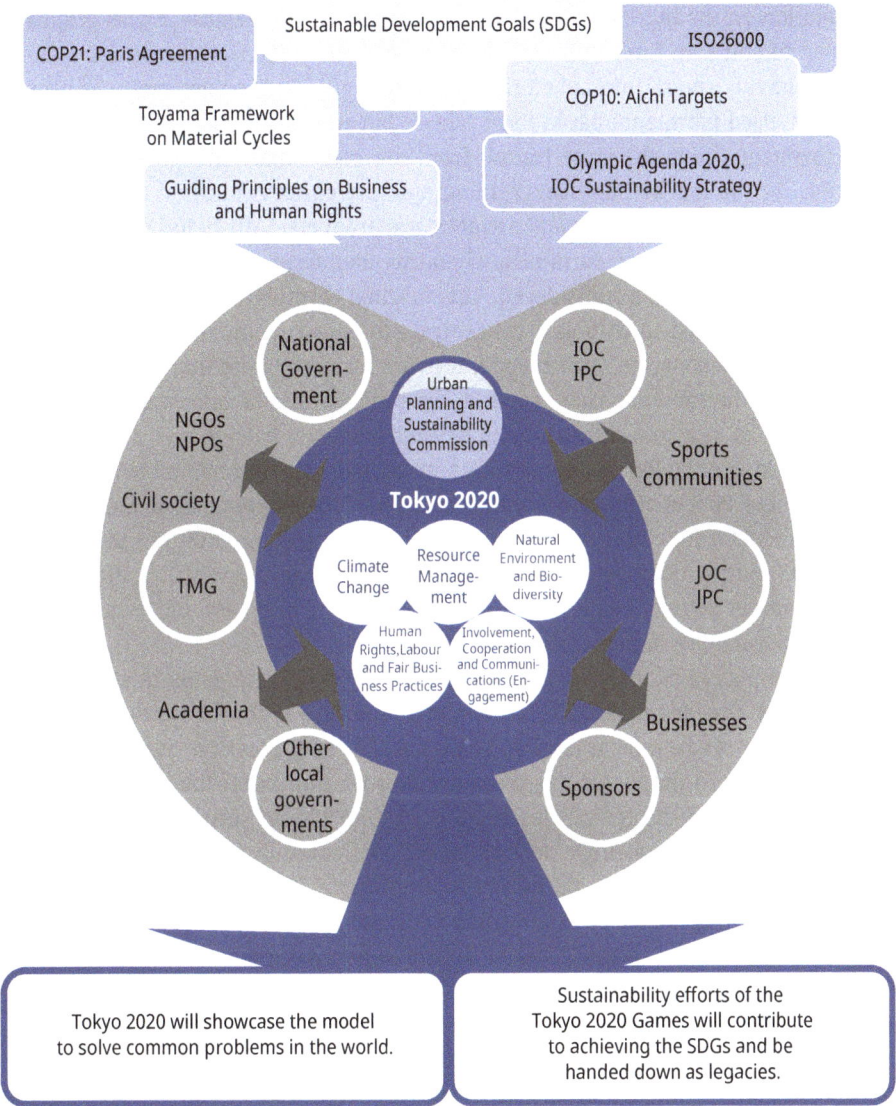

Figure 17.1: Sustainable Sourcing Code (3rd Edition) – public handout to potential material suppliers from welcome packet of the organizing committee, January 2016.

17.3 Introduction to SMD-70 concrete form release

One solution on the market today is SMD-70 Concrete Form Release. It was originally designed to solve an Environmental Protection Agency (EPA) violation for a client who was having runoff into Lake Erie in Ohio in the 1990s The customer was using a commonly used petroleum-based form release but was negatively affecting the local fish ecosystem due to the large run off from concrete builds. Though SMD-10 is still available on the market, and several boutique pre-casters continue to say SMD-10 makes the best concrete blocks and architectural products; SMD-10 has the same logistical footprint as standard form oils, when the amount of energy spent to ship the product to the job site is considered. This original formulation has now been perfected and stabilized to ship as a concentrate to construction sites throughout the world as SMD-70 concrete form release. It is interesting to note that SMD-70, though derived from the raw materials grown in the United States of America, is commonly shipped abroad as a concentrate due to its environmental friendliness, ability to support the making of higher quality concrete, and smaller environmental footprint, as it is shipped as a 20% concentrate, eliminating 80% of the shipping costs, thereby making this product competitive, even in some of the lower income countries in Southeast Asia. SMD Products Company spent just over 10 years to develop SMD-70, after having seen the market potential experienced by selling SMD-10 to several of North Americas largest construction companies for several years.

The complexity of logistics is removed as SMD-70 ships globally in standard 55-gallon drums. One drum of SMD-70, when mixed on site, will produce the equivalent of 5 drums of standard form release. The ability to reduce logistic costs by not shipping water around the world has created a positive impact on the overall environmental footprint of the form release industry. One 20-foot shipping container can hold over 40 drums on a single-floor shipment configuration, and a grand total of 78 drums when double stacked in the same container. A typical customer will buy 40 drums in a container and have 2.5-gallon cans and jugs for smaller projects, which may be loaded on the second level of the container. All the raw materials used in SMD-70 are regionally sourced in the mid-western United States of America, from the original soybean oil component to the proprietary blend of chemicals used in the creation of SMD-70. The industry was misled by some legacy manufacturers who conducted a negative campaign against form releases that have a mixture of water, and it has taken years for SMD-70 to catch on as "the form release" of choice for several job sites. The below image shows why the legacy manufacturers pushed hard to prevent SMD-70 from becoming the new standard on the market. SMD-70 quickly proved to major contractors that it provided a better release.

The unique concept of having water as the delivery agent is the key to the product. This is the secret sauce of SMD-70. Unlike traditional petroleum-derived release agents who use petroleum as the delivery agent that evaporates quickly, leaving the fatty acid (lubricant) on the form material, SMD-70 does something previously unheard-of: SMD-

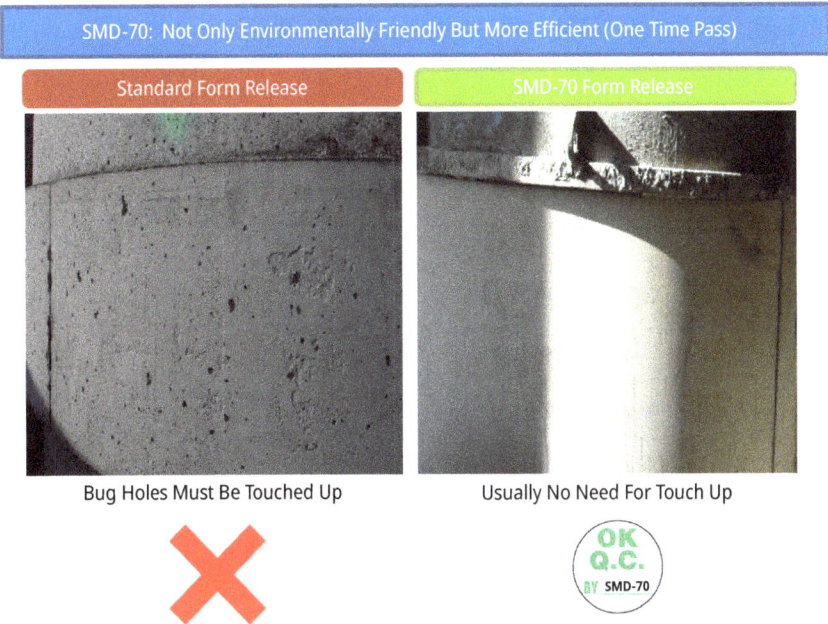

Figure 17.2: Concrete product using traditional release agent on left and using SMD-70 on the right. (Courtesy, SMD Products Company, Inc.).

70 is blended with water that remains as an emulsion. This means the fatty acid molecule is encapsulated by water. The water is used to deliver the product as a fine mist, but never allowing the water to enter the molecule. This is the key point and leads to better concrete with fewer bug holes. Please see Figure 17.2 above. Though users have been confused by the legacy suppliers to believe that when you add water, the product becomes a water-based product, this is not the case with SMD-70 as the water is only used as the delivery mechanism. For this reason, SMD-70 was the product of choice, used in the building of the Tokyo Olympic Stadium and the Olympic Village. Please see figure 17.3, which highlights the Tokyo 2020 olympics project, courtesy of Saitama Ryokoku.

During the 2021 AFCC Biobased Economy Conference, the effective use of bio-based form releases was discussed in detail. The consensus of the panel discussion was that one does not need to give up concrete strength or the quality of the concrete product when using the latest technologies that are on the market. Sophisticated engineering companies are using artificial intelligence to determine the concrete strength using very powerful image-based algorithms that rely on deep learning to find areas of weakness in concrete samples. Figure 17.4 below shows one such solution from Aiforia that is applying its innovative use of artificial intelligence (AI) using image recognition, combined with deep learning algorithms, to the oncology space to potentially solve some of the greatest challenges at some of the world's largest medical institutions. The process is also being used in the areas of sustainable construction and sustainable fish farming.

Figure 17.3: Report of document sharing construction project details (Courtesy, Saitama Ryokoku).

We believe the use of AI, especially the type of AI deployed by Aiforia will continue to help advance the construction industry.

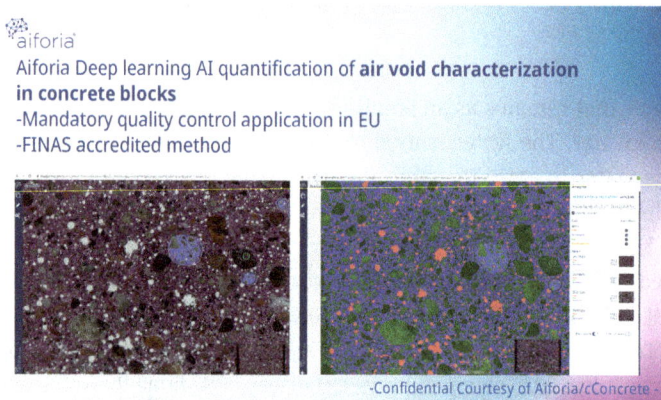

Figure 17.4: Color image of how AI is applied in the concrete industry. Image courtesy of Aiforia.

The structural integrity of concrete is becoming more of a recent talking point as several recent mishaps have occurred with infrastructure that has started decaying. Western nations are now looking into perfecting the process to make better concrete. As we know, there are seven steps for making concrete: (1) Make Cement (2) Set forms (3) Apply form release (4) Place rebar (5) Pour concrete (6) Let concrete set (7) Cure the concrete. It is now becoming more evident that the timing to apply the rebar and the process of curing concrete play important roles in the process. Historically, rebar was placed and then sprayed with form release, as the form release would quickly evaporate and lose its lubricating properties. SMD-70 has been shown to be the ideal solution for this timing problem, as it can be applied and will not evaporate (only the delivery agent, "the water" evaporates), allowing for more time to place the rebar with no loss of lubricating features. Please see image below (Figure 17.5) where one can observe the proper placement of rebar. We observed this first hand when we noticed and observed the use of SMD-70 by some of the suppliers to Singapore's world-renowned HDB flat construction, where torrential rains and tropical heat would sometimes wash away legacy form releases, when SMD-70 would remain in its useful state even after some of the harshest weather patterns had taken their toll on the construction.

Figure 17.5: Typical image of rebar being placed in form prior to concrete pour. (Photo courtesy of SMD Products Company, Inc.).

Research out of Japan is now being validated by the Japanese Government for use in major infrastructure projects where we are seeing the curing benefits of a derivative of SMD-70. A nationally recognized university in Japan has conducted testing, which showed that the application of SMD-70 in its concentrated form will create a better environment for curing the concrete when compared to the environmentally damaging methods of water bath, plastic, and heating. This was validated, and can be seen in the below Figure 17.6, where it is shown that under the Comprehensive Strength Test, Air Permeability Test, and Neutralization Acceleration Test, there is no worse outcome when compared to traditional curing methods. Though there is still work to be done, it seems clear that using a concentrated version of SMD-70 may solve some of the environmental footprint issues that occur downstream after the concrete is poured. This is an evolving project that may open up new uses for SMD-70, here again showing that bio-based solutions can help not only with the release part of the project but the downstream curing part as well.

Examination	Curing Method (When Compared to Whats Commercially Available))			
	SMD-70 100%	SMD-70 75%	SMD-70 50%	SMD-70 33%
Comprehensive Strength Test	Effective	Effective	Effective	Effective
Rebound Hammer Test	No consistent effect was observed during this review			
Air Permeability Test	Effective	Superior Effect	Superior Effect	Not Very Effective
Neutralization Acceleration Test	Superior Effect	Superior Effect	Effective	Not Very Effective

Figure 17.6: Diagram of test results carried out in collaboration with Academia. Test data may be received upon request to SMD Products Company, Inc. (Courtesy of Saitama Ryokoku & University Professor).

17.3.1 Tokyo Olympics & major projects

As noted in the previous section, SMD-70 was found to be the best solution for the building of the Tokyo Olympic Stadium and the Athlete Village. The global concern regarding concrete strength and concrete quality, combined with the demand for an environmentally friendly solution, created the unique environment whereby SMD-70 became the only viable candidate. Thanks to the partnership of one of the largest construction companies in Japan with the exclusive partner in Japan for SMD-70, the company can proudly say it took part in the behind the scenes work of building a successful 2020 Tokyo Olympic games.

In parallel to the Tokyo Olympic infrastructure building, the positive publicity helped SMD-70 become the obvious choice for Japan's largest dam, built in the North of the Tokyo Kanto-region. SMD-70, being a bio-based product, is fully biodegradable, and allows it to be used in complex waterways, where the ecological system is of such great concern when building this 87 meter-tall dam. Images can be seen below Figure 17.7 and on the publicly available website set up for the public to review this important project.

Figure 17.7: Japan's largest dam project in an environmentally important area, north of Tokyo. Photo's courtesy of Saitama Ryokoku, artist rendition courtesy of the publicly available project website.

17.4 Conclusion

As we started in Tanzania, went to Ohio, and finished with SMD-70 from the heartland of the United States of America, it is quite apparent: Biobased Products are not the compromise, but the solution. The aspect of forming better concrete-finished product with less need for touch-up will reduce the overall project labor costs. The ability to ship as a concentrate reduces the logistics costs by eighty percent. The opportunity to potentially offer stronger concrete, thanks to the use of SMD-70 as a curing agent, is under investigation, but preliminary results have led to the Japanese National High-way Board (NEITS) granting the SMD-70 derivative with recognition for use on core infrastructure projects. All these benefits alone justify the immediate use of SMD-70, the bio-based solution for the concrete industry. On top of all these obvious benefits, when one considers the fact that the solution is sustainable, biodegradable, and en-vironmentally friendly, it is clear that sometimes the best and easiest solution to tackle a problem is right in front of you, and not as complicated as one may have thought. Over the coming years, we can all look forward to observing the benefits that SMD-70 and other bio-based materials will have on the global construction indus-try. The authors would like to encourage governments and institutions to create a free and open space for products like SMD-70 to be used, not because they are from a small business, but because they are the best solutions for the business.

In conclusion, the authors would like to share some highlights and closing points from the presentation given at the 2021 AFCC Biobased Economiy Conference in figure 17.8. We stand committed to help provide the best solution for all stakeholders with no compromises.

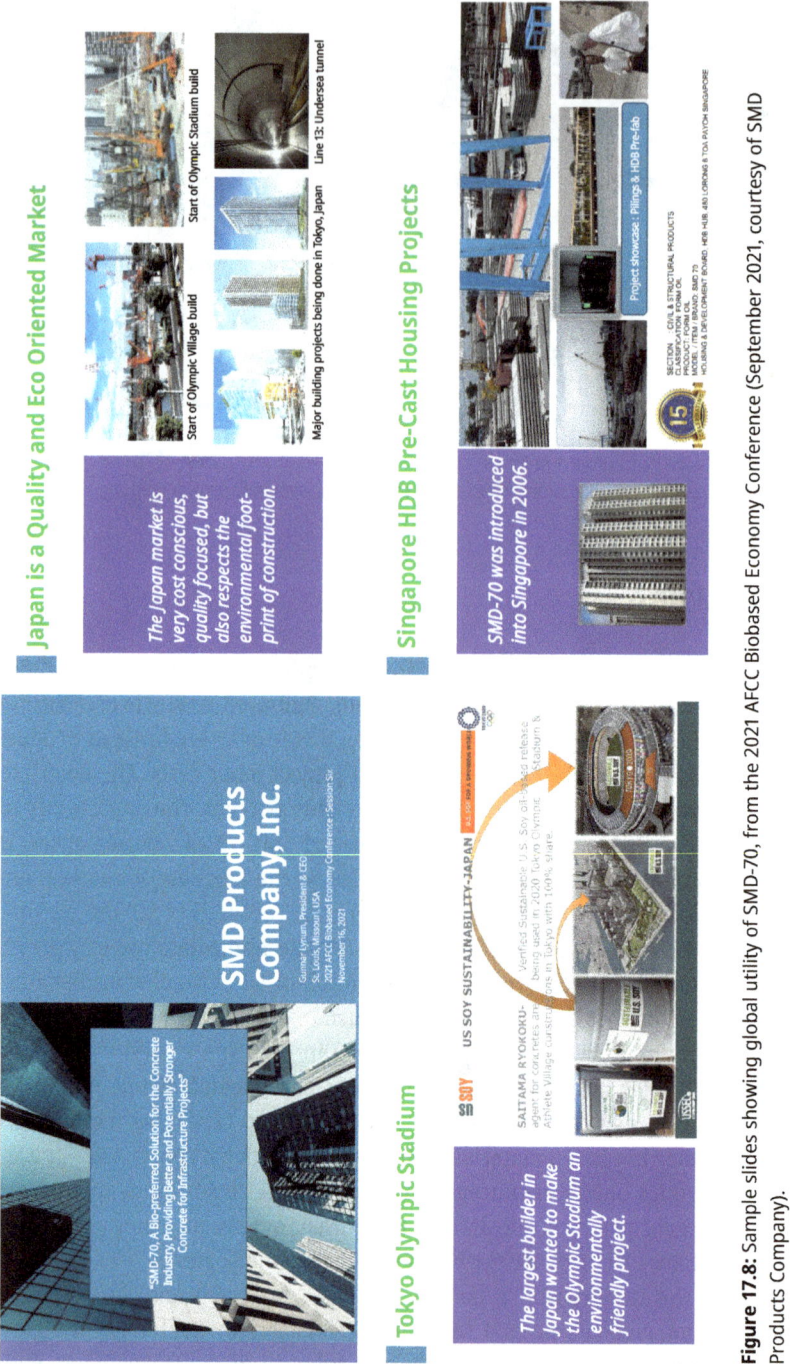

Figure 17.8: Sample slides showing global utility of SMD-70, from the 2021 AFCC Biobased Economy Conference (September 2021, courtesy of SMD Products Company).

References

[1]   de la Torre I. The OGAP Core Team. 2022, May 27. https://www.olduvai-gorge.org/.

[2]   Worth K. COP26 opens in Glasgow with calls for ambitious solutions to tackle climate emergency. 2022, May 27. https://news.un.org/en/story/2021/11/1104522.

[3]   Day M. Everything about concrete. 2022, May 27). https://www.everything-about-concrete.com/.

[4]   Weber B. About new 7 wonders. 2022, May 27. https://world.new7wonders.com/.

[5]   Giatec Scientific. 2017, July 28. The history of Concrete. https://www.giatecscientific.com/education/the-history-of-concrete/.

[6]   Hospitality Net. Courtyard by Marriott Cincinnati Downtown. 2022, May 30. https://www.hospitality net.org/announcement/41006628/courtyard-by-marriott-cincinnati-downtown.html.

[7]   Geo Engineer. The history of hoover dam. 2020, June 30. https://www.geoengineer.org/news/history-of-the-hoover-dam.

[8]   Ellis LD, et al. Toward electrochemical synthesis of cement – An electrolyzer-based process for decarbonating $CaCO_3$ while producing useful gas streams. 2019.

[9]   Alternative Fuels & Chemicals Coalition. 2022, May 30. https://www.altfuelchem.org/2021-afcc-workshop-sessions.

[10]   University of Cambridge. 2022, May 30. https://www.cam.ac.uk/.

[11]   Markets & Markets. Precast Market Research Report. March 2021.

[12]   Environmental Protection Agency Fact Sheet. 2022, May 30. https://www.epa.gov/stationary-sources-air-pollution/fact-sheets-consumer-and-commercial-product-section-183e-voc-rules.

[13]   Worth K. COP26 opens in Glasgow with calls for ambitious solutions to tackle climate emergency. 2022, May 27. https://news.un.org/en/story/2021/11/1104522.

[14]   United States Department of Agriculture, Bio-Preferred website. 2022, June 25. https://www.biopre ferred.gov/BioPreferred/.

[15]   Tokyo 2020 Website. 2022, March 31. https://www.tokyo2020.jp.

Jonathan Cristiani*, Grace Dearnley, Justin Distler,
Andrew Doerflinger, Vincent Mazzoni

Chapter 18
Green hydrogen production, use, and project delivery

18.1 Introduction

As interest in bio products rises with the progress of the clean energy transition, hydrogen has become an increasingly attractive fuel, feedstock, and energy carrier due to its compatibility with both traditional and decarbonized processes and systems. Hydrogen is a fundamental building block of the natural world that provides a diverse array of sources for production of molecular hydrogen, such as biomass, waste, fossil resources, and water-splitting technologies. Hailed for its flexibility, hydrogen can be used in a range of markets such as transportation fuels, energy storage, and chemical precursors.

Hydrogen is valued for its high energy content on a mass basis, which makes it favorable as a fuel for both aerospace and terrestrial transportation. "Green" hydrogen produced via electrolysis and integrated with renewable energy resources has gained significant attention in the sustainability movement for its potential as a decarbonization tool in historically carbon-intensive markets. It shows potential for use as feedstock to reduce carbon intensity in the "hard-to-abate" cement, steel, chemical and petrochemical industries, as a zero-carbon fuel in fuel cell electric vehicles, and to support the production of synthetic fuels and chemical products.

The broad field of hydrogen production and use has wide implications for the bio products industry, particularly in the marriage of renewable net-zero carbon resources such as biomass with renewable electricity. Despite the challenges associated with transportation and storage, low-carbon hydrogen can be viewed as a powerful energy carrier and feedstock in the successful implementation of integrated biorefineries. It is also highly complementary in less apparent ways, such as the decarbonization of the multitude of transportation systems associated with biomass feedstock and bio product logistics. This chapter provides an overview of the many hydrogen production and

Acknowledgements: Contributors and Reviewers: Jacob Clanton, Jack Donoghue, Michael Clifford, Michael Goff, and Frank Jakob.

*Corresponding author: Jonathan Cristiani, Black & Veatch Corporation, 11401 Lamar Avenue, Overland Park, Kansas 66211, United States, e-mail: CristianiJM@BV.com
Grace Dearnley, Justin Distler, Andrew Doerflinger, Vincent Mazzoni, Black & Veatch Corporation, 11401 Lamar Avenue, Overland Park, Kansas 66211, United States

https://doi.org/10.1515/9783110791228-018

utilization alternatives, as well as a discussion of hydrogen infrastructure projects and their synergies with the bio products sector.

18.2 Hydrogen production

Numerous processes have been developed to produce hydrogen, and all are used today with varying levels of scale and technological maturity. Each of these hydrogen production processes relies upon different technologies and reaction pathways. Determining a best-fit process for any given hydrogen production operation requires evaluation of a range of factors from affordability and spacing constraints to technological shortfalls and advantages. Each production type holds a unique role in the overall hydrogen production market.

18.2.1 Electrolysis

Electrolysis is the process of splitting water into hydrogen and oxygen using electricity in an electrochemical cell. Electrolyzers come in a variety of capacities and chemistries, but the fundamental concept remains the same. Electrolyzers, like fuel cells and batteries, have electrodes (i.e., anodes and cathodes) separated by an electrolyte. The combination of electrodes and electrolyte – the stack – varies by the type of chemical reactions taking place. Electrolysis is considered "green" when the electricity consumed is provided by renewable energy resource(s). Here, four types of electrolyzers are examined: proton exchange membrane (PEM), alkaline water electrolysis (AWE), anion exchange membrane (AEM), and solid oxide electrolysis (SOE).

PEM electrolyzers exchange a proton (H^+ ion) through the polymer electrolyte between the electrodes. In a PEM electrolyzer, water is split into oxygen and hydrogen, with the hydrogen ions traveling from the anode to the cathode and exiting out of the cathode side of the stack. Oxygen exits out of the anode side of the stack. Electrocatalysts help lower the activation energy required for the splitting of water. Although high electrolyzer capital cost is a barrier to adoption, recent research and development initiatives have optimized the catalytic activity of the cell while minimizing the amount of expensive electrocatalysts, thereby lowering the cost of PEM electrolyzers [1].

AWE electrolyzers fundamentally function similarly to PEM electrolyzers; however, the hydroxide anion (OH^-) ion transported in the electrolyte and travels from the cathode to the anode. The hydrogen then exits out of the cathode side while the oxygen exits out of the anode side of the stack. Since AWEs have a lower current density, they typically require a larger footprint compared to PEMs, though the technology is considered more mature for large-scale hydrogen production [2].

The AEM electrolyzer is an emerging technology that offers construction and operational advantages over both PEM and AWE types. Similar to PEM, AEM electrolyzers use an electrolyte membrane, but instead of passing a proton across the electrolyte, the OH⁻ anion is passed, similar to AWE. AEMs also use less expensive (nonplatinum group) electrocatalysts than PEM and can potentially operate at higher operating pressures to reduce system size [3].

SOE stacks have high conversion efficiencies compared to PEM/AWE, primarily because they operate at higher temperature (i.e., 600–850 °C) where thermodynamics and reaction kinetics are favored. Additionally, SOEs can be used for the direct electrochemical conversion of steam, carbon dioxide (CO_2), or both, into hydrogen, carbon monoxide (CO), and/or synthesis gas (i.e., syngas, which is a mixture of hydrogen and CO). SOEs are further capable of converting captured CO_2 and water into synthetic natural gas (i.e., methane or CH_4), gasoline, methanol, or ammonia. This type of electrolyzer consists of two porous electrodes surrounding a dense ceramic electrolyte capable of conducting oxide ions (O^{2-}). In addition to the efficiency benefits, SOE technology is also characterized by low-cost materials of construction compared to PEM electrolyzers, particularly in their use of non-noble metal electrocatalysts. Additionally, many SOEs in development are also thought to show greater promise than existing, commercialized electrolysis technologies for reversible operation.

Table 18.1 shows the relative performance characteristics of each of the electrolysis technologies considered.

18.2.2 Reforming

Reforming technologies use a hydrocarbon feedstock to produce a syngas, which is then converted and purified to achieve a hydrogen product. Traditionally, natural gas or coal is used as feedstock. Reforming reactions take place in reformers, usually in the presence of catalysts, at high temperatures and net endothermic conditions. However, reforming systems can have exothermic steps as part of their reaction mechanism, depending on the reaction taking place. The resulting syngas contains CO, which is then converted to hydrogen and carbon dioxide through a water gas shift (WGS) reaction. Finally, the hydrogen is purified, typically through a pressure swing adsorption (PSA) process to remove any unreacted components and impurities (primarily water, CO, CO_2, and nitrogen depending on the process) from the hydrogen product. There are three primary reforming technologies: steam methane reforming (SMR), partial oxidation (POX), and autothermal reforming (ATR).

SMR is the most mature hydrogen production process, having been widely used for industrial-scale applications throughout the twentieth and the beginning of the twenty-first centuries. In an SMR process, natural gas is reacted with steam at temperatures of approximately 810–870 °C in a catalyst-filled tube in the reformer to produce syngas. The SMR process is highly endothermic, requiring large amounts of heat

Table 18.1: Electrolysis technology performance comparison.

Attribute	Proton exchange membrane (PEM)	Alkaline water electrolysis (AWE)	Solid oxide electrolysis (SOE)	Anion exchange membrane (AEM)
Technology maturity	Commercially available	Commercially available	Developmental	Developmental
Technology scale	Small to large-scale	Small to large-scale	Small-scale	Laboratory-scale
Key features	Solid polymer electrolyte that is highly conductive for positively charged ions and resistive to negatively charged ions [6]	Positively charged anode and negatively charged cathode immersed in liquid electrolyte [6]	Electrolyzes water as steam and includes two porous electrodes on either side of a dense ceramic electrolyte capable of conducting O^{2-} ions [4]	Solid polymer electrolyte that is highly conductive for negatively charged ions and resistive to positively charged ions [3]
Key characteristics	– Smaller footprint – Modest current density – Faster dynamic response, high turndown capabilities – Higher pressure operation – Higher purity product gas (99.99%) [7]	– Large footprint – Low current density – Lower capital expenses (CAPEX) and operating expenses (OPEX) compared to PEM – Older, most established technology [6]	– High conversion efficiency and current density – Lower CAPEX compared to PEM/AWE – Abundant raw materials for cell components [4]	– Smaller footprint – Lower CAPEX compared to PEM/AWE – Higher pressure operation – Higher purity product gas (99.99%) [3]
Disadvantages	– Higher CAPEX and OPEX compared to AWE [6] – Expensive platinum electrocatalyst [4]	– Less suitable for space-constrained facilities [6] – Lower gas purity, relative to PEM (>99.5%) [7] – Requires bulk chemical storage of hazardous, corrosive electrolyte	– New emerging technology – Small stack capacities (< 10 kW) [2] – More complex thermal management [2]	– New emerging technology – Lower current density compared to PEM – Decreased durability – Increased degradation rates [3]

Table 18.1 (continued)

Attribute	Proton exchange membrane (PEM)	Alkaline water electrolysis (AWE)	Solid oxide electrolysis (SOE)	Anion exchange membrane (AEM)
Normalized degradation [1]	1.5% per year	0.9% per year	4.2% per year	Unknown
Life expectancy and stack replacement schedule [9]	20 + Years [5] 20,000–60,000 h [7]	20 + Years [8] 60,000–90,000 h [7]	Unknown Life < 10,000 h [7]	Unknown Life and stack replacement
Stack energy efficiency [2] [7]	54–71% 47–61 kWh/kg	51–71% 47–66 kWh/kg	< 94% > 36 kWh/kg	58–62% 53–58 kWh/kg

Table Notes:
1) Estimated based on 95% annual capacity factor and assuming moderate operating conditions (i.e., stable temperature/loading and minimal cycling). No specific formula for degradation exists and is based on a complex set of factors that have been modeled and discussed in the literature.
2) Efficiency calculated based on lower heating value of hydrogen.

input. This heat is supplied to the SMR by fired burners surrounding the tubes, with the heat from the exiting flue gas recovered in the convection section of the SMR. The convection section preheats the natural gas and steam for the process, along with usually producing steam to recover as much heat as possible. Hot syngas from the SMR is cooled in a waste heat boiler and then passes through a series of WGS reactors to convert the CO in the syngas to hydrogen. Finally, it is purified (typically in a PSA system) to achieve the final desired hydrogen purity. Primarily due to the high temperatures required for SMR, the hydrogen production process produces large amounts of CO_2 and yields hydrogen with a high carbon intensity. Hydrogen produced through SMR is often referred to as "gray" hydrogen. However, if the SMR process is coupled with carbon capture, utilization, and sequestration (CCUS) technologies to reduce the emissions of CO_2, the product is referred to as "blue" hydrogen.

POX reforming also uses a natural gas feedstock but employs a different reaction pathway than SMR to produce syngas. Rather than using steam as a reactant, a limited amount of oxygen is supplied to partially oxidize natural gas to supply heat to the endothermic reforming reactions. Oxygen can theoretically be supplied from air but is usually supplied as a high purity oxygen stream produced from an air separation unit (ASU). The amount of oxygen supplied is limited in stoichiometric quantity to about 30–35% of that required for complete oxidation of the hydrocarbon feedstock to CO_2 and water. The CO is then converted to hydrogen in a WGS reaction and purified in a PSA. Since the POX process is exothermic, fired heaters are not required as with SMR, thereby requiring smaller reactor sizes. However, the POX reaction pathway produces less hydrogen than SMR for the same quantity of fuel input due to the lower amount of hydrogen produced from the steam. Additionally, using air as the oxygen source creates a product with large amounts of nitrogen, increasing the purification costs (via nitrogen rejection in a PSA) of the final hydrogen product [10].

ATR can be thought of as a hybrid process between SMR and partial oxidation, employing aspects of both reaction pathways and using both steam and oxygen as inputs to the catalyzed process with natural gas. Natural gas and steam are preheated in a fired heater before combining and entering the reformer burner, where it is mixed with an oxygen input (which can be air or a pure oxygen stream). The oxygen partially oxidizes the hydrocarbon components in the natural gas in an exothermic process, reducing the energy input to achieve the required high temperatures of the reforming process. The hot, partially oxidized stream flows downward over a catalyst bed that helps drive the reforming reactions. Exiting the reformer, the syngas is cooled in a waste heat boiler and is then upgraded and purified through a WGS reaction and PSA process. Using air as the oxygen source will result in a syngas product with a high nitrogen concentration, which must be removed during purification to achieve the desired hydrogen purity.

Table 18.2 shows the relative performance characteristics of each of the reforming processes considered.

Table 18.2: Reforming processes: performance comparison.

Attribute	Steam methane reforming (SMR)	Partial oxidation (POX)	Autothermal reforming (ATR)
Feedstocks	Natural gas, steam	Natural gas, oxygen (pure or air)	Natural gas, steam, oxygen (pure or air)
Required heat input	High	Little to none	Low to intermediate
Hydrogen production per unit of fuel	Highest	Lowest	Intermediate
Reformer Size	Largest	Smallest	Intermediate
Primary impurities removed	CO_2, CO, water	CO_2, CO, water, nitrogen (if air used)	CO_2, CO, water, nitrogen (if air used)

18.2.3 Gasification

Gasification refers to the conversion of a solid biomass to syngas through a series of reactions, which can broadly be separated into four stages. The first of these stages is a drying phase (typically occurring at temperatures in the range of 100–320 °C), in which the moisture in the biomass fuel evaporates. The second stage is pyrolysis, in which the volatile matter trapped in the solid matrix of the fuel is volatilized and released as a vapor. Pyrolysis occurs at temperatures ranging from just above drying to

590 °C. The third stage of the gasification process is the gasification of the remaining solids, which are mostly fixed carbon by reaction with water, hydrogen, or CO_2 to produce CO, hydrogen, and CH_4. These reactions occur at temperatures above 650 °C and are endothermic. The fourth stage is a combustion stage, where oxygen reacts with carbon and CO to produce CO_2. This reaction is exothermic and produces the heat required for the endothermic gasification reactions. The ordering and speed of these stages varies, based on the gasifier reactor design and operating conditions.

Following the gasification process, the cleaning and treating processes for the resulting syngas depend upon the operating characteristics of the gasification system as well as the composition of the biomass. To achieve the final hydrogen product, the gasification-produced syngas, often, must undergo an acid gas removal process, and then proceed to a syngas upgrading and purification process similar to that described in the reforming section.

In an entrained flow gasifier, solid feedstocks (e.g., coal or petroleum coke) are ground to a fine consistency and injected into the gasifier with water or steam and oxygen or air. There are generally no limitations on the feedstocks to an entrained flow gasifier, but high ash and high moisture feedstocks increase the oxygen consumption more than in other gasification types because of the high operating temperatures of entrained flow gasifiers. Additionally, biomass cannot be used in an entrained flow gasifier due to its inferior friability, unless when blended or pretreated.

In a fluidized bed gasifier, the feedstock is introduced near the bottom of the gasifier and is fluidized by gas feedstocks (typically steam and oxygen or air) to the gasifier. This method may result in poor carbon conversion, as smaller particles tend to become entrained in the gas flow and will exit the gasifier before they have the chance to be gasified. Relative to entrained flow gasifiers, fluidized bed gasifiers operate at lower pressures and temperatures, have longer fuel residence times, and have lower capacities. The temperature of the gasifier is typically kept below the ash softening point to prevent agglomeration and bed defluidization. Specific types of fluidized bed gasifiers include circulating fluidized bed and bubbling fluidized bed.

In a moving (or fixed) bed gasifier, the feedstock is introduced at the top of the gasifier and slowly descends by gravity toward the bottom of the gasifier. Liquids can be injected into the gasification zone of moving bed gasifiers. The most common arrangement for gas flow through moving bed gasifiers is countercurrent to the bed flow, where the feedstock at the top of the gasifier is dried and then pyrolyzed by heat exchange with the exiting gases. This typically generates a syngas with a high CH_4 content. Compared to entrained flow gasifiers, moving bed gasifiers operate at lower temperatures, have much longer fuel residence times, and have lower capacities. There are many small-scale fixed bed gasifiers that focus on biomass or coal.

Indirectly heated gasifiers do not use partial oxidation to provide the heat required to drive the gasification reactions. Instead, heat is added to the gasifier indirectly by a heated medium, by heat transfer through the vessel walls, or by electric current. Indirectly heated gasifiers can also be entrained flow, fluidized bed, or moving bed; however, they have special importance in smaller biomass applications where ASUs are not economically feasible. Indirectly heated gasifiers can produce a medium heating value syngas that is mostly free of nitrogen (a portion of the fuel-bound nitrogen will end up in the syngas).

Table 18.3 shows the relative performance characteristics of the first three gasifier technologies presented.

Table 18.3: Gasifier technology performance comparison.

Attribute	Entrained flow	Fluidized bed	Moving (or fixed) bed
Solid feedstock particle size	< 0.127 mm	6.35–12.7 mm	6.35–50.8 mm
Residence time	Seconds to minutes	Minutes	Hours
Relative operating temperature	High	Low	Low
Relative operating pressure	High	Low	Low

18.2.4 Methane pyrolysis

Methane pyrolysis is a process that has been discussed since the 1970s, when the temperatures needed were thought to be too high for commercial production without the use of catalysts. CH_4, usually in the form of natural gas, is heated to high temperatures (>1,000 °C) in the absence of oxygen, splitting the carbon and hydrogen atoms through radical chain dehydrogenation and methylation, or "cracking" [11]. The hydrogen atoms naturally coalesce to form hydrogen gas, while the carbon forms solid carbon. Through various solid-gas separation and filtration processes, the solid carbon and any other stray gases can be removed from the hydrogen stream, producing a high-purity hydrogen gas known as "turquoise" hydrogen. Recovered carbon can be further processed and sold for a variety of end use applications (e.g., steel making, battery electrodes, composites, etc.) to decrease total cost of producing hydrogen by this method. However, if methane pyrolysis became a significant source for hydrogen production, the supply of co-product solid carbon would quickly outpace the global carbon demand. To tackle this, additional markets would be needed [12].

The temperature required to crack the CH_4 can be reduced if a catalyst is used. Many different catalysts can be used, with transition metals such as nickel, iron, and cobalt being the most common. While these catalysts can reduce the required temperatures to as low as 500 °C, product carbon tends to coat the reactor, thereby deactivating the active sites on the catalysts and reducing the rate of reaction to zero over time. One solution to this is to use carbon-based catalysts, which are usually less active than the metal catalysts and require higher temperatures but tend to have longer catalyst lifetimes. In both cases, the catalyst must be regenerated. Currently, the only large-scale regeneration processes produce carbon emissions through the combustion or gasification of the catalyst, reducing the benefits gained from the clean production of hydrogen [13].

Even with the high temperatures required for noncatalytic thermal decomposition, methane pyrolysis is a competitive production method from an energy requirement standpoint. In SMR, the energy to convert liquid water into steam and react the steam and CH_4 is $63 \frac{kJ}{mol\,H_2}$. According to standard reaction enthalpies, using methane pyrolysis requires $37.8 \frac{kJ}{mol\,H_2}$. This energy could be produced by combusting natural gas or other low-carbon fuels, generating as little as $0.05 \frac{mol\,CO_2}{mol\,H_2}$, much less than the $0.43 \frac{mol\,CO_2}{mol\,H_2}$ generated from SMR. If part of the hydrogen produced is combusted instead, carbon emissions could be prevented altogether [14].

18.3 Hydrogen storage and transportation

Due to the fact that hydrogen is the lightest element, it can be challenging to store large quantities in terms of both mass and volume. Methane is about eight times denser than hydrogen at standard conditions on a gravimetric basis, so the pressures and temperatures required to store hydrogen in an economical manner are more extreme than that of natural gas. Two potential hydrogen storage methods are compressed gaseous hydrogen storage and cryogenic liquid hydrogen storage. Selection of storage type may depend on a variety of factors, including the amount of hydrogen stored and the relative cost of each storage option.

For over-the-road hydrogen transportation, hydrogen storage density relative to the weight of the transporting vehicle is especially important. Liquid hydrogen deliveries provide the densest hydrogen state, however the production of liquid hydrogen as well as the handling can make it cumbersome for short ranges with low capacities. For most hydrogen transportation applications, the distance traveled and quantity delivered between the production and the consumption of hydrogen will drive the economics of liquid hydrogen, high-pressure compressed hydrogen, or low-pressure compressed hydrogen delivery. Over long distances, transportation through pipelines is possible, and is discussed below.

18.3.1 Compression

Compressed hydrogen storage is the most common method of storage for today's industrial hydrogen consumers. Depending on the amount of hydrogen being stored, pressures can range from 140–690 bar, with small cylinders used in the transportation sector being more suitable for the high end of the range than the large bulk tanks for industrial users. Depending on the pressure and storage volume, many smaller vessels may be more economical than one large bulk tank. Because of its very low molecular weight and small molecule size, hydrogen presents an issue with leakage, and some compressed storage applications may require special materials to line the inside of the vessel for leakage prevention.

To compress and store hydrogen for later use, a combination of compressors, heat exchangers, tanks, and balance of plant piping are required. Depending on the final storage pressure, there may be several intercooled compressor stages. The balance between capital and operations and maintenance (O&M) costs would need to be optimized to determine the best means of compression for any given situation. Lower pressure storage would require less power consumption to achieve the storage pressures, but at a cost of a greater number of tanks and a larger footprint to hold the same mass of hydrogen. Higher pressure storage would require greater power consumption to achieve the storage pressures, but offers the benefit of fewer tanks and smaller footprint.

18.3.2 Liquefaction

Hydrogen liquefaction is more energy intensive than compressed storage but may be an attractive option, depending on the amount of hydrogen storage needed. The larger the quantity of hydrogen being stored, the more economical liquefaction becomes, relative to compressed storage on a mass basis. Consider the density of liquefied hydrogen compared to compressed hydrogen: density of liquefied hydrogen is approximately 70.8 $\frac{kg}{m^3}$, while density of compressed hydrogen ranges from 2.56–50$\frac{kg}{m^3}$, depending on the pressure. The storage volumes for liquefied hydrogen would be much smaller than the storage volumes for compressed hydrogen of the same mass. However, liquefied hydrogen requires far more complex auxiliary equipment.

To liquefy hydrogen, extremely cold temperatures (i.e., – 250 °C) need to be maintained. This requires a vapor-compression cycle with liquid nitrogen as a refrigerant. Boil-off compressors are also required to reliquefy the hydrogen that boils off during storage. However, depending on the amounts of storage required, liquefaction can still be more economical than compressed storage. This is particularly true at large scales.

When considering liquefaction equipment, one must evaluate thermal cycling and ramp time. Cycling from ambient to extremely low temperature thermally stresses the equipment. The equipment associated with liquefaction is designed for a certain number of thermal cycles over its lifetime, and frequent cycling will significantly reduce the useable life of the equipment. Ideally, the liquefaction equipment is run continuously to minimize thermal cycles and maximize the life of the equipment. Additionally, the startup/shutdown times associated with the liquefaction train are in the 4–8 h range (from cold standby) or up to 24 h (from ambient temperature). Liquefaction is best suited for continuous operation or seasonal operation at a minimum. Daily cycling of the liquefaction equipment is not considered feasible; however, designing a system for very low turndown may offer some additional operating flexibility [15].

Current methods for liquefying hydrogen use a vapor-compression cycle, which requires various equipment including trains of intercooled compressors, brazed aluminum heat exchangers, parallel closed-loop nitrogen refrigerant cycles, and boil-off compressors. The amount of equipment required makes the process of liquefying hydrogen quite energy intensive.

18.3.3 Pipeline transportation

Pipelines are the most cost-efficient way to transport large quantities of hydrogen over long distances. Hydrogen gas pipelines are considered mature technologies and can typically cost approximately up to 10 percent more than a traditional natural gas transmission pipeline. For dry hydrogen service, the use of carbon steel is perfectly acceptable for the typical temperatures/pressures associated with most low-carbon hydrogen projects. In instances where corrosive contaminants or moisture condensate are present, a stainless-steel pipeline material would be selected instead, which can increase costs [16].

One attractive option is to blend hydrogen into the existing United States (US) natural gas pipeline network, which includes over 644,000 km of infrastructure. It is estimated that at typical pressures and diameters associated with natural gas pipelines, approximately 19 metric tons of hydrogen could be stored per linear mile. Hydrogen is generally thought to be limited to 5–10% blending throughout most of the US, primarily due to safety and pipeline integrity concerns. While greater percentages may be possible if natural gas pipelines supporting infrastructure and the end users of the blended gas are evaluated converted for use with hydrogen, these costs and the required modifications are the subject of much research and development [17].

18.3.4 Geophysical storage

Geophysical storage takes advantage of existing geological formations such as salt caverns, rock caverns, and depleted gas fields, which present an opportunity to store large volumes of hydrogen in existing features. Conceptually, hydrogen is compressed and stored in an existing geological formation and then withdrawn for later use. In practice, the application of geophysical hydrogen storage is incredibly site-specific.

Salt caverns present the most suitable geological storage feature, followed by rock caverns and then, depleted gas fields as the least suitable of the three. Since hydrogen is the lightest gas, it has the fastest molecular velocity compared to any other gas at the same conditions. Depending on the geological feature, upgrades such as a liner may need to be added to minimize leakage. Another consideration associated with geological storage is contamination. Depending on the geological formation, other compounds such as CH_4 or water may be present. Additional clean up equipment upon discharge of hydrogen may be required depending on the geographic location and the hydrogen user quality requirements.

Geophysical storage presents an attractive method to store large quantities of hydrogen for seasonal variations but is highly dependent on the location. Upgrades to the geological formation or additional clean up equipment may drive the effective cost above that of traditional compressed storage or liquid storage [15].

18.3.5 Metal hydrides and chemical hydrides

Hydrogen can also be stored in metal hydrides or chemical hydrides that function via chemical bonding. Metal hydrides store hydrogen in a solid material, while many chemical hydrides are stored in a liquid phase. The chemical bond is much stronger than adsorption bonds and thus, often, requires more energy in synthesis and dehydrogenation. However, the strong chemical bonding allows hydrogen to be stored at a higher density at ambient conditions relative to physical hydrogen storage as compressed gas or cryogenic liquid. Releasing the hydrogen from the hydride carrier also requires energy or releases energy, depending on the chemical bond. This typically warrants balance of plant thermal management equipment and can further lower the round-trip efficiency of the storage. Despite benefits of metal hydrides having a high volumetric energy density, their low gravimetric energy density typically limits their applicability in certain end uses.

Current hydride storage techniques are typically more expensive compared to gaseous and liquid storage methods. However, due to the high hydrogen density at ambient conditions, there is still significant interest in both metal and chemical hydride storage. A few specific chemicals that are produced from hydrogen and could be considered chemical hydrides are discussed in greater detail in Section 3.

18.4 Power to products and chemicals

Power to products and chemicals (also known as "Power-to-X" or "P2X") uses electricity from renewable energy resources to produce products such as hydrogen, ammonia, methanol, and renewable natural gas (RNG). Hydrogen is used as feedstock in many industrial chemical processes; thus, when produced using electrolysis paired with renewable electricity, the use of "green" hydrogen in these processes decarbonizes the final product. The most common chemical, ammonia, is discussed first. Then, an important companion process to deploying hydrogen derived products – CCUS – is discussed, followed by several products thus enabled.

18.4.1 Green ammonia

Ammonia is considered a leading hydrogen carrier chemical that can overcome the challenges associated with storage and transportation of hydrogen, while taking advantage of existing infrastructure. It stores hydrogen in an energy-dense form as NH_3. Although ammonia has historically been used exclusively as a precursor chemical feedstock for a variety of applications (e.g., fertilizers, pharmaceuticals, cleaning products), its role as a hydrogen carrier is also envisioned to support a variety of new applications, such as transportation fuel, energy storage, and power generation. Thus, after ammonia is synthesized, it can be transported from regions with abundant renewable energy resources to those without such resources. Once there, it can be converted back to hydrogen via cracking, for the end user to use directly.

While ammonia was first discovered in the late nineteenth century, the modern ammonia process that is attributed to commercial-scale production was developed by German chemist, Fritz Haber, in the early twentieth century [18]. This process produced ammonia from the reaction of nitrogen gas and hydrogen gas over a catalyst at high temperature and high pressure, per the following chemical equation:

Ammonia synthesis reaction:

$$3H_2 + N_2 \rightarrow 2NH_3$$

The patents to Haber's process were purchased by BASF, and the process was further commercially developed, including catalyst improvements contributed by company chemist, Carl Bosch, from whom, the second name in the Haber-Bosch ammonia process originated. The first commercial production of ammonia using the Haber-Bosch process began at a German BASF plant in 1913. The Haber-Bosch ammonia process has been modernized to increase process efficiency and ammonia production. Ammonia production technology is well proven and commercial designs range from about 45 metric tons per day (TPD) to 3,000 TPD production rates. Ammonia is typically produced using hydrogen generated from natural gas, which is readily available in large quantities and allows for large-scale ammonia production.

18.4.1.1 Synthesis and purification

"Green" ammonia synthesis requires hydrogen from electrolysis and nitrogen. Nitrogen can either be purchased from a pipeline, if nearby, or obtained from the air by a PSA unit or a cryogenic air separation unit. To produce ammonia, hydrogen and nitrogen are compressed in a molar ratio of 3:1 to a pressure of about 150 bar. The mixture then goes through a feed/effluent heat exchanger to heat the gas before it goes to the ammonia converter, which typically has three catalyst beds in the pressure vessel, with the gas flowing in an axial/radial direction. Between the beds, heat exchangers heat the incoming feed and cool the gas coming out of each catalyst bed. The gas exiting the converter is typically at a temperature of 370–475 °C and is cooled when it is used to produce steam. The gas then passes to a waste heat boiler, boiler feed water (BFW) heater, and then the feed/effluent exchanger. It is then cooled against cooling water, and then, a refrigeration condenser to condense more ammonia.

The refrigerant used in the process is an open-loop ammonia system driven by a refrigeration compressor. The effluent from the refrigeration condenser then goes to a vessel to separate the liquid ammonia from the unreacted gases. The unreacted gases are recycled back to the converter for near-complete conversion of the hydrogen/nitrogen fed to the system. The only unreacted gas is the relatively small amount that is dissolved in the liquid ammonia and is flashed off at lower pressure. Utilities required for ammonia production include water for steam production and cooling water. The synthesis of ammonia is exothermic, and steam is produced downstream of the converter. This steam is used to drive the compressor required to compress the hydrogen/nitrogen to the required pressure.

18.4.1.2 Storage and transportation

Pure ammonia is a gas at atmospheric pressure/temperature and can be stored as a liquid at moderate pressure or as a liquid at – 33 °C under atmospheric pressure. Gas storage is favorable for small-scale applications where large quantities of product are not required. Ammonia production plants typically store ammonia in the refrigerated liquid phase, as the increased density allows them to store larger volumes for offtake and distribution of their product. The ammonia liquefaction process is also less expensive and challenging when compared to hydrogen liquefaction, because only a temperature of – 33 °C needs to be reached for liquefied ammonia storage, whereas liquefied hydrogen requires a temperature of – 253 °C. Thus, the roundtrip energy efficiency of producing ammonia and converting it back to hydrogen is greater than for hydrogen liquefaction, which when coupled with the superior volumetric energy density of ammonia is a strong enabler for use in energy applications.

18.4.2 Carbon capture

CCUS is an important technology in the hydrogen economy as it allows decarbonization of conventional, fossil-based hydrogen production methods by capturing CO_2 generated from the process for either use or sequestration. Furthermore, captured CO_2 can be used as a feedstock for the production of chemicals such as methanol in combination with low-carbon hydrogen. For purposes of this chapter, post-combustion capture and direct-air capture (DAC) technologies are discussed.

18.4.2.1 Post-combustion capture

Post-combustion technologies remove CO_2 from the flue gas produced out of combustion of a fuel such as in a natural-gas or coal-fired power plant. The two primary methods of post-combustion CO_2 capture include solvent absorption and solid adsorption. Other emerging post-combustion technologies include chemical looping, separation membranes, and cryogenic separation.

The solvent absorption technology, as shown in Figure 18.1, uses a solvent with a high affinity for CO_2 (typically amine-based) to selectively capture CO_2 in a multi-tray or packed absorber column. The rich amine is regenerated in a stripper tower where the CO_2 is released, and the amine is recirculated to the absorber. Amine absorption systems are widely demonstrated with well-understood costs and reliability [19]. However, because of the large capital investment, solvent absorption systems are best suited for large-scale CO_2 capture implementation.

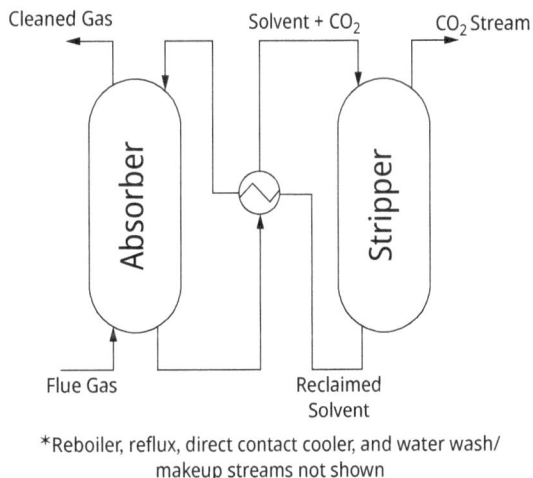

Cleaned Gas   Solvent + CO_2 CO_2 Stream

Absorber   Stripper

Flue Gas   Reclaimed Solvent

*Reboiler, reflux, direct contact cooler, and water wash/ makeup streams not shown

Figure 18.1: Representative solvent absorption CO_2 capture process flow diagram.

Solid adsorption systems capture CO_2 with a selective adsorbent material that bonds to CO_2 in the flue gas. As the flue gas passes over the sorbent surface, the sorbent selectively retains CO_2 until it reaches saturation, as which point it must be regenerated to continue removal. Regeneration can occur via a temperature-swing, pressure-swing, or vacuum-swing process. To maintain a semicontinuous operation, the gas flow is typically switched to a secondary column, while the first column is regenerated. Adsorption systems are a newer technology compared to absorber systems and, generally, are more easily modularized, with the potential for reduced operating expenses compared to absorber systems. For this reason, adsorption systems are better suited for smaller-scale operations. Figure 18.2 depicts a high-level process flow arrangement for a solid adsorption system.

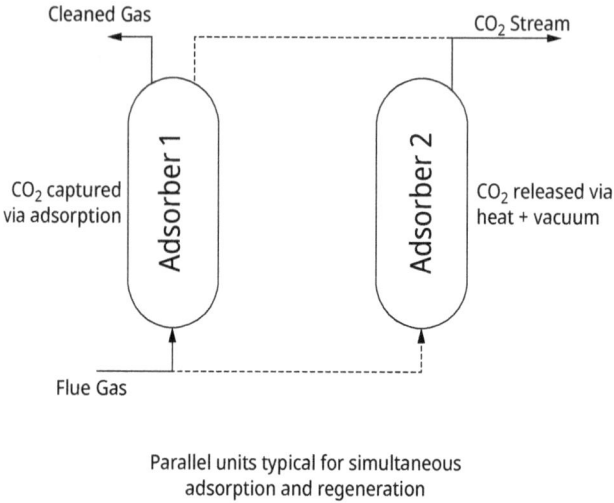

Cleaned Gas CO_2 Stream

Adsorber 1

CO_2 captured CO_2 released via
via adsorption   heat + vacuum

Adsorber 2

Flue Gas

Parallel units typical for simultaneous
adsorption and regeneration

Figure 18.2: Representative physical adsorption CO_2 capture process flow diagram.

18.4.2.2 Direct-air capture

Direct air capture (DAC) technology captures CO_2 directly from the atmosphere using solid and liquid sorbents. The solid DAC technology uses sorbent filters that chemically bind with the CO_2. The system operates by adsorbing the CO_2 and desorbing it at low pressure and temperatures between 80 °C and 100 °C. The liquid DAC technology uses two closed chemical loops, wherein the first loop contacts air with an aqueous solution that captures the CO_2; the second loop releases the CO_2 from the solution and operates at high temperatures between 300 °C and 900 °C [20].

Captured CO_2 is sequestered in underground geological formations such as saline formations, oil and natural gas reservoirs, unmineable coal seams, organic rich shales, and basalt formations. The type of formation selected must be both safe and environmentally sustainable. To minimize the storage volume required, the CO_2 is stored underground as a supercritical fluid above temperatures of 30 °C and pressures above 70 bar. To prevent leakage, the formation is sealed using structural trapping, residual trapping, solubility trapping, or mineral trapping [21].

18.4.3 Green methanol

Alcohols, such as methanol, are organic compounds with at least one hydroxyl functional group that acts as essential building block for the chemical industry. Currently, methanol is produced primarily using fossil resources; however, pathways for low-carbon methanol production using "green" or "blue" hydrogen in combination with captured CO_2 or other "renewable" carbon sources (e.g., biomass) are of significant interest [22]. As with low-carbon ammonia, methanol is viewed as a promising hydrogen carrier that can take advantage of existing infrastructure for storage and transportation.

Methanol is produced using syngas feedstock, which is a composed of CO_2, CO, and hydrogen. Traditional sources of syngas are produced using autothermal reforming, steam methane reforming, and coal gasification [23]. In a conventional gas-phase process, the syngas is fed to a methanol converter where the CO, CO_2, and hydrogen are reacted in a highly exothermic manner over a catalyst in a gas-phase fixed-bed reactor at pressures between 40 to 120 bar and at 200–320 °C [24]. Once crude methanol is generated in the converter, it is distilled to remove water and byproducts of the synthesis process. Methanol is typically stored as a liquid at ambient conditions and transported to end users by ship, rail, and truck. Pipeline transportation of methanol is also possible [22].

18.4.4 Methanation and renewable natural gas (RNG)

Methanation is the hydrogenation of either CO or CO_2 to produce CH_4 and water vapor. When using hydrogen produced via electrolysis in the hydrogenation of CO or CO_2, the CH_4 produced is referred to as RNG. Both reactions are exothermic and are often used to produce steam that is needed in other unit operations in the overall RNG production facility. Methanation thermodynamic equilibrium experiments suggest that higher pressures and moderate temperatures favor the production of CH_4. Syngas methanation projects have been developed primarily since the 1970s and 1980s, as a result of interest in the production of "substitute"/"synthetic" natural gas from coal, which has given rise to several demonstration projects as well as a couple of commercialized concepts [25].

The most noteworthy commercial methanation scheme is the adiabatic fixed-bed concept. Fluidized bed methanation (FBM) concepts have also been pursued in the past but have found limited success due to issues such as catalyst attrition, mechanical stresses during transients, and de-fluidization. Biological methanation technologies are also being developed as a means of using microorganisms to produce CH_4 at low temperatures and in stirred tank or trickle-bed reactors [25].

In a traditional process, RNG is produced from the cleaned and shifted syngas via the hydrogenation of CO and CO_2 in a series of fixed-bed reactors according to the exothermic methanation reactions shown in the following equations:

CO methanation reaction	CO₂ methanation reaction

$$CO + 3H_2 \rightarrow CH_4 + H_2O \; \Delta H = \frac{-206.2 \; kJ}{mol} \qquad CO_2 + 4H_2 \rightarrow CH_4 + 2H_2O \; \Delta H = \frac{-165 \; kJ}{mol}$$

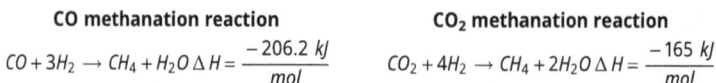

To protect the methanation catalyst from sulfur poisoning, the feed is first passed through a sulfur guard bed to remove traces of sulfur compounds. The syngas is then mixed with recycled gas to control the maximum temperature rise of the reactor and then passed to the methanation reactor(s). The exothermic methanation reaction results in a high outlet temperature, which allows generation of superheated high-pressure steam in downstream heat exchangers. After cooling, the partially methanated syngas passes through additional methanation reactors in series, for complete conversion of the CO into CH_4. The process stream leaving the final methanation reactor is cooled and sent to the final drying unit and compressed to meet pipeline specifications for natural gas [26].

In an RNG production operation that uses FBM, some noteworthy simplifications can be made. According to the Paul Scherrer Institute (PSI), FBM can enable fewer tar removal units, eliminate the need for olefin and carbonyl sulfide hydrotreating, and replace the WGS unit and series of methanation reactors, with a single FBM reactor resulting in up to 33% reduction in capital costs [27]. As with many fluidized bed reactors, FBM allows reactions to proceed at suboptimal conditions: lower operating temperatures, less than stoichiometric hydrogen to CO/ CO_2 ratios, and less carbon deposition on catalyst surfaces [28, 29]. Prior research has also shown that the addition of three to four percent steam and maintaining a CO_2: CO ratio of less than 0.52 in the feed to an FBM process positively impacts the carbon deposition avoidance and increases the conversion of hydrogen, respectively [30]. However, some literature notes a couple of drawbacks for FBM compared with fixed beds for RNG production, such as limitations in scaleup and lower potential for the economic cogeneration of RNG and power [28, 31].

The RNG generated from methanation can be used onsite where it is generated for vehicle fuel, power generation, or thermal applications. More often, RNG is injected into a natural gas pipeline network for distribution to numerous users, assuming it meets the specifications of that particular gas utility or pipeline owner/operator [32].

18.4.5 Other fuels and chemicals

Liquid organic hydrogen carriers (LOHCs) are defined by reversible hydrogenation and dehydrogenation reactions. The benefit of LOHCs is that they are chemically stable and easy to be transported long distances because they exist in the liquid state at ambient conditions [33]. One of the more established examples of LCOHs is methyl cyclohexane (MCH) and toluene, where MCH is the hydrogenated form and toluene is the dehydrogenated form of the LCOH pair. Toluene, like most dehydrogenated LCOHs, is aromatic with a relatively low gravimetric hydrogen storage density. Once hydrogenated, MCH has a high volumetric hydrogen storage density of 47 $\frac{kg}{m^3}$. The process for dehydrogenation of

MCH and other cyclic hydrocarbons is well established in industry and uses noble metal catalysts such as platinum or palladium. The process is endothermic and requires a heat source to reach the high temperature requirements, which for MCH is 270–290 °C. To hydrogenate toluene into MCH, hydrogen is supplied at elevated temperatures (130–200 °C) and pressures (10–50 bar) to a reactor with toluene. The hydrogenation reaction also uses a noble metal catalyst; ruthenium is most commonly used for hydrogenated toluene [15].

Hydrogenation of other compounds including naphthalene and benzyl toluene is possible. As a result, other potential LOHC systems exist, including decalin, perhydro-dibenzyl toluene, perhydro-benzyl toluene, dodecahydro-N-ethyl carbazole, 1-methylperhydro indole, 2-methylperhydro indole, 1,2-perhydrodimethyl indole, perhydro-phenazine, and perhydro-2-(N-methyl benzyl pyriadine) [34]. However, these chemicals are still somewhat developmental and are in the early stages of commercialization; thus, their performance and economic competitiveness with toluene/MCH, as well as other hydrogen carriers, are not yet known.

18.4.6 End-use applications

P2X products synthesized using hydrogen have a wide range of end use applications. Ammonia is primarily used as a nitrogen fertilizer but does show potential as a fuel and energy carrier, as discussed previously. Similarly, methanol is a building block for a host of different products, including acetic acid, methyl methacrylate, silicone, olefins, and formaldehyde. These chemicals are used to produce a variety of products ranging from adhesives and paints to plastics and more. Additionally, methanol can be used as a fuel. RNG produced via methanation has similar end-use applications as fossil-based natural gas and is used as a fuel in both heat and power generation applications. LOHCs are expected to continue to be commercialized and eventually play a role in the large-scale transportation of hydrogen in a more stable form.

18.5 Project development and delivery

The development and delivery of hydrogen and P2X projects mirrors the steps that are undertaken in many different parts of large, complex infrastructure projects, such as in the power generation, chemical process, and the bio products industries. The development of low-carbon hydrogen and P2X infrastructure projects has increased dramatically, of late, as interest in hydrogen as a versatile fuel, feedstock, and energy carrier has increased manifold. The International Energy Agency (IEA), amongst other authorities, maintains a database of global infrastructure projects to produce hydrogen from low-carbon resources for use across the decarbonization landscape [35]. Some of the major commercial-scale projects that have reached a final investment decision (FID) since 2020 are summarized in Table 18.4.

"Green" hydrogen and P2X projects are mainly being pursued for their potential benefits in reducing key air emission criteria pollutants and greenhouse gas (GHG) emissions. These benefits are realized primarily in reducing or displacing the use of fossil-based

Table 18.4: Major global hydrogen and P2X infrastructure projects.

Project name	Location	Capacity, investment, and end use	Status
Air Products Net-Zero Hydrogen Energy Complex [36]	Edmonton, Alberta, Canada	520 ktpa $1.3B (CAD) Merchant, Power Generation	FID
Inner Mongolia Green Hydrogen Project [37]	Ordos, Inner Mongolia, China	67 ktpa Mobility, Chemicals	FID
HYBRIT [38]	Svartöberget Luleå, Sweden	58 ktpa 200 M (SEK) Iron/Steel	Construction
Advanced Clean Energy Storage [39, 40]	Delta, Utah, US	36 ktpa $500 M (USD) Energy Storage	Construction

fuels and feedstocks. However, GHG emissions reductions are highly dependent on the feedstock source, conversion process, transportation logistics, and end-use application. The lifecycle GHG emission avoidance of these products are computed based on a rigorous lifecycle analysis, thereby resulting in a carbon intensity score assigned to the specific fuel or feedstock. Most "green" hydrogen and P2X technologies offer a reduction in carbon intensity of greater than 50 percent, and their versatility makes them highly complementary with electrification and bio products initiatives of several industries.

18.5.1 Project development lifecycle

Major hydrogen and P2X projects often begin with several critical strategy and planning activities, which can vary quite a bit, depending on the target markets for product offtake, the geography, and set of enabling technologies that must be employed to ensure success. After the conception of the project, a developer will often focus on establishing the technical and financial viability via a feasibility study. Such endeavors focus on the market forces driving the end-product sales and the principal technologies, systems, and processes that are needed to support a profitable asset. For example, many "green" hydrogen project developers will first explore the design of an electrolysis plant to be integrated with a renewable energy asset, what the associated capital or operating and maintenance (O&M) costs are, and whether the "levelized" cost of hydrogen from the facility supports their target offtake markets and price points.

Once the feasibility of a project is established, the developer will typically turn his/her attention to advancement in the design to support financing, offtake contract negotiations, permits, and final budgetary authorization to proceed. This can involve an assortment of activities with some variation in roles, deliverables, and arrangement, but is often referred to in the chemical process industry as the front-end loading (FEL) process. FEL-1 through FEL-3 refer to a stage-gate procedure that includes a standardized set of technical deliverables (and associated improved accuracy in cost estimates at each stage) to advance the project through front-end engineering design, until it reaches the point where it can be firmly bid for full execution. Figure 18.3 illustrates the full lifecycle of an infrastructure project within the context of the broader decarbonization industry, including hydrogen, P2X, and bio products.

18.5.2 Technology scaling and project delivery

For projects that involve an advanced technology, it is critical that appropriate measures are taken to scale that technology, so it can be seamlessly integrated into the infrastructure asset when the project is ready for execution. For example, an RNG project may involve a cutting-edge gasification or carbon capture technology, or a "green" hydrogen project may involve an advanced electrolyzer technology; in both cases, they would need to be scaled and de-risked in parallel with their implementation into full-scale facilities. This requires a combination of strategic planning and research, development, and demonstration (RD&D) activities to ensure success. A developer often will need to make critical decisions on how to manage their intellectual property and determine a strategy for the commercialization of the advanced technology (i.e., spin-off, start-up, licensing). As the technology progresses through various technology readiness levels (TRLs) during RD&D, it will become critical to consider how the technology will be manufactured and integrated into pilot, demonstration, and commercial-scale facilities, including how each phase will be financed [41].

Upon successful scaling of the technology and during the late stages of an infrastructure project development, attention can be turned to the critical quality variables and associated costs for the manufactured product. These will have a significant impact on formulating a project execution strategy, including the form of contract with the engineering, procurement, and construction (EPC) contractor, the performance risks to be shared between the stakeholders of the project, and the associated impacts on cost and schedule. Figure 18.4 shows a few project delivery contracting alternatives and how risk transfer, flexibility, and cost/schedule can be balanced. However, note that this is a very simplified depiction of the various considerations, and the selection of a project execution strategy is a complex topic that should be carefully studied prior to moving forward with a specific approach [42].

Once a project is financed and permitted, and an EPC contractor is hired, it may proceed into detailed engineering, equipment/material procurement, and construction.

EXECUTION

BUSINESS CHOICES

Strategy

- Current state assessment
- Technology scenarios
- Market & regulatory view
- Provider landscape
- Economic modeling across business
- Project Roadmapping

Investment Planning

- Ranking of project investment options
- Co-optimize cost savings with CO_2 savings (as applicable)
- Identification and decision-making support for initial investments

Prelim. Feasibility & Design

- Project Feasibility
- Detailed site assessment
- Cost efficiency validation for project
- Technology Optimization
- Energy access (fossil / renewable)

Detailed Engineering

- Engineering designs
- Zoning and permitting
- Power/utility/transport requirements
- Procurement
- Qualify and manage subcontractors

Construction & Commissioning

- Union and Non-union
- Mobilization
- Site prep/civil works
- Mechanical
- Electrical
- Communications
- Commissioning

Operation & Maintenance

- Testing and training
- Startup
- Alarms & monitoring
- Infrastructure management
- Plant Improvement /Retrofits

Program Management

- Manage overall program of multiple concurrent initiatives globally
- Track & measure results of applicable initiatives
- Apply lessons learned across projects
- Continual update of investment prioritization as technology advances

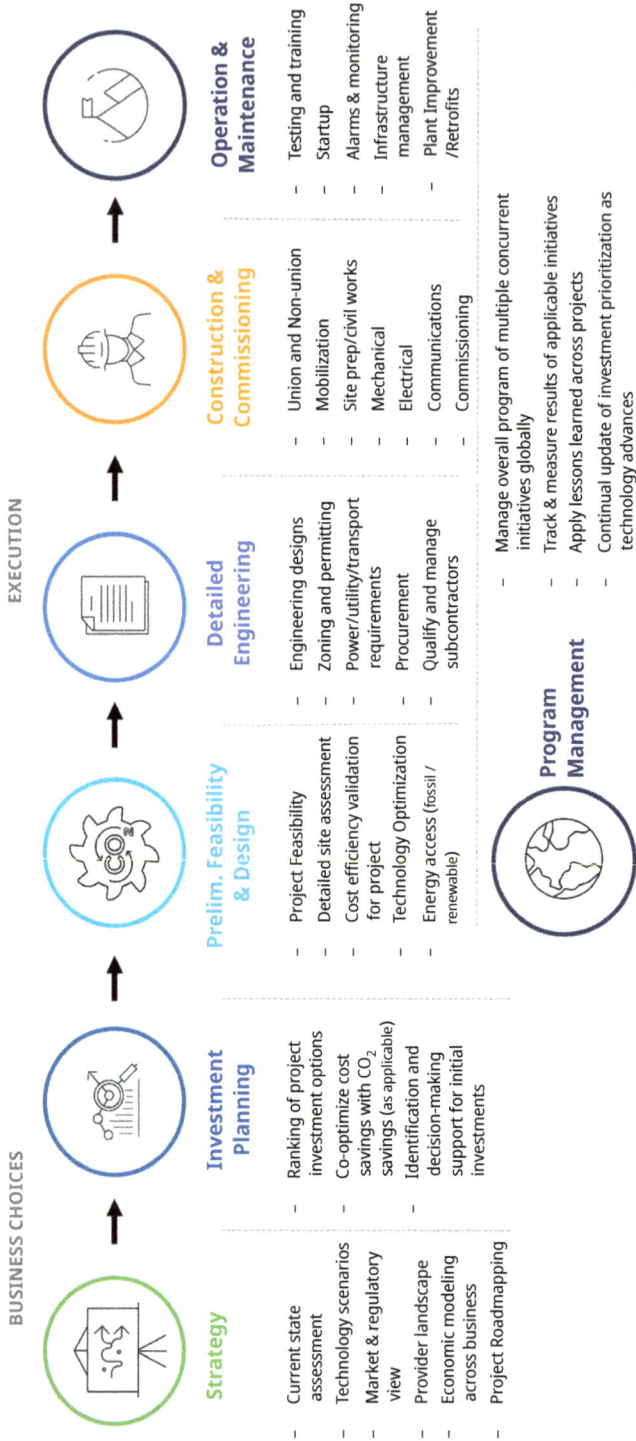

Figure 18.3: Infrastructure project development lifecycle.

Figure 18.4: Project delivery contracting alternatives.

Throughout execution, the roles of the project owner, owner representatives, the EPC contractor, and all other stakeholders will need to be carefully managed. The EPC contractor will have rigorous procedures for execution of engineering, procurement, construction, and then, commissioning and facility turnover that will be administered through meetings and various project control tools, in which the owner and their representatives play a key monitoring and approval role. Environmental health and safety, quality management, and risk management best practices will also be employed by project management professionals in a synergistic manner throughout the execution.

18.5.3 Operations and maintenance (O&M)

After the project owner has accepted the handover from the EPC contractor, the longest portion of the project lifecycle begins via operations. This initially entails training of operators and maintenance staff so that a safe and profitable work environment can be maintained throughout the asset life. For complex hydrogen and P2X projects, operations can sometimes rely on numerous automated functions, but will frequently require ongoing asset monitoring and intervention, particularly as equipment begins to age and failures are experienced. A carefully executed management plan is key to successful operations and planned/unplanned maintenance outages. As new technologies are developed and commercialized, it is also advantageous to explore opportunities for plant improvements and retrofits to ensure that the asset profitability is maximized. At the end of the project life, which is typically 20–30 years for hydrogen and P2X projects, the plant will be retired and decommissioned, after which the major equipment can be recycled or repurposed and the site can be used for new means.

18.6 Conclusions

Hydrogen is one of the most abundant and diverse chemical substances in the universe, in terms of its production and consumption. These fundamental qualities have driven significant interest in hydrogen as a means for the decarbonization of numerous industries across the chemical, energy, and heavy industrial landscape. The synergies between hydrogen and the bio products industry are innumerable and vast, with respect to the supplantation of fossil-based resources in favor of the union of renewable carbon from biomass and/or carbon capture with renewable electric power resources. The delivery of infrastructure associated with low-carbon hydrogen and bio products also has many noteworthy similarities, further affirming their cross-applicability.

References

[1]   Vichard L, Harel F, Ravey A, Venet P, Hissel D. Degradation prediction of PEM fuel cell based on artificial intelligence. Int J Hydrogen Energy 2020;29(45):14953–63.
[2]   Brauns J, Turek T. Alkaline water electrolysis powered by renewable energy: A review. Processes 2020;8:248.
[3]   Pozio A, Bozza F, Nigliaccio G, Platter M, Monteleone G. Development perspectives on low-temperature electrolysis. Enea 2021;1:66–72.
[4]   Hauch A, Küngas R, Blennow P, Hansen AB, et al. Recent advances in solid oxide cell technology for electrolysis. Science 2020;370:6513.
[5]   Kuckshinrichs W, Ketelaer T, Koj TC. Economic analysis of improved alkaline water electrolysis. Front Energy Res 2017;5.
[6]   Christensen A. Assessment of Hydrogen Production Costs from Electrolysis: United States and Europe. Washington: International Council on Clean Transportation 2020.
[7]   Schmidt O, Gambhir A, Staffell I, Hawkes A, Nelson J, Few S. Future cost and performance of water electrolysis: An expert elicitation study. Int J Hydrogen Energy 2017;52(42):30470–92.
[8]   Felgenhauer M, Thomas Hamacher T. State-of-the-art of commercial electrolyzers and on-site hydrogen generation for logistic vehicles in South Carolina. Int J Hydrogen Energy 2015;5 (40):2084–90.
[9]   Marcelo C, Fritz DL, Mergel J, Stolton D. A comprehensive review on PEM water electrolysis. Int J Hydrogen Energy 2013;12(38):4901–34.
[10]   Hydrogen production: Natural gas reforming. Office of Energy Efficiency & Renewable Energy. (Accessed June 2022 at https://www.energy.gov/eere/fuelcells/hydrogen-production-natural-gas-reforming).
[11]   Chen CJ, Back MH, Back RA. Mechanism of the thermal decomposition of methane. Am Chem Soc Symp Ser 1976;1–16.
[12]   Dagle RA, Deagle V, Bearden MD, Holladay JD, Krause TR, Ahmed S. An Overview of Natural Gas Conversion Technologies for Co-Production of Hydrogen and Value-Added Solid Carbon Products, USDOE Office of Energy Efficiency and Renewable Energy, 2017.
[13]   Sánchez-Bastardo N, Schlögl R, Ruland H. Methane pyrolysis for zero-emission hydrogen production: A potential bridge technology from fossil fuels to a renewable and sustainable hydrogen economy. Ind Eng Chem Res 2021;60:11855–81.

[14] Ahmed S, Aitani A, Rahman F, Al-Dawood A, Al-Muhaish F. Decomposition of hydrocarbons to hydrogen and carbon. Appl Catal A: Gen 2009;359:1–24.

[15] Andersson J, Grönkvist S. Large-scale storage of hydrogen. Int J Hydrogen Energy 2019;23 (44):11901–19.

[16] Tan-Peng C, Kelly B. Hydrogen Delivery Infrastructure Options Analysis. US Department of Energy Hydrogen Program, 2006. (Accessed June 2022 at www.hydrogen.energy.gov/pdfs/progress07/iii_a_ 1_chen.pdf).

[17] Domptail K, Hildebrandt S, Hill G, Maunder D, Taylor F, Win V. Emerging Fuels – Hydrogen State of the Art, Gap Analysis, and Future Project Roadmap. Chantilly: Pipeline Research Council International Inc, 2020.

[18] Venkat P, Richardson J. Introduction to Ammonia Production. New York: American Institute of Chemical Engineers, 2016.

[19] Meeting the Dual Challenge: A Roadmap to at-Scale Deployment of Carbon Capture, use, and Storage, Chapter 5 – CO_2 Capture, National Petroleum Council, 2021.

[20] Budinis S. Direct air capture: a key technology for net zero. International Energy Agency, 2022: 16.

[21] Carbon Storage FAQs. National Energy Technology Laboratory 2022. (Accessed June 2022 at https://www.netl.doe.gov/carbon-management/carbon-storage/faqs/carbon-storage-faqs).

[22] Kang S, Boshell F, Goeppert A, Prakash SG, Landälv I. Innovation Outlook: Renewable Methanol, Abu Dhabi: International Renewable Energy Agency; 2021.

[23] Methanol Production. Methanol Institute 2016. (Accessed June 2022 at https://www.methanol.org/ wp-content/uploads/2016/06/MI-Combined-Slide-Deck-MDC-slides-Revised.pdf).

[24] Syngas Conversion to Methanol. National Energy Technology Laboratory, 2022. (Accessed June 2022 at https://netl.doe.gov/research/coal/energy-systems/gasification/gasifipedia/methanol).

[25] Rönsch S, Schneider J, Matthischke S, et al. Review on methanation – from fundamentals to current projects. Fuel 2016;166:276–96.

[26] Schmider D, Maier L, Deutschmann O. Reaction kinetics of CO and CO_2 methanation over nickel. Ind Eng Chem Res 2021;60:5792–805.

[27] Biollaz SMA, Schildhauer TJ, Jansohn P. Status PSI – R&D on Fluidised Bed Methanation. Würenlingen: Paul Scherrer Institute, 2016.

[28] Rönsch S, Kaltschmitt M. Bio-SNG production – concepts and their assessment. Biomass Convers Biorefin 2012;2:285–96.

[29] Liu B, Ji S. Comparative study of fluidized bed and fixed bed reactor for syngas methanation over Ni-W/TiO2-SiO2 catalyst. J Energy Chem 2013;22:740–46.

[30] Liu J, Cui D, Yu J, Su F, Xu G. Performance characteristics of fluidized bed syngas methanation over Ni-Mg/Al2O3 catalyst. Chin J Chem Eng 2015;23:86–92.

[31] Heyne S. Bio-SNG From Thermal Gasification – Process Synthesis, Integration, and Performance, Chalmers University of Technology Sweden; 2013.

[32] Renewable Natural Gas. Environmental Protection Agency United States, 2022. (Accessed June 2022 at https://www.epa.gov/lmop/renewable-natural-gas).

[33] Kurosaki D. Introduction of Liquid Organic Hydrogen Carrier and The Global Hydrogen Supply Chain Project, Chiyoda Corporation, Advanced Hydrogen Energy Chain Association for Technology Development; 2018.

[34] Rao PC, Yoon M. Potential liquid-organic hydrogen carrier (LOHC) systems: A review on recent progress. Energies 2020;13.

[35] Hydrogen Projects Database. International Energy Agency, 2021. (Accessed June 2022 at https://www.iea.org/reports/hydrogen-projects-database).

[36] Air Products Announces Multi-Billion Dollar Net-Zero Hydrogen Energy Complex in Edmonton, Alberta, Canada. Air Products Inc. 2021. (Accessed June 2022 at https://www.airproducts.com/news-center /2021/06/0609-air-products-net-zero-hydrogen-energy-complex-in-edmonton-alberta-canada).

[37] China approves renewable mega-project for green hydrogen. The Straits Times, 2021. (Accessed June 2022 at https://www.straitstimes.com/business/economy/china-approves-renewable-mega-project-for-green-hydrogen).

[38] HYBRIT: SEK 200 million invested in pilot plant for storage of fossil-free hydrogen in Luleå. Vattenfall AB, 2019. (Accessed June 2022 at https://group.vattenfall.com/press-and-media/pressreleases/2019/hybrit-sek-200-million-invested-in-pilot-plant-for-storage-of-fossil-free-hydrogen-in-lulea).

[39] US DOE Closes $504.4 Million Loan to Advanced Clean Energy Storage Project for Hydrogen Production and Storage. Mitsubishi Power Americas, 2022. (Accessed June 2022 at https://power.mhi.com/regions/amer/news/20220609).

[40] World's Largest Green Hydrogen Hub Selects Black & Veatch as EPC Provider. Black & Veatch Corporation, 2022. (Access June 2022 at https://www.bv.com/news/hydrogen-hub-selects-black-veatch).

[41] Technology Readiness Level Definitions. National Aeronautics and Space Administration United States 2017. (Accessed June 2022 at https://esto.nasa.gov/files/TRL_definitions.pdf).

[42] Shahani G. Comparative analyses of common project execution alternatives. Hydrocarbon Processing, 2019. (Accessed June 2022 at https://www.hydrocarbonprocessing.com/magazine/2019/april-2019/project-management/comparative-analyses-of-common-project-execution-alternatives).

Section V: **Bioplastics and biopolymers**

Ashok Adur
Chapter 19
Renewable green biopolymers and bioplastics

19.1 Introduction

19.1.1 What are biopolymers and bioplastics?

Biopolymers are polymeric materials produced from renewable biomass, mostly plant-based sources, found in nature. Natural biopolymers are produced by the cells of living organisms. Examples of these include corn starch, straw, sawdust, woodchips, recycled food waste, as well as various vegetable fats and oils called lipids, from plants or animals [1]. As covered in earlier chapters in this book, some feedstocks for biopolymers are obtained by chemical processing directly from polysaccharides (e.g., starch, cellulose, chitosan and alginate) and proteins (e.g., soy protein, gluten and gelatin), or biologically generated by fermentation of sugars or lipids. Others are chemically synthesized or derivatized from these some sources like lactic acid or various cellulose derivatives. On the other hand, common traditional plastics, such as fossil-fuel plastics (also called petro-based polymers) are derived from petroleum or natural gas or, in a few cases, from coal. Either way, all polymers are made of repetitive units called monomers [2], so how can one distinguish between biopolymers and synthetic polymers? One way is based on their molecular weight distribution in their chemical structure. Most natural biopolymers typically contain similar sequences and numbers of monomers and thus, all have the same chain length and hence, mass or a very narrow molecular weight distribution (M_w/M_n) of 1. This phenomenon is also called monodispersity, in contrast to the polydispersity encountered in synthetic polymers. As a result, many, but not all, biopolymers have a dispersity of 1 [3].

Natural biopolymers can be classified into three main families according to the monomers used and the structure of the biopolymer:

- Polynucleotides [4], such as RNA and DNA, are long polymers composed of 13 or more nucleotide monomers.
- Polypeptides [5] and proteins are polymers of amino acids, and some major examples include collagen, actin, and fibrin.
- Polysaccharides are linear or branched polymeric carbohydrates, and examples include starch, cellulose, and alginate [6].

Ashok Adur, Everest International Consulting (LLC), Jacksonville, FL 32256, USA,
e-mail: aadur@outlook.com

https://doi.org/10.1515/9783110791228-019

From an organic chemistry perspective, biopolymers can be classified based on the type of repeat chemical bonds in each monomeric part or in the way each monomer unit is linked to the next one in the polymer structure:

- Polyesters: Examples are polylactic acid (PLA), polyhydroxyalkanoate (PHA) and poly-3-hydroxybutyrate (PHB)
- Polyamides: Examples are proteins and polypeptides like silk, collagen, fibrin, actin, and zein.
- Cellulosics: Examples include cellulose, starch, chitosan, and alginate.
- Unsaturated olefinics: Examples include pinenes and terpenes as well as natural rubber (polyisoprene)

Nature is far more complex than to allow for simple classifications like those above. Examples of biopolymers found in nature with complex structures are suberin, cutin and lignin, which are complex polyphenolic polymers, cutan, which is a complex polymer of long-chain fatty acids, and melanin, which is a broad term for a group of natural pigments found in most organisms, produced through a multistage chemical process involving oxidation of the amino acid tyrosine, followed by polymerization.

While most people think biopolymers and bioplastics are recent inventions, excavations have shown that people in Mesoamerican cultures (Olmec, Maya, and Aztecs) used natural latex and rubber to make balls, containers, and waterproof clothes, around 1500 BCE [7].

Many natural biopolymers are further processed and chemically modified to make them useful for various applications. Also, a few industrial polymers produced today use bio-based monomers, while some petro-based polymers and a few bio-based polymers use bio-based plasticizers, lubricants, and other additives, which are compounded in to modify the properties of both bio-based and petro-based polymers (plastics and rubber/elastomers) and make them suitable for large-scale applications.

Biopolymers like thermoplastics or thermosets that can be processed are called bioplastics. A thermoplastic is a polymer material that becomes pliable or moldable at a certain elevated temperature and solidifies upon cooling, so it can be shaped using polymer processing techniques such as injection molding, blow molding, compression molding, calendaring, or extrusion processes, while a thermoset is produced by irreversible chemical bonds during the curing process and does not melt when heated, but typically decomposes and does not reform upon cooling. While all bioplastics are biopolymers, all biopolymers are not bioplastics. For example, wool and silk are biopolymers but not bioplastics. So, bioplastics are a subset of biopolymers but they are not synonymous, although many people wrongly use both terms interchangeably.

This chapter covers biopolymers that are used as plastics, with emphasis on those that are not covered elsewhere in other chapters in this book. A few non-bio-based polymers that are biodegradable and/or compostable, even though they are sourced from fossil fuels like polyvinyl alcohol (PVOH), poly(butylene succinate) (PBS), and polybutylene adipate-terephthalate (PBAT). PBS and PBAT are covered in another chapter. Some bio-feedstocks that are not covered in other chapters in this book will also be covered in this chapter, especially those that are used for bioplastics. In addition, natural biopolymers that are chemically modified like cellulose acetate, cellulose butyrate, and other cellulose derivatives as well as vulcanized natural rubber that are of industrial importance will also be covered briefly. These sections will be followed by a discussion of limitations of biopolymers and challenges to growth in the future. Despite these limitations, biopolymers do have a bright future, especially if some of these limitations can be overcome. This chapter will end with a perspective on future trends and prospects for growth.

In most cases, polymers (plastics and rubber) whether they are bio-based or not, are rarely used without the addition of additives like plasticizers, lubricants, antioxidants and stabilizers, fillers, colorants, reinforcements, etc. in order to meet various customer requirements all the way downstream in the value chain. These additives are added during a compounding process step. Such compounds based on biopolymers as well as bio-composites, which are compounds where reinforcements are used with biopolymers, will be covered in a separate chapter, whether the polymer and/or the reinforcement is bio-based in the composite.

There are numerous academic and other research institutions, primarily funded by governments and foundations that publish valuable information on biopolymers. A vast majority of these biopolymers may not get commercialized. The reasons for this are numerous. In many cases, the researchers who have worked on conducting such research either are not interested in overcoming or are not able to overcome some of the important but necessary challenges to commercialization that exist. These include how to overcome some of these limitations of their biopolymers besides challenges of starting a new business as a startup like scale-up processes, financing, time needed to develop new products, applications and markets, adequate capital till their sales grow to generate sufficient profits, or the contacts needed. In some cases, it is critical for a new entrant to meet an unmet need, to replace an existing product with a new entrant, or to offer better performance per unit net price to succeed. Thus, some of these biopolymers that are unlikely to be commercialized will be referenced, but the emphasis will be on commercialized or products likely to be commercialized. (Many already are.)

19.1.2 Global production capacities of bioplastics

With a mounting number of materials, applications, and products, along with development of technology, the number of manufacturers, converters, and end users are also rising steadily. Significant financial investments have been made in production and marketing to guide and accompany this development, especially in Europe.

Figure 19.1 shows the estimated global production capacities of bioplastics in 2021, by region. The Asian region shows the largest production capacity, followed by Europe, and then, North America. However, on a per capita basis, due to the much higher population in Asia, the growth rates there will continue to be higher than the rest of the world.

Figure 19.2 shows the breakdown of biodegradable plastics, by market segment, based on estimates. For biodegradable plastics, packaging is the largest segment with flexible packaging being almost double that of rigid packaging, followed by agriculture and horticulture and consumer goods.

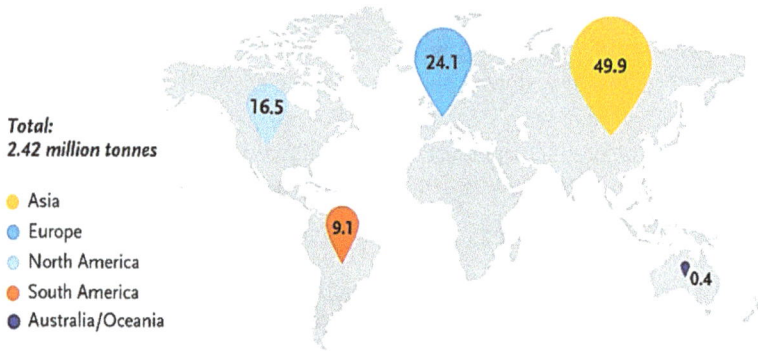

Figure 19.1: Global production capacities of bioplastics, by region.

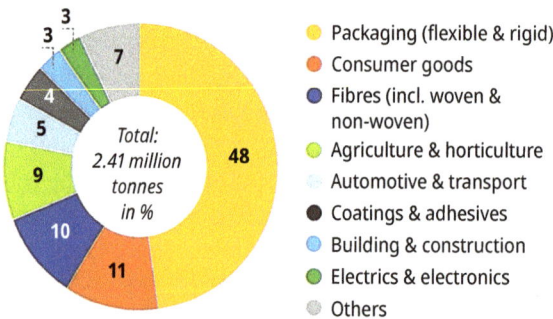

Figure 19.2: Biodegradable plastics, by market segment.

Figure 19.3 shows the global production capacities of bioplastics separated by each of the major specific biopolymers, in 2021, for various market segments.

This set of latest market data from all three figures are from www.bio-based.eu/ and demonstrates the contributions of the industry in moving towards a sustainable future with reduced environmental impact. Their forecast also predicts that the budding bioplastics industry will continue to show enormous economic potential growth, over the coming decades [8].

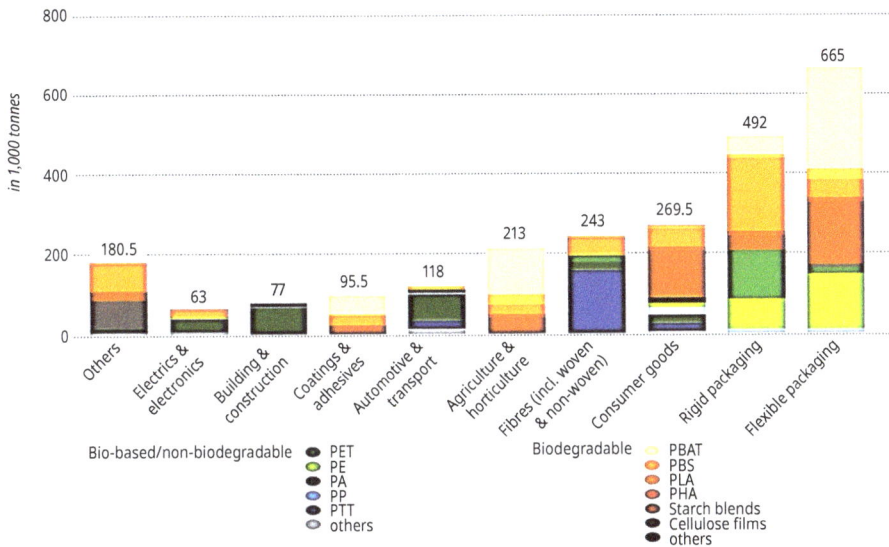

Figure 19.3: Biodegradable plastics in 2021, by market segment and biopolymer type.

The factors driving market development are both internal and external to the biopolymer industry. Obviously, the biopolymer industry has an economic and professional bias to drive market development and increase the market share and acceptance of biopolymers and bioplastics globally. Internal factors are based on the growth of technology that is being developed along with a number of players at all stages of the value chain, covered in the first paragraph of this section. There are external factors too, that make bioplastics the choice. This can be demonstrated by the escalating rate of consumer acceptance, as the general population has become more environmentally conscious. The extensively publicized effects of climate change, mounting prices of coal, oil and natural gas and materials derived from fossil fuels, and the increasing dependence on fossil resources also contribute to bioplastics being viewed favorably [8].

According to the latest Eurobarometer Survey conducted by the European Commission in 2020, about 90% of European customers want to buy products that have minimal impact on the environment. The fact that bio-based plastics display clear advantages over conventional petro-based plastics makes them attractive to environmentally conscious customers. However, the prices of bio-based products are higher than those that are petro-based.

Bioplastics are believed to efficient and technologically mature materials by European Bioplastics e.V. because they are able to improve the balance between the environmental benefits and the environmental impact of plastics. Life cycle analyses (LCA) demonstrate that bioplastics can significantly reduce carbon dioxide (CO_2) emissions compared to conventional plastics (depending on the material and application). Furthermore, the increasing use of biomass in bioplastic applications has two clear advantages: renewability and availability [8]. However, the LCA data also show that other factors such as energy and water requirements of bioplastics are much higher over the full "cradle-to-grave" life cycle.

According to a recent market study by Markets and Markets™, Inc., the global bioplastics and biopolymers market size was projected to grow at a compound annual growth rate of 23% from USD 11 billion in 2021 to USD 30 in 2026 [9].

Companies like NatureWorks (USA), Braskem (Brazil), BASF (Germany), Total's Corbion and Synbra (both in Netherlands), Novamont (Italy), Biome Bioplastics (UK), Mitsubishi Chemical Holding Corporation (Japan), CJ Bio (Korea), Biotec and KFUR Kunstoff (both in Germany), Cardia BioPlastics and Plantic Technologies (both in Australia), Toray Industries (Japan), Futerro (Belgium), PTT Global Chemical Public Company Limited (Thailand), and Zhejainag HiIsun Biomaterials and Tianan Biologic Materials (both in China) are among those producing these bio-based polymers.

19.1.3 Are biopolymers and bioplastics green, environmentally friendly, and/or sustainable?

The general belief is that all bioplastics are "green" and that anything produced from a renewable source is green. Is this a scientific fact? It is debatable, because the word "green," from an environmental perspective, is used without a strict definition. There are standardized LCA tests that have been developed and used. The results of these tests will surprise many people. The common perception that bioplastics are derived from biological sources and are, therefore, biodegradable and compostable, is not a fact. While some of these bioplastics are biodegradable and a few are compostable, one cannot make such a generalization for all bio-sourced materials. Those interested in understanding the differences between biodegradable, compostable and oxo-biodegradable may refer to this well written article [10]. This outcome of biodegradability is achievable because the microorganisms in nature the environment are able to break down the chemical structures, making the bioplastic biodegradable under certain conditions. This is a result that is considered to be eco-friendly, compared to the traditional plastics that we throw away every day. Biodegradability of a polymer depends on the chemical structure of the polymer but not on the feedstock source used for the monomers' collection.

Bio-based plastics can also be divided into four principal groups based on the source of their feedstock:

A. Bioplastics that are based on renewable resources and are also biodegradable, like starch plastic, cellulose polymers, proteins, lignin and chitosan plastics, polylactic acid (PLA), polyhydroxy alkanoates (PHAs), but also polyhydroxybutyrates (PHBs), polyhydroxyvalerate (PHV) and their copolymers in different percentages of two different monomers called poly(3-hydroxybutyrate-co-3-hydroxyvalerate) or (PHBV);

B. Biopolymers such as PVC, PE, PP, PET, nylon, and polyamides (PA), named as bioplastics because the starting monomers are obtained from biological resources;

C. Bioplastics based on petroleum resources, which are 100% biodegradable, like poly-caprolactone (PCL), polybutylene succinate (PBS), polybutylene adipate (PBA) and its copolymers with synthetic polyesters like polybutylene adipate-terephthalate (PBAT) and polyvinyl alcohol (PVOH), although some of these have also been recently shown in the lab to be produced from bio-based sources; and

D. Bioplastics obtained by using monomers from mixed biological and petroleum resources like polyesters obtained with petroleum-derived terephthalic acid and biologically derived ethanol, 1,4-butanediol and 1,3-propanediol, such as polybutylene terephthalate (PBT), polytrimethylene terephthalate (PTT), polyethylene-co-isorbite terephthalate (PEIT), polyurethane (PUR), and epoxy resins (thermoset plastic). Major biodegradable polymers that also fall within this group include PLA; PHAs, including PHB and related copolymers, cellulose diacetate, regenerated cellulose, copolyesters such as PBAT and polybutylene succinate adipate (PBSA), PBS, PCL, polyglycolic acid (PGA); and starch compounds (mixtures of starch with other biodegradable polymers such as PBAT or PLA) [11].

Most of these in Group A as well as bio-polybutylene succinate (bio-PBS), bio-polyethylene furanoate (bio-PEF), bio-polyethylene terephthalate (bio-PET), bio-polybutylene adipate terephthalate (bio-PBA) will be discussed in this book in Chapter 20.

On the other hand, there are a few polymers that are derived from petro-based sources but are biodegradable and some are even water-soluble, especially when heated. Examples include PVOH, which is produced from vinyl acetate, PCL, which is produced from ε-caprolactone, a cyclic monomer, and PBS and some of its copolymers produced by polycondensation reactions of glycols, such as ethylene glycol and 1,4-butanediol, with aliphatic dicarboxylic acids, such as succinic and adipic acid [12]. Caprolactone as well as the monomers for PBS can be bio-based in smaller quantities or petro-based in larger quantities.

Biopolymers that do decompose in compost can be marked with a "compostable" symbol, under European Standard EN 13,432 (2000) [13]. Packaging marked with this symbol can be put into industrial composting processes and will break down within six months or less. An example of a compostable polymer is PLA film under 20 μm thick. Films that are thicker than that do not qualify as compostable, even though they are "biodegradable," because they take too long to decompose [14]. In fact, many biodegradable and compostable bottles will not degrade much and will behave similar to petro-based nondegradable plastics, unless both sufficiently high temperature and the right range of humidity as well as the right bacteria and/or fungi are present. Too much or too little humidity or insufficient temperature can slow down the composting rate. Certain biodegradable plastics need very specific bacteria in order to decompose. In Europe, there is a home composting standard and associated logo that enables consumers to identify and dispose of packaging in their compost heap [14]. In Europe, most countries have readily accessible large-scale composting facilities. The ASTM D6400 industrial composting testing and certification standard was adopted in North America in 2002, but in the USA and most other countries around the world, such composting facilities are not available everywhere and cannot be readily accessed. In such circumstances, recycling is the best option.

Biopolymers can be sustainable, carbon neutral, and are always renewable, because they are made from plant materials that can be grown indefinitely. The growth cycle for each plant source may be different. If these plant materials come from agricultural nonfood crops, the use of biopolymers would create a sustainable industry. However, when we evaluate all materials for being green or sustainable, it is important to also determine other important factors such as water and energy usage, as well as what happens when the biopolymer degrades. In many cases, biopolymers form carbon dioxide when they degrade and use a lot more energy and water than petro-based polymers as comparable "cradle-to-grave" lifecycle studies have shown.

There are some bio-sourced plastics that are produced by a fermentation process from sugar (sugarcane, beets, corn, or some other sugary bio-source, which could be from food crop), converted to ethanol, then, to ethylene and polyethylene that is undistinguishable from the polyethylene produced from ethylene from natural gas or crude oil. However, if we pursued this approach, food prices to the public would increase. To avoid using bio-feedstocks that come from food crops, other bio-sources such as agricultural or food waste need to be used. A lot of work is ongoing to do just that.

Some "biopolymers" are produced from a combination of a bio-feedstock and a non-bio-feedstock, yet are claimed to be biopolymers. Hence, it is critical that we stay away from such generalizations like all "biopolymers or natural polymers are biodegradable or compostable," "all synthetic or oil and gas or coal-based products live forever or at least a few thousand years and are not biodegradable and definitely not compostable." Each biopolymer or bioplastic material or composite or article has to be judged not on such generalizations, but on scientifically based LCA testing.

19.2 Bio feedstocks for biopolymers

19.2.1 Introduction

Many of the common bio-feedstocks for producing biopolymers are covered in Chapters 9, 20 and 25 in much more detail. Other bio-based chemicals of interest for use in biopolymers covered in this chapter are listed below:

19.2.2 Castor oil and castor oil derivatives

Castor oil is extracted from seeds of the castor plant and is a vital industrial raw material.

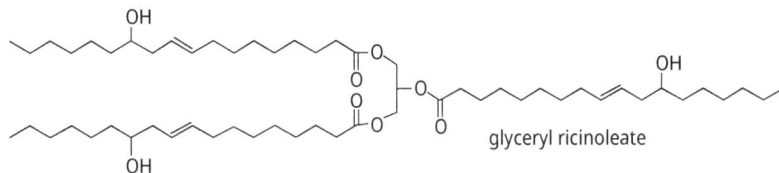

glyceryl ricinoleate

The unique aspect of castor oil is its high hydroxyl fatty acid composition of over 90%, much higher than other vegetable oils. Various derivatives like ricinoleic acid (12-hyroxyoctadec-9-enoic acid), undecenoic acid (10-undecinoic acid), sebacic acid (octanedicarboxylic acid), azelaic acid (1.7-nonanedioic acid), and oleic acid (9-octadec-9-enoic acid) are produced from this glyceryl ricinoleate [15, 16]. Castor oil is used to produce polyols that are used in the production of many different specialty polyurethanes. India is the largest producer of castor seeds and accounts for over 80% of the global production, followed by China and Brazil.

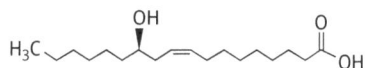

ricinoleic acid
12-hydroxyoctadec-9-enoic acid

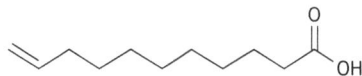

undecenoic acid
(10-undecinoic acid)

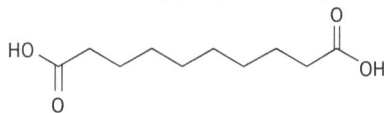

sebacic acid
(octanedicarboxylic acid)

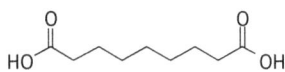

azeliac acid
(1,7-nonanedioic acid)

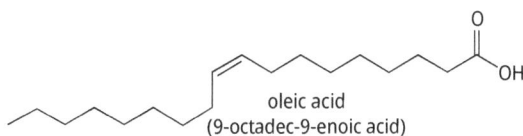

oleic acid
(9-octadec-9-enoic acid)

Other vegetable oils are also used to produce polyols, but the polyurethanes produced from them have different properties. Other polyols can obviously be produced from petro-based sources. Producers of polyols include BASF, Cargill Inc., MCNS, Emery Oleochemicals, Croda, Alberdingk Boley, Jayant Agro-Organics Limited, Maskimi, Stahl, Polylabs, Xuchuan Chemical, Vertellus, NivaPol, MCPU Polymer, Global Bio-Chem Technology Group, and EDB Poliois Vegetais. The overall global market is estimated to be about USD 7.6 billion in 2021, with the bio-based segment about 11% of the total but growing at double-digit rates [17].

Castor oil is converted also to ricinoloeic acid (12-hydroxyoctadec-9-enoic acid) by transesterification. From ricinoloeic acid, undecenoic acid (10-undecinoic acid) is produced by thermolysis. This 11-carbon acid can then be used to produce polyamide-11. Ricinoloeic acid is also converted to sebacic acid (octanedicarboxylic acid) and, then, is used to produce polyamide copolymers like 6–10, 5–10, 4–10, and 10–10.

Similarly, oleic acid (9-octadec-9-enoic acid) is produced from peanut oil. Oleic acid is converted to nonanoic acid (1-octanedioic acid) which can be polymerized to produce the same polyamide as sebacic acid. Oleic acid can also be converted to azeliac acid (1,7-nonanedioic acid) and in a few more steps, eventually to polyamide-9,9 and PA-9 T.

1,12-dodecanedium acid is typically derived from an oil transformation process, but can also be obtained from palm kernel oil. This 12-carbon acid is used to produce polyamide-12 and its copolymers such as PA-6,12.

19.2.3 Bio-ethylene

While ethylene itself is not found in nature, it is produced on a large scale from ethane, which is separated from natural gas and also by catalytic cracking of longer alkanes present in natural gas as well as in petroleum, for producing different types of polyethylenes (HDPE, LDPE, LLDPE, VLDPE), polypropylene, and ethylene copolymers with other olefins and comonomers like vinyl acetate and other unsaturated esters. However, due to the push for using bio-based sources, ethylene is also produced from glucose or sucrose, which is extracted from sugarcane, sugar beets, starch crops from corn, wheat or other grains, and lignocellulosic materials. The 12–13% solution of sucrose is anaerobically fermented in order to obtain ethanol. The ethanol is then distilled in order to remove water, giving an azeotropic solution of at 95.5 volume percentage of ethanol and water [18]. The bio-ethanol is dehydrated to obtain ethylene gas.[18–21]

19.2.4 Bio-propylene

There are multiple routes to produce bio-propylene (Bio-PP) that have been documented by Venkatraman [22]. One way is by fermentation of glucose to iso-butanol, which is then converted to bio-butylene and, later, to Bio-PP. The first company that produced

Bio-PP on a pilot-plant scale was Braskem; however, their process route has not been disclosed but is believed to be based on metathesis of ethylene and butene producing propylene monomer. The most promising route to obtain propylene is believed to be through methanol, which is further processed to obtain propylene monomer *via* Lurgi's methanol-to-propylene (MTP) process or UOP's methanol-to-olefins process [21], using the same industrial plants as for petrochemical methanol. Another route to bio-propylene could be producing 1,2-propanediol or acetone through fermentation, followed by conversion to 2-propanol and then, dehydration to propylene [18–20]. The thermochemical route uses a wide variety of bio-based feedstocks including corn, grass, and agricultural wastes, which are carbon-rich and can be gasified to produce syngas. Bio-diesel and vegetable oils can also be used in the production of green propylene. Production cost is a very important factor in the choice of production route. The capital costs for the fermentation route are significantly lower compared to the gasification route, but the fermentation route is usually much slower [21].

19.2.5 Bio-caprolactone

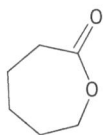

ε-caprolactone

Caprolactone can be prepared from petro-based sources [23] or from renewable resources *via* chemical treatment of saccharides. The bio-caprolactone production process involves two steps: In the first step, the saccharide is converted to ethanol and acetic acid by fermentation. In the second step, ethanol is converted to cyclohexanone by using chromic acid. Then, the generated cyclohexanone is oxidized by a Bayer–Villiger reaction to get ε-caprolactones covered in a Daicel patent[24], and also covered in a review by Grandhe Rani and Suraj Sharma [25].

19.2.6 Bio-glycolic acid

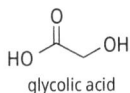

glycolic acid

Glycolic acid is mainly produced chemically from petrochemical resources by a process that requires toxic formaldehyde. However, from a sustainable point of view, in a biotechnological production route, it is also produced commercially by three to four different fermentation methods listed [26]. Glycolic acid is used to produce PGA, which is covered in the next section.

19.2.7 Bio-butadiene (Bio-BD), Bio-butylene glycol (Bio-BG), and Bio-1,4-butanediol (Bio-BDO)

These chemicals that are usually produced from crude oil distillates are now being produced also from plant-based sources.

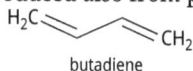

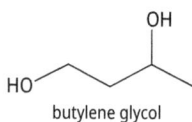

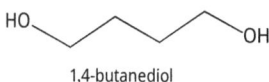

H_2C ⟍⟋⟍ CH_2
butadiene

OH
HO ⟍⟋⟍⟋
butylene glycol

HO ⟍⟋⟍⟋⟍ OH
1,4-butanediol

Genomatica, a company based in San Diego, California, USA, developed manufacturing processes that enable the production of intermediate and basic chemicals from renewable feedstocks. It has made progress since its first product, bio-BDO. BDO is an industrial chemical that can be used as a solvent, in its own right. The largest use of BDO, around 40%, is in producing tetrahydrofuran (THF), which is then used to manufacture polytetramethylene ether glycol (PTMEG), from which Spandex™ fibers are produced. BDO is also used as an intermediate to produce polyurethanes. Another use for BDO is in making single-use compostable bags in Europe. As a biotechnology partner to the chemical industry, Genomatica's business model is as a technology enabler and not as a producer of large-volume chemicals. Its patents cover the use of fermentation and other biotech technologies to produce bio-BD and bio-BG. These technologies have been licensed to several companies in the US, Europe, and Japan that have built plants in their home countries as well as in China and elsewhere [27].

19.2.8 Bio-hexamethylenediamine (HMDA), bio-caprolactam, and bio-adipic acid

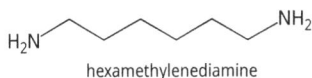

hexamethylenediamine

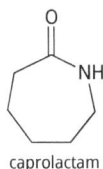

caprolactam

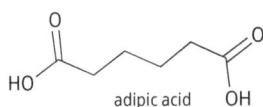

adipic acid

In 2014, Genomatica announced that it had entered the bio-based nylon intermediates market focusing on HMDA, caprolactam, and adipic acid [27]. This company would develop complete process technologies for the bio-based production of these intermediates, which will then be licensed to major firms in the nylon value chain. These three intermediates are used primarily in the production of nylon 6 (or polyamide-6) and nylon-6,6 (PA-6,6). At that time, these two polyamides were 100% produced from petro-based intermediates at a value of over $18 billion per year. Genomatica's announcement claimed that they had issued eight patents. The current status will be discussed later in the section on bio-polyamides.

19.2.9 Bio-1,5-pentamethylene diamine

1,5-Pentamethylene diamine

A Japanese patent application disclosed a biomass resource-derived 1,5-pentamethylene diamine to produce a polyamide resin composition having a good color tone. The 1,5-pentamethylene diamine which contains a total of the contents of an acetamide, 2,3,4,5-tetrahydropyridine, 2-piperidone, α-amino-ε-caprolactam, lysine, and ornithine has impurities that are less than or equal to 150 weight ppm [28]. This diamine is then used to produce several different bio-based polyamides that have five methylene groups and are covered in the bio-polyamides section later in this chapter.

19.2.10 Natural rubber

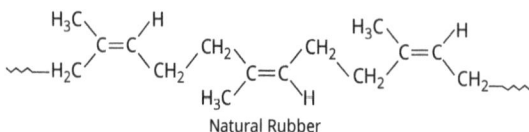

Natural Rubber

Natural rubber is obtained from rubber trees in the form of latex. Such rubber trees are grown in the tropics in countries like Thailand, Indonesia, Malaysia, Sri Lanka, as well as parts of equatorial countries in Africa and Latin America. The raw latex is not easy to use and does not have desirable properties to be used directly, except for coatings. Chemically, it is an emulsion of polymers of isoprene. Rubber latex is a sticky, milky, and white colloidal emulsion, which is allowed to drip off by making incisions in the bark and collected as fluid in vessels by a process called "tapping." The latex itself is difficult to use directly and has limited utility, except to be used as water-proof coating on fabric. Hence, the latex is allowed to coagulate in the collection cup. These coagulated lumps are collected and processed into dry forms for sale. Natural rubber is used extensively in many applications and products, either alone or in combination with other materials. In most of its useful forms, it has a large stretch ratio and high resilience and is also water-proof. Significant progress was made after the invention of vulcanization of natural rubber, which will be discussed in a section on "Vulcanized Natural Rubber" later in this chapter.

19.3 Biopolymers (not covered elsewhere)

19.3.1 Introduction

While many biopolymers are covered in the chapter by Drs. Kumbhar and Ghosalkar, and some by Dr. Kulkarni, some are not. This section will cover those biopolymers. This list includes bio-polyethylene (Bio-PE), Bio-PP, PCL, PGA, bio-polytrimethylene terephthalate (Bio-PTT), bio-polyamides (Bio-PAs), cellulose derivatives like cellulose acetate, and starch compounds (mixtures of starch with other biodegradable polymers such as PLC or PLA and with synthetic polymers like polyolefins). Other renewable resin chemistries such as collagen and chitosan are not covered because their roles are not well developed, at least in industrial plastics. Other bio-based polymers have also been developed like bio-poly(acrylic acid) (Bio-PAA), bio-poly(methyl methacrylate) (Bio-PMMA), bio-polystyrene (Bio-PS), bio-poly(vinyl acetate) (Bio-PVA), and bio-based poly(vinyl chloride) (Bio-PVC) but with insufficient quantities being produced, if at all, are also not covered here.

19.3.2 Bio-polyethylene

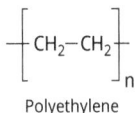

$$\left[CH_2{-}CH_2 \right]_n$$

Polyethylene

Polyethylene (PE) is the largest volume plastics material used in the world. The starting monomer used for synthesis is ethylene, which is also used to produce other polymers like PVC and PS, and it is generally obtained from petroleum feedstock by distillation, especially in Europe. Almost everywhere else, it is separated from natural gas or by cracking of natural gas liquids. To obtain ethylene from biological resources, instead, a lot of research and development work has been put in. The only approach that has been successful so far has been to obtain the ethylene monomer by dehydration of bio-ethanol, obtained from glucose. Once ethylene gas is produced from a bio-source, it is indistinguishable from the ethylene produced from a petro-source and the same polyethylene grades can be produced from ethylene, irrespective of source. Consequently, the subsequent polymerization of bio-ethylene monomer is exactly the same as if the ethylene is derived from petroleum or natural gas, and the corresponding bio-polymer is identical in its chemical, physical, and mechanical properties to fossil-based PE, also with regards to the mechanical recycling processes [18, 19, 29].

By various polymerization methods that are used for polymerizing non-bio-ethylene, various types of bio-PE can be produced: bio-high density polyethylene (bio-HDPE) with a low degree of short-chain branching (about seven branches per 1,000 carbon atoms), bio-linear low density polyethylene (LLDPE) with a high degree of short-chain branching, and bio-low density polyethylene (bio-LDPE) with a high degree of short-chain branching + long-chain branching (about 60 branches per 1,000 carbon atoms). Ethylene copolymers with other olefins like propylene, butene, hexene, octene-1, and 4-methyl pentene-1 can be produced, and based on the level of the comonomer, could be LLDPE grades or more elastomeric very low-density polyethylene (VLDPE), also called flexomers, or even ethylene-propylene rubber. Also, other ethylene copolymers with polar comonomers could be produced to produce ethylene-vinyl acetate (EVA), ethylene-methyl acrylate (EMA), ethylene-ethyl acrylate (EEA), ethylene-butyl acrylate (EBA), ethylene-acrylic acid (EAA), ethylene-methacrylic acid (EMAA), etc. as well as bio-polyethylene wax using bio-ethylene.

In 2010, Braskem, the company based in Brazil, became the first company to produce bio-PE on a commercial scale[18–20, 29] for applications in food packaging, cosmetics, personal care, automotive and toy applications. Even though in 2007, a joint venture between Dow and Crystalsev [29–31] to produce bio-LLDPE was announced, it did not go anywhere. In 2020, Dow said that their bio-PE produced from tall oil, a waste product from paper production is being used by Thong Guan to produce cling film [31]. Solvay, Nova Chemicals, and other chemical companies, which are in the bio-PE market, were supposed to contribute to the production of about 20% of the world's bio-based plastics production. At that time, Braskem was expected to cover 10% of the worldwide plastic market by the year 2020 [29], but that did not happen. Since then, Braskem has acquired many other chemical operations especially polymer production facilities, most of which are not bio-based. Today, bio-polyethylene is a small part of Braskem's polymer production.

When production of bio-ethylene began, it was not considered to be cost-competitive when compared to petrochemical-derived ethylene [29]. As bio-ethanol production has ramped up, the gap has narrowed somewhat, but even today, the price of ethanol is about 35% higher than that of crude oil. The price of 1 kg of bio-PE is currently about 30% higher than petrochemical PE.

In May 2021, Neste, Toyota Tsusho Corp., and Mitsui Chemicals, Inc. announced a joint effort to increase the industrial-scale production of renewable chemicals and plastics, including polyethylene, from 100% bio-based hydrocarbons [32, 33]. Some reports indicate that bio-PE is also being produced or is likely to be produced by SABIC, Solvay, Neste, and Nova Chemicals.

The main selling point and advantage is that bio-ethylene is a drop in with no changes needed downstream for producing the same articles, films, bottles or parts using the same equipment and the same processing conditions compared to those for conventional petro-based polyethylene. However, bio-PE has two major challenges to larger adoption: higher cost and insufficient availability [25].

The global bio-PE market was estimated to be around USD 1.5 billion in 2021 and is expected to exceed USD 2.2 billion by 2029, generating compound annual growth rate (CAGR) values of 5.23% during the forecasted period in a market report by Data Bridge Market Research [34]. In reality, it is rare that either production or sales of such commodity resins grow in a linear fashion.

19.3.3 Bio-based polypropylene

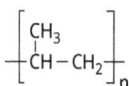

$$\left[\begin{array}{c} CH_3 \\ | \\ CH-CH_2 \end{array} \right]_n$$

Polypropylene

Polypropylene (PP) is the second largest plastic produced globally after polyethylene. Yet it is the most important organic building block for polyolefin production, as it has much wider set of critical applications in a multitude of end applications. Bio-PP is produced from bio-propylene, using conventional polymerization methods that are used for non-bio-based polypropylene production. Once the bio-propylene monomer is produced, polymerization technologies to produce high molecular weight polypropylene with identical properties on a commercial scale are exactly the same as those for petro-based PP. At this stage, it needs to be emphasized that green PE and PP may be bio-based and identical to their fossil fuel counterparts (assuming the same polymerization catalysts and process are used). However, they are not biodegradable and, of course, not compostable.

Braskem was the first company that produced bio-PP on a pilot-plant scale [29]. It made sense for Braskem was to leverage their R&D and capital investment in the bio-PE plant to the production of bio-PP. Yet, developing the process technology to produce bio-PP that can compete in the marketplace on cost parity with fossil fuel-based propylene has been challenging for potential bio-PP producers. So, in spite of the knowhow that Braskem had, commercial production of bio-PP did not take place.

Then again, bio-PP derived from bio-based naphtha has been on the commercial market for several years in a process led by SABIC and Dow at Terneuzen, the Netherlands, using Neste's ability to supply bio-based naphtha as feedstock. It is understood that they use the biomass balance approach to market their bio-derived polypropylene [30].

In 2019, Neste and Lyondell-Basell announced their first commercial production of both bio-PP and bio-LDPE. This joint project uses Neste's renewable hydrocarbons derived from sustainable bio-based raw materials like waste and residue oils. The project successfully produced several thousand tons of bio-based plastics, which are marketed under Circulen™ and Circulen™ Plus, the new family of Lyondell-Basell circular economy product brands, and are approved for the production of food packaging [35]. This achievement used the combination of Neste's unique renewable feedstock and Lyondell-Basell's technical capabilities and their cracker flexibility at their Wesseling, Germany site, which was converted directly into bio-based polyethylene and bio-based polypropylene. According to the announcement, an independent third party tested the polymer products using carbon tracers and confirmed that these products contained over 30% renewable content. Lyondell-Basell sold some of the renewable products produced in the initial trials to multiple customers like Cofresco to create sustainable food packaging materials and to brands like Toppits and Albal, which are Europe's leading supplier of household film [35].

19.3.4 Polycaprolactone (PCL)

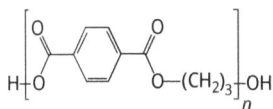

Polycaprolactone (PCL) is a synthetic biodegradable polyester that is partially crystalline but has a low melting point of about 60 °C and a glass transition temperature of around − 60 °C. It is an aliphatic polyester with a high tolerance to liquids, solvents, and gasoline. This polymer is prepared by ring-opening polymerization of ε-caprolactone, using a catalyst such as stannous octanoate [36]. The ε-caprolactone monomer could be produced from bio-based or from bio-petro sources.

In general, PCL is biocompatible, nontoxic, and cost-effective. Its biodegradability is due to its much lower glass-transition temperature (Tg = − 60 °C) than other biodegradable polymers, despite its high degree of crystallinity, typically 50%. However, unlike other biopolymers, many different grades of PCL can be produced by adjusting the polymer's molecular weight, cross-linking, and degree of crystallization, and by preparing PCL blends to vary mechanical, barrier, thermal, chemical resistance, and bio-degradability rates [25, 36]. This flexibility makes PCL more adaptable for a multiple range of end applications. PCL's exceptional mechanical, chemical, and outstanding bioresorbable properties have led to its complete commercial production for biomedical applications [21, 23, 37].

The biodegradation rate of PCL is very fast when exposed to air, but it is slower when not exposed to air. Biodegradation of PCL occurs by a specific thermotolerant microorganism called Aspergillus sp. at 50 °C [37, 38]. PCL is degraded by the hydrolysis reaction of its ester linkages in the human body and hence, has been accepted for use as a very important implantable biomaterial, specifically for use in the preparation of long-term implantable devices.

Ingevity offers PCL derivatives for soft thermoplastic polyurethane (PU) elastomers, adhesives, and coatings under the trade name CAPA. Some of the other commercial manufacturers and trade names are BASF's Capromer™, eSun, Daicel, and Perstorp's CAPA™ (previously Solvay). Dow's Tone™ has apparently been discontinued.

Some grades of PCL have elastic properties. PCL is used in various forms such as film, fibers, and micro-particle. In Sweden, bags were produced from PCL and PCL-starch bags, but they degraded very quickly, even before reaching customers [39].

Overall, PCL is biocompatible, nontoxic, and cost-effective. Consequently, PCL has been used in a wide variety of applications in the medical field, such as preparation and formulation for pharmaceutical purpose, nanoparticles coating, drug conjugates, as well as craniofacial bone repair, bone fixation, drug release, and even for *in vivo* applications in the biomedical industry. Another application is encapsulation of a variety of drugs within PCL beads for controlled release and targeted drug delivery. To lower the cost, PCL is often diluted with starch to obtain a good biodegradable product produced with mixed biopolymers. Products formulated with PCL and PCL-based products have exhibited promising results in controlled drug release, formation of scaffold, artificial organs, and nerve regeneration. Development of such advanced techniques encourages the progress in the field of bio-based polymer science [37, 38, 40]. Biodegradable PCL scaffolds also have been fabricated using 3D printing [41].

Despite all these advantages that PCL possesses for therapeutic applications such as biodegradability, biocompatibility, and high permeability, it also has some drawbacks like burst release, low encapsulation efficiency, and low bioactivity towards the tissue because of its hydrophobic nature. To overcome this, more research is needed and is being carried out for developing modified PCL and PCL-based materials for broadening the range of effective applications even further, in the biomedical field [37, 38, 40].

19.3.5 Bio-polytrimethylene terephthalate (Bio-PTT)

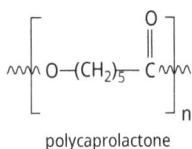

polycaprolactone

Polytrimethylene terephthalate (PTT) is a polyester that was polymerized and patented as long ago as 1941. It can be produced by condensation polymerization or by transesterification reactions from two monomers, 1,3-propanediol and either terephthalic acid or dimethyl terephthalate [42].

More economical and efficient methods to produce 1,3-propanediol in the 1980s by Degussa, *via* acrolein, and by Shell using the hydro formylation of ethylene oxide have greatly increased PTT's value as a commercial polymer. DuPont has since also commercialized the production of this polymer *via* a different bio-based method, using 1,3-propanediol generated by fermentation to produce bio-PTT with the trade name Sorona™. Similar to bio-PE and bio-PP, once bio-PTT is produced, it is indistinguishable from the same polymer produced from petro-based monomer. These developments may allow PTT to effectively compete against PBT and PET – two polyesters that have been far more successful than PTT to date, in some applications and markets. On March 20, 2009, the US Federal Trade Commission (FTC) approved a subclass to this polyester. The US Federal Trade Commission had last approved an extension for residential carpets in 1959. The PTT fiber used in Mohawk's SmartStrand carpet, and branded Sorona™ by Dupont can be labeled Triexta™. Mohawk Industries and DuPont had applied jointly for FTC approval of the Triexta polyester subclass in 2006 [43]. Other companies that are also involved with PTT include Teijin Frontier, Shenghong Group, and Glory. The factors for the rising demand for growth include wider diversity of products, increasing applications in the textile industry, as well as a larger number of applications for manufacturing monofilament, films, and engineering plastics. This demand was expected to boost the growth of the bio-PTT to 6.4%, to reach an estimated volume of about 320,000 tons by 2027, in spite of higher cost of raw materials and manufacturing as well as patented process that restrained growth [44]. Newer estimates by marketwatch.com take into account the lower volumes due to COVID but higher CAGR values after that, but overall attain similar forecasts for estimated volume of production and applications in the 325,000–350,000 ton range, by 2028 [45].

19.3.6 Polyglycolic acid (PGA)

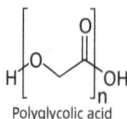

Polyglycolic acid

PGA, also known as polyglycolide, is a thermoplastic biodegradable polymer with a linear aliphatic polyester chemical structure. It can be polymerized from glycolic acid by either polycondensation or ring-opening polymerization. Other methods of production use the reaction of formaldehyde, carbon monoxide, paraformaldehyde, or trioxane or other related compounds in presence of an acidic catalyst. PGA has been known since 1954 as a tough fiber-forming polymer. However, due to its hydrolytic instability, its use was limited for several decades [46].

Such hydrolytic instability is because of the ester linkage in its main chain itself. When exposed to physiological conditions, polyglycolide is degraded by random hydrolysis, and it seems to decompose by certain enzymes, especially those with esterase activity. When this happens, the initial degradation product, glycolic acid, is nontoxic; this acid gets metabolized to oxalic acid, which limits its use. However, in the absence of this enzyme, PGA apparently does not decompose this way.

Currently, PGA, polyglycolide and its copolymers like with lactic acid called poly (lactic-co-glycolic acid, with ε-caprolactone called poly(glycolide-co-caprolactone), and with trimethylene carbonate) called poly(glycolide-co-trimethylene carbonate) are widely used as materials for the synthesis of absorbable sutures.[45–47] This traditional role of PGA and, now, its copolymers as materials for biodegradable sutures has led to their evaluation in other biomedical fields. Implantable medical devices including anastomosis rings, pins, rods, plates and screws have been produced with PGA.[46] It has also been explored for tissue engineering or controlled drug delivery.

DuPont's PGA grade called TL F 6267 is a crystalline polyester with a molecular weight of approximately 600, a specific gravity of 1.58 g/ml, and a melting point of 93–99 ° C. It is sold as a finely ground tan powder with an average particle size of 20 micron, which is insoluble in both water and most organic solvents, but is used as a dispersion in oil and gas applications, as time release agent for corrosion inhibitors, dispersants, and in decomposition inhibitors for lubricants in moving equipment in the wellbore and other channels of the formation. This product was developed by DuPont but is now available from The Chemours Company and is claimed to be useful in such applications [48].

Kureha Chemical Industries has commercialized high molecular weight polyglycolide for food packaging applications under the tradename of Kuredux™ produced at Belle, West Virginia, USA [49]. Its high degree of crystallization creates a tortuous path mechanism for low permeability and hence, its use as a barrier material. It is expected that the high molecular weight version will have to be used as an interlayer between layers of polyethylene terephthalate to provide improved barrier protection for perishable foods, including carbonated beverages and foods that lose freshness on prolonged exposure to air. This polyglycolide interlayer technology could also enable thinner plastic bottles, which still retain desirable barrier properties.

According to a recent market report from Global Industry Analysts, Inc., the global PGA market is expected to achieve sales of $453.4 million by 2027, while growing at a CAGR of just over 8.5% in the 2020–2027 period. While medical applications are projected to record an 8.5% CAGR and reach USD 185.2 million by 2027, growth in packaging applications was revised to 8.6% CAGR for the same period, from an earlier estimate [50].

19.3.7 Bio-polyamides (Bio-PAs)

A polyamide is a polymer with repeating units connected by amide bonds and is produced by the polymerization of a cyclic amide or reaction of a diamine with a dicarboxylic acid or acid derivative, like an anhydride or a dicarboxylic acid chloride. While there are several natural polyamides like silk, wool, and proteins, and nucleic acids like DNA and RNA, there are no known commercial paths from these to bio-PAs. Synthetic polyamides, especially polyamide-6 and polyamide-6,6, are the largest polyamides produced globally and are used in numerous applications but they are produced from petro sources.

This chapter is intended to provide a basic background on the origin and importance of bio-based polyamides, different synthetic routes of a wide variety of starting monomeric materials (that were covered in an earlier section) obtained from biosources, and a brief summary of properties and applications of some common aliphatic, semi-aromatic, and fully aromatic bio-PAs from both technical and commercial perspectives.

Just like bio-PE and other biopolymers, once the polymerization is carried out with a bio-sourced monomer or a non-bio-sourced monomer, the bio-PA formed from one source is indistinguishable from the other. With some bio-PA, the bio-sourced product is lower cost, while with others, it is not. Purchasers, as you would expect, purchase whichever has the lower cost, unless their customer specifies a bio-source. In some cases, only one of the comonomers is bio-sourced, while the other is petro-based.

What is common among bio-polyamides is that they all have all have an amide group in their repeat unit, but the differences amongst them is in the length and type of the rest of the repeat units. In general, the longer the aliphatic part within the repeat unit, the less crystalline and more hydrophobic the polyamide is. The lower crystallinity translates into a lower melting point and a softer polymer. If an aromatic part is in the repeat unit, the polyamide becomes more rigid and has a higher melting point.

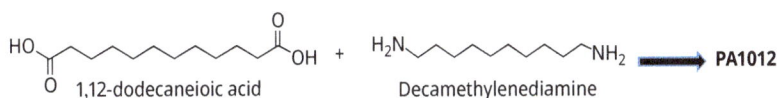

1,12-dodecaneioic acid   Decamethylenediamine   PA1012

The 11-carbon carboxylic acid discussed earlier in this chapter, undecenoic acid is used to produce polyamide-11. The first application of this approach was by Organico, in 1947, to produce polyamide-11 [51]. It is now sold by Arkema under the brand name Rilsan-11 [52]. Since then, other partly or fully bio-sourced polyamides have been commercialized for a very wide range of applications. Arkema also uses bio-based feedstocks to make other polyamides, some of which are thermoplastic elastomers (TPEs), all under the Pebax™ trade name [52].

Ricinoloeic acid is also converted to sebacic acid (octanedicarboxylic acid) to produce various polyamides that have 10 carbon atoms such as polyamide-10-10 and other polyamide copolymers like polyamide-6,10, polyamide-5,10, polyamide-10,10, and polyamide 4–10 [52–58].

Similarly, oleic acid (9-octadec-9-enoic acid) is produced from peanut oil, and then converted to azeliac acid (1,7-nonanedioic acid), and in a few more steps, eventually, to polyamide-9,9 and bio-based copolyamides, PA-6,9, PA-10,9, and PA-12,9 [47, 59]. Oleic acid can also be converted to nonanoic acid (1-octanedioic acid), which can be polymerized to produce polyamide-10,10.[52–60]

In 2014, Genomatica announced that it had entered the bio-based nylon intermediates market focusing HMDA, caprolactam, and adipic acid. This company has licensed its technology to produce various bio-monomers for producing bio-PA to several companies around the world. The list of such companies includes (Aquafil, BASF, DSM, Solvay, Evonik, Asahi Kasei, Covestro, Cargill, Helm, Daicel, and Braskem, as well as retail brands like H&M and Lululemon [27, 61].

Poly(pentamethylene oxamide) (PA-5,2), is a unique high molecular weight bio-based polyamide, prepared from dibutyl oxalate and renewable monomer of 1,5-pentanediamine, using a two-step polymerization method. The results of analytical chemical testing using Fourier transform infrared (FTIR) spectroscopy and proton nuclear magnetic resonance spectroscopy (H^+-NMR) showed that it was indeed PA-5,2. Other evaluations using methods including differential scanning calorimetry (DSC), thermogravimetric analysis (TGA), and water uptake measurements for the PA-5,2 showed that this bio-polymer has excellent properties like high temperature resistance, excellent crystallization performance, and low water absorption [62].

Cathay Industrial Biotech, a company listed on the Shanghai Stock Exchange, has been promoting its new renewable polyamides, PA-5,6, PA-5,10, PA-5,12, and PA-5,14 (as well as their copolymers) marketed under the Terryl™ brand for textile applications and as Ecopent™ for applications in downstream end markets such as automotive, electronics and electricals, cable ties, film, and other engineering materials. These bio-PA are based on Cathay Industrial Biotech's 100% bio-based 1,5 pentamethylenediamine monomer, C-BIO N5, produced under a proprietary technology process, using sugarcane for feedstock [63]. The company claims the 5-carbon diamine has the potential to substitute HMDA, a key raw material in the production of traditional petrochemical-based nylons, but many of the properties seem to be not the same as those of the corresponding 6-carbon-containing polyamides. The company is also working on PA-5,11 (which reportedly contains 31% renewable-based monomers) and PA-5,13 (28–100% renewable). Cathay Industrial Biotech has been producing other long chain α,Ω-dicarboxylic acids for polyamides, such as series of products from sebacic acid (DC 10) to octadecanedioic acid (DC 18) and including undecanedioic acid (DC 11), dodecanedioic acid (DC 12), brassylic acid (DC 13), tetradecanedioic acid (DC 14), and hexadecanedioic acid (DC 16). These longer chain acids can be used to produce more hydrophilic, softer, and lower melting bio-polyamides like PA-1012.

11-aminoundecanoic acid　　　　　　　　　　　　　PA 11

The company's dibasic acids are produced from paraffin wax using fermentation methods followed by purification, compared to multistep chemical processes used for petrochemical-based materials. The paraffin-based products are obviously not biobased, but the ones from sugar and other cellulosics are. Since paraffin wax comes from crude oil, which is a natural resource, the company considers it to be a natural product. But it is important to remember that just because a product is natural, it does not mean that it is ecofriendly or sustainable. It all boils down to how the product is collected, produced, and disposed of.

Polyamide-6,10 (PA- 6,10) is a semi-crystalline polyamide that is produced by the polymerization of HMDA with a dibasic acid, that is, sebacic acid, in this case. The melting point of PA-6,10 is 223 °C. Tests have shown that nylon-6,10 has lower moisture absorption than nylon-6 or nylon-6,12 and is stronger than nylon-11, nylon-12, and nylon-6,12. Nylon- 6,10 has excellent room temperature toughness even at low temperatures, much better than nylon-6 and nylon-6,6. Nylon-6,10 also has good resistance to most solvents and to dilute mineral acids. It also resists the environmental stress cracking action of salts like zinc chloride. It is believed that a patent assigned to DuPont that expires this year[64] covers their process for making partially aromatic polyamides in which an aromatic dicarboxylic acid component, at least 20–100% by weight of the dicarboxylic acid in the acid component is in the form of an alkylated ester and a diamine component, comprising a diamine having from 6–12 carbon atoms are admixed, in the presence of water and with heating, to form polyamide having 1–100% of N-alkylated amide and amine groups on a molar basis.

Table 19.1 on the next page 376 shows the renewable content of the newer bio-PAs and their performance comparison with those of the two largest volume petrochemical-based PA-6,6 and PA-6.

Polyamide 6,10 is particularly useful in the manufacture of products intended for use at elevated temperatures, or in applications where retention of flex modulus and tensile strength at elevated temperatures is required along with retention of good impact properties, at low temperature. Such requirements are targeted especially, therefore, for "under-the-hood" parts in the engine compartment of automobiles and for various electronic appliances. These polyamides can also be formed into films and fibers for other applications.

Table 19.1: Properties and percentage bio-Content of bio-PAs.

Polyamide	Specific gravity (g/ml)	Tensile modulus (GPa)	Tensile strength @ yield (MPa)	Moisture absorption (%)	Max. service temperature (°C)	Thermal coeff. of expansion (µm/m °C)	Coefficient of friction	% bio Content*
PA-6	1.14	2.59	71.0	4.60	123	82	0.277	100%
PA-6/30%GF	1.37	7.89	150.0	2.99	155	27	0.364	70%
PA-6,6	1.17	3.71	72.3	3.72	136	83	0.352	100%
PA-6,6/30%GF	1.38	8.46	155.0	2.02	165	32	0.310	70%
PA-6,10	1.26	6.32	114.0	1.39	132	68	0.400	100%
PA-6,10/30%GF	1.31	8.60	160.0	1.20	135	26	N/A	70%
PA 4.6	1.21	2.36	136.0	8.62	163	85	0.388	100%
PA-4,6/30%GF	1.27	9.00	120.0	4.87	163	30	0.300	70%
PA-11	1.04	1.17	35.2	1.24	102	98	0.350	100%
PA-12	1.36	4.78	42.8	1.12	120	239	0.312	100%
PA-5,6**	1.13	2.95	70.0	N/A	N/A	N/A	N/A	47%
PA-5,10	1.16	2.34	26.0	N/A	N/A	N/A	N/A	100%
PA-9 T	1.17	2.56	80.0	0.25	N/A	66	N/A	100%
PA-10,10	1.05	0.50	38.0	1.50	N/A	170	N/A	100%
PA-6,12	1.05	1.80	50.0	1.00	N/A	270	N/A	100%

*Bio Content (assumes all monomers used are bio-based)

**HMDA was not bio-based

However, nylon-6,10 and most of these other smaller-volume polyamides are much more expensive than both bylon-6 and nylon-6,6. Nylon 6–10 compounds are used in a wide range of applications including zip fasteners, electrical insulators, precision parts, and filaments for brushes. DuPont™ markets PA-6,6 under the Zytel™ and longer chain PA-6,10, PA-10,10, and PA-6,12 under the Zytel™ LC brand names [65]. They also market a specialty polyamide produced from 1,4-Benzenedicarboxylic acid, with 1,6-hexanediamine and 2-methyl-1,5-pentanediamine under the Zytel™ HTN [66]. On February 18, 2022, DuPont announced that it has entered into a definitive agreement with Celanese Corporation to divest a majority of the Mobility & Materials segment including the Engineering Polymers business line and select product lines within the Performance Resins and Advanced Solutions business lines for $11.0 billion in cash. Zytel falls within the sale, but details of whether their other products covered in this chapter are also included in the sale is not known, at this time [67].

A Japanese patent[68] describes polyamide having excellent heat resistance, water resistance, and impact resistance. With a glass transition temperature at 125°C, it can withstand sufficient heat even in the automobile engine compartment, which is called 'under-the-hood' in some applications such as electrical components (jacks, switches, terminals) and engine component housings and covers. This polyamide is produced from an aromatic dicarboxylic acid like a terephthalic acid and an aliphatic diamine. This patent is believed to be the one that discloses Kuraray's new semi aromatic Genestar™ polyamide-9 T. The 9-carbon diamine could be from a bio-based source. With a glass transition temperature at 125°C, it can withstand sufficient heat even in the automobile engine compartment, which is called 'under-the-hood' in some applications such as electrical components (jacks, switches, terminals) and engine component housings and covers. Polyamide-9 T has a material with low water absorption properties, even among other polyamides. This gives it excellent dimensional stability and stable mechanical properties over a wide range of temperatures. This material exhibits low outgassing during injection molding. It provides excellent formability with very few burrs created even for their high flow grade, compared to other long chain polyamides.

The same company was also granted an earlier patent [69] that has an alicyclic dicarboxylic acid like a 1,4-hexane dicarboxylic acid reacting with an aliphatic diamine to produce alicyclic-aliphatic polyamides, which are more flexible than those with the aromatic-aliphatic polyamides from the patent discussed in the previous paragraph.

In most cases, these bio-PAs like other polymers (plastics and rubber) are rarely used without the addition of additives and/or reinforcements, in order to meet various customer requirements all the way downstream in the value chain. For meeting specific applications, it may be necessary to reinforce these bio-PAs with glass fibers, glass beads, and carbon fibers to improve their mechanical and thermal performance. For applications like bearing materials where low friction and improved wear resistance are important, they can be compounded with additives like PTFE and molybdenum disulfide. Such compounds and reinforced composites of bio-plastics will be covered in later chapters in this book.

Rilsan™ Clear G850 Rnew and Rilsan™ Clear G120 Rnew are partially bio-based materials with excellent chemical and fatigue resistance, which will provide advantages in the optics, medical, and consumer electronics markets. The company has also added PA-10,10, PA-10,12, and PA-6,10 to its portfolio. These resins offer optimized properties between those of short- and long-chain polyamides, with the added benefit of having significant bio-sourced content, providing solutions for automotive, monofilament, and industrial applications. Pebax™ TPEs, which have been well known in the sports and medical industries for years, are now breaking ground in breathable, monolithic applications. New grades of Pebax TPEs are now used in the house wrap and breathable textile applications by producing tough and tear-resistant, waterproof barriers, while remaining breathable, plasticizer-free, and monolithic (nonporous) [70].

Polyamide 10/10 (PA-10,10) is the product of 1,10-decamethylene diamine and 1,10-decanedioic acid (sebacic acid). Both monomers can be extracted from castor oil beans, which can make it 100% bio-based. PA-10,10 has properties that are between those of higher performance polyamides such as PA-12 and lower cost PA-6 and PA-6,6. PA 10/10 absorbs less water than PA-6 and PA-6,6, and hence, its properties are less affected by humidity. PA-10,10 has excellent mechanical properties too, including high tensile strength, flexibility, toughness and good resilience, and abrasion resistance. It offers higher impact strength, and improved chemical resistance to greases, oils, fuels, hydraulic fluids, and salts over PA-6 and PA-6,6 with a higher melting point, and a much higher elongation at break.

This section ends with recently published data on the growth of the global market of bio-based polyamides to emphasize the economic importance of this family of biopolymers and bioplastics [71]. In recent times, production volume and applications of these bio-based polyamides have been rapidly growing. Besides Arkema, other producers of such polyamides include Radici, Evonik, DuPont, DSM, EMS-Grivory, DOMO Group, Cathay Industrial Biotech, BASF SE, Dupont, and Lanxess AG. Other players include AdvanSix, Honeywell International Inc., SK Group, and Asahi Kasei, who also produce some grades of these specialty polyamides. While these polyamides are produced from fully or partially bio-based sources, they are generally not biodegradable. The total global market for polyamides was around USD 30 billion growing at a CAGR of 4.5% to USD 43 billion by 2027. Within that the global bio-PA market was valued at about USD 130 million in 2021 and will grow with a CAGR of 10.58% from 2021 to 2027, based on research in a newly published report [72].

On the one hand, most governments globally are looking for ways to reduce their CO_2 emissions, and dependence on oil-based products and bioplastics is increasingly being viewed as one option to accomplish these goals. On the other, support from companies like Coca-Cola, Ford, Mercedes, Suzuki, Pepsi, H.J. Heinz, AT&T, and Toyota as well-known global conglomerates is providing market pull to the bio-sourced plastics industry. Prominent big brands that have introduced bioplastic packaging include

Danone™, Actimel, Activia, Volvic), Coca-Cola's PlantBottle), and Ecover™ cleaning products. Carrefour, Sainsbury, Billa, Spar, and Hofer which are some of the European supermarket chains offer different packaging products and/or shopping bags made of bioplastics. In the leisure/sport sector, Puma, Nike, and Adidas are among companies that use bioplastics compounds in their products. Ford, Toyota, Honda, Hyundai, Kia, Suzuki, and Mercedes are among the automotive manufacturers that have introduced various bioplastic components in several car models. Fujitsu is a well-known brand in the consumer electronics market that uses bioplastics in some of its products [73]. Applications in the packaging market are covered in more detail in Dr. Kulkarni's chapter 21.

19.3.8 Bio-polycarbonates (Bio-PCs)

Polycarbonates (PCs) are a group of thermoplastic polymers containing carbonate groups in their chemical structures. PCs used in engineering applications are strong, tough materials, and many grades are optically transparent. They are easy to work with and can be molded and thermoformed. Consequently, polycarbonates find many applications. Products made from polycarbonate can contain the precursor monomer, bisphenol A (BPA), which has been banned in most countries for applications where it is in contact with hot water or milk, due to endocrine disruptions and metabolic complications. Bio- PCs are eco-friendly bioplastics that can replace conventional polycarbonate, BPA. Until now, Mitsubishi Chemical Corp. of Japan is the only one that has successfully commercialized bio-PC sold as Durabio™, which is partially bio-based and a truly durable engineering plastic. It is made from isosorbide, which is derived from sorbitol, a bio-feedstock [74].

A more recent development is from Korea Research Institute of Chemical Technology (KRICT), which has developed another semi-bio-carbonate isosorbide with nanocellulose as a bio-derived reinforcing agent, with much higher tensile strength [74], using two patent applications from Samyung Corporation [75].

In addition, many other companies are evaluating the use of waste cooking oil as a source of producing feedstocks, especially for producing bio-based polycarbonates. This is a relatively new development and in the next two to five years, we should see some major announcements from at least four companies, because replacement of BPA has been a significant driver.

19.4 Non-bio-based polymers, but biodegradable and/or compostable

19.4.1 Introduction

There are a few polymers that are produced from fossil fuel sources but happen to dissolve in water and are biodegradable and/or compostable. Typically, these are not followed by most researchers who work with bio-based materials, because of their source is not bio-based. However, if the goal is finding suitable biodegradable and compostable products, they are, at least, worth taking a look at. A list of these materials includes polyvinyl alcohol (PVOH) and some of its copolymers, DBS, and PBAT. BDS and PBAT are covered by Dr. Ghosalkar in his chapter, because recently, these polymers have been prepared from monomers derived from bio-sources. Hence, these two polymers will not be covered in this chapter.

19.4.2 Polyvinyl alcohol (PVOH)

Polyvinyl alcohol

Polyvinyl alcohol, also known as PVOH or PVAL, is a synthetic polymer that is soluble in water. Sometimes it is mistakenly called PVA, which can be confusing, since it is also used for polyvinyl acetate from which it is produced. Unlike most vinyl polymers, PVOH is not produced by polymerization of the corresponding monomer, because vinyl alcohol monomer is thermodynamically unstable and tautomerizes to acetaldehyde. Instead, PVOH is prepared by hydrolysis of polyvinyl acetate, which is produced by polymerizing vinyl acetate.

PVOH solution has excellent film-forming, emulsifying, and adhesive properties. It has no odor and is not toxic, and is resistant to grease, oils, and solvents. Its film is ductile but strong, yet flexible. The thermoplastic film typically produced using a small amount of mixture of glycerin and water as the plasticizer is transparent and has excellent barrier properties to oxygen and flavor. However, PVOH film loses part of its barrier as a function of higher humidity because water acts as a plasticizer, which limits its use in many applications in packaging. When opaque films are produced, rather than using minerals like in polyolefin films, starch-polyvinyl alcohol blends are used.

PVOH is mostly used to produce water-soluble film in applications like detergent pods and in the field of biomedical engineering, such as artificial cartilage, drug delivery systems, microorganism enwrapping, cell microcapsulation, antithrombin materials, and biomedical sponges [76]. Over time, the PVOH dissolves as the wound heals.

In Japan and N. Korea, it has been used to produce Vinylon™ fiber for low-cost clothing, which is stiff and difficult to dye but much cheaper than nylon, cotton, or polyester. This polymer is not biodegradable due to chemical modification. It is also used to produce polyvinyl butyral (PVB). PVB has amazing adhesive properties and is a water-resistant plastic film. So, one of the largest applications for this PVB film is for laminating the inner layer between two glass or polycarbonate sheets used as safety windshield glass for vehicles.

Companies producing PVOH include Sekisui, Nippon-Gohsei, Kuraray, Sinopec, and Wacker Chemie. The global polyvinyl alcohol (PVA) market is estimated to be around USD 1.0 billion this year and is expected to be about USD 1.2 billion by 2025 [77].

19.5 Chemically modified natural biopolymers

19.5.1 Cellulose derivatives

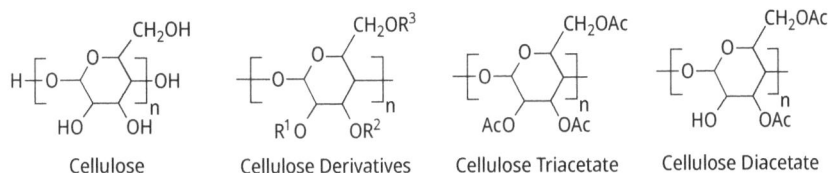

Cellulose | Cellulose Derivatives | Cellulose Triacetate | Cellulose Diacetate

Cellulose is the largest organic polymer found in nature. It has been covered in the chapter by Dr. Ghosalkar. Cellulose derivatives like cellulose acetate, cellulose butyrate, and ethyl cellulose have the general structure shown here. Amongst these three derivatives, cellulose diacetate has two acetate groups at the R2 and R3 position, while R1 has a hydrogen, while cellulose triacetate has three acetate groups replacing all three Rs, as shown here.

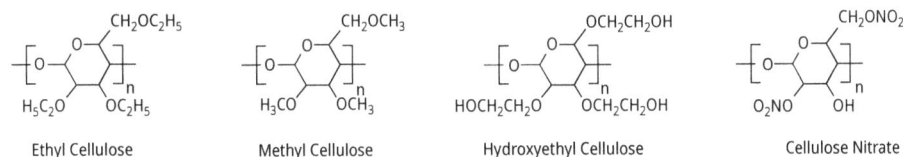

Ethyl Cellulose | Methyl Cellulose | Hydroxyethyl Cellulose | Cellulose Nitrate

The largest application of cellulose is in producing paper and paperboard. Compared to this application, other relatively smaller quantities of cellulose include converting it to semisynthetic cellulose derivatives, such as cellophane, rayon, cellulose acetate, and cellulose ethers. Cellulose acetate, which is widely used for industrial applications, is the largest cellulose derivative from a commercial perspective. When converted from cellulose, it can be either cellulose diacetate and cellulose triacetate or a mixture. This cellulose derivate has many important uses, which include fibers and threads for quality fabrics used in textiles, plastic films such as optical film for LCD technology and antifog goggles, consumer products such as cellulose-based filters, and labels.

Of the other derivatives, methyl cellulose (MC) is the most commercially important cellulose ether. It is water soluble and gels when exposed to heat. Unlike other water soluble polymers, films made from methyl cellulose have a unique characteristic in that they usually retain their strength and do not become tacky when exposed to humidity. Polymer films of both methyl and ethyl cellulose have excellent strength (60–70 MPa) and low elongation (5–15%) at room temperature, but their strength decreases rapidly with increasing temperature. These nonionic polymers also have excellent resistance to UV, oil, and solvent. These useful properties of this biopolymer and its ability to form gels easily have enabled numerous applications such as water-soluble films used for packaging products that dissolve in water like medical capsules, tooth pastes, detergent powders, bubble bath capsules and liquid products, rat poison and bread dough, as well as a thickener in food products for these cellulose ethers, especially for the methyl derivative [78].

Cellulose nitrate or nitrocellulose (called cellophane or celluloid) has been in use for over 160 years. It was really the first transparent plastic film invented. Due to its good tensile strength and insolubility in water and many other solvents, it gained acceptance in general packaging use and for celluloid film, until it was eventually replaced by cellulose acetate, polyolefin, polyester, and PVC films. Cellulose derivative manufacturers include AkzoNobel, Daicel, Eastman, Mitsubishi Rayon, Shin Etsu, Solvay, and Tembec.

19.5.2 Vulcanized natural rubber

Another natural polymer is natural rubber, which is obtained from rubber trees as latex. The raw latex is not easy to use and does not have desirable properties. Chemically, it is cis-1,4-polyisoprene. Natural rubber is used extensively in many applications and products, mostly after compounding with other additives and involving chemical modification, and very rarely by itself. In most of its useful forms, it has a large stretch ratio and high resilience and is also waterproof. The only reason this section exists is to illustrate the wide variety of bio-based applications.

To make it more useful, rubber is vulcanized or "cured" using sulfur or other cross-linkers, along with various additives. Vulcanization is the process of formation of cross-links between polymer chains for hardening rubber material [79]. Covalent cross-links provide a three-dimensional elastic network with "junctions." The degree of cross-linking can be controlled by modifying the formulation and the process conditions of the vulcanization process. The higher the degree of cure, measured by its cross-link density, the less elastic and more durable, it is.

There are numerous applications where the flexibility of rubber is advantageous. These include rubber bands, tubing, hoses, tires, and rollers for a wide range of devices ranging from domestic and office supplies to large industrial machinery. Its elasticity and resistance to impact even at low temperature makes it invaluable for various kinds of shock absorbers and for specialized machinery mountings vibration damping. Once natural rubber is fully cured, it becomes a thermoset and it cannot be melted nor can its shape be changed. Its relative good gas impermeability makes it useful in the manufacture of articles such as air hoses, balloons, balls, and cushions. The low water absorption and resistance of cured rubber to water and to most fluid chemicals has led to its use in rainwear, diving gear, and chemical and medicinal tubing and as a lining for storage tanks, processing equipment, and railroad tank cars. Rubber's electrical resistance, poor heat conduction and low water absorption are reasons why soft rubber goods are used for thermal and electrical insulation and for protective gloves, shoes, and blankets. Hard rubber is used for articles such as telephone housings and parts of radio sets, meters, and other electrical instruments, although in a lot of this type of applications, today, rubber is being replaced by TPEs. Another unique application not well known to the general public is its use for power-transmission belting, highly flexible couplings, and for water-lubricated bearings in deep-well pumps, due to the coefficient of friction of rubber, which is high on dry surfaces and low on wet surfaces [80]. Due to shortages of natural rubber, synthetic rubber was developed, primarily based on styrene-butadiene copolymers. Several other elastomeric polymers including TPEs have been also developed since then. The largest application for rubber has been its use in tires. Automotive tires are mostly produced using synthetic rubber, whereas tires for trucks need to have a significant ratio of natural rubber to synthetic rubber for their far superior tear and fatigue resistance. As with any thermoset, vulcanized natural rubber is not biodegradable. It can only be degradable by controlled pyrolysis.

19.6 Limitations of biopolymers and challenges to growth

The production and use of bioplastics instead of nonbiodegradable and oil-based plastics reduce emissions of carbon dioxide based on using plants that absorb it during their growth stage. The technology of bioplastics also provides products from renewable and/or biodegradable sources, and can offer a promising alternative for the destination of solid biomass residues for production of these materials. Hence, use of these products can be considered as one of the mitigating methods to minimize global warming [81].

However, bio-based products have several limitations that become challenges to growth and fast adoption on a really large scale, except in a few cases where they have found niches, many of which are growing at high growth rates or have grown already. Most biopolymers used have practical shortcomings that include high moisture permeability and consequent poor dimensional stability, poor oxygen and flavor permeability, fragility, meagre thermal stability resistance, poor mechanical properties especially when wet, vulnerability to biological, ultraviolet, and thermal degradation, which are critical for durable applications, and very narrow temperature window for plastics processing. For some applications, biopolymers are inherently too absorbent.

Other commercial issues that also need to be addressed and that prevent widespread adoption include:

– High costs of the biopolymer in comparison to synthetic plastics derived from fossil sources.
– Large-scale production of bioplastics for different applications is limited by limited rate of production.
– Cost of production of most bioplastics on average is at least 25% higher than conventional plastics that they are trying to replace.
– Ability to provide consistency in quality.
– Ability to supply globally prevents global consumer companies to switch to biobased products.

To address many of the functional disadvantages listed earlier in this section, it has been necessary to address each of these, to improve the range of applicability and processability. This is carried out by either polymer modification or compounding in various additives, many of which are bio-based. The bio-based additives, whether they are bio-fibers, bio-lubricants, or plasticizers, are also used in compounds where the carrier resin polymer is non-bio-based. Such compounds have found many different applications for a wide variety of end-use markets. The bio-based compounds where reinforcements are used are called bio-composites. Both bio-based plastics compounds and composites will be discussed in two separate chapters later in this book.

19.7 Will bioplastics solve the world's plastics problem?

Jim Robbins wrote a thought-provoking article entitled "Why Bioplastics Will Not Solve the World's Plastics Problem" in 2020 [82]. In his article, he has used an example of a large consumer company like Coca Cola, which has developed PlantBottle™ a new type of recyclable plastic container, produced from a blend of 30% of a biopolymer sourced from sugarcane and other plants and 70%from traditional oil-based plastic. It is understandable that the company and its competitors like Pepsi are responding to public pressure to solve the problem of "plastic pollution" in the oceans. It is an extremely difficult challenge to come up with an organic plastic that satisfies all the cost and product requirements from production to packaging to handling to storage to the consumer, while maintaining shelf life and, after its use, becomes biodegradable quickly as part of nature again. Bioplastics, which make up part of Coke's PlantBottle, have been touted as an important solution to the world's plastic pollution problem. Yet, we know for a fact that thicker the bioplastic, slower is the biodegradation rate. Also, when it is a minor volume phase the biopolymer is not on the surface and it is difficult for the bio-organisms to degrade it. Yet, the performance requirements for such a bottle of carbonated beverages needs to meet specific permeability and strength requirements, which dictates what the minimum thickness of such a bottle must be. So, while such efforts by companies need to be applauded for a start, they are not a solution! As my friend, Dr. Ramani Narayan, Distinguished Professor in the Department of Chemical Engineering and Materials Science at the School of Packaging, Michigan State University, USA, has said: "The concept that we could use it, throw it away, and it doesn't matter where you throw it, and it's going to safely disappear, that does not exist. Nobody could engineer something like that, not even nature."

Instead, many experts believe the solution to plastic waste, mainly, is not in developing better bioplastics, but instead in overhauling the world's economy to recycle far greater quantities of plastic than are currently being reused. A recent two-year study called "Breaking the Plastic Wave" by Pew Charitable Trusts and SYSTEMIQ, found that despite the efforts of industry, governments, and NGOs, the plastics problem is getting much worse [83]. It is a fact that despite the production of biopolymers and bioplastics growing at a higher rate, it will take a long time for them to match the volume of production of petro-based plastics.

There are other such presentations including one by Dr. Geoffrey Mitchell presented at the 8th World Congress on Biopolymers & Bioplastics, in 2018 [85]. He said that an immense global demand for sustainable materials exists, especially for those which biodegrade in sea water to avoid the current problems of plastics in the oceans. He rightly observed that "before we can transform the materials supply chain it is helpful to reflect on the particular properties which have made polyolefin-based polymers so successful." We all know that just having a suitable material is not sufficient; we need manufacturing processes that successfully produce objects, as well as the appropriate designs for those materials that meet all the performance requirements of that product for each stage of the value chain. We cannot generalize that all petro-based plastics can be replaced. For certain applications, high-performing polyolefins and other petro-based plastics are essential for applications like the continued safe supply of water and electrical power, and parts for automobiles and other transportation end uses that need lightweighting for better fuel efficiency. For other applications like some packaging but not all, we certainly can deliver on the promise of a greener future. It is that we need to be more discriminating in our use of materials through appropriate design based on the application requirements for material substitution. Dr. Mitchel cautioned us against seeing bioplastics as a straight swap for existing plastics. He proposed that "embracing the whole life cycle analysis approach to product design and material choice provides a sounder basis for future developments. It is timely that we embrace the concept of the circular economy whilst we deal with the plastics in the ocean disaster." [85]

The same 154-page Pew report concludes that the only solution to this proliferating problem, is an enormous overhaul of the global plastic ecosystem, so that these materials are reused and recycled using a circular economy [83]. Bioplastics have a smaller role in this but are not the ultimate solution. If all of the report's recommendations are adopted, "plastic waste could be reduced by 80 percent over the next two decades" is the claim. The report proposes a combination of all of the following recommendations, along with many other suggestions:

a) eliminating plastic packaging wherever possible, substituted with paper or compostable material;

b) designing products for effective recycling;

c) increasing mechanical recycling;

d) scaling up collection and recycling efforts in moderate- and low-income countries, where the vast majority of ocean plastic originates; and

e) ending exports of waste plastic, which would force countries where the waste is generated to come up with solutions to the plastics problem.

This prescription of a long-term comprehensive solution to address this issue makes a lot of sense, except for point (a) above, because numerous cradle-to-grave lifecycle studies have shown, one report after another, that plastics are greener and more sustainable than paper, paperboard, metal, and glass [84]. By the way, Dr. DeArmitt's unsponsored work (meaning unbiased by any outside organization for or against plastics) shows that plastics are a tiny 0.5% of global waste, and despite fake photographs in social media that tug on the viewers' emotions, there is no scientific evidence to prove that it is toxic [84]. Meanwhile many "pro-environmental" organizations push policy proposals to ban plastics, without studying scientifically obtained life cycle data based on either ideology or preconceived conclusions with the knowledge that most journalists, politicians, and indeed, the general population will not seek out such lifecycle data.

Plastics serve many useful purposes in our daily lives today. Imagine everything globally that uses plastics were eliminated today. We would consume more energy because transportation would be heavier. Shipping costs would jump by 50%. We would need food supply to grow at least 30%, if proper plastics packaging that ensure sufficient shelf life to food and medicine would disappear. What about our phones, computers, appliances, electronics, medical devices, and everything else we use every day? Yet, it is easy for journalists, politicians, bureaucrats, and especially "fake environmentalists" to blame someone, or in this case, a set of materials, instead of putting in the work to find out what scientific data has already been generated, which of those are facts, to determine what actions actually make sense, to allocate money to take actions and follow through on implementation to ensure this complicated issue is resolved!

As many others have pointed out, there are numerous cases of wooden and metal ships and many other materials, which have lasted over 500 or even thousands of years in the ocean. Yet, no government or organization in over 210 countries on this planet have said or done much about it. There are so many hazardous chemicals and metals like lead and mercury that are over a million times more dangerous that end up in our oceans, yet there is no action or even a serious discussion about these materials. Plastics are less than 1% of materials and the other 99% of materials are ignored. Yet, LCA data shows that usually plastics are the greenest solution; greener than alternatives such as paper, glass and metal in applications that they are used, in terms of carbon dioxide generation and consumption of water and energy, and usually the lowest cost option. Data shows that replacing plastics with these alternatives would generate 5–15 times the amount of CO_2. If we banned plastics, we would increase the amount of carbon dioxide in the atmosphere with other alternatives and consume more water, energy, and other resources, increase food spoilage, and it would cost us all a lot more money with every other alternative. But it is easy to blame materials that cannot talk, protest, or vote!

19.8 Future trends and prospects for growth

An article that provided a good review on future trends for bio-based polymers was published almost a decade ago in 2013 and talked about their high growth potential [86]. A recent report from ResearchAndMarkets.com that is more relevant today contends that the global bioplastics market volume in metric tons would grow from an estimated 2.2 million metric tons in 2021 to about 3.3 million metric tons by 2026, with a CAGR of 8.3% for this 5-year forecasted period. From a regional perspective, the bioplastics market volume is expected to grow at a CAGR of 8.4% from 0.93 million metric tons in 2021 to 1.4 million metric tons in 2026 in the same period, while the Asian bioplastics market catering to a much large population is estimated to grow from 0.6 million metric tons in 2021 to 0.97 million metric tons in 2026, at a CAGR of 9.6% — a higher growth rate but much lower per capita consumption [72]. Based on this report, the future is certainly bright for bioplastics!

Another recent review article that lists 532 biopolymer startups in several promising areas where new innovative developments are being carried out in new business ventures [87]. It highlights the top nine biopolymer trends and 18 promising startups. Even if only 50% of these succeed, they will have shown the way to others and portend a lot of promise to future growth. Some of the examples from this article showing examples of these innovations are:

– Protein-based biopolymers and some bio-PAs have advantages of being biocompatible and biodegradable along with possessing superior gas barrier and mechanical properties, compared to cellulosics. Hence, they are one of the most promising classes of biomaterials. Due to their possibility of interaction with a wide range of bioactive molecules, many protein biopolymers and PCLs find applications in the medical industry for tissue engineering and tablet coating.

– Recent advances in starch-based materials have enabled films to be made antimicrobial in order to prevent microbial growth for packaging of food products. Many startups are expected to adopt this technology to meet the growing demand for starch in applications including agriculture, packaging, cosmetics, medicine, and textiles.

– Pivot Materials, an American startup, uses bamboo and rice waste fiber among other bio-materials in their products. It specializes in manufacturing natural fiber-based, lightweight, sustainable, and durable composite plastics for injection and extrusion molding to improve specific material properties. The end markets for such sustainable plastics are suitable for home goods, packaging, and transportation.
– A Chinese startup called WAVE has developed a cassava-based polymer compound in pellet form, which is then converted to bags. The natural cassava starch is water-soluble, composts naturally, and is edible by land and aquatic animals. The startup company produces a bio-based plastic compound that safely returns to nature in the form of water and carbon dioxide, which, in turn, supports sustainable agriculture. This solution is an eco-friendly alternative to regular plastic, as the starch-based bags use only sustainable natural resources. One caveat is that the bags need to stay dry, otherwise they fall apart!

– Seaweed is used as a source material by two startup companies, UK-based SoluBlue and Norway-based B'Zeos.
– Solublue's product is suitable for FDA-approved food packaging because it absorbs moisture and prevents mold growth on fresh food, thereby extending its shelf life. This alternative to single-use petro-based plastics also captures a vast amount of carbon and locks it in the soil when composted on land. If dropped in the ocean, their seaweed biopolymer-produced film breaks down within weeks and can also be safely ingested by marine life.
– B'Zeos's bio-based packaging materials using seaweed biopolymers are targeted to replace disposable plastic articles in the food service industry. This startup partners only with seaweed cultivators and harvesters that work with sustainable and ethical practices. The eco-sourced seaweed uses their green processing technology to extract valuable compounds to create a packaging solution that is circular, home-compostable, and even edible.

– An Italian startup called Earth Bi uses biomass-derived natural waste as a source to produce biodegradable and compostable polymers. They also use renewable energy for their production process and do not require pure materials. Not only is their biopolymer product suitable for traditional processing of plastic without extensive modifications through its distinct heating process, they also claim it to be recoverable and recyclable.

– Starting from mushrooms, a startup called Mushroom Material based in New Zealand has developed bio-based foam to replace Styrofoam™, which is based on petro-based polystyrene. They take the vegetative part of the mushroom called mycelium and combine it with fibrous agricultural waste to create this unique product, which is waterproof, fire-resistant, durable, and compostable, even at home for sustainable packaging applications with the same protective properties as Styrofoam.

– An American startup called Verde Bioresins produces bio-PET under the brand name PolyEarthylene™. This is a plant-based sustainable and renewable biopolymer claimed to be environmentally friendly, economically more feasible, and lighter compared to petroleum-based polymers for applications in production of durable consumer goods. While they are not competitive with PET for applications like bottles, they do compete with PLA, PHA, green PE, and other bio-based materials.

– Another US-based startup called Biorgani is offering a bio-based resin called Faunnus™ for packaging. It is designed to be used as a blend with petro-based polyolefins, by matching the pure petro-based polyolefin's weight resistance and performance for applications like vegetable wraps in the produce section of groceries and for grocery bags with a lower carbon footprint. The same company also manufactures a "zero polyethylene resin" called Solum™ Compost Resin, which goes through conventional plastic processing equipment needing less temperature for processing and hence, saving energy and reducing carbon emissions during manufacturing, compared to petrobased polyethylene resin.

These are just a few examples of the progress being made, a little at a time, by many companies around the world to try and mitigate the issue of the existing and growing pile of plastics in landfills, oceans, and elsewhere. While plastics that end up in waste are less than 1% of total global waste by weight (the rest is wood, metal, glass, minerals, etc.), they seem to have more visibility. Every day, we are adding more plastics that are used globally, so there is crying need for finding sustainable alternatives as well as for solving this issue. Bio-based and biodegradable materials have shown strong growth across industries over the past years, mainly in packaging, driven by consumer market trends towards greener packaging and waste reduction. As several examples in this chapter have shown, there are numerous applications in the medical industry for drug delivery and tissue engineering and packaging, as well as in other industries. On our way to a more sustainable future, innovations in this field of bioplastics to improve biomaterials properties further, reduce their cost, and make production processes environmentally friendly are sorely needed. Although bioplastics represent just about 1% of the total amount of plastics today, as more innovative techniques are developed to overcome the various limitations listed earlier in this chapter, the field of bioplastics will grow faster, especially as the larger volumes make them more cost competitive; and as fossil fuel prices increase further and more bioplastics are produced from waste materials rather than from food crops or from sugarcane, which raises consumer prices for the growing global population.

References

[1]   Thielen, M,"Bioplastics – Basics. Applications. Markets", Polymedia Publisher GmbH, Mönchengladbach. Germany, 3rd Revised English Edition 2020, ISBN 978-3-9814981-4-1.

[2]   Young, RJ, "Introduction to Polymers", 1987, Chapman & Hall, London, United Kingdom, ISBN 0-412-22170–5.

[3]   "Natural Polymers and Biopolymers", at https://iscamapweb.chem.polimi.it/citterio/it/education/course/topics/

[4]   "Polynucleotides" at the US National Library of Medicine Medical Subject Headings (MeSH).

[5]   Hamley, I W, "Introduction to Peptide Science". Wiley, Hoboken, NJ, USA, 2021; ISBN 9781119698173.

[6]   "Polysaccharide – Definition, Examples, Function and Structure", at https://biologydictionary.net/polysaccharide.

[7]   Barrett, A, "The History of Bioplastics", Bioplastics News, at https://bioplasticsnews.com/2018/07/05/history-of-bioplastics/

[8]   "EUBP Facts and Figures – European Bioplastics Documents" at https://www.european-bioplastics.org/; www.bio-based.eu/markets/

[9]   https://www.marketsandmarkets.com/PressReleases/biopolymers-bioplastics.asp/

[10] Barrett, A, Bioplastics News, https://bioplasticsnews.com/2019/04/13/what-is-the-difference-between-biodegradable-compostable-and-oxo-degradable/

[11] (a) Isabelle Vroman and Lan Tighzert, "Biodegradable Polymers", at https://www.ncbi.nlm.nih.gov/pmc/articles/PMC5445709/; (b) European Bioplastics – Report Bioplastics. [(accessed on 25 May 2020)]; at https://docs.europeanbioplastics.org/publications/market_data/Report_Bioplastics_Market_Data_2019.pdf/

[12] Fujimaki T., "Processability and properties of aliphatic polyesters", "Bionolle", synthesized by polycondensation reaction. Polym. Degrad. Stab. 1998;59:209–214.

[13] "BS EN 13432:2000 Packaging" at https://www.en-standard.eu/bs-en-13432-2000-packaging.-requirements-for-packaging-recoverable-through-composting-and-biodegradation.-test-scheme-and-evaluation-criteria-for-the-final-acceptance-of-packaging/

[14] Ranjan, S, "Biopolymers & Their Impact", 2017, at https://polymeracademy.com/biopolymers-their-impact/

[15] D.S.Ogunniyi, "Castor oil: A vital industrial raw material", at https://www.sciencedirect.com/science/article/abs/pii/S0960852405002026/

[16] Vasishtha AK, et al., "Sebacic acid and 2-octanol from castor-oil", J. Am. Oil Chem. Soc. 67, 333–337 (1990), at https://doi.org/10.1007/BF02539685

[17] "Global Green and Bio Polyols Market Analysis 2022" by Syngene Research LLP, Bangalore, India, May, 2022, at https://www.marketresearch.com/Syngene-Research-LLP-v4190/Global-Green-Bio-Polyols–31908029/

[18] Valentina Siracusa and Ignazio Blanco, "Bio-Polyethylene (Bio-PE), Bio-Polypropylene (Bio-PP) and Bio-Poly(ethylene terephthalate) (Bio-PET): Recent Developments in Bio-Based Polymers Analogous to Petroleum-Derived Ones for Packaging and Engineering Applications", Polymers (Basel), Aug. 2020, 12(8): p. 1641; at https://www.ncbi.nlm.nih.gov/pmc/articles/PMC7465145/#B9-polymers-12-01641/

[19] Morschbacker A., "Bio-Ethanol Based Ethylene" Polym. Rev. 2009;49:79–84.; doi.org/10.1080/15583720902834791.

[20] Braskem Report. "Development of Bio-Based Olefins" (accessed on 22 August 2017); at http://www.inda.org/BIO/vision2014_659_PPT.pdf/

[21] Kahn, J, "Green PE: The Value Proposition and Success in the Marketplace and Green PP: Update on Commercialization", Bioplastics: The Reinvention of Plastics Conference, March 6, 2013, Las Vegas,

NV; at https://polymerinnovationblog.com/bio-based-polypropylene-multiple-synthetic-routes-under-investigation/

[22] Venkataraman, V, "Propylene Supply, Go Green, On Purpose Technologies for the Future", at https://www.slideshare.net›beroe_inc›propylene-supply-go/

[23] Chen GQ; Patel MK; Plastics derived from biological sources: Present and future: A technical and environmental review. Chem. Rev., 2012, 112 pp.2082–2099; doi.org/10.1021/cr200162d/

[24] Ohara, E; Kawazumi, K, United States Patent 6,936,724, "Process for producing epsilon-caprolactone", assigned to Daicel Chemical Co. issued August 30, 2005.

[25] Rani, GU; Sharma, S, "Biopolymers, Bioplastics and Biodegradability: An Introduction", in Reference Module in Materials Science and Materials Engineering, 2021, Elsevier Inc.

[26] (a) Koivistoinen OM, Kuivanen J, Barth D, Turkia H, Pitkänen JP, Penttilä M, Richard P. Glycolic acid production in the engineered yeasts Saccharomyces cerevisiae and Kluyveromyces lactis. Microb Cell Fact. 2013 Sep 23;12:82. doi.org/10.1186/1475-2859-12-82/ (b) Soucaille, P, Canadian Patent CA-2654182-A1, " Glycolic Acid Production By Fermentation from Renewable Resources", Owner: Metabolic Explorer; PCT/EP2007/055625 & WO2007/141316.

[27] https://greenchemicalsblog.com/tag/polyamides/; www.genomatica.com/; https://greenchemicals blog.com/ 2014/08/08/genomatica-enters-bio-polyamides-market/; https://www.icis.com/sub scriber/news/home/viewarticles/

[28] Murata Y; Kuribayashi T; Onishi, F; Hiura; T, "1,5-Pentamethylene diamine and Production Method Thereof", United States Patent Application 20160289165 Filed Nov. 18, 2014, published Oct. 6, 2016 Assigned to Toray Industries Inc. and Ajinomoto Co., Inc.

[29] Alvarenga, RAF; Dewulf J, Plastic vs. fuel: "Which use of the Brazilian ethanol can bring more environmental gains?" Renew. Energy. 2013;59:49–52. doi.org/10.1016/j.renene.2013.03.029. [CrossRef] [Google Scholar]

[30] de Jong, E; Stichnothe, H; Bell, G; Jørgensen H, "Bio-Based Chemicals A 2020 Update", at https://www.academia.edu/42073840/Bio_Based_Chemicals_A_2020_Update/

[31] (a) https://www.marketwatch.com/story/dow-chemical-crystalsev-to-form-brazil-joint-venture/ (b) https://www.prnewswire.com/in/news-releases/dow-and-thong-guan-introduces-sustainable-plas tic-stretch-cling-films-made-with-renewable-feedstock-in-asia-pacific-836547588.html/

[32] https://www.biofuelsdigest.com/bdigest/2021/05/23/neste-mitsui-chemicals-and-toyota-tsusho-to-start-japans-first-production-of-renewable-plastics-from-100-bio-based-hydrocarbons/

[33] McKeen, LW, "Plastic Films in Food Packaging – Materials, Technology and Applications", Ebnesajjad, S, Editor, Elsevier, 2013, Hardcover ISBN: 9781455731121, eBook ISBN: 9781455731152; McKeen, LW, "Permeability Properties of Plastics and Elastomers", (Third Edition), Elsevier, 2012.

[34] https://www.databridgemarketresearch.com/reports/global-bio-based-polyethylene-pe-market.

[35] "Neste and LyondellBasell produce bio-PP and bio-LDPE at commercial scale", https://www.bioplas ticsmagazine.com/en/news/meldungen/20190618Neste-and-LyondellBasell-produce-bio-PP-and-bio-LDPE-at-commercial-scale.php/; LyondellBasell, Neste launch new bio-based plastics (chron.com)

[36] McKeen, LW, "The Effect of Sterilization on Plastics and Elastomers" (Third Edition), 2012. Elsevier Inc.; https://doi.org/10.1016/B978-1-4557-2598-4.00012-5

[37] Dhanasekaran, NPD; Muthuvelu, KS; Arumugasamy, SK, "Recent Advancement in Biomedical Applications of Polycaprolactone and Polycaprolactone-Based Materials", in Reference Module in Materials Science and Materials Engineering, January 2022; doi.org/10.1016/B978-0-12-820352-1.00217-0.

[38] Arif, U; Haider, S; Haider, A; Khan, N; Alghyamah, AA.; Jamila, N.; Khan, MI; Almasry, WA; Kang, IK, "Biocompatible Polymers and their Potential Biomedical Applications: A Review", *Current Pharmaceutical Design*, **2019**, 25(34):3608–3619;

[39] Flieger M; Kantorová M; Prell A; Rezanka T; Votruba J, Folia Microbiol (Praha). 2003;48(1):27–44. doi. org/10.1007/BF02931273.2003; https://www.researchgate.net/publication/358654485/

[40] "Caprolactone", at https://www.sciencedirect.com/topics/chemical-engineering/polycaprolactone/

[41] Singh, MK; Singh, R;Dhami, M (2020 "Biocompatible Thermoplastics as Implants/Scaffold", Reference Module in Materials Science and Materials Engineering, 2020, Elsevier, https://www.scien cedirect.com/science/article/pii/B9780128203521000122/

[42] https://bioplasticsnews.com/polytrimethylene-terephthalate-ptt/

[43] "FTC Approves Federal Register Notice Establishing New Fiber Name and Definition" (Press release). Federal Trade Commission. 2009-03-20.] Triexta has several advantages over polyester, including better stain resistance and softness. [Simmons, Cheryl (8 August 2019). "Triexta PTT Carpet Fiber". The Spruce.; https://www.thespruce.com/triexta-ptt-carpet-fiber-2908799/

[44] https://www.databridgemarketresearch.com/reports/global-bio-based-polytrimethylene-terephthal ate-ptt-market/

[45] https://www.marketwatch.com/press-release/polytrimethylene-terephthalate-ptt-polymers-market-2022-size-2022-investigation-report-by-global-industry-share-demand-key-findings-regional-analy sis-key-players-profiles-future-prospects-and-forecasts-to-2028-2022-06-16/

[46] Gilding, DK; Reed, AM. "Biodegradable polymers for use in surgery – polyglycolic/poly (lactic acid) homo- and copolymers: 1", Polymer, 1979, 20 (12):1459–1464, doi.org/10.1016/00323861(79)90009-0.

[47] (a) Middleton, J; Tipton, A, "Synthetic biodegradable polymers as medical devices". Medical Plastics and Biomaterials Magazine, (March 1998); (b) Middleton, J; Tipton, A, "Synthetic biodegradable polymers as orthopedic devices", Biomaterials, Dec. 2000, 21(23):2335–46. (c) Pişkin E, "Biodegradable polymers as biomaterials", J Biomater Sci Polym Ed. 1995;6(9):775–95. doi.org/10.1163/156856295x00175/

[48] http://www2.dupont.com/Oil_and_Gas/en_CA/assets/downloads/DuPont_Polyglycolic_Acid_Sheet.pdf

[49] Kuredux® Polyglycolic Acid (PGA) Resin Archived 2020-12-09 at the Wayback Machine www.kureha.com, accessed 4 December 2021.

[50] Polyglycolic Acid – Global Market Trajectory & Analytics, ID: 5309880 Report April 2021, Research And Markets, Dublin, Ireland.

[51] "Polyamide 11", https://www.etigo.fr/en/blog/p-polyamide–11

[52] "Rilsan", "Pebax", Specialty Polyamides", all at https://www.arkema.com/global/en/search/?qc= search&q=polyamides

[53] Khedr, MSF, "Bio-based polyamide", Physical Sciences Reviews; https://doi.org/10.1515/psr-2020-0076/

[54] "Bio-based Polyamides – Current Trends and Overview" paper presented by Doug Weishaar, Evonik Corporation at the 10th Annual World Congress on Industrial Biotechnology, Montreal, Canada, June 16–19, 2013, https://www.bio.org/sites/default/files/legacy/bioorg/blogs/the_new_age_of_bio-based_polyamides_doug_weishaar.pdf/

[55] Humphreys, R, "Bio-Polyamides: Where Do They Come From?" https://polymerinnovationblog.com/bio-polyamides-where-do-they-come-from/

[56] Devaux, J-F; Le, G; Pees, B, "Application of Eco-profile Methodology to Polyamide 11", https://www.docsity.com/en/polyamide-11-and-its-properties-for-hose/538543.

[57] Thielen, M, "Basics of Bio-polyamides", Bioplastics Magazine, March 2010, 5, 50–53.

[58] "Bio-polyamides for Automotive Applications", https://www.tu-braunschweig.de/en/presse/wissen schaft-medien/homepage.

[59] Tao, L, "Preparation and properties of biobased polyamides based on 1,9-azelaic acid and different chain length diamines", Polymer Bulletin, 2020, 77, 1135–1156.

[60] Kuciel, S; Kuzniar, P; Liber-Knec, A, "Polyamides from renewable sources as matrices of short fiber reinforced biocomposites", Polimery, 2012, 57, 627–634.

[61] www.genomatica.com/press_releases/

[62] Mutua, FN; Cong Cheng, C; Dong, Y; Zheng, C; Zhu, B; He, Y, "Synthesis and Properties of Bio-Based Poly(pentamethylene oxamide)", Polymer Engineering and Science, 2018, Vol. 58, Issue 5; doi.org/10.1002/pen.24596

[63] https://www.cathaybiotech.com/

[64] Ng, H, "Partially aromatic polyamides and a process for making them", United States Patent 6,355,769, assigned to DuPont Canada, Inc., issued March 12, 2002.

[65] "Zytel®- A Portfolio of High-Performing Nylon Resin Products", https://www.dupont.com/products/zytel.html/

[66] "Zytel® HTN -A family of cost-effective, high-performance polyamides, https://www.dupont.com/products/zytel-htn.html/

[67] https://www.dupont.com/news/dupont-announces-agreement-to-divest-majority-mobility-and-materials-segment-to-celanese.html

[68] Tamura, K. "Polyamide", United States Patent 10,544,262, assigned to Kuraray Co. Ltd., issued Jan. 28, 2020.

[69] Takeda, H, "Polyamide Resin", United States Patent 10,465,071, assigned to Kuraray Co. Ltd., issued Nov. 5, 2019

[70] LaPree, J, "High-performance polymers in the spotlight";, Chem. Engg., July 1, 2015, 122(7):20–23, Publisher: Access Intelligence, LLC, Rockville, MD, USA

[71] Bio-Polyamide Market Research Report 2022 Professional Edition, https://galleonnews.com/uncategorized/283974/bio-polyamide-market-research-report-2022-professional-edition/

[72] "Global Bioplastics Technology Markets Report 2021–2026: Industry Landscape, History of Bioplastics, Recent Developments, Pros and Cons of Bioplastics", 2021, Research and Markets, Dublin, Ireland.

[73] "Global Markets and Technologies for Bioplastics", November 2021, BCC Research, Wellesley, Massachusetts, USA.

[74] (a) https://www.mcpp-global.com/en/mcpp-america/products/brand/durabioTM/; (b) Sasaki, H; Tanaka, T, United States Patent 8,889,790, "Polycarbonate resin composition and molded article thereof", assigned to Mitsubishi Chemical Corp., issued November 18, 2014.

[75] "South-Korea To Compete With Japan on Bio-Polycarbonate", at https://bioplasticsnews.com/2019/10/29/south-korea-japan-bio-polycarbonate/

[76] (a) Jin, SC; Chang, YJ; Kwon, YD, "Isosorbide-aromatic polycarbonate copolymer", Korean Patent Applications 10-2015-0004084; (b) Jin, SC; Chang, YJ; Rhee, HC; Kwon, YD, "[Poly(isosorbide carbonate- aromatic carbonate)]-[polycarbonate] block copolymer", Korean Patent Application 10-2015–0004085.

[77] Guo, D; Xu, K, "Application of Polyvinyl Alcohol in Biomedical Engineering"; (Article in Chinese) at www.pubmed.ncbi.nlm.nih.gov/16013269/

[78] "Polyvinyl Alcohol (PVA) Manufacturing Industry" report published by NIIR Project Consultancy Services (NPCS), New Delhi, India, www.entrepreneurindia.co.

[79] Heinze, T; El Seoud, OA; Koschella, A, "Cellulose Derivatives: Synthesis, Structure, and Properties" (Springer Series on Polymer and Composite Materials) 1st ed. 2018 Edition.; at https://polymerdatabase.com/polymer%20classes/Cellulose%20type.html

[80] "Polymers and Composites Material Consulting" at https://www.polymerexpert.biz/services/materials/

[81] "Natural Rubber", at https://en.wikipedia.org/wiki/Natural_rubber/

[82] Abe MM, Martins JR, Sanvezzo PB, et al. "Advantages and Disadvantages of Bioplastics Production from Starch and Lignocellulosic Components", Polymers (Basel). 2021;13(15):2484; doi.org/10.3390/polym13152484/

[83] Robbins, J, "Why Bioplastics Will Not Solve the World's Plastics Problem", August 31, 2020, https://e360.yale.edu/features/why-bioplastics-will-not-solve-the-worlds-plastics-problem/

[84] Reddy, S; Lau, W, "Breaking the Plastic Wave: Top Findings for Preventing Plastic Pollution", https://www.pewtrusts.org/en/research-and-analysis/articles/2020/07/23/breaking-the-plastic-wave-top-findings/

[85] De Armitt, C, "The Plastics Paradox: Facts for a Brighter Future", Phantom Plastics LLC, Terrace Park, OH, USA; ISBN 978-0-9978499-6-7.

[86] Mitchell, G; Mahendra, V; Pinheiro, J; Sousa, D; Abdulgahni, S; Dias, J; Faria, P; Gaspar, F; Mateus, A; "Challenges with Biopolymers" presented at 8th World Congress on Biopolymers & Bioplastics June 28–29, 2018, Berlin, Germany, https://www.longdom.org/proceedings/challenges-with-biopolymers-19169.html/

[87] Babu, RP; O'Connor, Kl; Seeram, R, "Current progress on bio-based polymers and their future trends", Prog Biomater. 2013 Mar 18;2(1):8. doi: 10.1186/2194-0517-2-8.

[88] "Innovation Map outlines the Top 9 Biopolymer Trends & 18 Promising Startups", at https://www.startus-insights.com/innovators-guide/biopolymer-trends-innovation/

Pramod Kumbhar*, Anand Ghosalkar, Yogesh Nevare

Chapter 20
Bio-derived polymers and plastics

20.1 Introduction

Bio-based polymers have received great attention in recent years due to their sustainability, the global movement towards circular economy, and uncertainties associated with cost and availability of fossil fuel [1]. The emerging demand for bio-based products has motivated researchers to develop novel technology solutions to produce a wide variety of biopolymers and improve their characteristics. Multiple bio-based polymers have been developed with a wide range of properties suitable for applications in different sectors, including food, pharmaceuticals, cosmetics, packaging, agriculture, and consumer goods [2]. Bio-based polymers have the potential to replace synthetic petroleum-derived polymers and can play a major role in the sector of polymer industries. For replacement of fossil-based molecules by renewable counterparts in an environment friendly manner, twelve principles of green chemistry should be followed along with sustainable chemical alteration and polymerization techniques. Use of renewable raw material along with different biocatalytic or chemocatalytic processes has been investigated to produce carbon-neutral products [3]. Similar to the twelve principles of green chemistry by Anastase and Warner [4], Slootweg and coworkers formulated the twelve principles of circular chemistry [5].

Biopolymers are generally categorized as: (i) Bio-based polymers directly extracted from biomass, such as starch, cellulose, and guar gum [2, 6]; (ii) Bio-based polymers obtained from bio-derived monomers such as poly(L-lactic acid) (PLLA), poly(butylene succinate) (PBS), Bio-poly(ethylene) (PE), Bio-poly(ethylene terephthalate) (PET); and (iii) Bio-based polymers produced directly by natural or genetically modified organisms such as poly (hydroxy alkanoates) (PHAs) [2]. The term bioplastic is sometimes used for fossil-based polymers that are biodegradable, such as poly (butylene adipate-co-terephthalate) (PBAT), poly (caprolactone) (PCL), and polyvinyl alcohol (PVA). However, use of this term is misleading because they are not produced from renewable feedstocks and do

Note: This chapter covers the following topics and complements subject matter covered in Chapter No. 19, 21, 22 and 23. To the extent possible, attempt is made to minimize duplication and repetition of topics covered in these chapters without sacrificing flow of the subject matter.

*Corresponding author: Pramod Kumbhar, Praj Matrix-R&D Center (Division of Praj Industries Ltd), GAT#1098,402,403, Urawade, Pirangut, Mulshi, Pune 412115, India, e-mail: pramodkumbhar@praj.net
Anand Ghosalkar, Yogesh Nevare, Praj Matrix-R&D Center (Division of Praj Industries Ltd), GAT#1098,402,403, Urawade, Pirangut, Mulshi, Pune 412115, India

https://doi.org/10.1515/9783110791228-020

not provide the benefits of greenhouse gas (GHG) reduction. Table 20.1 shows classification of different bio-based and fossil-derived polymers and categorizes them as per their biodegradation characteristics.

Table 20.1: Classification of bioplastics (see Chapter No. 19, 21, and 22 for acronyms used in Table and Figures).

Origin	Biodegradable	Nonbiodegradable
Bio-based	Starch and cellulose, guar gum, xanthan gum, PLA, PHA, Bio-PBS	Bio-polyamide, Bio-polycarbonate, Bio-PET, Bio-PE, Bio-PP
Fossil based	PBS, PBAT, PCL, and PVA	PET, PE, PP, PS

The word "bio-derived" denotes a polymer that is made fully or partially from renewable, biological, or organic waste biomass, and the word "biodegradable" denotes a material's ability to degrade into natural by-products such as CO_2, H_2O, and biomass by the action of microorganisms. The polymer that is bio-based and chemically analogous to their petrochemical-derived counterpart, for example, Bio-PE, Bio-PP, and bio-PET are known as "drop-in polymers" [2].

Like any chemical or biochemical, feedstocks selection plays an important role in biopolymer production on a large scale. Figure 20.1 shows the classification of feedstocks that can be used to produce different bio-based products. Extensive research has been done to expand feedstock options including first- (sugar and starch based) and second-generation feedstocks (biomass-based sugars and lignin) to produce biopolymers. Plant-based bio-oils, including waste cooking oils (triglycerides and fatty acids) and gaseous feedstocks like biogas (CH_4 and CO_2) produced from anaerobic digestion have also been demonstrated as sustainable sources for conversion into biopolymers.

Development and production of biopolymers is at different stages and currently, some of the biopolymers are successfully produced at commercial scale (Figure 20.2). Apart from the direct extraction of biopolymers from plant-based resources, the synthesis approaches mostly revolve around production of monomers from multiple bio-based feedstocks and subsequent polymerization of monomers *via* a single or multi step process. Today, almost every monomer required to produce drop-in polymers can be obtained from renewable feedstocks. Bio-based feedstocks also support direct synthesis of novel polymers like PHAs, using microbial fermentation [7, 8].

In this chapter, we provide an overview of the most commercially relevant biopolymers which can be used as bio-derived thermoplastics. Apart from chemical composition and mechanical properties, the chapter includes major feedstocks and technological approaches reported for production of biopolymers. A brief overview of EOL scenarios have been covered along with current global market and applications of bio-derived thermoplastics.

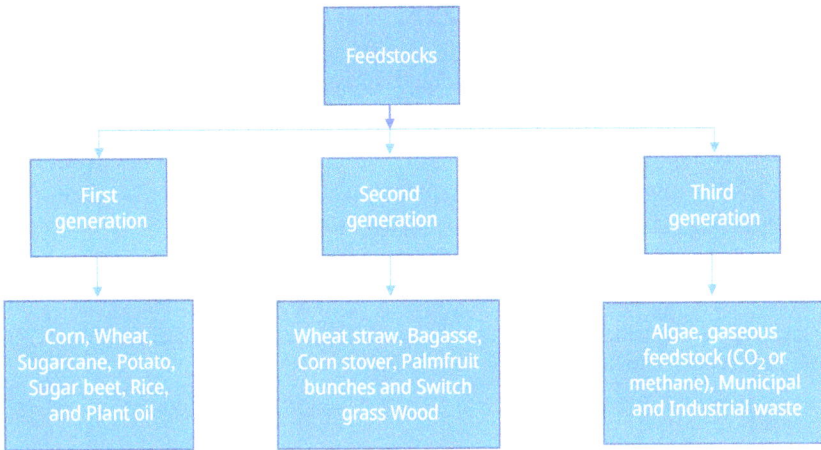

Figure 20.1: Classification of feedstocks.

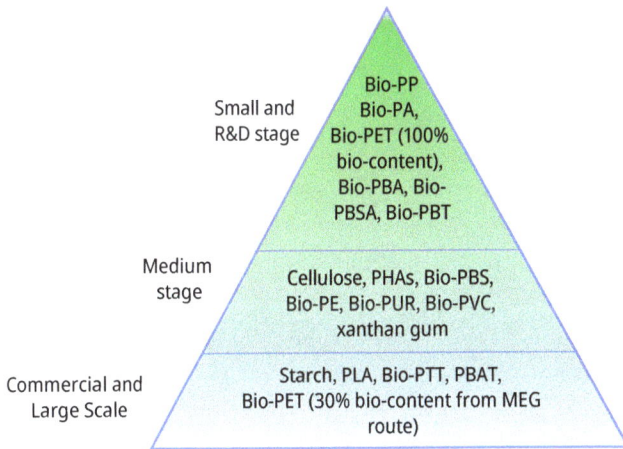

Figure 20.2: Development stages of bioplastic production.

20.1.1 Bio-derived thermoplastics – chemistry, properties, feedstock, technologies, and applications

The polymers that can be softened, reshaped, and molded into a new article by the application of heat and pressure are called thermoplastics. Due to their great versatility and distinctive properties, especially in the field of packaging (~ 40% thermoplastic), the thermoplastic market is growing continuously, since the 1950s. However, only a small amount of these materials is recycled, and the majority is landfilled. This has resulted in generation of huge amount of plastic waste in the form of macroplastics and microplastics, since they are not properly disposed at the end of their life. As a consequence, the development of biodegradable packaging has emerged as an attractive option to deal with the thermoplastic waste [9]. Several synthetic aspects such as atom-efficient methods, green catalysis and processes for the synthesis of bio-based thermoplastic, especially aliphatic polyester, have gained increased attention over past few decades [10–15]. The physical, chemical, and mechanical properties of some of the commercially synthesized bio-based polymers are shown in Table 20.2.

Table 20.2: Thermal and mechanical properties of commercial bio-based polymers.

Parameter	Starch	Cellulose	PHA	PLA	PBS	PEF	PET	PBAT
Glass transition temperature, Tg (°C)	35–85	493–518	2	55–60	−32	88	69–115	NR
Melting temperature, Tm (°C)	65–115	Not reported (NR)	160–175	125–178	114	210–230	245–265	115–125
Tensile strength (MPa)	2.1–46.8	15–90	15–40	10–60	34	90–100	70–177	21
Elongation at break (%)	1–1200	1–74	1–15	1.5–380	560	NR	70–180	670
Tensile/ Young's modulus (MPa)	1000–2900	3300–10,800	1000–2000	300–3800	NR	3600	1700–3500	126
Ref.	[16]	[16]	[8]	[16]	[17]	[18]	[16]	[19]

20.1.1.1 Starch

Starch is an important polysaccharide produced by plants and is a natural biopolymer with useful properties. Moreover, it is biodegradable, renewable, and amply available at low cost [20]. Starch can be produced from several sources including corn, wheat, potato, rice, tapioca, tam, and barley, but the major source for commercial production is maize [21]. Starch is made up of two D-glucose homopolymers: amylose (20–30%) and amylopectin (70–80%) (Figure 20.3) [9]; the composition of these components can affect the properties of starch-based films [21]. Starch constitutes a substantial portion of food waste and is used for nonsynthetic, starch-based bioplastics produced by direct processing of the starch into films [1].

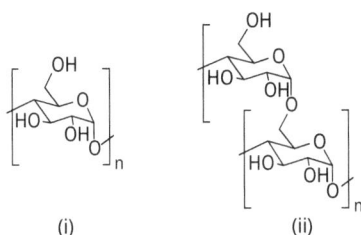

Figure 20.3: Chemical structure of starch: (i) amylose and (ii) amylopectin.

From the processing perspective, starch can be used as a substitute for polystyrene (PS) and exhibits properties like low water vapor barrier, high brittleness, and poor mechanical properties. Starch-based materials are highly brittle in nature and have high glass transition temperature (Tg) [22]. Tg is the temperature at which the material shows the transition from glassy or brittle material to flexible rubbery state. Starch shows Tg in the range of 35–85 °C and Tm ranges from 65–115 °C (Table 20.2), indicating a narrow processing window, which limits its applications [16]. Starch is gaining commercial interest due to its availability at low cost, biodegradability, and abundant availability of –OH chemistry. Owing to –OH chemistry, starch can be modified in a number of ways by chemical or melt processing in the presence of plasticizers [21, 22]. Melt processability of starch is limited by its hydrophilicity, thermal and mechanical characteristics, quick degradability, and strong intra- and intermolecular hydrogen bonding of the polymer chains. Therefore, the modifications are required to alleviate these drawbacks, while introducing new desirable qualities [23].

Thermoplastic starch (TPS) appears to be a good biopolymer because it can be processed using standard synthetic plastics manufacturing technology (extrusion, injection molding). TPS can be made using a plasticizer such as water, glycerol, or sorbitol in combination with proper temperature and pressure treatment. The plasticizer penetrates the starch granules, disrupting the initial crystallographic structure. The mechanical properties of TPS can be influenced by the temperature of starch production and water content, type and amount of plasticizers, and ancillary materials [24]. Plasticizers such as glycerol or glycol act as diluents and reduce the interaction between molecules; subsequently, they reduce tensile strength and improve the macromolecular flexibility, leading to an increase in elongation at break. The Tg of most TPS ranges from -75 to 10 °C. TPS has modest tensile strength, typically below 6 MPa. Synthetic polymers, other biopolymers, nano clay, and fiber can be added to increase tensile strength and water resistance [25].

Corn, potato, wheat, and tapioca are generally used as feedstock for starch production [9, 21]. More than 85% of the corn starch is produced by wet milling and very high purity starch (99.5%) can be obtained, as compared to dry milling. Wet milling aims to fractionate different components of the corn kernel and release starch granules with minimal mechanical damage.

The industrial process involves chemical, biochemical, and mechanical operations to separate pure starch fractions from corn kernels. Various products are isolated, including starch (about 68.0%), corn bran (12.0%), corn germ (7.5%), steep liquor (6.5%), gluten (5.6%), and others (0.04%) [26]. Different studies demonstrate the applicability of starch bioplastics in shopping bags, food packaging, agriculture, and medicine [9].

20.1.1.2 Cellulose

Cellulose is the most abundant, renewable natural polymer [24] that can be extracted from plant biomass or from specific cellulose-producing bacteria, also called bacterial cellulose (BC). Cellulose is also a polysaccharide chain of D-glucopyranose units linked by β-1,4-glycoside (Figure 20.4) [27]. Cellulose can be processed into food packaging materials or used as a nano-filler additive with other bioplastics to improve barrier and tensile properties for food packaging applications. Cellulose is a highly crystalline structured material, which is useful for film application. Since cellulose is insoluble in water as well as in common organic solvents, it is essential to modify cellulose into water-soluble derivatives. Cellulose acetate, which is a derivative of cellulose, is used to make cigarette filters, membranes, and in food packaging [21]. Due to acetylation, cellulose becomes hydrophobic and degrades extremely slowly, in the environment. Cellulose is an interesting material because of its outstanding mechanical properties and biodegradability. Yet, regenerated cellulose, which is popular as "viscose" or "rayon" textile fibers, constitutes 60% of seabed microplastics [1].

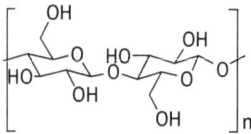

Figure 20.4: Chemical structure of cellulose.

Although around 100 years old, the viscose process for the production of cellophane (films/membranes) and rayon (fibers) is still in use [28]. The cellulose is alkalized and derivatized to cellulose xanthogenate by addition of CS_2, resulting in a metastable intermediate. Derivatized cellulose can be dissolved in an aqueous sodium hydroxide solution to form the so-called viscose. High purity cellulose is regenerated by removing the substituent and processed into fibers or films, in a wet process. The major disadvantage of the viscose process is requirement of sulfur and heavy metal compounds [29]. The CarbaCell process provides an alternative to viscose spinning technology and avoids use of hazardous sulfur-containing compounds and provides relatively high stability at room temperature [30]. In the Carbacell process, an aqueous or organic urea solution is used to convert the native cellulose into cellulose carbamate, which can be dissolved and processed identically to viscose. However, the CarbaCell process has not yet been industrially established due to certain requirements such as a catalyst, organic solvents, long reaction times, and high temperatures. Recently, techniques for *in vitro* cellulose synthesis were reviewed and critically compared, with a special focus on more recent developments [31].

The potential of cellulose can be further extended when bundled cellulose chains with highly ordered regions can be isolated as nano-particles, known as cellulose nanomaterials or nanocelluloses [32]. A number of nanocellulose forms like cellulose nano crystals (CNCs), cellulose microfibrils (CMF), cellulose nano fibrils (CNF), amorphous nanocellulose (ANC), cellulose nano yarn (CNY) can be produced using different methods and from various cellulosic sources [33–35]. Size, morphology, and other characteristics of each nanocellulose can be varied depending on the cellulose origin, isolation, and processing conditions. Although several methods have been investigated to synthesize nanocellulose, acid hydrolysis using sulfuric acid is the oldest process and is most widely used. A typical approach starts with alkali and bleaching pretreatments followed by acid hydrolysis. Cellulose nanocrystals, usually produced by acid hydrolysis are rod-like cylindrical nanoparticles with 4–70 nm width, 100–6,000 nm length, and 54–88% crystallinity index [36]. Some of the other processes reported for CNCs include mechanical treatment, oxidation methods [37], enzymatic hydrolysis [38], ionic liquid treatments [39], and subcritical water hydrolysis [40].

Currently, nanocellulose can be employed in several fields including continuous fibers and textiles, food coatings, barrier/separation membranes, antimicrobial films, paper products, cosmetics, biomedical products, supercapacitors, batteries, catalytic supports, electroactive polymers, and many more emerging applications [41, 42]. Despite significant industrial interest in R&D of nanocellulose as evident from nanocellulose patents, it has been produced at limited capacities and very limited technologies have transitioned from laboratory to pilot scale. Beyond pilot plant-scale capacities of CNCs include CelluForce (Canada, 1,000 kg/day), American Process Inc. (USA, 500 kg/day), Melodea/Holmen (Sweden, 100 kg/day), Blue Goose Biorefineries (Canada, 10 kg/day) [43, 44].

20.1.1.3 Guar gum

Guar gum is a naturally occurring seed gum made from the endosperm of *Cyamopsis tetragonolobus* [45, 46] or *Cyamopsis psoraloides* [46], a member of the Leguminosae family. As illustrated in Figure 20.5, the primary component of guar gum is a high molecular weight polysaccharide of galactomannans that is a linear chain of (1→4)-linked β-D-mannopyranosyl units with (1→6)-linked α-D-galactopyranosyl residues as side chains [45]. Guar gum is a high molecular weight hydrophilic polysaccharide that is soluble in water but often insoluble in organic solvents like ketones, esters, lipids, and alcohols. Guar gum's physical qualities are determined by its chemical and structural composition: The majority of it is composed of a polysaccharide called galactomannan, which has a mannan backbone and side groups of galactose in a ratio of about one galactose unit for every two mannose units (Figure 20.5) [47]. The structure and behavior of guar gum have been studied using a variety of methods, including physical (optical rotation, stress-strain measurement, and x-ray analysis,), chemical (acid hydrolysis, ethylation, periodate oxidation, and creation of tolyl sulphonyl derivatives), enzymatic (by means of selective enzyme hydrolysis), and analytical (FTIR, NMR, XRD, DSC, TGA, SEM, MALDI-TOF-MS, and chromatography) [48].

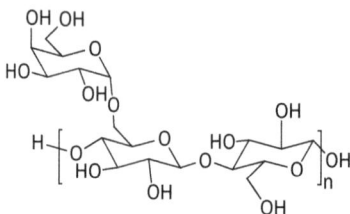

Figure 20.5: Chemical structure of guar gum.

When guar gum is dispersed, it swells and liquefies in polar liquids, creating powerful hydrogen bonds, while it creates fragile hydrogen bonds in nonpolar liquids [49]. Guar gum is used for a variety of purposes in the textile, culinary, petrochemical, mining, and paper sectors, where it serves as a suspending, emulsifying, gelling, and stabilizing agent. Because of its low degree of viscosity and hydration, guar gum has a limited range of applications in its basic forms. Its industrial applications are expanded by chemical modification, which leads to the formation of new types of superabsorbent with enhanced qualities. Guar gum-based films provide excellent barrier properties, high mechanical strength, and antibacterial or microbial resistant abilities [46]. Although raw gum has significant limitations and cannot always meet the needs of a particular application, incorporating newer functional groups into raw gums results in hybrid derivatives. The chemical, derivatization, and grafting copolymerization methods are used for the modification of guar gum, which helps eliminate their deficiencies that may restrain its use in a wide range of applications [48]. A major industrial application for guar is in oil and gas well fracturing.

Typically, guar gum is obtained by grinding the endosperm portion of a leguminous plant, *Cyamopsis tetragonolobus*, which has drought-resistant characteristics. The guar gum producing plants are mostly grown in India, Pakistan, and semi-arid zones in Brazil, Southern US, Australia, and South Africa due to favorable climatic conditions required for growth [45]. Processing involves removal of guar seeds from the pods followed by dehusking and gum extraction. The germ portion of the guar seed is a major by-product of guar gum processing and is used as cattle feed. Due to its low cost, it is mostly popular in the food industry and also has some beneficial properties for health like high fiber content; its consumption has been reported to reduce cholesterol levels.

20.1.1.4 Xanthan gum

Xanthan gum is also a natural polysaccharide known for inherent properties such as nontoxicity, biodegradability, and biocompatibility. Xanthan gum is mainly produced as an extracellular polymer by the *Xanthomonas campestris* bacteria [50, 51]. Xanthan gum is a long-chain pentasaccharide comprising glucose, mannose, and glucuronic acid that are repeated in the molar ratios of 2:2:1. A D-glucuronic acid unit is sandwiched between two D-mannose units that are linked to an alternate glucose unit in the main chain in the trisaccharide side chain, whereas the main chain is made up of β-D-glucose units linked at the 1 and 4 positions, resembling the structure of cellulose (Figure 20.6) [52, 53]. The D-mannose unit that is associated to the main chain is linked to an acetyl group, while the D-mannose unit at the end is linked to a pyruvic acid residue. The presence of acetic and pyruvic acid imparts an anionic character to xanthan gum. Stability of xanthan gum is affected by the structure of xanthan chain, especially the type of functional group at the outer mannose unit. Among acetyl and pyruvyl units, greater acetylation leads to more stability.

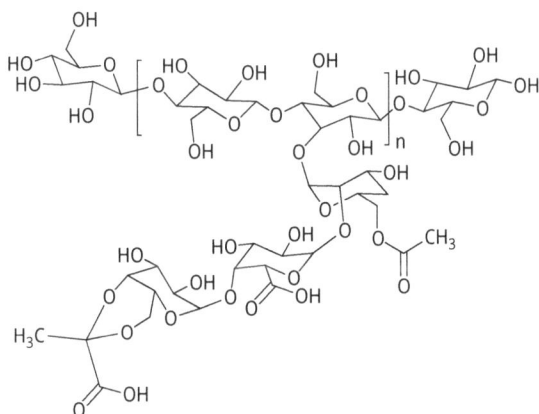

Figure 20.6: Chemical structure of xanthan gum.

Xanthan gum is produced on a commercial scale by bacterial fermentation of sugar-based feedstocks like glucose or sucrose by *Xanthomonas campestris*. Industrial feedstocks like molasses, cheese whey, and starch derived from corn and cassava have also been used for the production of xanthan gum [54]. Xanthan gum is recovered from the fermentation broth by steps of cell separation followed by alcohol precipitation and drying [55]. Being a fermentation-based product, selection of microbial strain, media, and fermentation conditions directly affect the composition of the functional groups in xanthan gum [39]. The concentration of pyruvic acid and acetyl depends on the type of bacteria and other aspects of the fermentation process like medium composition, variations in pH, temperature, and agitation rate [55]. The viscosity of xanthan gum solutions is affected by pyruvyl content, which has a significant impact on properties like gel formation.

Xanthan gum, mainly, has applications in food, cosmetics, cleaning products, paints, ceramic glazes, inks, and formulas for water-based drilling fluids. *The* most important properties of xanthan include its high viscosity at low concentrations, high solubility in hot and cold water, viscoelastic behavior, resistance to enzymatic degradation, high stability in a wide range of pH, temperature, and salt solutions, interaction with other polymers, as well as its straightforward processing mechanism [52]. Being a bio-based polymer, xanthan gum has also been investigated for matrices for tablets, nano- or microparticles, tissue engineering, and transdermal patches. However, the native xanthan gum has limitations with regards to its poor thermal and mechanical stability, inadequate water solubility, unusable viscosity, and susceptibility to microbial contamination. These limitations can be overcome by physical or chemical modification, or cross-linking of xanthan gum by esterification, etherification, amidation, oxidation, and acetylation of xanthan gum [56].

20.1.1.5 Poly (hydroxy alkanoates) (PHAs) (see chapter 22 for a more complete review)

PHAs are biosynthetic, biodegradable, thermoplastic aliphatic polyesters obtained from renewable resources (Figure 20.7) [21]. PHAs can be synthesized directly by microbial fermentation of different kinds of sugars, glycerol, fatty acids, and gaseous feedstocks like biogas (methane) [57, 58] Various bacteria, for instance, *Cupriviadus, Pseudomonas, Bacillus,* and methanotrophic strains have been widely explored for PHA synthesis.

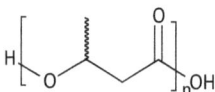

Figure 20.7: Chemical structure of poly (3-hydroxy butyrate) (PHB).

PHAs are classified as short-chain (3–5 carbon atoms), medium-chain (6–14 carbon atoms), or long-chain (>14 carbon atoms) lengths, based on the number of carbon atoms in the PHA repeating unit of monomer [59, 60]. Different material properties of PHAs can be adjusted by varying the chain length of comonomer or side chain functionality, for example, rigid and brittle PHB (Figure 20.7) and softer and flexible poly (3-hydroxybutyrate-co-3-hydroxyhexanoate) (PHBH). The flexibility to change the monomer chain length increases the design space for PHAs as compared to other biopolymers. PHAs would be most suitable replacements for packaging materials of fossil-based plastics such as PE and PP due to their good mechanical [24] properties and barrier properties to O_2 and CO_2 [21, 61]. Most PHAs degrade rapidly as compared to PLA and, hence, are most attractive for applications in which rapid biodegradation is desired [1].

PHA feedstocks can include first-generation or second-generation cellulosic sugars, vegetable oils, organic waste, municipal solid waste, and fatty acids. In addition to glucose, other lignocellulosic monomer sugars, such as xylose, arabinose, mannose, galactose and rhamnose, and biomass derived lignin can also be employed as PHA feedstocks. Furthermore, it is possible to produce PHAs directly from gaseous substrates like methane. An American company, Newlight Technologies' Air Carbon technology uses methane-based feedstocks for the production of PHB [62]. During fermentation, PHA's are produced by bacteria under nutrient-limited conditions. Some of the microbes can accumulate PHA intracellularly up to the 80% of the cell volume using different fermentation strategies [63]. The intracellular PHA granules can be extracted using different extraction methods including solvent extraction, chemical digestion assisted by surfactant and alkali, and biological extraction assisted by enzymes [64]. With increasing focus on sustainable processes, biological extraction methods (called bio-extraction methods) are gaining more attention in PHB extraction. With advances in synthetic biology and bio-processing technologies, different types of PHAs can be produced with improved functional properties. Detailed studies regarding recent trends and future perspectives of microbial PHA production from different feedstocks, including lignocellulosic biomass, can be referred from recent reviews [60].

The primary reason for the limited production of PHAs is the higher cost of production. To reduce the costs further, it is required to use low-cost feedstocks or different agro-industrial waste materials as a substrate for PHA production [21].

20.1.1.6 Poly (L-lactic acid) (PLLA)

PLLA is a synthetic, compostable, transparent, and aliphatic thermoplastic polyester obtained from different renewable resources by a combination of fermentation and chemical route [21]. PLA (Figure 20.8) can be prepared by ring opening polymerization of cyclic diester of lactic acid, namely, L-lactide (3,6-dimethyl-1,4-dioxane-2,5-dione) or direct polycondensation of lactic acid either in melt or solution. The former method is used successfully for the production of PLA on industrial scale. PLA can be used as potential substitute for the fossil--based polyolefin films and polystyrene foams including single use plastics. The presence of methyl group and small repeating unit makes PLA more brittle and slow to crystallize. Physical and mechanical properties of PLA can be further improved by blending with the other bio-based polymers, plasticizers, and nucleating agents before processing of the materials [1].

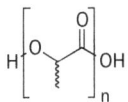

Figure 20.8: Chemical structure of poly (lactic acid).

Both sugar and starch-based feedstocks are being used today to produce lactic acid, which is subsequently polymerized to PLA. Different carbohydrate sources can be used, including sugarcane juice or sugar industry waste like molasses, starch-rich materials like corn, cassava, or rice, and lignocellulosic materials like agricultural and forestry residues [65]. Fermentation pathway from methane to lactic acid is also being developed [66]. The fermentation process is based on a methanotrophic organism that feeds off methane and, through synthetic biology, has been modified to convert methane to lactic acid. Production of high purity L-Lactic acid is critical for the polymerization process. Optically pure L (+) lactic acid is produced by fermentation of sugars by lactic acid bacteria or genetically engineered yeast species [67]. The advantage of the yeast-based process lies in the direct production of lactic acid at lower pH. Bacterial production of lactic acid commonly involves the generation of by-product, gypsum (calcium sulfate), which poses disposal challenges. L-Lactide is a widely used intermediate for the industrial production of PLLA. Lactide is obtained via a two-step process consisting of firstly polymerizing lactic acid under reduced pressure to produce an oligomer or prepolymer with molecular weight range of about 500–5,000 Da, and then depolymerizing the produced prepolymer under reduced pressure, in the presence of a catalyst (Figure 20.9) [68].

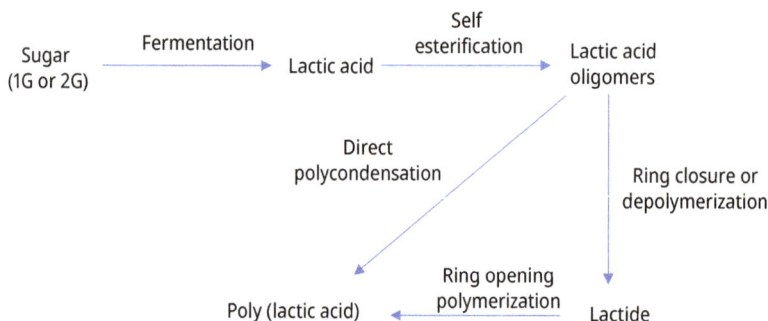

Figure 20.9: Process paths to prepare poly (lactic acid).

Different catalytic processes for the synthesis of lactide have been extensively discussed in the literature [69, 70]. Chemical and optical purity of lactide is most desired for synthesis of high quality and high molar mass PLLA [71, 72]. The conventional method of lactide production involves multistep synthesis, and purification by crystallization leads to high energy consumption, relatively lower yield, and lower optical purity [73, 74]. Lactide to PLA polymerization proceeds through a ring-opening polymerization process catalyzed by an organometallic catalyst at high temperatures and under vacuum [69]. This multistep process is commonly employed for PLA production, despite challenges related to higher capital and operating expenditure related to a multistep process. Research is ongoing to develop novel synthetic approaches to lactide and improve it to the desired quality [75–77].

To improve the crystallization kinetics for certain fabrication procedures and applications, D-lactic acid units are added to L-PLA. Tailoring D-isomer content in PLA has shown to significantly affect the melting temperature (Tm) of PLA. The optical comonomers alter the natural helical shape of the PLA by adding "kinks" and "defects" to the crystal arrangement, which lowers the Tm and the amount of crystallinity that can be achieved as well as slows down the crystallization process. D-lactide and L-lactide in equimolar concentrations are known as PDLLA or poly (rac-lactide) [68, 78].

Today, PLA is used in a variety of applications like packaging and food service ware and is increasingly favored for more durable applications in automotive, electronics, and textiles. PLA products with different molecular weights are available today. They are suitable in a variety of polymer processing applications like injection molding, fiber spinning, extrusion/thermoforming, and as a nucleating agent. PLA-based products comply with industrial composting standards (ASTM D6400, EN13,432) and have been approved for food contact applications in the USA, Europe, and China [79].

20.1.1.7 Bio-poly (butylene succinate) (Bio-PBS)

Bio-PBS (Figure 20.10) is an aliphatic polyester that can be prepared by step-growth polymerization of succinic acid and 1,4-butane diols (BDO) in the presence of a catalyst at high temperature and vacuum. Although PBS is generally produced from fossil-based feedstock, both of its monomers, succinic acid and butanediol, can be produced by the fermentation of renewable feedstock (Figure 20.11) [21].

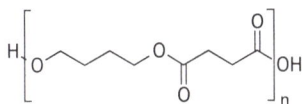

Figure 20.10: Chemical structure of PBS.

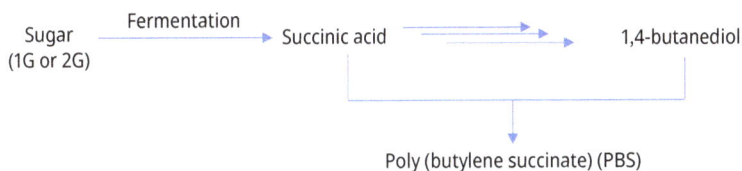

Figure 20.11: Process pathways to prepare poly (butylene succinate).

Bio-PBS has good thermal and mechanical properties that are analogous to PE and PP and is useful in numerous applications such as shopping bags, food packaging, agriculture mulch film, and hygiene products. PBS is a white, semi-crystalline thermoplastic, with a Tm of ~ 115 °C and Tg of about – 30 °C. It is characterized by good thermal stability and decent mechanical properties like tensile yield strength of 30–35 MPa, (similar to that of polypropylene), Young's modulus of 300–500MPa, and high flexibility [80]. PBS has longer chain ester linkage as compared to PLA and PHB and, thus, has more flexible molecular structure. Good oxygen barrier properties of PBS can substitute PP and its application in films, bags, or boxes, for both food and cosmetic packaging.

For the production of Bio-PBS, sugar-based feedstocks are being used for the production of monomers like succinic acid as well as butanediol. PBS synthesis can be done into two steps: the first stage is the esterification of succinic acid and BDO to obtain oligomers. The second step is the polycondensation of oligomers to achieve high-molecular-weight PBS [17]. The synthesis of high-molecular-weight aliphatic polyesters by conventional polycondensation is difficult due to the simultaneous competing reactions of condensation and degradation [81]. For this reason, to obtain polyesters with useful mechanical properties, it is necessary to introduce side chains with aromatic units and chain extenders to increase the molecular weight or, more frequently, catalysts to accelerate the kinetics [82].

In 2016, Novamont, (Italy) inaugurated the first bio-based plant for the industrial-scale production of 1,4-butanediol directly from sugar, using a genetically engineered bacteria [83]. Although, bio-based succinate production suffers from poor productivity and high downstream processing costs, several companies are trying to scale up bio-succinate production processes [84]. Mitsubishi Chemical (Japan) has industrialized succinic acid derived from biomass in conjunction with Ajinomoto to put bio-based PBS onto the market. In parallel, DSM and Roquette are jointly developing a feasible fermentation process to produce succinic acid, 1,4-butanediol, and successful achievement of bio-based PBS [85]. The high cost of bio-based PBS production demands further developments to make it more competitive with petrochemical-based PBS.

20.1.1.8 Bio-poly (ethylene furanoate) (Bio-PEF)

Bio-PEF is a bio-based, highly recyclable biopolymer with superior performance properties as compared to petroleum-based packaging materials. Bio-PEF (Figure 20.12) has a semi aromatic structure and shows similar properties to that of PET. Bio-PEF can be prepared by polycondensation of bio-based mono- ethylene glycol and 2,5-furane dicarboxylic acid (FDCA) or through ring-opening polymerization of cyclic PEF oligomers, which can decrease reaction times and improve the control over molecular weight [18].

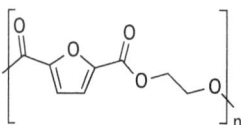

Figure 20.12: Chemical structure of PEF.

There are many similarities between Bio-PEF and PET in terms of synthesis by poly-condensation, properties, degradation, and EOL solutions. Bio-PEF has high Tg, higher tensile strength, and higher gas diffusion barrier properties, which may be advantageous for long-shelf-life packaging. Bio-PEF is thermally sensitive and needs more cautious processing but it also biodegrades relatively quicker (after 9 months) under suitable industrial composting conditions [1]. Molecular modification and copolymerization of PEF with different comonomers are being explored to tune Tg, mechanical strength, and degradability of PEF. FDCA, which is used a monomer for the synthesis of PEF is produced in two steps. The first step involves dehydration of fructose to 5-HMF, and in the second step, 5-HMF is oxidized to FDCA (Figure 20.13) [60].

Fructose ⟶ 5-HMF ⟶ 2,5-FDCA ⟶ PEF

Figure 20.13: Process path to prepare poly (ethylene furanoate).

Besides biomass, CO_2 is the only renewable carbon source for material synthesis. CO_2 can be converted into value-added products or monomers such as oxalic acid and glycolic acid and their polyesters. Avantium is a Dutch technology company in renewable chemistry focused on development of building blocks (monomers) from renewable sources (plant-based sugars and CO_2). Several aspects of Avantium YXY® Technology (Bio- PEF) technology, the process of bringing a new monomer and polymer to market, including process and application development, life cycle assessment (LCA), by-product valorization, EOL scenario, and recyclability and circular economy have been reviewed recently [18]. However, the production of FDCA as a cost-effective monomer remains a challenge [86].

20.1.1.9 Bio-poly (ethylene terephthalate) (Bio-PET)

Typically, PET is prepared by polycondensation of terephthalic acid and ethylene glycol in the presence of a catalyst at high temperature and under high vacuum [1]. Bio-based PET (Figure 20.14) can be produced using ethanol, which is first converted into mono ethylene glycol followed by reaction with petroleum-derived terephthalic acid (Figure 20.15) [87]. Bio-PET is an analogous drop-in to petroleum-based PET, and hence exhibits all the equivalent properties, which makes it suitable for multiple applications for instances in beverages and in textile markets [1]. Polyesters like PET can be recycled with constant molecular weight of the recycled polymer after multiple mechanical recycling, without the need to go back to monomers or polymer precursors [18]. These drop-in polymers are partly bio-based and nonbiodegradable and are chemically equivalent to the petroleum-based polymer.

Coca-Cola Company adopted sustainable packaging in 2009, with the launch of PlantBottle™, which is made using fully recyclable PET plastic bottles made partially

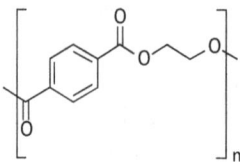

Figure 20.14: Chemical structure of PET.

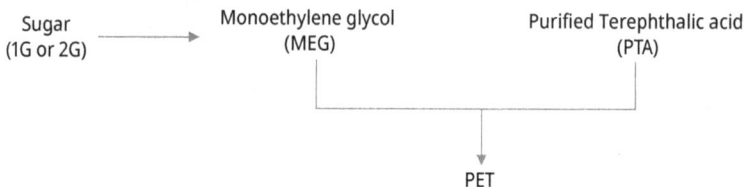

Figure 20.15: Process path to prepare poly (ethylene furanoate).

from plant-based source. Sugarcane was used as feedstock for Bio-PET, which was employed widely in soft drink and water bottles, beer bottles, mouthwash bottles, peanut butter and salad dressing containers, ovenable packaging films, and ovenable prepre-pared food trays. Similar to PET, its bio-based counterpart has the same clarity, strength / toughness, barrier to gas and moisture, and resistance to heat properties.

Both the monomers, ethylene glycol (also known as MEG in industry) and purified terephthalic acid (PTA), can be obtained from biomass or carbohydrates by catalytic reactions.

US Patent 7,960,594 described a process wherein ethylene glycol is produced from cellulose. The process involves catalytic degradation and hydrogenation reactions under hydrothermal conditions. Avantium has developed a single-step hydrogenolysis process to convert glucose to MEG [88].

About 70% of terephthalic acid used in PET manufacturing is produced from the oxidation of *para*-xylene [89]. Bio-based para xylene can be produced by bio-isobutanol in a three-step reaction involving dehydration, oligomerization, and dehydrocyclization [90]. Several cutting-edge technologies for synthesis of MEG and PTA have been reviewed and can be referred for details.

20.1.1.10 Poly (butylene adipate-co-terephthalate) (PBAT)

Although this chapter is focused on bio-derived thermoplastics or polymers derived from renewable resources, the term bioplastics has also been used for fossil-based plastics that are biodegradable. PBAT was synthesized by conventional step-growth polymerization (polycondensation) of 1,4-butanediol, terephthalic acid, and adipic acid monomers in the presence of zinc, tin, or titanium catalysts. The reported studies suggest that the condensation reaction is usually performed at high temperature > 190 °C, under high vacuum, and it requires long reaction time. The structure of PBAT includes terephthalic acid, adipic acid, and 1,4-butanediol as comonomers. Bio-based BDO produced by industrial fermentation, can directly replace petrochemical-derived BDO in PBAT. Adipic acid produced from castor oil has been used as a monomer to make copolyesters [91]. PBAT is a degradable aliphatic-aromatic copolyester sold as Ecoflex by BASF (Germany) and under other trade names by several suppliers in Asia. BASF, Germany produced Ecoflex, a PBAT-based product, with highest production capacity of 60,000 tons/year. Kingfa, China, Novamont, Itali DuPont, and Jinhui, China supply PBAT under the trade names Ecopond, Origo-Bi, Biomax, and Ecowords, respectively, with a production capacity of 20,000–50,000 tons/year [19]. PBAT (Figure 20.16) is widely used in agricultural mulch films, which can degrade in soil over a period of > 9 months [1].

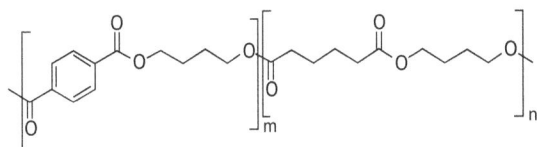

Figure 20.16: Chemical structure of PBAT.

PBAT is known for being flexible and tough, which makes it ideal for combinations with other bio-based polymers that have high modulus and strength, but are very brittle. PBAT is widely used in applications like garbage bags, wrapping films, and disposable plastic products (lunch boxes, dishes, cups, etc.,) [19].

20.2 End-of-life scenario of bio-derived thermoplastics (see also chapter 21, 22, and 23 for additional details)

The longevity of polymeric materials and the lack of proper collection and recycling technologies at the end of product life have led to continuous accumulation of plastic-related waste and incorrect product disposal practices. Extensive usage of single-use plastics and disposable PPEs during the COVID-19 outbreak further worsened the plastic waste problem. Most fossil-based plastic materials do not have pathways or mechanisms for natural degradation, and the stability provided by the polymeric structure further leads to their accumulation in the environment. Oceans, freshwater lakes, and soil tend to accumulate plastics, which are constantly damaging the environment and ecosystem [1]. There have been several reports on accumulation of microplastics in the environment, since the word was first coined in 2004. Microplastics have been found everywhere including in deep oceans; in Arctic snow and Antarctic ice; in shellfish, table salt, drinking water and beer; and drifting in the air or falling with rain over mountains and cities [92]. According to European Chemical Agency, any type of plastic particle less than 5 mm is considered as microplastic.

Plastic trash can be managed in one of three ways: it can be recycled for reuse, managed through landfilling and incineration, or directly disposed into the environment. To avoid leakage of contaminants in the environment or the formation of microplastics, complete digestion of polymeric materials is necessary. Biodegradation is an EOL option that can be achieved for easily hydrolysable polymers, such as aliphatic esters like PLA and PHA. The terms "biodegradation" and "composting" refer to the metabolic breakdown of polymeric material into CO_2, H_2O, and other inorganic compounds by a variety of microorganisms under favorable conditions. Disintegration by different physical techniques like fragmentation and particle size reduction can augment biodegradation process and enzymatic breakdown [1]. Degradation of polymeric materials by oxidative processes commonly known as oxo-degradation has been banned in the European Union and many other countries due to lack of evidence of the complete degradation of oxo-degradation particles and products by biological systems[1].

Another most sought-after EOL option for polymeric compounds is recycling for new products. The easiest, most affordable, and most popular type of recycling is mechanical recycling, which often involves separating plastic waste by polymer type, removing labels, washing, mechanically shredding, melting, and remolding into new shapes. Re-extrusion of different plastic has been attempted; however, mechanical recycling is currently not feasible for bioplastics at commercial scale. Deterioration in quality has been observed when PLA and PHA are mechanically recycled, with loss of tensile strength and molecular weight. To create polymers of the desired quality, plastic products must first be depolymerized into their monomeric subunits. Recycling by chemical processes like solvolysis and thermolysis are other EOL alternatives for polymeric materials. The only extensively used technology for treating plastic solid waste on a significant scale is mechanical recycling. The primary procedures involve washing away organic waste, followed by shredding, melting, and remolding the polymer, which is frequently merged with virgin plastic of the same type to create a substance with advantageous qualities for manufacture. As each type of plastic reacts to the process differently based on its chemical composition, mechanical behavior, and thermal qualities, mechanical recycling technologies have limitations [93, 94]. Enhancing chemical recycling efficacy [95, 96] and selectivity through catalyst development, reducing the requirement for sorting through compatibilizer design, and expanding recycling beyond thermoplastics are three current research approaches that show significant promise in augmenting plastic recycling methods. The reason that chemical recycling with thermolysis offers a recycling approach through decomposition of a polymer to lower-molecular-weight streams is that most polymers are immiscible with one another, yielding phase-separated mixes with diminished characteristics. The properties of recycled material may change ever so slightly when one type of plastic is contaminated with another, which could make it harder to use. By creating compatibilizers that regulate the phase behavior of polymer mixtures, this issue can be solved [93]. For plastics with petrochemical origin, recycling is well-established, but, for bioplastics, an efficient and effective recycling technology has not yet been developed. For majority of bioplastics that are in use at reasonable commercial scale today (starch, PLA, PBS, PHA, etc.) segregation may be required, based on their chemical make up for recycling into new products. Landfilling is another commonly used option for most of the single-use plastic products. Most of the fossil-derived plastics do not degrade in soil and, therefore, the sizes of landfills increase every year. It is imperative now to develop sustainable materials like bio-based plastics that can fulfil the growing demand, but, at the same time, are benign to the environment and human health. Figure 20.17 shows the ideal scenario to achieve circularity of plastic-based products, with different feedstocks and production processes. Among all EOL options, biodegradation is the most efficient method for treating bioplastics with bio-based origin. Bioplastics are expected to have little impact on the environment, because their biodegradation does not induce the release of any toxic compounds. However, widespread use of existing and newly developed bioplastics need regulations that verify and certify important properties like origin and bio-based content, biodegradability, and compostablity to ensure their safety in use and the lack of any harm to the environment [21]. Experts from various fields and countries have been collaborating with global organizations to develop standards that may be used and implemented to examine important characteristics of various bioplastics, under repeatable conditions. The International Organization for Standardization (ISO), the European Committee for Standardization (CEN), and the American Society for Testing and Materials (ASTM) are the principal organizations involved in developing these standards.

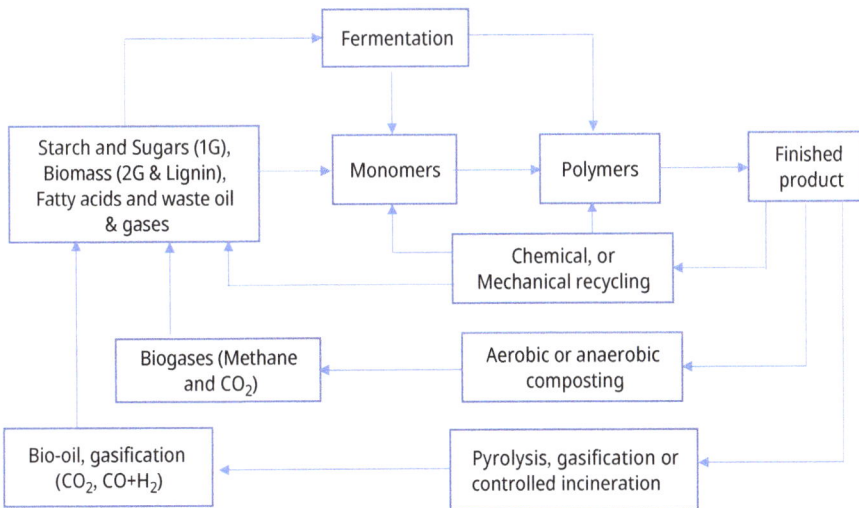

Figure 20.17: Circular economy of bio-plastics.

The most popular and often-cited plastics standards from various organizations are included in Table 20.3 [21]. Critical evaluation of sustainability of different bioplastics is another important aspect and multiple LCA studies have been published with different boundary conditions. There are international standards in place such as ISO 14,040 to guide the LCA structure, conduct, and assumptions made with limitations. EN 16,760 specifically provides guidelines for LCA requirements and guidance for bio-based products, excluding food, feed, and energy. However, heterogeneity in LCA approaches and assumptions persist and require a systematic framework for bioplastic LCAs that can improve meaningfulness for these studies and inform decision makers better [97].

Table 20.3: Standard test methods for aerobic and anaerobic biodegradation of plastics and bio-based content.

Parameter	Test method	Disposal	Ref.
Aerobic biodegradation	ASTM D6400-04, EN13432, ASTM D5338-15, ASTM D5338-98, ASTM D5511-18 and ISO 15,985 ISO 14,855, ASTM D6400-19, ASTM D7081 and ASTM D6868, ISO17,556, ASTM D5338-98, ASTM D6002-96 and AS 4736–2006, Soil (ISO 17556, ASTM D5988, ISO 11,266), water (ISO 14,851, ISO 14,852)	Compostable plastics or OK biodegradable or OK compost	[96, 98–101]
	ASTM D5929	Natural or home composting	[99]
	EN 13432	Industrial composting	[98]
	ASTM D6691- 17, ASTM D7473-12, ASTM D7991-15, ASTM D7081-05, ISO 14,851, ISO 14,852 and ASTM D6692	Marine biodegradation	[99, 101]
	ASTM D6400, ASTM D6868	US biodegradable product institute (BPI)	[96]
	EN17033	Biodegradation of mulch film (>90% degradation within 2 years)	[98]
	ASTM D5526-18, and ASTM D7475-20	Landfill biodegradation	[99]
Anaerobic biodegradation	ASTM D5210-92, ISO 14,853, ISO 13,975	Sewage sludge biodegradation	
	ASTM D5511 and ISO 14,853, ISO 15,985	Anaerobic digestion biodegradation	[102]
	ASTM D5526-18, ASTM D7475-20, ASTM D5526-18	Accelerated landfill biodegradation	[99]

Table 20.3 (continued)

Parameter	Test method	Disposal	Ref.
OK bio-based polymer	CEN/TS 16,137 and ASTM 6866		[96]
	ASTM 6866	USDA Certified Bio-based product	[102]
	ASTM 6866, CSN/TS 16,137	Din Gepruft bio-based polymer (DIN CERTCO)	

20.3 Global market and application of bio-derived thermoplastics (some of the market and applications are also covered in chapters 19, 21, 22 and 23)

The global production of 100% bio-based polymers is currently more than 2.4 million tons with two-thirds of it contributed by biodegradable polymers [103, 104]. In comparison to current fossil-based plastic production of 380 million tons, the fraction of biopolymers is very little and has significant potential to increase, over the years. Sustainability is the main driver for increase in demand for biopolymers. In the last couple of decades, it has become clear that biopolymers cannot be sold at much higher prices as compared to petrochemical counterparts; however, there is continued interest from large corporations as biomaterials can help in achieving emission reduction targets and help in transitioning to net zero commitments by 2050. A recent survey by McKinsey in ten countries has found that consumers see sustainability as increasingly important, and the vast majority of consumers showed willingness to pay more for sustainable packaging. Further, policy makers and regulatory bodies across the globe are pushing for sustainability through initiatives such as the "European Green Deal" and "Single-use plastic directives" of the European Union. Two key regulatory objectives are reducing CO_2 emissions and environmental leakage of plastics that do not biodegrade. Both the objectives can be addressed through biomaterials, and increased demand and recycling-oriented regulations might create further incentives. Food packaging and fast-moving consumer goods are the largest markets for short-lived to medium-lived plastics and, therefore, also for bioplastics. Global bioplastics production capacity is set to increase significantly from around 2.41 million tons in 2021 to approximately 7.59 million tons in 2026 [105, 106]. Out of the current total capacity in 2021, up to 40% biopolymers belong to both bio-based and biodegradable (PLA, PHA, starch blends, and cellulose films). Among different biopolymers, PLA is the most established and accepted material, with total global production capacity beyond 300 Kilotons/Annum (KTA), and another 100 KTA capacity expansion is under progress. The demand for PLA has steadily increased beyond 190 KTA, with greater than 60% demand coming from the packaging sector alone. It is estimated that by 2030, the potential demand for PLA may cross 600 KTA. From approximately 380 million tons of total plastics consumption, replacement of even 1% by bioplastics would require about 1.4 million tons of additional capacity. Improvements in chemical recycling and availability of biodegradation facilities can further pave the way for bioplastics, such as PHAs that are more hydrolysable and, therefore, more compatible with existing composters [99]. Table 20.4 shows the current production capacities and applications of main biopolymers in use, today. According to report by World Economic Forum, investor interest in environmental, social, and governance assets is growing, with 86% believing that these will be better long-term investments. Up to $30 trillion are now spent on sustainable assets globally (one-third of total investments), with plastic and climate change topping the list of sustainable investor interests [107]. The Alliance to End Plastic Waste, comprising major chemical companies and brand owners, has promised to spend a total of $1.5 billion for sustainable plastic-related projects. Companies from multiple sectors like Nestle from food and nutrition, and Toyota and

Table 20.4: Overview of bio-based polymers.

Bio-polymers	Producer	Cost US$/kg	Capacity (tons/year)	Applications	Ref.
Starch	Plantic® Australia, Mater-Bi® and Novamont Italy, BIOPAR® Portugal, Biofase® Mexico and Sonaly® Canada, Ecofoam National Starch	0.5–2	20,000	Filler, thickening, stiffening, gluing, and binding agent, precursor to dextrins, glucose, ethanol, and sugars for processed food, in the production of biofuel and as adhesive in the papermaking process. Agriculture, medicine, engineering, food, and packaging, food packaging, shopping bags, agriculture, medicine, disposable domestic utensils (cup and plates), even automotive industry, controlled release of fertilizer, efficient bone cement, and drug release film in medical industry.	[109–111]

Table 20.4 (continued)

Bio-polymers	Producer	Cost US$/kg	Capacity (tons/year)	Applications	Ref.
Cellulose	Eastman Chemical Company produced cellulose acetate	1.8–4		Drug delivery, cell and tissue growth, scaffold, implants, clothing, decorative, and automobile industries.	[109, 112]
PHA	Shenzhen Ecoman Biotechnology, China Danimer Scientific USA Kaneka, Japan Bio-on, Italy Tianan Biological material polyone, China Biomer, Germany Mitsubishi Gas, Japan PHB Industrial, Brazil Metabolix, USA P&G Chemicals, USA/ Japan Telles LLC, USA Jiangsu Nantian Graup, China Goodfellow Cambridge Ltd, UK Tepha Inc, USA Cofco, China Mango Materials, USA Newlight Technologies, USA PhaBuilder, China RWDC Industries Ltd. USA Bluepha Co. Ltd, China	2.4–5.5	100–10,000	Packaging film and food services (shopping bags, paper coatings, cups diapers, carpets, PHA coated boxes, paper, sheets, boards), agricultural (biodegradable plastic film, seeds encapsulation, encapsulation of fertilizers, and controlled release of pesticides and insecticides), Medical application, drug delivery (controlled delivery of antibiotics), tissue engineering (scaffolds, pins, films, sutures), and wound dressing (swabs, fleece, lint). Biofuels or fuel additives, protein purification specific drug delivery, healthy food additives, fine chemical industries, printing, and photographic industries.	[113–115]

Table 20.4 (continued)

Bio-polymers	Producer	Cost US$/kg	Capacity (tons/year)	Applications	Ref.
PLA	Bionolle Starcla^{TM} - Showa Denko Japan, BIOFRONT-Teijin Japan, Cargill-Dow LLC, IngeoTM-Nature Works USA, WeforYou Austria, Total-Corbion Netherlands/Thailand	4–6	140,000	Food packaging, Cups, fiber, tissue engineering scaffolds, bio-absorbable medical implants, dermatology, cosmetics, targeted delivery of bioactive compounds, fixation rods, plates, pins, screws, and sutures, suitable for bone, cartilage, ligament, skin, blood vessels, nerves, construction, electronics, textile, and agriculture applications.	[79, 109, 116]
PBS	DSM and Roquette, Mitsubishi and Ajinomoto, Japan, BASF, Germany Hexing Chemicals, China IPC-CAS, China IRE Chemical, Korea Mitsubishi Gas Chemical, Japan Showa, Japan	4–10	3000	Food packages, bottles, supermarket bags, flushable hygiene products, mulch film and compost bags, agricultural mulch film, dishware, fishery, forestry, construction, electronics, drug encapsulation systems, orthopedic, and coffee capsules.	[17, 109, 117]
PEF	Avantium, BASF,	0.15–1.75	1000	Film, bottles, and fibers for textiles.	[18, 118]
PET	Coca-Cola-Gevo Venture, PepsiCo-Virunt Venture Toyota Tsusho Corporation, Japan			Fibers, packaging, bottles, containers, and films.	[119]
PBAT	BASF, Germany Kingfa, China Novamont, Italy Tunhe, China Xinfu, China Jinhui, China	3.8–5.8	20,000–60,000	Rigid and flexible packaging, shopping and garbage bags, sacks, cutlery, mulch film, fibers, paper coating, and foodservice.	[19]

Peugeot Citroën SA from the automotive sector have pledged to introduce renewable polymers in their products [108].

20.4 Summary and future perspectives

Significant technological advancements have been made in the last few decades in developing bio-based alternatives of fossil derived polymer. With ever-increasing focus on circular economy and sustainable products, biopolymers can provide multiple options with similar or better thermal and mechanical characteristics. However, in comparison to fossil-based polymers, bio-derived polymers are mostly produced in small quantities at relatively high cost, and in some cases, do not always have substantial environmental benefits. To increase production of biopolymers to a commercial scale, a few limitations need to be addressed including, the cost of production, availability of raw material, EOL status, biodegradability, and compatibility of the end products. Substantial growth in biopolymer sales would also depend on their engineering application and expansion in consumer goods. Low-cost feedstock is the key to further expand production of bio-derived polymers and besides first-generation feedstocks, lignocellulosic materials, including agricultural and other bio-wastes including gaseous streams, present abundant and viable feedstock from a future sustenance perspective. It is expected that bio-refinery processes will improve in efficiency, and with application of green chemistry principles (such as using nontoxic chemicals and reducing the energy demand), supply of monomers or polymer building blocks will be increasingly available at cost-competitive and sustainable manner. Advances in synthetic biology are expected to improve monomer yields, biopolymer diversity, and synthesis rates (e.g., PHAs, Xanthan Gum). New approaches of enzyme discoveries like metagenomic approaches will improve biomass deconstruction rates and will also help in biological depolymerization for recycling. In the long run, just the biodegradability of biopolymers will not be sufficient, and recycling technologies including chemical and biological will be required to handle mixed waste streams of poor quality as well as recycled material from multiple sectors. Apart from recycling of polymeric materials, consumer preferences towards renewable materials and strict usage of renewable energy for polymer and plastic production remain essential strategies to mitigate plastic waste and carbon emissions. Harmonization of bioplastic waste and disposal regulations and homogenization of methodology standards to make LCAs more consistent and comparable will help in identifying sustainability limitations of different biopolymers.

References

[1]   Rosenboom J-G, Langer R, Traverso G. Bioplastics for a circular economy. Nat Rev Mater 2022;7:117–37. https://doi.org/10.1038/s41578-021-00407-8.

[2]   Moshood TD, Nawanir G, Mahmud F, Mohamad F, Ahmad MH, AbdulGhani A. Sustainability of biodegradable plastics: New problem or solution to solve the global plastic pollution? Curr Res Green Sustain Chem 2022;5:100273. https://doi.org/10.1016/j.crgsc.2022.100273.

[3]   Llevot A, Dannecker P-K, von Czapiewski M, Over LC, Soeyler Z, Meier MAR, et al. Renewability is not enough: Recent advances in the sustainable synthesis of biomass-derived monomers and polymers. Chem – A Eur J 2016;22:11510–21. https://doi.org/10.1002/chem.201602068.

[4]   Anastas PT, Warner JC. Green Chemistry: Theory and Practice, Oxford University Press; 1998.

[5]   Keijer T, Bakker V, Slootweg JC. Circular chemistry to enable a circular economy. Nat Chem 2019;11:190–95. https://doi.org/10.1038/s41557-019-0226-9.

[6]   Coppola G, Gaudio MT, Lopresto CG, Calabro V, Curcio S, Chakraborty S. Bioplastic from renewable biomass: A facile solution for a greener environment. Earth Syst Environ 2021;5:231–51. https://doi.org/10.1007/s41748-021-00208-7.

[7]   De Donno Novelli L, Moreno Sayavedra S, Rene ER. Polyhydroxyalkanoate (PHA) production via resource recovery from industrial waste streams: a review of techniques and perspectives. Bioresour Technol 2021;331:124985. https://doi.org/10.1016/j.biortech.2021.124985.

[8]   Bugnicourt E, Cinelli P, Lazzeri A, Alvarez V. Polyhydroxyalkanoate (PHA): Review of synthesis, characteristics, processing and potential applications in packaging. Express Polym Lett 2014;8:791–808. https://doi.org/10.3144/expresspolymlett.2014.82.

[9]   Morinval A, Averous L. Systems based on biobased thermoplastics: From bioresources to biodegradable packaging applications. Polym Rev 2021;1–69. https://doi.org/10.1080/15583724.2021.2012802.

[10] Albertsson A-C, Varma IK. Aliphatic Polyesters: Synthesis, Properties and Applications. Degrad. Aliphatic Polyesters, Berlin, Heidelberg: Springer Berlin Heidelberg; vol 157, 2002, pp. 1–40. https://doi.org/10.1007/3-540-45734-8_1.

[11] Pang K, Kotek R, Tonelli A. Review of conventional and novel polymerization processes for polyesters. Prog Polym Sci 2006;31:1009–37. https://doi.org/10.1016/j.progpolymsci.2006.08.008.

[12] Maisonneuve L, Lebarbé T, Grau E, Cramail H. Structure-properties relationship of fatty acid-based thermoplastics as synthetic polymer mimics. Polym Chem 2013;4:5472–517. https://doi.org/10.1039/c3py00791j.

[13] Stempfle F, Ortmann P, Mecking S. Long-chain aliphatic polymers to bridge the gap between semicrystalline polyolefins and traditional polycondensates. Chem Rev 2016;116:4597–641. https://doi.org/10.1021/acs.chemrev.5b00705.

[14] Debuissy T, Pollet E, Avérous L. Biotic and abiotic synthesis of renewable aliphatic polyesters from short building blocks obtained from biotechnology. Chem Sus Chem 2018;11:3836–70. https://doi.org/10.1002/cssc.201801700.

[15] Nomura K, Binti Awang NW. Synthesis of bio-based aliphatic polyesters from plant oils by efficient molecular catalysis: a selected survey from recent reports. ACS Sustain Chem Eng 2021;9:5486–505. https://doi.org/10.1021/acssuschemeng.1c00493.

[16] Filiciotto L, Rothenberg G. Biodegradable plastics: Standards, policies, and impacts. Chem Sus Chem 2021;14:56–72. https://doi.org/10.1002/CSSC.202002044.

[17] Xu J, Guo BH. Poly(butylene succinate) and its copolymers: Research, development and industrialization. Biotechnol J 2010;5:1149–63. https://doi.org/10.1002/biot.201000136.

[18] de Jong E, Visser HA, Dias AS, Harvey C, Gruter G-JM. The road to bring FDCA and PEF to the market. Polymers 2022;14:943. https://doi.org/10.3390/polym14050943.

[19]   Jian J, Xiangbin Z, Xianbo H. An overview on synthesis, properties and applications of poly(butylene-adipate-co-terephthalate)–PBAT. Adv Ind Eng Polym Res 2020;3:19–26. https://doi.org/10.1016/J. AIEPR.2020.01.001.

[20]   Apriyanto A, Compart J, Fettke J. A review of starch, a unique biopolymer – structure, metabolism and in planta modifications. Plant Sci 2022;318:111223. https://doi.org/10.1016/j.plantsci.2022.111223.

[21]   Melchor-Martínez EM, Macías-Garbett R, Alvarado-Ramírez L, Araújo RG, Sosa-Hernández JE, Ramírez-Gamboa D, et al. Towards a circular economy of plastics: An evaluation of the systematic transition to a new generation of bioplastics. Polymers 2022;14:1203. https://doi.org/10.3390/ polym14061203.

[22]   Janssen L, Moscicki L. Thermoplastics starch. Biodegrad Polym Their Pract Util 2009;1–29.

[23]   Ojogbo E, Ogunsona EO, Mekonnen TH. Chemical and physical modifications of starch for renewable polymeric materials. Mater Today Sustain 2020;7–8:100028. https://doi.org/10.1016/j. mtsust.2019.100028.

[24]   Peelman N, Ragaert P, Ragaert K, De Meulenaer B, Devlieghere F, Cardon L. Heat resistance of new biobased polymeric materials, focusing on starch, cellulose, PLA, and PHA. J Appl Polym Sci 2015;132:42305. https://doi.org/10.1002/app.42305.

[25]   Zhang Y, Rempel C, Liu Q. Thermoplastic starch processing and characteristics – a review. Crit Rev Food Sci Nutr 2014;54:1353–70. https://doi.org/10.1080/10408398.2011.636156.

[26]   Zhang R, Ma S, Li L, Zhang M, Tian S, Wang D, et al. Comprehensive utilization of corn starch processing by-products: A review. Grain Oil Sci Technol 2021;4:89–107. https://doi.org/10.1016/J. GAOST.2021.08.003.

[27]   Babu RP, O'Connor K, Seeram R. Current progress on bio-based polymers and their future trends. Prog Biomater 2013;2:8. https://doi.org/10.1186/2194-0517-2-8.

[28]   Klemm D, Heublein B, Fink HP, Bohn A. Cellulose: fascinating biopolymer and sustainable raw material. Angew Chemie Int Ed 2005;44:3358–93. https://doi.org/10.1002/ANIE.200460587.

[29]   Rose M, Palkovits R. Cellulose-based sustainable polymers: State of the art and future trends. Macromol Rapid Commun 2011;32:1299–311. https://doi.org/10.1002/MARC.201100230.

[30]   Turunen O, Mandell L, Eklund V, Ekman K, Huttunen JI. Procedure for producing soluble cellulose carbamate. FI61033, 1982.

[31]   Lehrhofer AF, Goto T, Kawada T, Rosenau T, Hettegger H. The in vitro synthesis of cellulose – a mini-review. Carbohydr Polym 2022;285:119222. https://doi.org/10.1016/J.CARBPOL.2022.119222.

[32]   Foster EJ, Moon RJ, Agarwal UP, Bortner MJ, Bras J, Camarero-Espinosa S, et al. Current characterization methods for cellulose nanomaterials. Chem Soc Rev 2018;47:2609–79. https://doi.org/10.1039/C6CS00895J.

[33]   Phanthong P, Reubroycharoen P, Hao X, Xu G, Abudula A, Guan G. Nanocellulose: Extraction and application. Carbon Resour Convers 2018;1:32–43. https://doi.org/10.1016/J.CRCON.2018.05.004.

[34]   Pires JRA, Souza VGL, Fernando AL. Valorization of energy crops as a source for nanocellulose production – current knowledge and future prospects. Ind Crops Prod 2019;140:111642. https://doi.org/10.1016/J.INDCROP.2019.111642.

[35]   Salimi S, Sotudeh-Gharebagh R, Zarghami R, Chan SY, Yuen KH. Production of nanocellulose and its applications in drug delivery: A critical review. ACS Sustain Chem Eng 2019;7:15800–27. https://doi.org/10.1021/ACSSUSCHEMENG.9B02744/ASSET/IMAGES/MEDIUM/SC9B02744_0021.GIF.

[36]   Naz S, Ali JS, Zia M. Nanocellulose isolation characterization and applications: A journey from non-remedial to biomedical claims. Bio-Design Manuf2019 23 2019;2:187–212. https://doi.org/ 10.1007/S42242-019-00049-4.

[37]   Sun B, Hou Q, Liu Z, Ni Y. Sodium periodate oxidation of cellulose nanocrystal and its application as a paper wet strength additive. Cellul2015 222 2015;22:1135–46. https://doi.org/10.1007/S10570-015-0575-5.

[38]   Enzymatic preparation of nanocrystalline and microcrystalline cellulose, TAPPI JOURNAL May 2014 n.d. https://imisrise.tappi.org/TAPPI/Products/14/MAY/14MAY35.aspx (accessed August 11, 2022).

[39]   Lazko J, Sénéchal T, Bouchut A, Paint Y, Dangreau L, Fradet A, et al. Acid-free extraction of cellulose type i nanocrystals using brønsted acid-type ionic liquids. Nanocomposites 2016;2:65–75. https://doi.org/10.1080/20550324.2016.1199410.

[40]   Novo LP, Bras J, García A, Belgacem N, Curvelo da AAS. A study of the production of cellulose nanocrystals through subcritical water hydrolysis. Ind Crops Prod 2016;93:88–95. https://doi.org/10.1016/J.INDCROP.2016.01.012.

[41]   Moon RJ, Schueneman GT, Simonsen J. Overview of cellulose nanomaterials, their capabilities and applications. Jom 2016 689 2016;68:2383–94. https://doi.org/10.1007/S11837-016-2018-7.

[42]   Thomas B, Raj MC, Athira BK, Rubiyah HM, Joy J, Moores A, et al. Nanocellulose, a versatile green platform: From biosources to materials and their applications. Chem Rev 2018;118:11575–625. https://doi.org/10.1021/ACS.CHEMREV.7B00627/ASSET/IMAGES/MEDIUM/CR-2017-00627Z_0029.GIF.

[43]   Xie H, Du H, Yang X, Si C. Recent strategies in preparation of cellulose nanocrystals and cellulose nanofibrils derived from raw cellulose materials. Int J Polym Sci 2018;2018. https://doi.org/10.1155/2018/7923068.

[44]   Trache D, Tarchoun AF, Derradji M, Hamidon TS, Masruchin N, Brosse N, et al. Nanocellulose: From fundamentals to advanced applications. Front Chem 2020;8:392. https://doi.org/10.3389/fchem.2020.00392.

[45]   Mudgil D, Barak S, Khatkar BS. Guar gum: Processing, properties and food applications – a review. J Food Sci Technol 2014;51:409–18. https://doi.org/10.1007/S13197-011-0522-X/TABLES/3.

[46]   Sharma G, Sharma S, Kumar A, Al-Muhtaseb AH, Naushad M, Ghfar AA, et al. Guar gum and its composites as potential materials for diverse applications: A review. Carbohydr Polym 2018;199:534–45. https://doi.org/10.1016/j.carbpol.2018.07.053.

[47]   Theocharidou A, Mourtzinos I, Ritzoulis C. The role of guar gum on sensory perception, on food function, and on the development of dysphagia supplements – a review. Food Hydrocoll Heal 2022;2:100053. https://doi.org/10.1016/J.FHFH.2022.100053.

[48]   Thombare N, Jha U, Mishra S, Siddiqui MZ. Guar gum as a promising starting material for diverse applications: A review. Int J Biol Macromol 2016;88:361–72. https://doi.org/10.1016/j.ijbiomac.2016.04.001.

[49]   Adimule V, Kerur SS, Chinnam S, Yallur BC, Nandi SS. Guar gum and its nanocomposites as prospective materials for miscellaneous applications: A short review. Top Catal 2022;1:1–14. https://doi.org/10.1007/S11244-022-01587-5/TABLES/2.

[50]   García-Ochoa F, Santos V, Casas J, Gómez E. Xanthan gum: production, recovery, and properties. Biotechnol Adv 2000;18:549–79. https://doi.org/10.1016/S0734-9750(00)00050-1.

[51]   Becker A, Katzen F, Pühler A, Ielpi L. Xanthan gum biosynthesis and application: A biochemical/genetic perspective. Appl Microbiol Biotechnol 1998;50:145–52. https://doi.org/10.1007/S002530051269.

[52]   Bhat IM, Wani SM, Mir SA, Masoodi FA. Advances in xanthan gum production, modifications and its applications. Biocatal Agric Biotechnol 2022;42:102328. https://doi.org/10.1016/j.bcab.2022.102328.

[53]   Abu Elella MH. Synthesis and potential applications of modified xanthan gum. J Chem Eng Res Updat 2021;8:73–97. https://doi.org/10.15377/2409-983X.2021.08.6.

[54]   Habibi H, Khosravi-Darani K. Effective variables on production and structure of xanthan gum and its food applications: a review. Biocatal Agric Biotechnol 2017;10:130–40. https://doi.org/10.1016/J.BCAB.2017.02.013.

[55]   Palaniraj A, Jayaraman V. Production, recovery and applications of xanthan gum by xanthomonas campestris. J Food Eng 2011;106:1–12. https://doi.org/10.1016/j.jfoodeng.2011.03.035.

[56]   Patel J, Maji B, Moorthy NSHN, Maiti S. Xanthan gum derivatives: Review of synthesis, properties and diverse applications. RSC Adv 2020;10:27103. https://doi.org/10.1039/D0RA04366D.

[57] Winnacker M. Polyhydroxyalkanoates (phas): Recent advances in their synthesis and applications. Eur J Lipid Sci Technol 2019;1900101. https://doi.org/10.1002/ejlt.201900101.

[58] Pérez V, Mota CR, Muñoz R, Lebrero R. Polyhydroxyalkanoates (PHA) production from biogas in waste treatment facilities: Assessing the potential impacts on economy, environment and society. Chemosphere 2020;255:126929. https://doi.org/10.1016/J.CHEMOSPHERE.2020.126929.

[59] Kopf S, Åkesson D, Skrifvars M. Textile fiber production of biopolymers – a review of spinning techniques for polyhydroxyalkanoates in biomedical applications. Polym Rev 2022:1–46. https://doi.org/10.1080/15583724.2022.2076693.

[60] Isikgor FH, Becer CR. Lignocellulosic biomass: A sustainable platform for the production of bio-based chemicals and polymers. Polym Chem 2015;6:4497–559. https://doi.org/10.1039/C5PY00263J.

[61] Intasian P, Prakinee K, Phintha A, Trisrivirat D, Weeranoppanant N, Wongnate T, et al. Enzymes, in vivo biocatalysis, and metabolic engineering for enabling a circular economy and sustainability. Chem Rev 2021;121:10367–451. https://doi.org/10.1021/acs.chemrev.1c00121.

[62] Newlight technologies, Inc. | AirCarbon n.d. https://www.newlight.com/ (accessed July 3, 2022).

[63] Blunt W, Levin D, Cicek N. Bioreactor operating strategies for improved polyhydroxyalkanoate (PHA) productivity. Polymers (Basel) 2018;10:1197. https://doi.org/10.3390/polym10111197.

[64] Haddadi MH, Asadolahi R, Negahdari B. The bio extraction of bioplastics with focus on polyhydroxybutyrate: A review. Int J Environ Sci Technol 2019;16:3935–48. https://doi.org/10.1007/S13762-019-02352-0.

[65] Abedi E, Hashemi SMB. Lactic acid production – producing microorganisms and substrates sources-state of art. Heliyon 2020;6:e04974. https://doi.org/10.1016/j.heliyon.2020.e04974.

[66] Henard CA, Smith H, Dowe N, Kalyuzhnaya MG, Pienkos PT, Guarnieri MT. Bioconversion of methane to lactate by an obligate methanotrophic bacterium. Sci Rep2016 61 2016;6:1–9. https://doi.org/10.1038/srep21585.

[67] Miller M, Suominen P, Aristidou A, Hause BM, Van Hoek P, Dundon CA. Lactic acid-producing yeast cells having nonfunctional l-or d-lactate: Ferricytochrome c oxidoreductase gene. US8137953B2, 2006.

[68] Drumright RE, Gruber PR, Henton DE. Polylactic acid technology. Adv Mater 2000;12:1841–46. https://doi.org/10.1002/1521-4095(200012)12:23<1841::AID-ADMA1841>3.0.CO;2-E.

[69] Cunha BLC, Bahú JO, Xavier LF, Crivellin S, de Souza SDA, Lodi L, et al. Lactide: Production routes, properties, and applications. Bioengineering 2022;9:164. https://doi.org/10.3390/bioengineering9040164.

[70] Meng X, Yu L, Cao Y, Zhang X, Zhang Y. Progresses in synthetic technology development for the production of l-lactide. Org Biomol Chem 2021;19:10288–95. https://doi.org/10.1039/D1OB01918J.

[71] Di Lorenzo, Maria Laura Androsch R. Synthesis, Structure and Properties of Poly(lactic acid), Cham: Springer International Publishing; vol 279, 2018. https://doi.org/10.1007/978-3-319-64230-7.

[72] Masutani K, Kimura Y. Chapter 1. PLA Synthesis. From the Monomer to the Polymer. In: Jiménez A, Peltzer M, Ruseckaite R, editors. RSC Polym. Chem. Ser., Cambridge: Royal Society of Chemistry; 2014, vol 12, pp. 1–36. https://doi.org/10.1039/9781782624806-00001.

[73] Yoo DK, Kim D, Lee DS. Synthesis of lactide from oligomeric PLA: Effects of temperature, pressure, and catalyst. Macromol Res 2006;14:510–16. https://doi.org/10.1007/BF03218717.

[74] Yarkova AV, Novikov VT, Glotova VN, Shkarin AA, Borovikova YS. Vacuum effect on the lactide yield. Procedia Chem 2015;15:301–07. https://doi.org/10.1016/j.proche.2015.10.048.

[75] Ghadamyari M, Chaemchuen S, Zhou K, Dusselier M, Sels BF, Mousavi B, et al. One-step synthesis of stereo-pure l,l lactide from l-lactic acid. Catal Commun 2018;114:33–36. https://doi.org/10.1016/j.catcom.2018.06.003.

[76] Xu Y, Fang Y, Cao J, Sun P, Min C, Qi Y, et al. Controlled synthesis of l-lactide using Sn-beta zeolite catalysts in a one-step route. Ind Eng Chem Res 2021;60:13534–41. https://doi.org/10.1021/acs.iecr.1c02448.

[77] Dusselier M, Van Wouwe P, Dewaele A, Jacobs PA, Sels BF. Shape-selective zeolite catalysis for bioplastics production. Science 2015;349:78–80. https://doi.org/10.1126/science.aaa7169.

[78] Baratian S, Hall ES, Lin JS, Xu R, Runt J. Crystallization and solid-state structure of random polylactide copolymers: poly(l-lactide-*co*-d-lactide)s. Macromolecules 2001;34:4857–64. https://doi.org/10.1021/ma001125r.

[79] Balla E, Daniilidis V, Karlioti G, Kalamas T, Stefanidou M, Bikiaris ND, et al. Poly(lactic acid): a versatile biobased polymer for the future with multifunctional properties – from monomer synthesis, polymerization techniques and molecular weight increase to PLA applications. Polymers 2021;13:1822. https://doi.org/10.3390/polym13111822.

[80] Celli A, Colonna M, Gandini A, Gioia C, Lacerda TM, Vannini M. Polymers from Monomers Derived from Biomass. Chem. Fuels from Bio-Based Build, Blocks, Weinheim, Germany: Wiley-VCH Verlag GmbH & Co. KGaA; 2016, pp. 315–50. https://doi.org/10.1002/9783527698202.ch13.

[81] Sugihara S, Toshima K, Matsumura S. New strategy for enzymatic synthesis of high-molecular-weight poly(butylene succinate) via cyclic oligomers. Macromol Rapid Commun 2006;27:203–07. https://doi.org/10.1002/MARC.200500723.

[82] Takiyama E, Fujimaki T, Seki S, Hokari T, Hatano Y. Method for manufacturing biodegradable high molecular aliphatic polyester. US 5310782, 1994.

[83] Sisti L, Totaro G, Marchese P. PBS makes its entrance into the family of biobased plastics. Biodegrad Biobased Polym Environ Biomed Appl 2016;225–85. https://doi.org/10.1002/9781119117360.CH7.

[84] Aeschelmann F, Carus M. Biobased building blocks and polymers in the world: Capacities, production, and applications–status quo and trends towards 2020. Ind Biotechnol 2015;11:154–59. https://doi.org/10.1089/ind.2015.28999.fae.

[85] Sanford K, Chotani G, Danielson N, Zahn JA. Scaling up of renewable chemicals. Curr Opin Biotechnol 2016;38:112–22. https://doi.org/10.1016/J.COPBIO.2016.01.008.

[86] Gallezot P. Conversion of biomass to selected chemical products. Chem Soc Rev 2012;41:1538–58. https://doi.org/10.1039/C1CS15147A.

[87] Thuo M, Gregory P, Banerjee S, Du C. 1 Introduction: Biopolymers and Biocomposites. In: Biopolym. Compos., De Gruyter; 2021, pp. 1–26. https://doi.org/10.1515/9781501521942-001.

[88] Van Der Waal JC, Gruter GJM. Process for preparing ethylene glycol from a carbohydrate. WO 2016/114658 A1, 2016.

[89] Li M, Niu F, Zuo X, Metelski PD, Busch DH, Subramaniam B. A spray reactor concept for catalytic oxidation of p-xylene to produce high-purity terephthalic acid. Chem Eng Sci 2013;104:93–102. https://doi.org/10.1016/J.CES.2013.09.004.

[90] Peters MW, Taylor JD, Jenni M, Manzer LE, David H. Integrated process to selectively convert renewable isobutanol to p-xylene. US20110087000A1, 2012.

[91] Kanwal A, Zhang M, Sharaf F, Li C. Polymer pollution and its solutions with special emphasis on poly (butylene adipate terephthalate (PBAT)). Polym Bull 2022:1–28. https://doi.org/10.1007/s00289-021-04065-2.

[92] Lim XZ. Microplastics are everywhere – but are they harmful? Nature 2021;593:22–25. https://doi.org/10.1038/D41586-021-01143-3.

[93] Garcia JM, Robertson ML. The future of plastics recycling. Science 2017;358:870–72. https://doi.org/10.1126/science.aaq0324.

[94] Lamberti FM, Román-Ramírez LA, Wood J. Recycling of bioplastics: Routes and benefits. J Polym Environ 2020;28:2551–71. https://doi.org/10.1007/S10924-020-01795-8.

[95] Häußler M, Eck M, Rothauer D, Mecking S. Closed-loop recycling of polyethylene-like materials. Nature 2021;590:423–27. https://doi.org/10.1038/s41586-020-03149-9.

[96] Niaounakis M. Recycling of biopolymers – the patent perspective. Eur Polym J 2019;114:464–75. https://doi.org/10.1016/j.eurpolymj.2019.02.027.

[97]   Spierling S, Knüpffer E, Behnsen H, Mudersbach M, Krieg H, Springer S, et al. Bio-based plastics – a review of environmental, social and economic impact assessments. J Clean Prod 2018;185:476–91. https://doi.org/10.1016/J.JCLEPRO.2018.03.014.

[98]   Hottle TA, Bilec MM, Landis AE. Sustainability assessments of bio-based polymers. Polym Degrad Stab 2013;98:1898–907. https://doi.org/10.1016/j.polymdegradstab.2013.06.016.

[99]   Meereboer KW, Misra M, Mohanty AK. Review of recent advances in the biodegradability of polyhydroxyalkanoate (PHA) bioplastics and their composites. Green Chem 2020;22:5519–58. https://doi.org/10.1039/d0gc01647k.

[100] Fakirov S. Biodegradable Polyesters, Weinheim, Germany: Wiley-VCH Verlag GmbH & Co. KGaA; Vol 5, 2015. https://doi.org/10.1002/9783527656950.

[101] Helanto K, Matikainen L, Talja R, Rojas OJ. Bio-based polymers for sustainable packaging and biobarriers: A critical review. BioResources 2019;14:4902–51. https://doi.org/10.15376/biores.14.2. Helanto.

[102] Greene JP. Sustainable Plastics, Hoboken, NJ, USA: John Wiley & Sons, Inc.; 2014. https://doi.org/10.1002/9781118899595.

[103] Global production capacities of bioplastics 2021-2026 Global production capacities of bioplastics 2021 (by material type) Global production capacities of bioplastics 2026 (by material type) n.d. (accessed August 11, 2022)

[104] PlascticsEurope. Plastics-the Facts 2021 An analysis of European plastics production, demand and waste data n.d. (accessed August 11, 2022)

[105] Plastics – the Facts 2021 • Plastics Europe. n.d. https://plasticseurope.org/knowledge-hub/plastics-the-facts-2021/ (accessed August 11, 2022).

[106] Global bioplastics production will more than triple within the next five years – World Bioeconomy Forum. n.d. https://wcbef.com/news-bulletin/global-bioplastics-production-will-more-than-triple-within-the-next-five-years/ (accessed August 11, 2022).

[107] The business case for investing in sustainable plastics | World Economic Forum. n.d. https://www.weforum.org/agenda/2020/01/the-business-case-for-investing-in-sustainable-plastics/ (accessed July 29, 2022).

[108] Bioplastics Market Size, Share & Growth Analysis Report n.d. https://www.bccresearch.com/market-research/plastics/global-markets-and-technologies-for-bioplastics.html (accessed July 29, 2022).

[109] Gross RA. Biodegradable polymers for the environment. Science 2002;297:803–07. https://doi.org/10.1126/science.297.5582.803.

[110] Singh AA, Genovese ME. Green and Sustainable Packaging Materials Using Thermoplastic Starch. In: Sustain. Food Packag. Technol., Wiley; 2021, pp. 133–60. https://doi.org/10.1002/9783527820078.ch5.

[111] Shafqat A, Tahir A, Mahmood A, Tabinda AB, Yasar A, Pugazhendhi A. A review on environmental significance carbon foot prints of starch based bio-plastic: a substitute of conventional plastics. Biocatal Agric Biotechnol 2020;27:101540. https://doi.org/10.1016/j.bcab.2020.101540.

[112] Mishra S, Singh PK, Pattnaik R, Kumar S, Ojha SK, Srichandan H, et al. Biochemistry, synthesis, and applications of bacterial cellulose: A review. Front Bioeng Biotechnol 2022;10:https://doi.org/10.3389/FBIOE.2022.780409.

[113] Koller M, Mukherjee A. A new wave of industrialization of PHA biopolyesters. Bioengineering 2022;9:74. https://doi.org/10.3390/bioengineering9020074.

[114] Sharma V, Sehgal R, Gupta R. Polyhydroxyalkanoate (PHA): Properties and modifications. Polymer 2021;212:123161. https://doi.org/10.1016/j.polymer.2020.123161.

[115] Chen G-Q. A microbial polyhydroxyalkanoates (PHA) based bio- and materials industry. Chem Soc Rev 2009;38:2434–46. https://doi.org/10.1039/b812677c.

[116] Xiao L, Wang B, Yang G, Gauthier M. Poly(lactic acid)-based biomaterials: Synthesis, modification and applications. Biomed Sci Eng Technol 2012;902. https://doi.org/10.5772/23927.

[117] Aliotta L, Seggiani M, Lazzeri A, Gigante V, Cinelli P. A brief review of poly (butylene succinate) (PBS) and its main copolymers: Synthesis, blends, composites, biodegradability, and applications. Polymers 2022;14:844. https://doi.org/10.3390/polym14040844.

[118] Loos K, Zhang R, Pereira I, Agostinho B, Hu H, Maniar D, et al. A perspective on PEF synthesis, properties, and end-life. Front Chem 2020;8:585. https://doi.org/10.3389/fchem.2020.00585.

[119] Siracusa V, Blanco I. Bio-polyethylene (bio-PE), bio-polypropylene (bio-PP) and bio-poly(ethylene terephthalate) (bio-PET): Recent developments in bio-based polymers analogous to petroleum-derived ones for packaging and engineering applications. Polymers 2020;12:1641. https://doi.org/10.3390/polym12081641.

Sandeep Kulkarni

Chapter 21
Green bioplastics in sustainable packaging for the emerging bio economy

21.1 Introduction

The chapter will cover the following topics, namely, definition of bioplastics, sources of bioplastics, end of life options for bioplastics, global standards for biodegradation, common types of bioplastics in packaging, drivers for bioplastic use in packaging, some key considerations for developing bioplastic packaging for wider use, challenges, and opportunities. As some of these topics are also covered in other chapters in the book, the coverage here will attempt to minimize duplication as much as possible, yet have a good flow.

21.2 Definition of bioplastics

Bioplastics, according to the most widely accepted definition, are plastics that are partially or fully bio-based and/or are biodegradable. There are two key terms in this definition as outlined below:

a. A bio-based bioplastic has some or all of its carbon derived from a renewable source.
b. Biodegradable bioplastics are those that degrade into carbon dioxide (CO_2), methane (CH_4), water (H_2O), and biomass through biological action in a defined environment and in a defined timescale. The defined environments typically include composting and anaerobic digestion. (See chapter 19 and 23 by Adur for further details.)

It should be noted that the term "biodegradable" is an open-ended term and does not mean much unless it is accompanied by a description of the conditions of degradation (temperature, humidity, and nature of microbes) as well as the time frame of degradation. As will be discussed in section 4 of this chapter, the unqualified use of "biodegradable" as a marketing claim for packaging is misleading, and is either restricted or banned by many global trade authorities and governments.

The two important considerations related to bioplastics are where they come from/originate ("beginning of life" or source) and what happens to them at the end of their useful life ("end of life"). Table 21.1 provides examples of some common commercial bioplastics, categorized by their beginning and end of life. See chapter xxx for acronyms used in the Table.

Sandeep Kulkarni, KoolEarth Solutions, Inc., Alpharetta, GA, USA, e-mail: skulkarni@koolearthsolutions.com

https://doi.org/10.1515/9783110791228-021

Table 21.1: Beginning and end of life of some commercial bioplastics.

	Biodegradable	Non-biodegradable
Bio-based (corn, sugarcane, stover, wheat straw, etc.	PLA, PHA, PBS	PET, PEF
Non-bio-based (petroleum based)	PBAT	Not classified as bioplastic

As can be seen from this table, different bioplastics can have different combinations of beginning-of-life and end-of-life features. A common misconception is that all bioplastics are also biodegradable. However, this is not true. As indicated in the table, bioplastics such as PET and PEF can be derived from biological/renewable sources but are not biodegradable (at least in any reasonable timeframe).

The following sections describe the beginning-of-life and end-of-life considerations in more detail.

21.3 Sources of bioplastics (where do bioplastics come from?)

Bioplastics can come from variety of plant-based and other renewable sources, called "feedstocks." These feedstocks are generally classified into first, second, and third generations.

First-generation (or Gen1) feedstocks include crops (such as corn and sugarcane) traditionally grown for human consumption. The use of such feedstocks for producing bioplastics has given rise to concerns that this will result in reduced availability as food sources and also that it will result in land being used exclusively for growing crops for bioplastic production. These concerns can be largely mitigated by sustainable sourcing of feedstocks and sustainable farming practices. It should also be noted that as per information summarized by European Bioplastics [1] the feedstock currently used to produce bioplastics relies on less than 0.02 percent of the global agricultural area, compared to 91 percent of the area used for pasture and the production of food and feed. Despite the predicted continued growth in the bioplastics market at the current stage of technological development, the share of global agricultural area used to grow feedstock for the production of bioplastics will only slightly increase – to still below 0.06 percent in 2026. This clearly demonstrates that there is no competition between food/feed and industrial production.

Second-generation feedstocks (Gen2) are ones that are traditionally not used for human consumption. These are the byproducts remaining after harvesting of the primary food crop, and include materials such as corn stover, wheat straw, rice husks, and oat hulls.

Third-generation (Gen3) feedstocks are non-plant-based sources such as algae and waste gases, such as CO_2 and methane from steel mills or landfills.

Due to the concerns and perception issues with Gen1 feedstocks, there is a great deal of focus and interest in using Gen2 and ultimately Gen3 feedstocks for bioplastics

21.4 End-of-life options for bioplastics (where do they end up and what happens to them?)

The options for processing bioplastic-based packaging, after the end of its useful life, depends (as shown in the table in Section 1) on whether the bioplastic is biodegradable or non-biodegradable. For non-biodegradable bioplastics (such as bio-PE, bio-PET, and PEF), the only end-of-life option available is recycling. These materials should be recycled through the appropriate recycling stream available for their petrochemical counterparts. For instance, bio-PET can be recycled along with conventional PET (petroleum-based) recycling processes.

For biodegradable bioplastics, there are few different end-of-life options available.

21.5 Composting

Composting is defined as the biological degradation process of heterogeneous solid organic materials under controlled moist, self-heating, and aerobic conditions to obtain a stable material that can be used as organic fertilizer [2].

As is clear from the above definition, composting is a specific type of biodegradation that takes place under controlled conditions and in a fixed time frame (usually 90–120 days). Compost is rich in nutrients. It is used, for example, in gardens, landscaping, horticulture, urban agriculture, and organic farming. The compost itself is beneficial for the land in many ways, including as a soil conditioner, a fertilizer, as a source of vital humus or humic acids, and as a natural pesticide for soil. Composting can be carried out on an industrial scale as a manufacturing process or on a smaller scale (such as home/backyard composting).

21.5.1 Industrial composting

These processes use well defined temperature, humidity, and processing conditions. The following describes the properties of an ideal compost pile:
- Carbon-to-nitrogen ratio is roughly 30:1.

- Presence of oxygen in the decomposition process is necessary to oxidize the carbon.
- Moisture levels of 40–50% are necessary to promote growth of natural microorganisms that break down the waste.
- Temperature in the range of 120–170 °F during the active composting period.

There are three major industrial composting processes used by compost manufacturers in North America:

21.5.1.1 Windrow composting

In this process, the organic waste is laid in long piles called windrows. The piles are aerated periodically by either manually or mechanically turning them. The height of the pile is large enough to generate the required temperatures, yet is small enough to allow air flow. This process produces compost within 90–120 days.

21.5.1.2 Aerated static pile (ASP) composting

In this process, organic waste is mixed in a large pile. To aerate the pile, layers of loosely piled bulking agents (e.g., wood chips, shredded newspaper) are added so that air can pass from the bottom to the top of the pile. The piles can also be placed over a network of pipes that deliver air into or draw air out of the pile. Air blowers might be activated by a timer or temperature sensors. The ASP process produces compost in 3 to 6 months.

21.5.1.3 In-vessel composting

In-vessel composting can process large amounts of waste without using as much space as the windrow method and it can accommodate virtually any type of organic waste. This method involves feeding organic materials into a drum, silo, concrete-lined trench, or similar encasing. This allows control of the environmental conditions such as temperature, moisture, and airflow. The compost is mechanically turned or mixed to make sure the material is aerated. This method produces compost in just a few weeks. It takes a few more weeks or months until it is ready to be used because the microbial activity needs to stabilize, and the pile needs to cool.

As discussed above, depending on the type of the industrial composting process, the timing for completion of the composting process can be as short as 90 days and as long as 180 days. Most compost manufacturers attempt to produce the compost in as short time as possible, since this allows them to increase their throughput, and hence revenue. This can pose a challenge for packaging targeted to be compostable in an industrial composting process, as the packaging may not completely degrade within the short time span and may get rejected or screened out of the final compost product.

Another consideration is that compostable packaging by itself provides no value (adds no nutrients) to the final compost product. The value of compostable food packaging comes from fact that the packaging can divert more food waste/food scraps to composting facilities, since food waste is desirable for composters. Also, it keeps food waste and food scrap from ending up in landfills where it can generate methane, a highly potent greenhouse gas. A study published by the Sustainable Packaging Coalition (SPC) in 2017 [3] examined a variety of scenarios, such as a farmer's market, a concert event, an all-day festival, a quick-service restaurant, and a national grocery chain.

21.5.2 Home/backyard composting

Such a composting can refer to a variety of processes used by households to compost their food waste, yard trimmings, and other organics. These systems can range from simple bins or drums to more advanced systems where the organic waste is turned regularly and where high temperatures are employed. Unlike in the case of industrial composting, there are no standard processes or practices employed, and one home composting system can be completely different from another. Further, there is typically no control of temperature and humidity in the waste pile in home composting systems. Finally, most home composting systems are unable to achieve the high temperatures employed in industrial composting piles, making the process of composting slower and longer.

This lack of control and lack of standardization in home composting systems makes it very difficult for compostable packaging to degrade in home composting environments within a reasonable time frame. For this reason, packaging that is designed to be compostable in an industrial composting environment, cannot typically be composted in home composting systems. Furthermore, as will be discussed in the subsequent section, there are almost no standards and certifications available for home compostable packaging, making it very difficult to design packaging to be home compostable.

21.5.3 Marine and soil/land degradation

Over the past decade there has been a significantly heightened awareness and concern about litter and pollution due to plastic packaging (particularly single-use packaging), both on land and in the oceans. One of the approaches in addressing this issue has been development of packaging that could potentially biodegrade in marine as well as in land environments. The thought process behind this approach is that if any packaging is inadvertently and improperly disposed on land or is washed into the ocean, it will degrade within a relatively short time frame (not harming marine life or life on land).

The operative word in the above discussion is "**inadvertently**" since the ability of a package to degrade, either on land or in the ocean should not be taken as a license to litter.

Like the situation with home composting, there are almost no global standards and certifications for marine and soil/land degradable packaging. The primary reason for this is that soil and marine conditions vary tremendously across the globe. Marine temperatures, pH, types of microbes, and so on can be very different for different oceans and marine bodies, making it extremely difficult to define a standard set of marine conditions for testing degradation of packaging

21.6 Global standards and certifications for biodegradation

21.6.1 Need for standards

As mentioned earlier, the term "biodegradable" is an open-ended term that does not convey much information and can in fact be misleading to a consumer. Many suppliers of packaging or other plastic items have used a "biodegradable" claim/label, without any qualification, in an attempt to portray that the packaging is environmentally friendly. This approach is in fact considered "Greenwashing" (disinformation dissemination by an organization so as to present an environmentally responsible public image), and as a result many countries and regions across the world have either severely restricted or banned outright, the unqualified use of the term "biodegradable" in marketing claims. For instance, the US Federal Trade Commission states the following in its Green Guides:

> Marketers may make an unqualified degradable claim only if they can prove that the "entire product or package will completely break down and return to nature within a reasonably short period of time after customary disposal." The "reasonably short period of time" for complete decomposition of solid waste products is one year. Items destined for landfills, incinerators, or recycling facilities will not degrade within a year; so unqualified biodegradable claims for them shouldn't be made.

A company can be subject to significant financial penalties as well as loss of credibility and market share for violating these guidelines.

To address and eliminate Greenwashing and to further provide an avenue for legitimate biodegradation claims, few different global standards, certifications, and testing protocols have been developed. The following sections outline these standards and certifications.

21.6.2 Standards and certifications

ASTM D6400 is the primary standard used in the United States. This standard covers plastics and products made from plastics that are designed to be composted under aerobic conditions in municipal and industrial aerobic composting facilities where thermophilic conditions (i.e. microbes that live at high temperatures between 60 and 108 °C) are achieved.

To be considered compostable by ASTM D6400, the product (packaging, plastic item, etc.) must be tested and they should demonstrate the following:
1. Disintegration – The product must disintegrate into a defined particle/fragment size over a 12-week period.
2. Biodegradation – 90% of the organic carbon must be converted to carbon dioxide by the end of the test period when compared to the positive control (cellulose).
3. No adverse effect on the quality of the compost – No Eco toxicity.

The **Biodegradable Products Institute (BPI)** based in the US provides industrial compostablity certification for packaging that meets the ASTM D6400 standard, based on lab testing for disintegration, biodegradation, and eco toxicity. The following is an example of the type of label and communication that can be used on packaging that has been certified by BPI, as explained in the BPI CERTIFICATION MARK USAGE REQUIREMENTS Version 2 – September 2022 [4]:

COMPOSTABLE
IN INDUSTRIAL FACILITIES

Check locally, as these do not exist in many communities. **Not suitable for backyard composting.** CERT # SAMPLE

BPI ®

The **EN-13,432** is the primary standard used in Europe. This standard specifies that packaging may be deemed to be compostable only if all the constituents and components of the packaging are compostable. During the certification process, an assessment is made not only of the basic materials but also of the various additives and product properties.

The key requirements for the EN-13,432 certification are:
- Chemical composition
- Biodegradation
- Disintegration
- Quality of final compost and eco toxicity

TUV Austria, a leading Testing, Inspection, Certification & Training company based in Austria, provides industrial compostability certification for packaging that meets the EN-13,432 standard, based on lab testing (either by TUV itself or by another independent lab). The following is an example of the type of label and communication that can be used on packaging that has been certified by TUV Austria, as outlined on the TUV Green Marks website [5]

In addition to industrial compostability certification, TUV Austria also provides certification for home compostability (called OK Compost Home). The OK compost HOME certification program does not explicitly refer to a specific standard but details all the technical requirements that a product must meet in order to obtain the certification. In fact, the requirements of the OK compost HOME program, defined in 2003, have served as the basis for the drafting of several standards such as:
- Australia: AS 5810 (2010) – Biodegradable plastics – Biodegradable plastics suitable for home composting
- France: NF T 51,800 (2015) – Plastics – Specifications for plastics suitable for home composting
- Europe: prEN 17,427 (2020) – Packaging – Requirements and test scheme for carrier bags suitable for treatment in well-managed home composting installations

The following label communicates that packaging has met the TUV Austria requirements for home compostability:

In addition to these composting standards, TUV Austria also provides certifications for biodegradation in soil, marine, and freshwater environments. These labels are shown below:

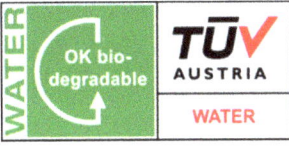

These last three certifications are not common, given the fact that there currently are no globally accepted standards for biodegradation in soil, marine, or natural freshwater environments. TUV provides these certifications based on testing and standards developed by them internally; however, these methods and standards are not yet widely accepted. Also, there is a risk that such labels can encourage consumers to dispose packaging improperly on land or in oceans, leading to increased littering. Nonetheless, as mentioned earlier, the ability of a plastic or packaging to biodegrade on land or in marine environments in a reasonably short time frame can be a useful in avoiding/managing any plastics that might **inadvertently** end up on land or in oceans as litter.

21.7 Examples of bioplastics used in packaging

The global production capacity for bioplastics in 2021 was roughly 2.5 million tons, as discussed in Chapter xxx by Adur. Further, this capacity is expected to grow to 7.5 million tons by 2026. This chapter will discuss the packaging applications of some of these bioplastics (note that the chemistry, production, and properties of these bioplastics are discussed in Chapter xxx by Adur and will not be repeated here)

21.7.1 Polylactic acid (PLA)

Polylactic acid (PLA) is a compostable polyester made from renewable feedstock. With lactic acid as the raw material, PLA is produced by fermentation of glucose or sucrose, and is refined to a high purity.

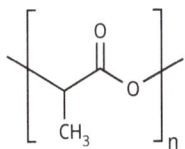

Figure 21.1: Structure Of PLA.

Over the past decade or so, PLA has found several applications in food and food service packaging industry, such as:
– Cups and lids for hot and cold beverages
– Carry-out food containers
– Single-use plates and cutlery (forks, spoons, knives etc.)

In addition to these rigid packaging applications, PLA has also been used in flexible packaging applications such as packaging for snack foods (chips, confectionery, and so on).

For both rigid and flexible packaging applications, it is necessary to compound PLA with appropriate additives to allow it to be processed using existing processing equipment (since current equipment is optimized for "conventional" plastics such as PE and PP).

As has been discussed in Chapter xxx, PLA inherently exhibits low barrier to penetration of gases such as oxygen and water vapor. For this reason, many of the flexible packaging applications of PLA require that it be used a part of a multilayered structure where the non-PLA layers can impart moisture vapor and oxygen barrier to the packaging structure. By appropriately selecting the non-PLA barrier layers, it is possible to maintain the compostability of the package while providing appropriate protection to the food being packaged.

It should be noted that PLA is only compostable in an industrial composting environment, and not in home composting or other uncontrolled environments (such as soil, ocean). The reason for this is that the PLA requires elevated temperature as well as specific moisture and microbial conditions for biodegradation, and these conditions are only found in an industrial composting situation.

One of the areas where PLA-based packaging has found good success is in situations where PLA-based packaging can drive increased diversion of food waste into composting (and hence away from landfills). For instance, in the case of some Quick Service Restaurants (QSRs) and sports arenas, it has been possible to collect food-serviceware (plates, bowls, and clamshells) along with leftover food scraps and send the combination to a local composting facility. However, it is important to ensure that the compostable packaging used in such cases is certified compostable by a recognized body (such as BPI in the US) and demonstrated to degrade completely within the 60–90 day timeframe used by most composters. This avoids the contamination of the compost by any packaging that has false/misleading compostability claims.

PLA has also found application as a liner for fiber-based coffee cups as an alternative to polyethylene, which is currently used in most coffee cups. Many of these PLA-lined cups are certified as compostable in industrial composting. However, currently there is a major lack of collection infrastructure as well as lack of widespread availability of industrial composting facilities (particularly those that accept compostable packaging). This has led to a large number of compostable cups ending up in landfills.

An example of the use of PLA in the flexible packaging arena is the compostable SunChips bag introduced by PepsiCo in early 2010. This bag was composed of a 3-layer all-PLA structure, with metallization on the innermost layer for moisture vapor and oxygen barrier. This bag was certified compostable by BPI. An image of the bag is shown below

While the bag was initially favorably received in the market, consumers discovered that the bag was extremely "noisy." This was due to the higher stiffness of the PLA structure (versus the traditional BOPP film construction), which resulted in a loud crackling sound when the bag was handled or pressed. Due to the noise issue as well as the higher cost of PLA, PepsiCo/Frito-Lay pulled the compostable bags out of the market towards end of 2010.

In late 2021, PepsiCo/Frito-Lay introduced an industrially compostable bag for its Off the Eaten Path brand. These bags are available at Whole Foods and select retailers. In order to promote composting of these bags, Frito-Lay has teamed up with Terracycle; so consumers can either mail in the bags or locate a local composting facility where they can drop off the bags. The bags are presumably made of PLA although Frito-Lay has not provided any details of their construction.

An example of the use of PLA in rigid packaging applications is yogurt cup by Danone for its Activia product line. This PLA-based yogurt cup was introduced in Germany in 2011 and Danone used Ingeo PLA from NatureWorks for this cup. The new yogurt cup was the result of a close cooperation between Danone, WWF (World Wildlife Fund)-Germany, and NatureWorks. Danone claimed (based on a LCA study) that this move resulted in 25% reduction of carbon footprint and 43% less fossil resources use, versus their conventional polystyrene cups.

One application area where PLA is notably absent is bottles for beverages such as water. The reason for this is that PLA bottles can be easily confused with PET bottles by a consumer (since both are clear). The recycling of PLA bottles along with PET causes significant disruption to the PET recycling process, and for this reason there was significant pushback from PET recyclers to the initial introduction of PLA water bottles in the market. Many of the large PLA producers, such as NatureWorks, have therefore not marketed their resins for use in beverage bottles.

21.7.2 Polybutylene succinate (PBS)

Polybutylene succinate (PBS) is polyester produced by copolymerization of butane diol and succinic acid:

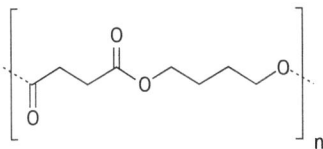

Figure 21.2: Structure of PBS.

While succinic acid is typically derived from renewable sources (usually corn), currently most butane diol is based on petrochemical sources. For this reason, currently available grades of PBS are only partially bio-based. There are, however, some companies (such as Genomatica) that have developed butane diol from biological sources and are in the process of scaling up production. However, given the price premium for bio-sourced butane diol versus its petroleum-based counterpart, it is unlikely that 100% bio-based PBS will be commercially available unless there is strong pull from major consumer brands.

Similar to PLA, PBS is compostable in an industrial composting environment. In addition, some grades of PBS, specifically those based on polybutylene succinate adipate, are also home compostable.

Some key advantages of PBS in flexible packaging applications are:

– Excellent heat sealability with low heat seal temperature
– Better oxygen barrier than LDPE
– Good to retain aroma, such as limonene
– Suitable for food contact, such as weak acidic, fatty, and oily food
– As sealant, this can be used by coextrusion or dry lamination

PBS has found commercial applications in compostable coffee pods/capsules and coffee pouches, and as liner for coffee cups. While PBS exhibits fairly good barrier to oxygen ingress into a package, it has very low barrier against water vapor. Most food packaging applications typically require a high degree of moisture barrier in order to prevent the food from getting soggy. For this reason, PBS film is typically laminated to a film with high moisture barrier (such as NatureFlex regenerated cellulose film from Futamura), along with a layer of vapor-deposited aluminum oxide (in a process known as metallization)

PBS was one of the technologies selected as the winner of the NextGen Cup Challenge (https://www.closedlooppartners.com/nextgen/impact/packaging-innovations/cup-challenge/), launched in 2018 by Starbucks and McDonald's, together with other leading global food service and coffee brands. The goal of the challenge was to identify and develop recyclable and compostable liner technologies to replace the polyethylene coating in current coffee and cold-drink cups. As a result of the development work, Starbucks conducted a market trial with cups lined with the bioPBS coating (from PTTMCC Biochem) in early 2020 [4]

The following picture shows some current applications of PBS:

Figure 21.3: Some PBS applications.

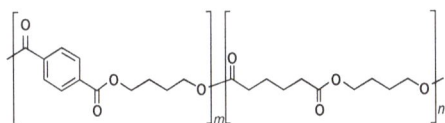

Figure 21.4: Structure of PBAT.

21.7.3 Polybutylene adipate terepthalate (PBAT)

PBAT is a copolymer of adipic acid, butane diol, and terepthalic acid. It should be noted that the structure of PBAT is a random copolymer of the blocks shown in the figure above. PBAT is currently derived completely from petroleum (non-renewable) sources and yet is industrially compostable (some grades are also home compostable).

One of the most well-known commercial PBAT resins is Ecoflex, marketed by BASF, for use in flexible packaging as a replacement for polyolefin films (PE, PP). However, typically PBAT is not used by itself but is blended with PLA to provide properties such as flexibility and toughness. Also, the incorporation of PLA makes the blend partially bio-based/renewable. Ecovio from BASF is an example of a PBAT/PLA blend available commercially. PBAT can also be blended with other materials such as thermoplastic starch (such as the MaterBi resin from Novamont).

As discussed in the case of PBS, PBAT/PLA films are typically combined with other barrier films as well as a layer of metallization in order to achieve high moisture barrier packaging. An example of such a structure is ana Ecovio film, laminated with NatureFlex film:

With appropriate selection of the film layers and the laminating adhesive, it is possible to achieve a high-barrier structure for demanding applications and yet have the package be fully compostable (industrially and potentially home compostable).

In addition to flexible packaging applications, PBAT/PLA blends have also been evaluated as paper coatings for applications such as liners for paper cups. However, there are currently no commercial cups that use the PBAT/PLA coatings.

21.7.4 Polyhydroxy alkanoate (PHA)

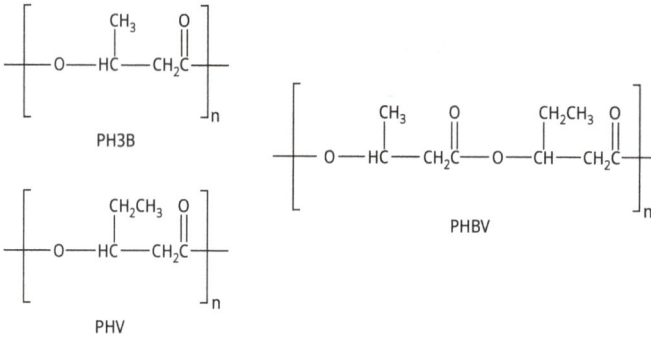

Figure 21.5: Structure of PHA.

PHAs are a unique family of bio polyesters produced by a bacterial synthesis process, starting from non-edible oils and other non-food feedstocks. See chapter xxxx for details.

PHA is the only bioplastic that is biodegradable under a wide range of conditions – industrial composting, home composting, and soil and marine degradation. As the chemistry and properties of PHAs are described in detail in Chapter xxx, these will not be repeated here.

One of the leading companies developing PHAs is Denier Scientific. Danimer and PepsiCo announced a joint venture a few years back to develop flexible packaging based on Danimer's Nodax PHA. Similarly, Mars Wrigley announced a 2-year partnership with Danimer Scientific to develop home compostable snack and candy packaging for Mars.

An example of a flexible packaging structure containing PHA is shown below:

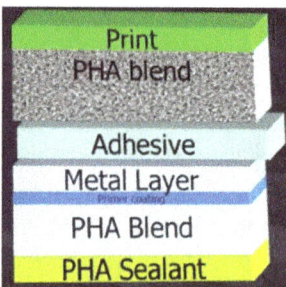

Figure 21.6: PHA flexible film packaging structure.

In addition to Danimer, several other companies such as Full Cycle Bioplastics, RWDC, Mango Materials, and so on are also working on developing PHAs from variety of sources such as used municipal solid waste (MSW), used cooking oil, and methane from landfills.

5Gyres, a well known non-profit organization that focuses on reducing plastics pollution (by conducting primary research on plastics in the ocean), has recently completed an extensive study on the degradation of various PHA articles (such as films, forks, bowls, and so on) in actual ocean environment. The results of this study are expected to be published in the coming months and this will help establish the time frame for degradation of PHA-based food service items in the ocean.

21.7.5 Bio-polyethylene terepthalate (bio-PET)

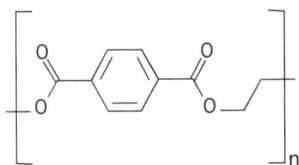

Figure 21.7: Structure of Bio-PET.

Bio-PET, like conventional PET polymer, is produced by the copolymerization of ethylene glycol (EG) with terepthalic acid (TA). In the case of bio-PET, however, either one or both monomers (EG and TA) can be derived from renewable sources.

Currently, the very limited commercially available bio-PET resin is partially bio-based where the ethylene glycol monomer has been derived from plant sources. One of the most prominent examples of a package, based on this partially bio-based PET, is the Coca Cola Plant Bottle that was first introduced into the market in 2009.

As discussed earlier, this bottle was partially bio-based (up to 30% bio content), where the EG monomer was derived from ethanol produced from sugarcane grown in Brazil. Coca Cola announced in 2015 that they had developed a fully bio-based PET bottle but this bottle was not commercially launched.

In addition to Coca Cola, several other global beverage brands (PepsiCo, Suntory, and Nestle Waters) have ongoing initiatives as well as alliances to develop 100% bio-based PET bottles.

Bio-PET (whether partially or fully bio-based) is identical in every way to petroleum-based PET and can be recycled using standard PET recycling processes.

21.7.6 Bio-Polyethylene (Bio-PE)

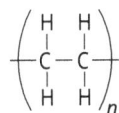

Figure 21.8: Structure of bio-PE.

Bio-PE is identical to petroleum-based PE, except that it is derived from renewable sources. Currently, the only commercial producer of bio-PE is Braskem, based in Brazil. Braskem manufactures its bio-PE by polymerizing ethylene derived from sugarcane ethanol.

Bio-PE can be used as a "drop in" replacement for conventional PE in all applications. The use of bio-PE, instead of petroleum-based PE, allows a brand owner to reduce the Greenhouse Gas (GHG) impact of their packaging and also to claim renewable content in their packaging.

Given the fact that bio-PE is indistinguishable from petroleum-based PE, it can be recycled through established PE recycling processes

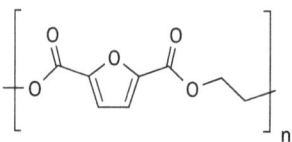

Figure 21.9: Structure of PEF.

21.7.7 Polyethylene furanoate (PEF)

PEF is a polyester synthesized by the copolymerization of ethylene glycol (EG) and furan dicarboxylic acid (FDCA). While the ethylene glycol monomer can be derived from either petroleum or renewable sources, FDCA monomer can be obtained only from natural sources such as certain carbohydrates. In fact, FDCA was identified by the US Department of Energy as one of 12 priority chemicals for establishing the "green" chemistry industry of the future. See chapter 2 for details.

FDCA exhibits significantly higher barrier to oxygen and carbon dioxide gases when compared to PET (as much as 6–8 times better gas barrier). This results in significantly longer shelf life for juices and carbonated drinks packaged in PEF bottles as against those packaged in PET containers, for the same weight of the bottle. Alternatively, significantly lighter weight PEF containers can provide similar level of shelf life as their heavier PET counterparts. PEF also has superior mechanical properties (tensile strength, hoop strength, and so on) when compared to PET, which in turn allows lighter weight bottles to be produced while maintaining good mechanical properties and stability.

Given the benefits of PEF over PET, there has been significant interest in developing this resin as well as beverage bottles from this material. Currently, two leading companies, Avantium and Corbion, are working on developing and scaling up the production of PEF as well as its applications. However, the higher cost of PEF when compared to PET has somewhat hampered the speed of development of PEF applications in beverage packaging. One of the approaches currently being explored is to create a multilayered bottle, with PEF as a thin layer "sandwiched" between thicker PET layers. This is expected to keep the cost increase reasonable, as against a bottle made fully of PEF, while still obtaining some level of shelf life (barrier) improvement.

One of the concerns with replacing PET with PEF in beverage bottles is the impact of PEF in the PET recycling process. PEF bottles will need to be collected and recycled through the existing PET recycling stream since a dedicated recycling process for PEF will not be economically (and logistically) viable, at least initially when the volumes of these bottles will be relatively small. Initial studies conducted by PEF producers as well as by some brand owners have indicated that small percentage of PEF bottles in the PET stream (likely 10% or less) will likely have little or no impact on the quality of the recycled PET.

21.8 Sustainability benefits of bioplastics

Bioplastics can provide several sustainability benefits for a consumer packaged goods (CPG) company. Some of these benefits/attributes have been discussed in other chapters; so this section will briefly touch on 4 key drivers from a packaging standpoint.

21.8.1 Greenhouse gas (GHG) reduction

Most major CPG companies have announced targets for reducing their GHG emissions across the board. Packaging is typically a fairly large contributor to the carbon footprint of a company. The use of bio-based bioplastics can help significantly reduce the GHG impact from packaging. The reason for this is that renewable sources for bioplastics, such as Gen1 or Gen2 feedstocks, are in fact "carbon sinks" i.e. they absorb and sequester CO_2 during their growing process. The benefit of this carbon sequestration gets "passed on" to the bioplastics produced from these renewable sources. In fact, the use of bioplastics is by far the largest "lever" a company can have for achieving 20% or higher GHG reduction from its packaging.

21.8.2 Decoupling from fossil sources

There are growing concerns with the use of nonrenewable fossil resources (such as petroleum) for the production of plastics, given the finite quantities of these resources as well as the pollution and emissions involved in the production of conventional plastics. The use of bio-based bioplastics allows decoupling of plastic production from fossil sources. As discussed earlier, various types of renewable feedstocks (Gen1, Gen2, or Gen3) can be used for producing bioplastics, and this allows appropriate selection of feedstocks, based on region/geography to avoid large transportation distances (for instance sugarcane and bagasse in Brazil, corn and corn stover in mid-western US, wheat straw in western Europe).

21.8.3 End-of-life for non-recyclable (or hard to recycle) plastic packaging

Compostable bioplastics are a viable solution for plastic packaging that is not currently recyclable or is difficult to recycle for either or both of the following reasons:
a. Packaging that is composed of multiple layers of plastic (such as snack food pouches, sachets etc.) that cannot be recycled through existing recycling streams
b. Packaging that does not have a viable collection and recycling infrastructure (such as single-use cups, plastic cutlery, straws etc.)

As discussed earlier in this chapter, there have been some notable initiatives and successes in the use of biodegradable bioplastics for packaging applications, such as the fully compostable SunChips bag from PepsiCo. The Seattle Mariners stadium (T-Mobile Park) in Seattle, Washington, was able to achieve over 80% waste diversion through converting all food serviceware (forks, spoons, cups, etc.) to PLA-based compostable plastics, collecting these items, and sending them to a composting facility.

21.8.4 Performance benefits

Some bioplastics can have significant performance advantages over comparable petroleum-based plastics. As discussed, PEF, which can be completely derived from plant sources, has 8 times better oxygen barrier as well as 6 times better CO_2 barrier when compared to PET. This can significantly increase the shelf life of juices and carbonated beverages, thereby reducing spoilage and waste.

21.9 Myths/misconceptions regarding bioplastics

While packaging based on bioplastics has some key benefits as outlined in the previous section, there are some widespread misunderstandings with regard to how bioplastics are perceived by consumers. Many of these misconceptions have in fact been perpetrated by misguided (or in some cases, downright unethical) suppliers of bioplastic-based packaging in an attempt to gain market share by portraying their packaging as "environmentally friendly." Some of these misconceptions/myths, as well as the actual facts are listed below.

Myth: All bioplastics will breakdown and "disappear" if discarded on the ground, for instance on roadsides or in open fields, thereby solving land litter problem.

Truth: As discussed in a previous section, compostable bioplastics will typically degrade ONLY in specific composting environments (elevated temperatures). Non biodegradable plastics (PET, PEF) will not degrade under ANY conditions.

Myth: All biodegradable bioplastics will break down and "disappear" in marine environments (thereby solving marine litter issue).

Truth: Currently, ONLY one class of biodegradable bioplastics (PHA) appears to have fairly rapid marine degradability (6 months), based on lab studies. There are currently studies underway to determine the rate of biodegradation of various PHA articles, such as bags, cutlery, and so on, in actual marine environments. Other biodegradable plastics, such as PLA, can take significantly longer to degrade in a marine environment.

Myth: Packaging based on bioplastics is always the most sustainable option.

Truth: While bioplastics do offer several sustainability benefits, they may not be the most sustainable choice for all packaging applications. For instance, as discussed earlier, replacing PET bottles with PLA bottles can be detrimental to the PET recycling stream.

Biodegradable bioplastics cannot be and should not be considered a panacea or a silver bullet to solve the plastic litter issues (either on land or in the oceans). Plastic litter can only be addressed through consumer education, proper waste management, and through diversion of plastic "waste" either into recycling or composting processes.

21.10 Key consideration in developing packaging based on bioplastics

Development of packaging based on bioplastics (either for a completely new packaging application or as a replacement of existing petroleum-based plastic packaging) requires significant upfront thought and analysis. Some of the considerations a brand must keep in mind are:

Goal/objective: It is important to clearly define what the sustainability goal, or combination of goals, is (GHG reduction, end-of-life, renewable content) and how bioplastics can help achieve these.

Packaging format and performance: It is critical to define the desired packaging format (rigid, flexible) and product protection needed (moisture barrier, oxygen barrier, CO_2 barrier). As discussed previously, a single bioplastic may not be able to meet the performance requirements, and in such a case, a combination of bioplastics (along with a non-bioplastic such as aluminum) may be needed.

End-of-life: End-of-life scenarios for the package (composting/recycling) should be clearly defined, along with strategies regarding how the packaging can be appropriately collected and disposed off.

Value chain partnerships: In order to successfully commercialize packaging based on bioplastics, a brand needs to work closely with bioplastic resin suppliers, compounders, converters, and so on. This collaboration is critical since many bioplastics can perform very differently, compared to conventional plastics, in their processing characteristics, package formation, filling, and so on.

Supply chain development: Given the fact that most bioplastics are produced from renewable/biological sources, new supply chains may need to be developed to produce the intermediates needed for production of these bioplastics. For instance, for the production of the Plant Bottle, Coca Cola had to work with partners to develop a completely new supply chain for the production of bio-based ethylene glycol (EG).

Regulations and plastic bans

Growing public awareness and concern regarding plastic pollution, in particular, marine litter, has prompted several countries worldwide to implement sweeping legislations and bans focused on plastics, specifically single-use plastics (such as straws, cutlery, cups, lids, and so on). For instance, India has announced bans of various single-use plastics. Similarly, the state of California in the US has passed a law that will ensure a 25% drop in single-use plastic by 2032. The law also requires that at least 30% of plastic items sold or bought in California are recyclable by 2028, and establish a plastic pollution mitigation fund. Other states in the US, as well as some countries in Europe, have announced bans against PE shopping bags.

Many of these regulations/bans specifically exempt compostable bioplastics such as PLA. This has prompted brand companies to switch to PLA-based straws, cups, lids, and other single-use items. This has led to significantly higher demand for PLA resin over the past few years. It is worth noting that many of the countries banning conventional single-use plastic packaging have little or no infrastructure for collecting and composting PLA-based (or other compostable plastic) items. As a result, it is very unlikely that the litter and pollution issues will be solved simply by switching from conventional plastics to compostable bioplastics for single-use applications.

21.11 Summary

Bioplastics are an important and growing segment of the flexible and rigid packaging industry worldwide to address the need to mitigate greenhouse gas emission, municipal waste, and the attendant marine pollution. Progress is evident on several fronts, including the availability of more number of bioplastic candidates, government and legislative actions, biodegradability standardization, increased awareness and acceptance by consumers, and major brand owners investing to develop and commercialize ecofriendly green packaging solutions.

References

[1]   https://www.european-bioplastics.org/faq-items/how-much-agricultural-area-is-used-for-bioplastics/.
[2]   Lobo MG, Dorta E. Postharvest Technology of Perishable Horticultural Commodities, 2019. pp. 639–66.
[3]   https://greenblueorg.s3.amazonaws.com/smm/wp-content/uploads/2018/05/Value-of-Compostable-Packaging.pdf.
[4]   https://bpiworld.org/resources/Documents/BPI%20Documents%202020/BPI%20Certification%20Mark%20Usage%20Requirements.pdf.
[5]   https://www.tuv-at.be/green-marks/.
[6]   https://stories.starbucks.com/stories/2020/starbucks-trials-a-nextgen-cup-solution/.

Ray Bergstra

Chapter 22
PHAs: the all-purpose green bioplastic for emerging bio economy

22.1 Introduction

Polyhydroxyalkanoates, or PHAs as they are most widely known, are one of the most promising biopolymer technologies that are coming into commercial prominence as the globe struggles with greenhouse gas emissions and plastic waste. It is a bioplastic. But what does this mean and why are PHAs considered to be so important?

Fundamentally, PHAs can address problems with plastic waste and, as is widely covered in today's media, the product sector generally referred to as "bioplastics" or biomaterials is expanding. Government and industry are led by strong consumer and public pressures to develop alternatives to petroleum-based plastic products.

Any discussion of bioplastics often includes the term "single-use plastics" or SUPs, which generally refer to "disposable" plastics such as materials used for takeaway or fast food and consumer goods packaging. The desire to replace existing plastic products has been taken on by packaging technology companies, which are responding in areas such as coated paper and paperboard, renewable plastics containers, biodegradable bottles, and a range of other approaches. Plastic straws, beverage containers, cutlery, and food containers are all targeted for replacement with other materials deemed to be more sustainable.

The sustainability movement has also significantly influenced corporate governance and the financial industry, as a 2020 survey by McKinsey found that the majority of company executives and investment professionals agreed that environmental, social, and governance (ESG) programs significantly enhance shareholder value [1]. And, it is well known that the world's major brand owners whose packaging materials are the leading contributors to plastic waste, are all active in trying to implement sustainable technologies and practices in order to reduce waste and improve the environmental sustainability of their operations. These companies include an ever-increasing number of major brands across a range of industries, such as Coke, Pepsi, Unilever, Nestle, Mondelez, Mars, and others.

In this chapter, we will first provide an overview of the bioplastics markets and the technologies, and then move on to technical details of PHAs, their role in the emerging bio economy, the advantages and disadvantages of PHAs, commercial developments, and what can be expected of PHAs as a viable alternative to petroleum-based plastics.

Ray Bergstra, TerraVerdae Bioworks Inc., Edmonton, Canada, e-mail: rjbergstra@terraverdae.com

https://doi.org/10.1515/9783110791228-022

22.2 What are bioplastics

There are numerous types and approaches to bioplastics that are best described by European Bioplastics (EUBP), an organization that represents the bioplastic industry in Europe. Publications by EUBP [2] highlight that bioplastics are not a single material, but rather a family of plastic materials that are either bio-based, biodegradable, or both.

Bio-based polymer plastics are those that are produced from biologically sourced carbon, such as starch, sugar, cellulose, or vegetable oils, where the carbon in these natural substances can be converted into plastics. Within this category are two production approaches. Some bio-based plastics are produced via chemical synthesis using bio-based monomers, while others (including PHAs) are produced directly by microbes through their natural metabolic processes.

Biodegradable plastics degrade naturally in the environment without leaving behind residues. Typically, this implies microbial degradation where microbes consume the plastic material and convert it into carbon dioxide and other natural carbonaceous materials that do not contribute to pollution.

It should be noted that throughout the literature, the terms biopolymer and bioplastic are often used interchangeably, and this is mainly because in the context of functional materials, they quite often mean the same thing.

Each bioplastic can be categorized as shown in Figure 22.1, as published by European Bioplastics. Pease see Chapter XX for further information and acronyms used in Figure 22.1 and Table 22.1. There are four quadrants in the chart that outline the relationship between a material's bio-based and biodegradability characteristics:
- Petroleum- or fossil-based and non-biodegradable, which includes common conventional plastics such as polyolefins and PET
- Petroleum-based and biodegradable, such as PBAT and PCL which are chemically synthesized but biodegradable polyesters
- Bio-based and non-biodegradable polymers where conventional type polymers are produced using bio-based monomers such as ethylene, derived from corn ethanol
- Bio-based and biodegradable – the most environmentally sustainable bioplastic category that includes PHAs, thermoplastic starch (TPS), some PBS polymers, and PLA.

Further to the categorization of biopolymers and bioplastics, a comparison of the different types is important for determining how they fit into the marketplace. A brief summary is provided in Table 22.1.

Each type of bioplastic has a place in the market. In some instances, the sustainability goals of a company can by met by using bio-based plastics that are not rapidly biodegradable in order to achieve specific performance characteristics, such as gas barrier properties for food packaging or mechanical strength and durability for a car part. In other applications, fast biodegradability may be the top priority, such as in

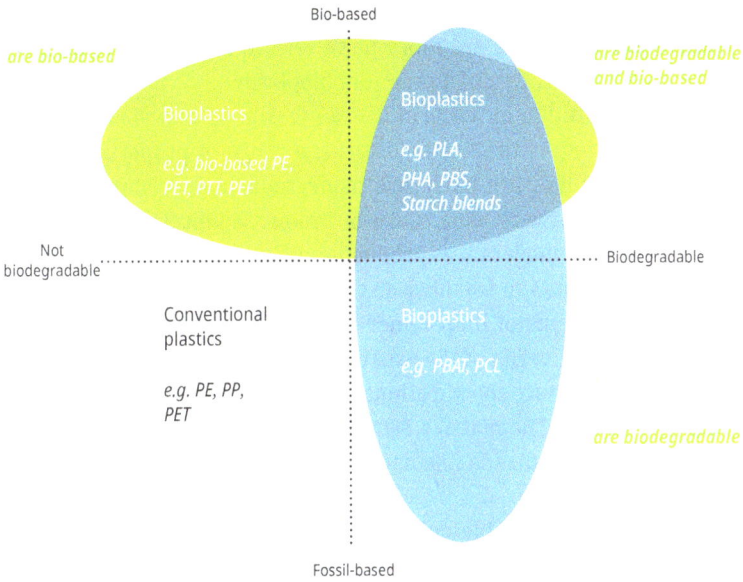

Figure 22.1: Categories of Bioplastics.
Source: European Bioplastics, used with permission.

Table 22.1: Comparison of Bioplastics.

Technology (pure polymer)	Carbon source	Mechanical properties	Biodegradability	Availability	Relative cost
PHA	Cereal sugars, vegetable oils, organic waste	Good	Yes	Limited	High
Thermoplastic starch	Cereal starches	Poor	Yes	Limited	Low
Bio-based PE	Sugar cane, cereal sugars	Good	No	Limited	Moderate
PBAT	Petrochemical	Good	Yes	Moderate	High
PBS	Cereal sugars, petrochemical	Good	Yes	Moderate	High
Polyethylene furandioate	Cereal sugars, Petrochemical	Good	No	Limited	High
PLA	Cereal sugars	Good	Low	Moderate	Moderate
Conventional polyolefins	Petrochemical	Good	No		Low

horticulture, where seedling containers can be planted and left to biodegrade in the ground. And of course, some markets are more price sensitive than others and are less prepared to take on the higher cost of a premium bioplastic.

PHAs occupy a leading position in terms of overall feasibility and viability. They can be produced from a variety of organic feedstocks; a range of properties can be built into the polymer design; they are bio-based, biodegradable, biocompatible, and compostable; they can address a very wide range of product applications; their production costs are expected to be moderate as production volumes grow.

As the bioplastic sector works to establish itself as a significant segment in the overall plastics industry, each type of biopolymer has challenges to overcome. First and foremost, in all cases, is the inertia of the existing industry where supply systems, processing technology, and product specifications are firmly ensconced at low cost and high production efficiencies. The massive $600 billion plastics market and the associated infrastructure represents formidable barriers to any proposed changes to the industry.

New bio-based and/or biodegradable polymers also face upfront financial challenges. Investments in the billion dollar range are required over what is usually more than a decade to develop a single new polymer production technology and adapt that biopolymer into final plastic products that meet the detailed and often complex requirements of different applications in existing markets. Securing these investments is further complicated by the fact that biopolymer production processes are inherently more costly than petrochemical processes, simply based on feedstock costs, limited scale, and availability. Nonetheless, biopolymer market feasibility is gaining increased support from the overwhelming global market demand for plastics and other materials, as they can be sustainably produced, minimize GHG generation, reduce impacts on climate change, and do not persist in the environment.

Governments and industry are responding to strong consumer and public pressures to develop alternatives to petroleum-based plastic products. Support is justified by more comprehensive and inclusive estimates of product costs, where direct feedstock and production costs of biopolymers are balanced by the social and direct expenses related to waste management and environmental consequences of petrochemicals.

EUBP provides projections of global biopolymer production capacities based on the impacts of known investments in each area. This is discussed further in Chapters 19, 20 and 21. The data indicates that in the near term, the largest growth will come from new bio-based and biodegradable polymers such as PHAs, but capacity in bio-based, durable, and non-biodegradable polymers will continue to build industry share.

Another critical observation is that while significant jurisdictions in North America (Canada, California, and other states), Europe, and China have made commitments to eliminate the use of conventional single-use plastics, the production of 7.5 million tons of all bioplastics by 2026 would represent a market share of less than 2%. If one were to take a position that a significant impact from any new technology is not realized until it has a 10% share, the bioplastics industry will require an exponential growth rate in financial infusion to see environmental impacts within the next 10 to 15 years.

22.3 PHAs technology: market drivers and opportunities

In terms of the potential for bioplastics, high level assessments of some key applications are provided in Table 22.2 (note Chapter 21 has further details on markets). These estimates are general, and are made using an overall aggregation of reports, press releases, and government policy announcements.

Agricultural plastic, adhesives, sealants, and general packaging are important markets for biopolymers, including PHAs. Plastics for agricultural and food applications are a natural fit for biodegradable polymers, based on their short-term use, probability of mixing with food waste, and the existing trend toward composting or biodigestion of waste organics. Used packaging that can be streamed into waste food processing represents a significant streamlining for waste management and reduced transporting of waste materials to landfills or incineration.

Multipurpose consumer goods packaging is a dominant use of plastics. It remains to be seen how manufacturers of consumer goods adjust their practices and demand for plastics to meet sustainability goals that may be diverse and unique to each company, industry, or within geopolitical jurisdictions.

Table 22.2: Market size estimates.

Market size	Total addressable	Serviceable by PHAs
Worldwide $800 B		,000 Tons
Agricultural [3]	$55 b / 20 m T	20,000
Adhesives/sealants [4]	$50 b / 12 m T	6,000
Packaging [5]	$300 b / 150 m T	120,000

Various market reports continue to paint a positive picture of the real growth in the bioplastics market, both due to the announced investments and production plant constructions, as well as growth trends in relevant markets. As already noted, bio-based and/or biodegradable plastic production capacity is projected to grow from 2.1 million tons in 2020 to 7.5 million tons by 2026. Other reports indicate that the agricultural mulch film market is projected to grow by 6.5% annually, which represents $200 million in new annual sales. Another interesting example is the 3D printing plastics market, which has been projected to grow by more than 25% annually from $600 million in 2021 to $6 billion by 2026. And finally, the global flexible packaging market is valued at $160 billion. It is a sector where major brand owners are increasingly making commitments and have the resources to drive growth in sustainable plastic packaging technology and reduce the negative environmental impact of their product packaging.

Where does PHA technology fit into this picture? At a high level, PHAs are of interest for two primary reasons.

1. They can be naturally produced using renewable carbon sources, which of course means that PHAs do not contribute to introducing new petroleum-based carbon into the Earth's environment.

2. PHAs will naturally degrade under a range of conditions with no residues, which means that even if a PHA material ends up in the environment, the long-term impact can be managed and is minimal.

Many review papers have been written over the years that discuss the history and fundamentals of PHAs and several recent ones are referred to here that provide excellent assessments of their market opportunities. Koller and Mukherjee's recently published paper, "A New Wave of Industrialization of PHA Biopolyesters" [6] includes the history of the development of PHAs as well as a summary of the commercial activities of a range of companies producing PHAs and related products.

Dublin-based O'Conner and Padamati published an extensive PHA technology review in 2022, entitled "A Review on Biological Synthesis of the Biodegradable Polymers Polyhydroxyalkanoates and the Development of Multiple Applications" [7]. The paper provides a thorough description, with more than 300 literature references of the biological processes, performance properties, PHA-based biomaterials, processing for applications, and end-of-life management.

Wageningen Food & Biobased Research and Invest-NL published a report entitled "Paving the way for bio-based materials – a roadmap for the market introduction of PHAs," which discusses all aspects of what is required to expand the PHA industry. The report focuses on a number of commercial or near-commercial grades of PHA, and includes the potential carbon sources, microorganism for each type of PHA, production processes and unit operations, typical properties, market shares, production capacities, as well as an assessment of where each type of PHA technology can have a best fit as replacement of petrochemical plastics.

The BioSinn report entitled "Products for which biodegradation makes sense" [8] published in 2021 by the Nova Institute (funded by Germany's Ministry of Food and Agriculture) includes fact sheets for 25 applications where biodegradation is a viable end-of-life option. The applications listed focus on specific products where the probability of loss to the environment is very high and even 100% by design. This includes products such as organic waste bags, geotextiles, tea bags, fertilizer coatings, brush bristles, and marine ropes. The report estimated the market potential to be in the 1 million tons per year range in the EU alone.

From this perspective, PHA technology can readily address many market needs by providing bio-based plastic materials and packaging products designed to function in industrial settings, and then either biodegrade in-place or be easily composted or biodigested, as required by the user. These four recent reports highlight how PHAs are ideally placed to provide sustainable plastic materials.

When discussion the management of waste plastics, recycling continues to be promoted by the existing industry. Unfortunately, recycling has been shown to be largely ineffective as more than 85% of the world's plastics end up in landfills and the value of the collected plastics is very low due to contamination and poor performance. The objective of plastics recycling is to blend the collected materials with virgin plastics, but the recycled content reduces the final product quality. This limit on recycled content therefore perpetuates the use of petroleum plastics, and we have ended up with a build-up of the collected and unused, dirty plastics.

Nonetheless, recycling efforts will continue despite the shortcomings of the current systems due to the effort already invested, along with ongoing support from existing producers with very large resources, like ExxonMobil, Sinopec, Dow, etc.

PHA and some other bioplastic technologies have significant potential to help manage plastic waste over the long term in terms of reducing green house gas emissions (GHGs), enabling alternatives to landfills, and minimizing the impact of waste plastic contamination of the environment (land and water).

- Companies and jurisdictions have ongoing efforts in place to improve the sustainability of their operations and waste management systems and to stabilize the global carbon cycle.
- GHG reductions targets are global, and bio-based plastic use can contribute by growing the circular carbon economy.
- There are a range of plastic applications where biodegradation in place provides operational benefits for industries – agriculture and forestry in particular – where used plastics would not need to be recovered and disposed.
- Plastic products, especially consumer packaging, often become part of mixed waste streams, and the contaminated plastics cannot be efficiently recycled. On the other hand, biodegradable plastics can be included in mixed organic waste streams and sent for composting instead of filling landfills.

The level of interest in PHAs from the market has never been higher and the primary reason is their robust sustainability profile. Being bio-based is favorable due to the increase in regulatory pressure from various government policies related to reducing the generation of waste plastics and GHGs. Compared to burning of fossil fuels, plastics are not a major contributor of GHG emissions, but a 300 million ton global plastics market represents a major portion of the carbon economy. PHA's strong environmental story includes how waste management practices of bioplastics can contribute to reduced GHG emissions. As landfills reach capacity (and new landfills are increasingly difficult to commission) and recycling continues to be problematic, more and more waste plastics are likely to be incinerated for energy recovery, but this practice would generate CO_2 emissions in the 1 trillion ton range every year. In the case of PHAs, incineration results in the return of CO_2 to the environment –the essence of the circular carbon economy [9].

Globally, the current PHA manufacturing capacity is approaching 100,000 ton per year, whereas just 2% of the plastic market is in range of 6 million tons. It will take decades to build PHA production capacity to a point where its replacement of petroleum plastics begins to have an environmental impact, considering renewability, biodegradability, and compostability. Nonetheless, during this time, PHAs and other bioplastics will begin to demonstrate that they are viable and sustainable options to the petroleum plastic manufacturing and waste management systems that exist today.

22.3.1 What are PHAs?

PHA are polyhydroxyalkanoates that are polymerized linear hydroxy acids. There are many versions of PHAs under development as the industry seeks out PHAs with optimal properties and there are technically an infinite variety of hydroxy acids that can be polymerized. Even the well-known polylactic acid (PLA) and poly-caprolactone (PCL) belong to the PHA group of materials.

> [Defining polymers in chemical terms is not straight forward; it is not simple to name a polymer and automatically communicate its function and usefulness. For example, the simplest (chemically speaking) polymer is polyethylene, and it comes in three main forms. The three forms have different functions based on their elasticity, strength, and other properties – high density, low density, and linear low density. Similarly, polyesters generally mean polyethylene terephthalate (PET) but there are other polyesters like PBT (polybutylene terephthalate) that is used for entirely different applications due to its toughness. The same goes for nylons or polyamides – there are numerous types, and each has its preferred uses that range from clothing to industrial materials to military-grade protective equipment.]

So, what are we really talking about when we discuss PHAs? In the industry today, PHAs refer to linear polyesters that are
1. derived from hydroxy fatty acids as monomers
2. naturally produced using microorganisms that consume natural carbon sources (sugars, CO_2, methanol, or methane,) and/or nutrients (vegetable oils, starches, etc.), and biologically convert the carbon source into polymers of hydroxy acids.

The most common variations of PHA are shown in the Figure 22.2, which shows how the length of the polymer side chains can be varied through the biological design of the production process. If the side chain is 1 or 2 carbons in length, the PHA is referred to as a short chain length PHA, and if between 3 and 8 carbons, it is known as medium chain length PHA.

It is important to note the origin of this conventional nomenclature is the fatty acid industry. The chain length indicated for PHAs is not the length of the polymer chain; rather, it is the length of the acid monomer. Historically, natural acids with 6–12 carbons are known as the medium chain length fatty acids; less than 6 are the short chain length acids; and more than 12 are the long chain length fatty acids.

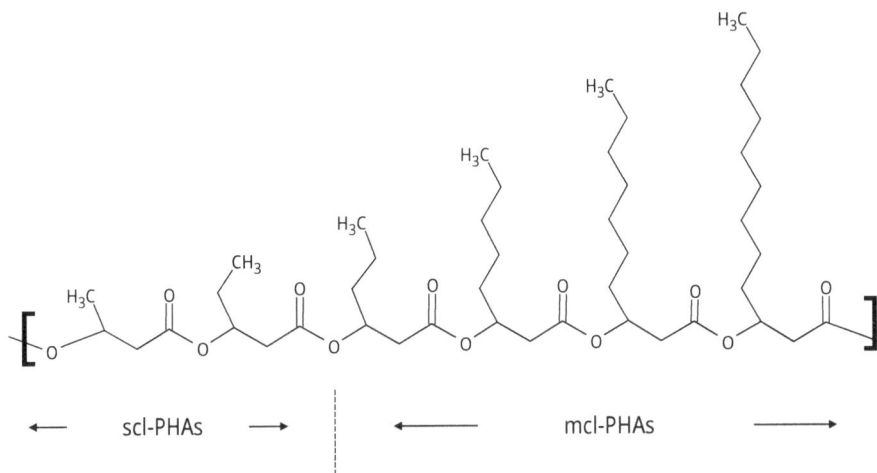

Figure 22.2: PHA Polymer Structure.

The most common PHA is polyhydroxybutyrate, which is an scl-PHA normally known as PHB (Figure 22.3). A quick comparison of the mechanical properties of PHB and mcl-PHAs highlights the range in properties that are achievable from PHAs, overall. PHB is rigid and strong, while mcl-PHAs are softer and flexible. This range of properties forms the basis of the bioplastic product development work discussed later in this chapter, where commercial ventures have worked to take advantage of the structural options available in order to produce high quality and functional PHA plastics.

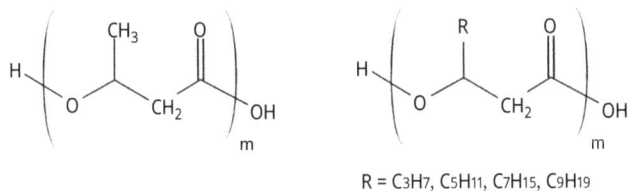

$R = C_3H_7, C_5H_{11}, C_7H_{15}, C_9H_{19}$

Figure 22.3: Structure of PHB compared to mcl-PHAs.

PHB polyhydroxybutyrate	mcl-PHA medium chain length polyhydroxyalkanoates

– rigid, crystalline, high strength and toughness
– bio-based and readily biodegradable and
  compostable
– often compounded with other biopolymers
  for conventional plastics applications
	– flexible, reduced crystallinity, softer
	– bio-based, readily biodegradable and
	  compostable
	– reduced melting temperature
	– targeted for coatings, adhesives, and sealants

22.3.2 Origins of PHA technology

Maurice Lemoigne was a French researcher who is largely credited with the first discovery of biodegradable plastic based on his work with *Bacillus megaterium*. He was able to isolate poly-3-hydroxybuturate (PHB), but this work was relatively unknown until the 1950s when researchers in the United States and Great Britain once again found PHB, and there was interest in its biodegradability. The oil crisis in the 1970s generated a high level of attention on bio-based technologies as alternatives to petroleum-based products. However, when crude oil prices dropped back to more acceptable levels, PHA interest faded. It was in the 1990s when a slowly growing but sustained interest in environmentally friendly products caught hold – first within the research community, but gradually moving into industrial operations. Polylactic acid (PLA) was the first bio-based polyester to achieve industrial scale commercial status, and now in 2022, we are indeed seeing more and more ventures moving forward with major investments and commercial plays in PHA.

Dr. Guo Qiang (George) Chen [10] is a well-known leader in developing PHA technology since the early 1990s. His research has focused on microbial materials, PHA metabolic engineering, and PHA biomaterials application. Dr. Chen's affiliation with research conducted in Graz, Austria during the early 1990s forms the foundation for PHB production using *c. necator*. After joining Tsinghua University in 1994, he has been actively promoting microbe-derived material industries related to the production of PHB, PHBH, mcl-PHAs, and P3HB-4HB in China. Dr. Chen has produced 200 international publications and his technologies and patents have been utilized by numerous companies that have succeeded in scaling up production of PHAs.

Another leading commercial development of PHA production technology is a convoluted tale. ICI, once a global leading and broad-based chemical company, was an original developer of microbial technology for the production of PHBV. A key aspect of this material was the advantage of PHBV being of mixed monomers (hydroxybutyric acid and hydroxyvaleric acid). During ICI's demerging activities in the early 1990s, this PHA technology and production of BIOPOL® was sold to Zeneca, and then a few years later to Monsanto. In 1999, this business was sold to Metabolix, a company that was a global leader in PHA innovation during the 2000s. Full-scale production of PHAs from corn sugars went ahead via a joint venture with ADM, but this effort was abandoned and in 2016, the PHA production and the related technology was sold to CJ Bio. In 2022, CJ Bio announced its plan to go ahead with full commercial production of P3HB4HB.

Another major PHA technology that is being commercialized is known as Nodax®, an scl/mcl PHA developed by Dr. Isao Noda at Proctor & Gamble. This PHA combines some of the properties of PHB with the elastomeric properties of mcl-PHAs by using vegetable oils as a feedstock, resulting in a range of lengths of the side chains. The Nodax technology and tradename are currently owned and marketed by Danimer Scientific. A similar PHA, known as PHBH – a copolymer of hydroxybutyric acid and hydroxy hexanoic acid, is produced by Kaneka.

A leading PHA production process developer is Dr. Bruce Ramsay of Queen's University in Canada and the founder of PolyFerm Canada, an mcl-PHA technology company. Dr. Ramsay has more than 80 publications related to microbiological systems, the majority of which have made a significant contribution to mcl-PHA technology in the areas of metabolic pathways, feeding strategies, and PHA performance.

Koller and Mukherjee provide a list of other biopolymer companies that have been focused on process development for the production of basic PHA polymers from organic waste. While these processes are attractive from a waste utilization standpoint, feedstock aggregation is often a challenge and the overall yields tend to be low, making it difficult to meet the demands of many industry sectors where biopolymers are required in high volume and at low cost in order to make an impact.

22.3.3 Common types of PHAs

The industry leading PHA polymers are discussed briefly below.

22.3.3.1 PHB

Poly-3-hyroxybutyrate is the most widely available and most well-known of the PHAs. There are numerous companies that are manufacturing small commercial quantities, including TianAn, Tianjin, Biomer, and others. For high volume and efficient production, the preferred bacterium is *cupriavidus necator*, with glucose as the feedstock.

There are numerous other routes to PHB production, including using methanotrophs, such as Newlight Technologies using methane, or TerraVerdae Bioworks, which has scaled up Canada's National Research Council technology for PHB production using *Methylorubrum extorquens* and renewable methanol as feedstock. As noted previously, there are numerous other early-stage PHB production ventures that are producing PHB from organic waste streams for pulp mills, waste cooking oils, and food waste, although these processes often require recombinant microbes that are undesirable systems for PHA applications in personal care markets.

Pure PHB is highly crystalline, which on the positive side, results in high strength, impact resistance, and good barrier properties, but the material is generally not sufficiently elastic to be used on its own in high volume flexible packaging applications. Therefore, it is normal to compound PHB with other biopolymers, biomaterials, or additives to enhance the processing and mechanical property requirements while maintaining the biodegradability and compostability of the resulting compounded resin.

22.3.3.2 PHBV

PHBV is the term used for polyhydroxybutyrate-valerate copolymer although it is important to note that this is a random copolymer according to the microbe's metabolic pathways. PHBV is commonly accepted to have a desirable blend of properties where the highly crystalline properties of pure PHB are appropriately compromised by the valerate content, resulting in a lower melting temperature and higher elasticity. This makes the material more amenable for flexible packaging applications as long as the hydroxyvalerate content is sufficient (more than 15%). Currently, PHBV, with elevated valerate content, is not widely available in the industry.

22.3.3.3 mcl-PHA

Pure mcl-PHAs have seen limited industrial activity to date even though their preparation has been known for decades. Using natural organic acids as feedstock, with *Pseudomonas putida* as the microbe, Polyferm Canada has developed the production process and licensed the technology to TerraVerdae Bioworks, which is developing applications such as coatings and adhesives, where lower melting points, low crystallinity, and high tackiness provide a range of performance advantages.

22.3.3.4 PHBH/PHBHHx

Similar to PHBV, random, naturally produced copolymers of hydroxybutyrate and hydroxyhexanoate or other mcl hydroxyalkanoates have a long history, which includes the Nodax technology. The microbe is recombinant *c. necatur* and the feedstock is mainly vegetable oil. These copolymer PHAs have a range of desirable properties in areas of elasticity, processability and melting temperature. Because of this, PHBHs have received a high level of commercial interest and indeed, Danimer, RWD, and Kaneka are three companies that are moving forward to achieve industrial-scale production capacity.

22.3.3.5 P3HB-4HB

P3HB-4HB is another copolymer that exhibits high elasticity. This PHA was developed by Metabolix and is now being commercialized by CJ Bio. It is produced using recombinant *e. coli* using sugar feedstock, along with butanediol, which introduces the 4-HB component. (See Figure 22.4). The random copolymer produced has sufficient disruption of the inherent crystalline nature of P3HB such that P3HB-4HB is amorphous, which is an important advantage for adhesive and coating applications.

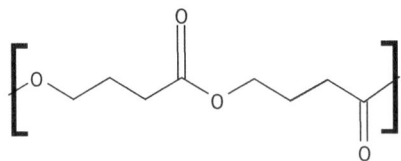

Figure 22.4: Poly-4-hydroxybutyrate (P4HB).

22.3.3.6 Compounded PHA resins

Developing a single new polymer material nearly always requires additional design and engineering to produce a new plastic products with the desired functional properties that the market can use. This is a major challenge for new bio-based PHA plastics, as this emerging industry is competing against 100 years of petroleum plastics technology development, evolution, and advancement. Therefore, in addition to PHA polymers, current offerings to industry often involve PHA compounded resins, particularly with rigid PHAs like PHB and PHBV. Compounding PHB with other bioplastics such as PLA, PBAT, PBS, and thermoplastic starches provide opportunity to balance the properties of the final compounded resins, with PHB/PHBV providing improved biodegradability, bio-based content, barrier properties, depending on the targets and expertise of the compounder.

22.4 Commercialization

In terms of manufacturing feasibility, there are two main points that support the use of PHAs. First, PHAs are produced using industrial fermentation technology – similar general processes and equipment that produce foods, ingredients, bio-based fuels, and chemicals. When establishing specific costs for PHA, high production processes are key factors facing the industry. Governments and the industry nonetheless have generally come to accept that optimized fermentation and the related bioprocessing using natural plant-based materials as the carbon source represent the future for non-petroleum-based fuels, chemicals, and polymers.

The commercial prognosis for PHAs continues to improve, with favorable tail winds from myriad fronts, such as improved technologies, large scale fermentation processes, and investments in major companies, as shown in Table 22.3.

Table 22.3: Snapshot of global bioplastic producers.

Company	Primary technology	Commercialization stage
Natureworks	PLA	Full commercial
BASF	PBAT	Full commercial
Kaneka	PHA	Small commercial
Danimer	PHA	Small commercial
Mitsubishi	PBS	Small commercial
Novamont	TPS	Small commercial
Avantium	PEF	Pilot
Total Corbion	PLA	Full commercial
TianAn Biopolymer	PHB	Small commercial
RWDC Industries	PHA	Pilot
Tianjin Guoyun Biomaterials	PHB	Small commercial
CJ Bio	P3HB4HB	Small commercial
Newlight Technologies	PHB	Small commercial
TerraVerdae Bioworks	PHB, mcl-PHA	Pilot

PHB production capacity is the most advanced within the field of PHAs, led by plants in China, such as TianAn and Tianjin. One of the largest producers of other PHA resins is Danimer Scientific, with a reported 65 million pounds (30,000 tons) annual capacity from two US plants, but other PHA ventures are moving forward quickly.

RWDC Industries announced a funding round of $95 million in 2021. According to the press release, RWDC intends to expand its production of PHA in Athens, Georgia, to 50,000 tons per year, as well as establish a production facility in Singapore.

CJ Bio announced the commissioning of 5,000 ton per year production capacity in Indonesia during 2022. CJ anticipates sufficient demand to plan for a total of 300,000 tons per year by 2030 of both its amorphous P3HBP4HB and scl-PHA.

In 2020, Kaneka announced the completion of 5,000 ton per year PHBH production plant in Japan as well as further plans for an additional 20,000 tons of annual capacity. Kaneka projected market demand to support 100,000–300,000 tons per year by 2030.

TerraVerdae has announced an MOU to acquire Polyferm's mcl-PHA technology and is expected to produce this PHA at demonstration scale by 2024.

There are many other PHA production technologies under development that are gaining support for commercialization, particularly from the standpoint of organic waste utilization using carbon sources, such as food waste and pulp mill streams. Examples include Nafigate, Bluepha, Genecis Bioindustries, Bosk Bioproducts, and Full Cycle Bioplastics.

The commercial activities of PHA received important support from GO!PHA [11], a PHA industry organization co-founded by Anindya Mukherjee, Jan Ravenstijn, and Rick Passenier – it has been working to educate the market on how PHAs can help solve waste plastic problems. This effort is supported by both PHA technology developers and final product manufacturers. GO!PHA published an industry review of the PHA world in 2022, "Mimicking Nature The PHA Industry Landscape Latest Trends and 28 producer profiles," which detailed the state of the industry in areas of applications and general growth [12]. Among other activities, GO!PHA has published numerous articles that discuss how PHAs fit into the demand for bio-based and biodegradable plastics. One of the main objectives is to align government objectives for clean technology policy, reduce the perceived risks with the financial sector, and to demonstrate to major global corporations and brand owners that PHAs will help them meet ESG goals (environmental social governance).

The work is based on a recognition that while PHAs are polymers and plastics, companies that been processing polyolefins for decades need to recognize and understand that some adjustments will be required in logistics, operations, and material cost as they look to implement PHA-based products. During transitions to a new product technology, some level of short-term impact can be expected in terms of efficiencies and revenues as existing operations adjust.

PHAs and PHA-based resins are developed so that they can be processed using the same equipment that is currently in use by polymer and plastic processors. Equipment operating conditions may require adjustment and optimization, but the plastic product supply industry is not expected to require major investments into new equipment in order to transition to PHAs. This is a critical message from GO!PHA supporters to the incumbent plastics industry. As a result, there are currently many longer-term strategic moves being played out as exemplified by large global brand owners like Coke, Pepsi, Nestle, Mars, and others.

22.5 Summary and conclusions

The future for PHAs can be summarized by a SWOT (Strength-Weakness-Opportunity-Threats) analysis.

STRENGTHS
- Bio-based and inherently sustainable: What does this mean in practical terms? If we consider that 50% of the worlds demand for plastics were to be met by PHAs, this will require nearly 200 million tons of polymers, which at a high level, would likely require around 400 million tons of starch, cellulose, and natural oils to produce. If purpose-grown crops are used and estimating an average production in the 5 ton per hectare range, these natural organic feedstocks might require in the neighborhood of 80 million hectares, which represents around 1.6% of the existing arable land globally. This level of arable land use for non-food applications can be managed without a significant impact on food supply and should be achievable with overall beneficial results, based on the potential reduction of GHG emissions in the 100 million ton CO_2 eq. range per year.
- Non-polluting at end-of-life: There is a plethora of data supporting the inherent microbial degradation of PHAs, which is a critical piece to reducing the impact of plastic waste in the environment. While the timelines for biodegradation will depend on the specific PHA products, PHAs are biodegradable within months and/or years time frames, whereas petroleum-based plastics degrade only over hundreds of years. The long-term impacts fully support a cleaner environment, particularly the oceans, through use of biodegradable PHAs.
- Offers a range of sustainable solutions: PHAs can be biodigested, composted, left to degrade in the environment by design (i.e., mulch films, erosion control products, etc.), depending on the application and regional waste management systems. Even if PHAs end up in an incineration stream, the escaped CO_2 simply returns into the natural carbon cycle and does not represent the additional petroleum-derived CO_2 into the Earth's environment since the PHA is produced from CO_2 taken from the environment via photosynthesis.
- Production feasibility has been demonstrated: The recent advancements of production capacity are just the first steps to demonstrate feasibility. As this activity continues to gain momentum, efficiencies and streamlining will only further support the feasibility of PHAs.
- Brand owners with major existing market presence support product development: The combined financial impact of major global corporations supporting sustainable plastics like PHA is more than sufficient to facilitate market demand and product roll-out from perspective of sustainable commercial operations.
- Solid financial backing: As noted above, with major global corporations and governments in support, the investments to grow the capacity of a new PHA plastics industry are more than sufficient. Even if a few of the global chemical companies,

(think of Dow, ExxonMobil, and Sabic) were to take on the challenge by each investing $10 billion over 10 years, this would be sufficient to build production capacity to the 10 million ton per year range.

- First-generation PHA polymer and resins are currently being introduced to the market: Refer to previous section regarding Danimer, CJ Bio, TianAn, RWDC, and Kaneka.
- Able to address diverse product applications: The existing variations of PHA polymer chemistry is just the first generation of the potential sophistication of PHAs. As the industry grows, new "by design" PHAs will be developed that provide tailored performance. While there are likely to be certain specific applications where highly engineered chemical plastics are required, PHAs' potential has already been demonstrated to include a broad spectrum of applications, particularly from high volume market sectors such as flexible and rigid packaging, coatings, films, adhesives, etc.

WEAKNESSES
- Constrained by limited chemical manipulations: There is the potential that chemical modifications to achieve performance requirements in areas of processability, barrier properties, durability, etc. could compromise biodegradability, which is one of the inherent strengths of PHAs.
- Native properties can be challenging to improve: Biological systems (fermentations) are inherently more difficult to control and manipulate than chemical processes since the microorganism needs to survive and thrive during production – altering temperature, pH, and chemical reagents are usually not viable options.
- Higher costs, compared to current market pricing for most plastics: This is strictly from a product cost perspective, as the world currently operates based on the low-cost structure of the petrochemical industry.
- Challenging to achieve the high-performance capabilities of existing fossil-based polymers, such as polyurethanes, polyamides, and engineered polyesters.

OPPORTUNITIES
- Provide solution to organizations and jurisdictions to meet sustainability goals.
- Enable a wide range of industries to implement sustainable (low carbon footprint) products, including very high volume (and waste-generating) applications, such as single use plastic packaging.
- Supply technology with direct benefits to GHG reductions, healthier soil, and cleaner water
- Reduce municipality waste management costs related to transportation to land fills.

THREATS
- Idealistic policies that do not favor any manufactured polymers, even if they are bio-based. Some markets and policy makers view all manufactured plastics as

potentially harmful to the environment and are pushing agendas supporting natural materials only, such as those based on cellulose and starch.

– Poorer economies not willing to make long-term investments into new plastics technology.
– Existing plastics industry sways government policy away from bioplastics, based on a strong lobbying position to maintain existing industries.
– Plastics industry does not have experience with PHAs and is resistant to change.

Overall, positive attributes and impact of PHAs are very encouraging and bode well for their commercialization. This positive picture is further boosted by increased awareness by end use customers for bioplastics, especially in packaging. A strong push to address global warming, greenhouse gas emissions, and the circular economy are very favorable for the strong growth of PHAs. However, PHA industry growth will not take place without business potential and the related financial support. Recent manufacturing capacity increases and investment announcements provide the real evidence that PHAs can achieve a significant position within the emerging bio economy.

References

[1]   McKinsey & Company. The ESG premium: new perspectives on value and performance, 2020.
[2]   https://www.european-bioplastics.org/news/publications/.
[3]   US Dep of Agriculture Statistics, StatsCan. https://saiplatform.org/wp-content/uploads/2019/06/190528-report_use-of-plastics-in-agriculture.pdf, https://www.fortunebusinessinsights.com/agricultural-films-market-102701.
[4]   https://www.fortunebusinessinsights.com/industry-reports/adhesives-and-sealants-market-101715.
[5]   US EPA. Plastics Industry Association, https://www.statista.com/topics/5266/plastics-industry/#dossierSummary__chapter1.
[6]   Koller M, Mukherjee A. A new wave of industrialization of PHA biopolyesters. BioEng 2022;9:74.
[7]   Dalton B, Bhagabati P, De Micco J, Padamati RB, O'Connor K. A review on biological synthesis of the biodegradable polymers polyhydroxyalkanoates and the development of multiple applications. Catalysts 2022;12:319.
[8]   https://renewable-carbon.eu/publications/product/biosinn-products-for-which-biodegradation-makes-sense-pdf/.
[9]   https://www.european-bioplastics.org/bioplastics/environment/.
[10]   currently Professor, School of Life Sciences, Tsinghua University.
[11]   www.gopha.org.
[12]   Ravenstijn J. GO!PHA, https://renewable-carbon.eu/publications/product/mimicking-nature-the-pha-industry-landscape-latest-trends-and-28-producer-profiles/.

Section VI: **Sustainable green biocomposites**

Ashok Adur
Chapter 23
Bio-based plastics compounds for the emerging bio economy

23.1 Introduction

Concerns about global warming and other environmental challenges among the global population have resulted in growth of both interest and demand for a clean and pollution-free environment, with reduction of carbon dioxide resulting from fossil fuel usage. Based on this, a lot of attention has been focused on research and development to replace petro-based (which includes oil and gas) commodity plastics with biodegradable materials sourced from biological and renewable resources. Biopolymers, which are polymers derived from natural sources, either chemically from a biological material or biosynthesized by living organisms, are claimed to be suitable alternatives for addressing these issues [1]. Polymers (plastics and rubber, thermoplastic or thermosets) whether they are bio-based or not, are rarely used without the addition of additives like plasticizers, lubricants, antioxidants and stabilizers, fillers, colorants, reinforcements, etc., in order to meet various customer requirements all the way downstream in the value chain. Some bio-based plasticizers are covered in Chapter 16. Bio-compounds are compounds that contain one or more biopolymers. These additives are added during a compounding process step to overcome limitations and to increase the range of performance characteristics required in the target end-use applications. This chapter covers several emerging and commercial bio-compounds, while the addition of reinforcements to form plastics bio-composites is covered in the next chapter.

23.2 Processing to produce bio-based compounds

Similar to other polymers with high viscosity, compounding of biopolymers needs sufficiently high shear and temperature to bring down viscosity to a range where adequate mixing is easier, whether the ingredients to be mixed in are in liquid, powder, or pellet form. The process of adding additives to withstand thermoplastic processing, to make the polymers processable downstream, and to modify their properties into useful articles is called compounding. When one polymer is blended with another polymer to obtain a combination of properties, it is called a polymer blend, especially

Ashok Adur, Everest International Consulting LLC, Jacksonville, Fl 32256, USA, e-mail: Aadur@Outlook.com

https://doi.org/10.1515/9783110791228-023

if it forms multiple phases. This is typical either if the viscosities of the two polymers at the mixing temperature are vastly different and mixing is inadequate or if the two polymers are chemically incompatible. Incompatible polymers can be made compatible by adding small amounts of compatibilizer as additives during compounding [2]. If the powders are not fed at a constant rate that can keep up with the production rate, special accessory equipment may be needed to ensure that it is uniformly introduced into the compounding equipment.

The major difference with bio-based ingredients in compounding is the need to ensure adequate moisture control, because most bio-based materials are very hygroscopic and absorb water. This can cause foaming, voids, and other issues. In a few end products, this sort of foaming and voids in the end product could be useful. Nevertheless, additional equipment to control moisture is needed in most applications. A few petro-based polymers like nylons and some polyesters also have this issue, but their moisture absorption levels are much less, compared to many biopolymers.

Examples of equipment that process compounds are no different from what is used for non-bio-based compounds, and includes Banbury mixers, Farrell continuous mixers (FCMs), various kinds of extruders that are single, twin, multiple screw, or oscillating types, heated roll mills, etc. Each type of equipment provides different shear profiles during mixing as well as different temperature buildup and residence times. It is important to choose the right equipment and processing conditions for each set of compounding, whether the product has biopolymers or bio-fibers, some of which are more sensitive to hot spots during processing, so screw optimization may be critical. In some cases, antioxidant additives are needed just to prevent degradation during compounding – especially true for polymers that are susceptible to oxidation. Typical compounding equipment converts the mixture of the polymer with the additives and reinforcement into either pellets (also called granules) or directly into sheet or profiles. The pellets are then molded by an injection molding or a blow molding process into films, sheets, parts, bottles, or other articles. The shape and size of the extruded sheets can be controlled to the required thickness and width and cut in the same process equipment into the desired lengths for the intended application, or further thermoformed or machined into other shapes in a secondary step. During extrusion, most fibers orient in one direction, typically the direction of the extrudate flow. This leads to different properties in the extrusion direction versus the perpendicular or transverse direction. This can happen a little during the compounding stage and a lot more at the subsequent extrusion stage or the injection molding stage. In injection molding, the design of the gate determines the direction of the fiber orientation. In order to obtain the same properties in both directions, various downstream methods can be used or in some cases, rather than feeding the fibers into the compounding equipment, a different set of processes are employed. Sometimes, depending upon the end application, the fiber is first converted into a cloth or a non-woven fabric, which is then, processed using a compression molding or similar process to convert the cloth or fabric into a two-dimensional sheet of the fiber-reinforced composite.

Compression or lamination molding and a variety of cast processes are more commonly used with thermoset type of plastics. Figure 23.1 shows processing steps for bio-fiber plastics compounds and composites.

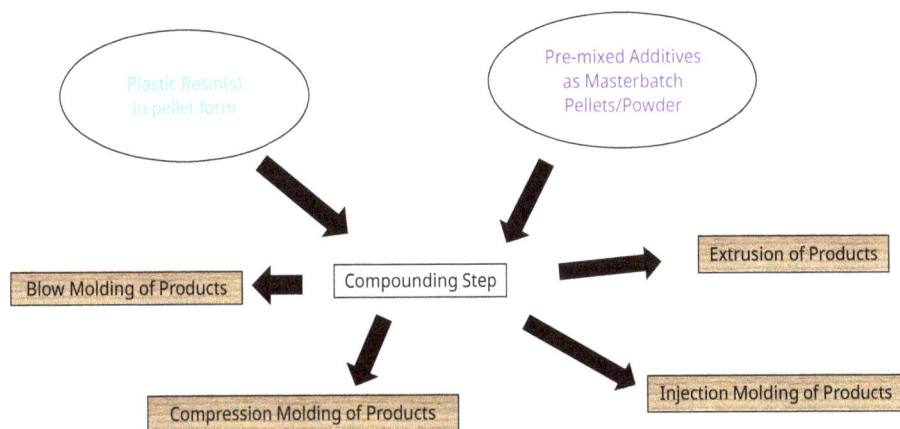

Figure 23.1: Processing steps for bioplastics compounds.

23.3 Unique challenges

In the chapter on biopolymers and bioplastics and in other chapters in this book, it has been made clear that biopolymers have the advantage of outstanding biodegradability and if some of them are not degradable or compostable, at least they are renewable, and in many cases, can be produced into useful form, starting from plant sources or even from waste products like industrial or consumer waste. However, they have characteristics and properties, as listed below, that pose challenges compared to conventional petro counterparts [3].

- some of them not degradable or compostable, but renewable
- poor mechanical properties
- higher sensitivity to moisture and water absorption
- poor resistance to fatigue life
- little chemical resistance
- low long-term durability
- issues related to production in large scale and supply chain
- limited processing capability

Most of these disadvantages of biopolymers can be overcome at least partially by compounding them with other polymers to form compounds and/or with reinforcing fillers or, in some cases, nanofillers. Nanofillers are fillers with at least one of its dimensions as low as 100 micrometers to qualify to be in the nanometer range [3]. Biocomposites are discussed in the next chapter.

Some bio-based polymers have thermal degradation points lower than those of petro-based polymers. In a few cases, their degradation points are lower than or very close to their melting point, so one needs to increase the processing window of the biopolymer by plasticizing it with a plasticizer. An extreme example of this is starch with water or other polar additives, such as glycerin, can be extruded, while dry starch just degrades and turns brown.

Another challenge that has to be overcome is that, as we have seen in the Chapter 19 on "Renewable Green Biopolymers and Bioplastics," just because a bioplastic is sourced from a bio-source like trees, plants, algae, or agricultural or food waste, it is not always biodegradable. When a nonbiodegradable plastic is added to biopolymer to form blends and alloys, only the biodegradable polymer can degrade or can be compostable. Whichever phase is larger in terms of volume, not weight, only one phase can biodegrade. If so, the nonbiodegradable plastic is encased inside the degradable biopolymer matrix and for biodegradability to occur. So, at the end of the bio-blend's life and if it is time to biodegrade it, the outer biopolymer degrades first, leaving bits and pieces of the other plastic to remain. On the other hand, if the nonbiodegradable polymer is the matrix as the larger volume phase, biodegradation slows down and what results over a longer period is a film or bottle with many holes in it.

In order to obtain better properties, it may be necessary in biopolymer blends to either chemically bond or "couple" the other polymer to the bioplastic matrix. This is carried out by modifying the chemistry of the biopolymer or the other polymer by some sort of radiation treatment (ozone or corona, or even high energy radiation like beta or gamma radiation) or chemical modification. Such modification is called "functionalization." The choice of such functionalization is based on the specific chemistry of the each of the major components of the blend. Such functionalization is used with conventional petro-based material multiphase systems too. Once compatibilized, the morphology of these blends changes to the extent that on a microscopic scale, the two phases intermingle enough that the domains of the minor phase get dispersed better and get smaller and, in a few cases, can be seen in photomicrographs to diffuse into each other's phases [2].

When such chemical bonding of one phase to the other occurs, it results in significant improvement in performance, opening up a wider range of applications but makes biodegradability even more challenging. Yet, from a net carbon dioxide balance perspective, such bio-compounds would be beneficial.

Thus, rather than concentrating on biodegradability or compostablity of bioplastics, this chapter will cover compounds made from biopolymers to turn them into useable bioplastics and bio-based compounds to include combinations of polymers where at least one of the combining materials is bio-based.

23.4 Bio-based compounds

23.4.1 Introduction

In this section, compounds of biopolymer are discussed to cover biopolymers that have been compounded to modify their properties, using various both bio-based and non-bio-based additives like impact modifiers, nucleating agents, and other polymers. There are, again, numerous academic publications and several patents granted. Some applications have been commercialized, and the use of such blends is growing fast globally.

However, it is important to remember that the biodegradability of articles or parts produced from such starch with nonbiodegradable plastics like polyethylene and polypropylene depends on several factors, such as the part's thickness, morphology of the blend in the part, and of course, on the specific micro-organism that will consume the part. The morphology of the blend in the part depends on factors such as the type of starch and particle sizes of both the starch and the other plastic (which is the dispersion of one phase in the other), the ratio of the starch to the nonbiodegradable plastic, how well the plastic is dispersed in the starch phase using compatibilizers (if any), and the equipment as well as the process conditions of the compounding process. The mechanical properties also depend on the morphology along with how well the starch is bonded to the other plastic. If the volume ratio of the plastic is higher than that of the starch, after biodegradation, you will be left with the film or other article with holes where the starch was. If this ratio is not low enough, you could be left with tiny pieces of plastic that are turned into microplastics. While there is a big hue and cry from some in the environmental community about microplastics, there is no data, yet, to show any harm from microplastics residue in humans or animals. However, appropriate formulation and compounding can surmount most issues. The optical properties of these blends differ from transparent individual plastic material. While starch blends are processed similar to mineral-filled plastics except for lower temperature and shear, almost all starch blends are opaque and not transparent, because it is difficult to match the refractive index of starch with other polymers.

23.4.2 Starch compounds: (see Chapter 20 for additional information)

Biodegradable compounds are disclosed in a patent where compositions comprising a biodegradable resin and starch, wherein fat or oil treated starch or its gelatinized product is utilized as the starch, and thereby, their production cost and biodegradability are improved, without losing their mechanical properties. This patent also covers biodegradable compositions comprising a gelatinized product of fat or oil-treated starch and a biodegradable resin and the process for the production of the biodegradable composition.

The process involves heating and kneading fat or oil-treated starch and a biodegradable resin in the presence of water or water and a plasticizer [4].

Biome Bioplastics Limited in the U.K. has developed a thermoplastic, biodegradable polymer blend having a high surface energy and good adhesion without the need for pretreatment, with starch in the range 10–60% by weight of the polymer blend, at least one further carrier polymer in the range 40–90% by weight of the polymer blend, and one or more process aids in the range of 0– 0.5% by weight of the polymer blend [5]. The same company also has developed copolymers of pyridinedicarboxylic acid (PDCA) and of furandicarboxylic acid (FDCA), both monomers derived from lignin-sourced furanic sugars. These copolymers are also biodegradable and useful in treating microbial infection.

Since starch-based plastic blends can be incorporated with various biopolymers and/or petroleum-based polymers to create unique bio-compounds, they can be used in a whole host of applications. These bio-compounds can then be extruded, blow molded, or injection molded using traditional processing equipment. As starch is relatively cheap, starch-based plastics compounds are more cost competitive than alternative bioplastics. If formulated appropriately, such compounds can be compounded to meet a wide range of physical properties that other bioplastics lack, such as tensile strength, low specific gravity, and heat resistance. To reduce cost further and increase sustainability, these starch composites can also be formulated by using recycled plastics. When starch is incorporated to replace traditional petro-based polymer resins, such starch-based compounds can reduce the carbon footprint, where they can replace petroleum-based polymers with natural ones. If the other polymer in such compounds is biodegradable or compostable as well, the compound also becomes highly degradable, that is, such compounds will also be very degradable, and if both polymers are compostable, the alloy or blend can be used as a compostable plastics material.

An analysis by Green Dot Plastics of a market research report stated that "starch blends are expected to account for the largest share in the market" from 2015 to 2020. In 2017, out of the 2.05 million tons of bioplastics produced worldwide, 18.8% were starch-based blended compounds. In fact, European Bioplastics stated that "Bioplastics are used in an increasing number of markets, from products, consumer electronics, automotive, agriculture/horticulture, and toys to textiles and a number of other segments." Additional opportunities will come from a continued capacity for growth and consumer demand for environmentally friendly products over and above the efforts to reduce pollution and to diminish our reliance on petroleum-based products in the future [6].

Rising demand results in growing supply and the Chinese government's increasing support for bioplastics as a way to improve the standard of living in the country's rural regions is just one example of this. Major indigenous producers of starch plastics and other types of bioplastics are rapidly changing marketplace dynamics. Municipal governments in central, eastern, and southern China are planning for strong growth in all sectors of this industry, and developing countries such as Thailand and Brazil are also emerging as bioplastic-producing powerhouses [7].

23.4.3 Polycaprolactone (PCL) compounds

Polycaprolactone (PCL) has a low degradation temperature. Hence, for it to be used, its processing temperature profile must be lowered to allow it to be manufactured into films, sheets, or articles molded with injection and blow molding. PCL is also much more expensive than starch. Hence, compounding PCL by adding lower cost biopolymers like starch as well as compounding in mineral fibers, animal fibers, and natural fibers as reinforcements and talc or calcium carbonate as low cost fillers makes PCL compounds more cost-effective. However, such mineral fillers are left behind after the biodegradable plastic part of such compounds gets degraded into carbon dioxide and water. Here are a few selected examples from literature:

- PCL is often used in biodegradable polymer blends such as PCL with wheat-based thermoplastic starch [8] and PCL/PLA blends[9]. Since PCL and PLA are both synthesized by ring-opening polymerization, PCL/PLA multiblock copolymers can readily be produced, to give biodegradable thermoplastic elastomers [10].
- There are many examples in academic and patent literature to produce PCL-based biodegradable blended plastics compounds such as starch/PCL, soy/PCL, chitin/PCL, lignin/PCL, cellulose acetate/PCL, PCL with cellulose derivatives like hydroxyethylcellulose acetate, cellulose acetate and ethyl cellulose, etc. It has been shown that the mechanical and thermal properties of blends of maleated PCL and starch polymer alloys, summarized as "PCL-g-MAH/starch" were studied to show significant improvement over similar blends without the maleation. What caused this improvement? The greater compatibility of PCL-g-MAH with starch resulted in much better dispersion and homogeneity of starch in the PCL-g-MAH compound (as determined by microscopic examination of the compound's morphology due to the formation of an ester carbonyl group) [11]. While Wu's title of his paper calls it a composite, from a polymer chemistry perspective, this is a polymer alloy compound and not really a composite.
- In Sweden, bags had been produced from PCL and PCL-starch bags, but they degraded very fast, even before reaching the customers [12]. However, a proprietary blend of PCL and starch has been successfully used by Yukong Company for making trash bags in Korea, but they use some polyethylene with a coupling agent in their patented formulation to obtain good enough mechanical properties along with biodegradability [13].
- Two synthesis routes to produce composites of graphene and PCL were investigated and the properties of the resulting composites compared [14]. In these composites, the graphene was also covalently bonded to the carrier biopolymer. Since *in vitro* cytotoxicity testing of the compounds showed good biocompatibility along

with good conductivity, products molded from these promising materials for applications such as conducting substrates for the electrically stimulated growth of cells claimed to be suitable for use in tissue engineering. This type of compounds has been used for several different applications in the biomedicine and medical device markets [15].

23.4.4 Polylactic acid (PLA) and polyhydroxy alkanoate (PHA) compounds

Both biopolymers, PLA and PHA, have been covered in Chapters 20, 21 and 22. PLA has several limitations such as slow bio-degradation rate, hydrophobicity, and low impact toughness when used alone. Also, PLA, by itself, or as a blend with starch has the major limitation of relatively low glass transition temperature (typically between 111 °F and 145 °F, depending upon the ratio of its dextro isomer to its levo isomer). This makes it fairly unsuitable for temperature applications. For example, when you put a PLA straw in a cup with boiling water, the straw will instantly deform, making it unusable. Therefore, PLA straws are only suitable for cold drinks, while hot drinks require the use of crystallized PLA as one of the options to extrude high-temperature straws. Parts produced from plastics with low heat distortion temperature can soften and deform in conditions like a hot car in the summer. In addition, PLA has higher permeability than other plastics, making both moisture and oxygen pass more easily through it than other plastics. This will lead to faster food spoilage, so PLA is not recommended for long-term food storage applications [16].

PLA and PHA are now well-known biodegradable biopolymers. Films produced from PLA offer sufficient gloss and transparency, comparable to polystyrene (PS) films. They have a decent level of heat resistance and mechanical strength and also enable the creation of biodegradable plastics with superior antibacterial and antifungal properties. Once such biopolymers are compounded, many of its limitations can be overcome. Blending such biopolymers with additives and other polymers are some of the common successful ways to improve properties, to overcome limitations, and to produce novel biopolymer blend compounds for target applications. For example, a Japanese patent disclosed compositions comprising PLA and 0.05–5 parts by weight of silicone oil per 100 parts by weight PLA, having significantly improved mold-release properties, elongation at break, impact strength, and tensile strength [17].

Another patent [18] discloses degradable materials, which include a nontoxic hydrolytically degradable polymer like PLA and a nontoxic modifier, wherein the modifier is compatible with the polymer, and the modifier is nonvolatile and non-fugitive. The modifier is selected from the group consisting of lactic acid, lactide, oligomers of lactic acid, oligomers of lactide, and mixtures thereof, – and more specifically, plasticizers and modifiers like acetyl tributyl citrate, lactide, glycolide, lactic acid esters, dimethyl adipate, diethyl adipate, caprolactone, acetyl triethyl citrate, bis 2-ethyl hexyl sebacate, and bis 2-ethyl hexyl adipate, dibutyl sebacate, and triethyl citrate. Also disclosed are processes for forming the various degradable materials, which include films, molded products, laminates, foams, powders, non-wovens, adhesives, and coatings. The disclosed materials and processes are particularly useful for the production of commercial and consumer products in high volumes, which are suitable for recycling after use or which are discarded into the environment in large volumes.

Another option to overcome the significant disadvantage of insufficient heat resistance of PLA for several applications is by compounding in inorganic fillers or polymers with higher heat deformation temperature (HDT). The addition of talc at about 1% improves heat resistance, as measured by HDT, by increasing the nucleation rate of its crystallization and its crystallinity [18]. However, it makes products molded or extruded from such PLA compounds too stiff and brittle. This lack of toughness of talc-nucleated PLA can be solved by adding compatible elastomers and/or plasticizers. If the elastomer is not compatible, functionalized products or coupling agents can be used to obtain compatibility. For example, the brittleness and low HDT can be overcome by the use of styrene-maleic anhydride (SMA) copolymer in combination with an epoxy functional styrene-acrylate oligomeric chain extender. Such a composition also often exceeds a threshold of 65 °C in heat deflection temperature, as shown in this patent by Avakian et al. [19]

Another significant disadvantage of the use of PLA is lack of flame retardance, which has been overcome by two different inventions. The first patent discloses the use of polyphosphonate-co-carbonate in combination with an impact modifier, a drip suppressant, and optionally, an epoxy functional styrene-acrylate oligomeric chain extender. The compound achieves a UL 94 rating of V-0 or V-1 and a Notched Izod value of more than about 876 Newtons/m. The compound also exceeds a threshold of 100 °C in HDT in one patent [20]. In the other patent, the same researchers at PolyOne, overcame the lack of flame retardance of PLA by using specific combinations of either polycarbonate or polyphosphonate-co-carbonate in combination with nonhalogenated flame retardants of polyphosphazene or phosphate ester, such as resorcinol bis(diphenyl phosphate) or metal hypophosphite, a drip suppressant, and, optionally, an inorganic synergist of either zinc borate or talc or both, and, optionally, other ingredients. The compound achieves a UL 94 rating of V-0 or V-1 at 1.6 mm [21].

At the recent SPE ANTEC held in Charlotte, N.C., USA, May 14–16, 2022, Prof. Raymond Pearson presented a paper entitled "Enhanced Compost Rate and Simultaneous Toughening by Multifunctional Additives in Polylactic Acid." In this work, a specially designed microencapsulated particle additive was added to simultaneously improve ductility and biodegradation behavior of PLA. This unique approach involves the use of encapsulation technology to create degradation-promoting additives. while limiting any breakdown of the matrix during melt extrusion and service life. These microencapsulated particles were dispersed within the PLA matrix by an extrusion process to 3D printer filament. The results showed that elongation at break was improved compared to PLA. with only a slight loss of yield strength. The degradation rate in the composting step of these compounds accelerated and decoupled from environmental conditions by embedding the degradant material into the PLA matrix itself. This rate was helped by encapsulation technology that isolated and protected the degradant. This technology of dual-use microencapsulated particles has the potential to broaden the uses of PLA by reducing total composting time by over 65% [22].

At the same conference, another paper was presented, which showed that adding small quantities of bio-based and fully biodegradable amorphous polyhydroxyalkanoate (a-PHA) copolymer increased the impact strength of PLA, without compromising the compostability and bio-based carbon content of the final product. These blends exhibited good toughness and clarity when they were injection molded and extruded into sheet and into blown film. What is significant is that, this technology fine-tunes the blend composition to achieve the targeted levels of toughness increases and flexural modulus reduction needed for the application. Such impact-modified PLA were highlighted for applications such as film, thermoforming, and injection molding using these new amorphous PHA commercial grades [23].

A South Korean patent disclosed PLA resin compositions and molded product comprising the same with excellent environmental friendliness and biodegradability had excellent heat resistance characteristics due to HDT and may be injection molded within a commercially reasonable cycle time due to high crystallinity and crystallization rate, making it suitable for the preparation of a molded product. They accomplished this by compounding a PLA resin, which includes a hard segment containing a PLA repeating unit and a soft segment containing a polyurethane polyol repeating unit in which polyether-based polyol repeating units are linearly linked *via* a urethane bond, a poly-D-lactic acid (PDLA) resin as a nucleating agent, and talc [24].

A series of highly flexible biodegradable poly(ethylene oxide)/poly(l-lactic acid) (PEO/PLA) (PELA) multiblock poly(ether-ester-urethane) polymers that perform as thermoplastic elastomers were synthesized and characterized [25].

PHA is a class of biopolymers, which are classified chemically as polyesters and can be rigid thermoplastic or elastomeric with melting points ranging from 40–180 °C. Polyhydroxy butyrate (PHB) and other PHAs with shorter aliphatic chains are much more rigid and not ductile, while the ones with longer chains are more elastomeric. Ideally, these two types can be blended to get to the desired set of mechanical properties. Alternatively, the rigid PHA can be made ductile as disclosed in a patent, by compounding it with macrocyclic poly(alkylene dicarboxylate) oligomer to produce a ductile biopolymer compound, compared to the original PHA without the macrocyclic poly(alkylene dicarboxylate) oligomer. Optionally, the compound can also include additional polymers and functional additives to modify physical properties of the compound to suit the required properties of the molded, or extruded plastic articles can be produced from the compound [26].

Table 3 of a report has summarized the progress of the large number of applications of PLA, PHA, PHB, and their blends with other biopolymers in various fields like medical and other biomedical devices as well as textiles, automotive spare and parts, packaging films, electrical appliances, commodity containers, floor mats, and mobile phone housings, from 1999 to 2013 [27].

An American patent assigned to Arkema Inc. disclosed the use of compatibilizers like their Lotader™ and Lotryl™ copolymers, which are functionalized olefin-methacrylic and olefin-acrylic copolymers, to compatibilize blends of polyolefins and a biodegradable polymer such as PLA and PHB. Such alloys had improved processability, melt strength, and melt elasticity [28].

A large number of compounding companies like Sukano, Americhem, Tosaf, AF-Color, Avient, and others have developed grades of masterbatches (also called concentrates) using PLA, PHA, PHB and other biopolymers as carrier resins for use downstream in extrusion and molding applications for a wide range of applications. Types of such masterbatches include color pigments, flame retardants, antioxidant packages, and other additives.

More than 100 eco-friendly companies around the world, large and small, are manufacturing various cutlery and other articles using PLA compounds, some of which also contain minerals like talc and calcium carbonate, which are added to provide additional stiffness, whiter color, and lower cost. A few biodegradable plastic applications compiled in an article in 2018[29] are listed below:

- Papelyco sells Plantable™ paper plates produced from bio-based plastic compounds that are not just biodegradable but also incorporate important minerals in the plate to provide their customers a new sapling with the nutrients it needs to grow. So, customers can use their plate, place it in the ground and a plant will literally grow out of it. This is a cool green invention and sustainable in several ways.
- A company called Ecoware in New Zealand has developed their biodegradable dishware and bioplastic packaging that has 80% lower carbon footprint than similar products produced from petro-based plastics. They claim to be a carbon neutral company, because they offset all of their energy output by investing in clean technology.
- The US-made line of compostable cutlery and packaging from a company called WorldCentric includes a very broad range of biodegradable bags, dishware, and cutlery for household and corporate use. Their products produced from bio-based plastics compounds are sourced from perennial plant fibers with long roots that help store carbon underground and are certified compostable disposables. In addition, 25% of their profits are donated to worthy causes.
- Compostable packaging is being brought to consumers by Be Green Packaging by partnering with companies such as Google, Whole Foods, Gilette™ from Procter and Gamble, Samsung, Virgin America, and others.
- A line of biodegradable flexible plastics like bags with zippers and garbage bags are produced by a company called Tipa. Such bioplastics are heavily preferred and used by consumers and have the potential to grow significantly to a more sustainable future.

23.4.5 Other compounds

Compounds comprising by weight: 70–96% of at least homopolyamides or copolyamides with 11 and/or 12 carbon atoms, 4 to 10% of a plasticizer, and 0 to 25% of an NBR or H-NBR elastomer, such that the sum of the amount of plasticizer and the elastomer is between 4% and 30% were patented [30]. Such compounds are used for pipes in the operation of offshore oil and gas fields due to their ageing resistance as well as in pipes in motor vehicles due to their temperature resistance, especially in the presence of automotive fluids for applications in the engine compartment. In some cases, such compounds are used as a layer in multilayer extruded pipes.

A patent disclosing bio-resin compositions containing cellulose acetate and a biodegradable secondary polymer in a ratio of at least 10:1 by weight of the total weight of the composition was issued in 2015 [31]. The secondary polymer is an aliphatic polyester, such as polybutylene succinate or an aliphatic co-polyester, such as polybutylene succinate adipate or an aliphatic/aromatic polyester, such as polybutylene adipate terephthalate. The secondary polymer may be present in an amount up to 10% by weight, more preferably up to 6% by weight, and even better in the range 2-3% by weight of the total weight of the composition, where the combined amount of secondary polymer and cellulose acetate may be up to 85% of the total weight of the composition.

Blends of thermoplastic starch with other polymers were reviewed by authors from Malaysia for applications mostly in cutlery and food packaging applications [32].

23.5 Limitations of bio-based plastics compounds and challenges to growth

Research on bio-based polymers has been active as evidenced by the large number of patents, technical publications, and new products. Globally, environmental and sustainability consciousness have risen to a stage where consumers desire eco-friendly products. This, in turn, has encouraged efforts for developing materials based on bio-based compounds for different end-use applications.

Conversely, these bio-based plastics compounds are not direct substitutes for petro-based plastics compounds without any negative issues or problems. They have several drawbacks and limitations such as poor moisture resistance (hydrophilicity), supply logistic issues, low thermal stability, flammability, poor electrical properties, and highly anisotropic properties. They also have challenges related to various parts of the value chain including extraction, processing, surface modification, machining, manufacturing, and characterization [33]. Consequently, due to challenges related to the properties of these materials compared to synthetic petro-based polymers, such as their resistance to chemicals or weather, applicability and use have not dramatically increased, yet, in many end-use markets compared to petro-based polymers, as was originally forecasted. For demanding applications as well as for those where durability is essential, such drawbacks will preclude biopolymers, and bioplastics as well as products produced from their compounds cannot displace incumbent non-bio-based products.

As stated earlier, another major issue is that since these biopolymers are all found in nature, there is a lot of inconsistency from one batch grown at one location to another grown elsewhere. This occurs due to differences in their growth due to differences in soil, temperature, and rainfall or water available at each location. Hence, consistency of mechanical performance is more difficult to maintain compared to non-bio-based polymers. Some of these issues can be mitigated by innovative compounding techniques that we have discussed in this chapter as well as by improving quality control of incoming raw materials and outgoing quality assurance of products.

Besides the limitations that are discussed earlier in this chapter, there are challenges in bio-based plastic compounds and products produced from them in trying to make them compatible with existing industry practice, equipment, capital investment, etc. In some cases, some modifications in process conditions or reduction of production rate may be needed. Lowering of processing temperature, on the one hand, can reduce energy consumption, while the reduction of production rate can lead to higher costs, on the other.

However, such bio-compounds may be 100% bio-sourced but suffer from the drawback that life cycle analysis (LCA) results show they consume a lot more of two precious resources, namely energy and water, than petro-based composites and sometimes cost more, based on data from LCA! [34, 35]. One prominent scientist-researcher, who has reviewed over 400 scientific articles on the subject of plastics, bioplastics, and their degradation chemistry, states in his book, "Every LCA study I have found clearly states that these biodegradable polymers are worse for the environment than standard plastics like PE and PP. So, they are not green at all, despite the marketing claims of the manufacturers"[36].

One more factor that needs to be considered for the long-term viability of bio-based polymers and compounds, as Michael et al. have pointed out in 2011 very eloquently, is the expectation that there will be feedstock competition as global demand for food and energy increases over time. Currently, a large number of renewable feedstocks used for manufacturing bio-based monomers and polymers often compete with requirements for food-based products [37]. With the current invasion of Ukraine by Russia and the resultant looming food shortage in many parts of the world, this factor is even more relevant. The expansion of first-generation bio-based fuel production, especially, could place unsustainable demands on biomass resources and is as much a threat to the sustainability of biochemical and biopolymer production as it is to food production [37].

Ultimately customers want value, and to get there for bioplastics in general and bio-based plastic compounds in particular, these products need to demonstrate value to break into the marketplace. Instead of trying to replace existing products that are not bio-based, the use of bio-based products needs to be more discriminatory, rather than seeing bioplastics as a straight swap with existing plastics, by embracing circular economy and whole LCA in making decisions on materials, according to Prof. Mitchell [38].

23.6 Conclusions

While biopolymers and bioplastics compounds have several limitations, it is clear from this chapter that interest in them is very high, not just related to research and creative ways to overcome some of these limitations, but also to move forward with scaling up these innovations to commercial products. Issues related to moisture absorption and dimensional stability are handled by encasing the bio-layer in a plastic material that provides a moisture barrier to protect the layer. Another limitation of poor interfacial adhesion is overcome by the use of coupling agents and compatibilizers, using techniques covered in a couple of online courses and a blog [2, 39].

As more and more bio-based polymers and compounds are being adopted, applications will grow and the total volume of such biomaterials should see an attractive annual increase – estimated by several commercially available market reports to grow at CAGRs in the range of 7–12%.

References

[1] Fakirov S, Bhattacharyya D. Handbook of Engineering Biopolymers – Homopolymers, Blends and Composites, Hanser Publishers, 2007, pp. 1–1. ISBN: 1-68015-262-9, 1-56990-405-7.

[2] Adur A. Compatibilization of Polymer Alloys Blends and Composites. online course taught on specialchem.com, 2002–2004, available from SpecialChem, Paris, France.

[3] Díez-Pascual AM, Cinelli P. Synthesis and Applications of Biopolymer Composites, Switzerland: MDPI, Basel, 2019. ISBN 3039211323 & 9783039211326.

[4] Shitaohzono T, Muramatsu A, Hino J. Biodegradable compositions. United States Patent 5,691,403, issued November 25, 1997 assigned to Nihon Shokuhin Kako Co., Ltd. (Japan).

[5] (A) Law PW, Longdon T, Perez AD, Gomis B, Gomis M. Biodegradable material. United States Patent 9,205,963, issued December 8, 2015, assigned to Biome Bioplastics Ltd. (GB).(b) https://biomebioplastics.com/applications/.

[6] https://www.greendotbioplastics.com/starch-based-plastics/.

[7] Global bioplastics industry report 2021–2026 with focus on plastics made from renewable resources such as biomass or food crops. Research and Markets, https://www.prnewswire.com/news-releases/global-bioplastics-industry-report-2021-2026-with-focus-on-plastics-made-from-renewable-resources-such-as-biomass-or-food-crops-301458684.html/.

[8] Averous L, Moro L, Dole P, Fringant C. Properties of thermoplastic blends: Starch–polycaprolactone. Polymer 2000;4111):4157–67. https://www.sciencedirect.com/science/article/abs/pii/S0032386199006369/.

[9] Przybysz-Romatowska M, Haponiuk J, Formela K. Poly(ε-caprolactone)/poly(lactic acid) blends compatibilized by peroxide initiators: Comparison of two strategies. Polymers (Basel) 2020;12(1):228.

[10] Cohn D, Salomon AH. Designing biodegradable multiblock PCL/PLA thermoplastic elastomers. Biomaterials 2005; 26(15):2297–305.

[11] Wu C-S. Physical properties and biodegradability of maleated-polycaprolactone/starch composite. Polym Degrad Stab 2003;80(1):127–34.

[12] Flieger M, Kantorová M, Prell A, Rezanka T, Votruba J. Biodegradable plastics from renewable sources. Folia Microbiol (Praha) 2003;48(1):27–44. doi.org/10.1007/BF02931273/.

[13] Lee BH, Jung KS, Kim YW, Bang SG, Cho WY, Jo B, Cheon U, Ki N. Biodegradable plastic composition, method for preparing thereof and product prepared therefrom. United States Patent 5,861,461, assigned to Yukong Ltd., issued January 19, 1999.

[14] Sayyar SS, Murray E, Thompson BC, Gambhir S, Officer SL, Wallace GG. Covalently linked biocompatible graphene/polycaprolactone composites for tissue engineering. Carbon 2013;52:296–304.

[15] Dhanasekaran NPD, Muthuvelu KS, Arumugasamy SK. Recent Advancement in Biomedical Applications of Polycaprolactone and Polycaprolactone-Based Materials.In: Reference Module in Materials Science and Materials Engineering, 2022. doi: doi.org/10.1016/B978-0-12-820352-1.00217-0.

[16] Barrett A. Advantages and Disadvantages of PLA. June 9, 2020, https://bioplasticsnews.com/2020/06/09/polylactic-acid-pla-dis-advantages/.

[17] Tanifuji Y, Tokushige Y. Polylactic acid composition. United States Patent 5,691,398, issued November 25, 1997, assigned to Shin-Etsu Chemical Co., Ltd.

[18] Sinclair RG, Lipinsky ES. Degradable polymer composition. United States Patent 5,502,158, issued March 26, 1996 assigned to Ecopol, LLC, Golden, CO, USA.

[19] Zhu S, Avakian RW. Heat resistant polylactic acid compositions. United States Patent 8,765,865, issued July 1, 2014, assigned to PolyOne Corp.

[20] Zhu S, Avakian RW. Flame retardant polylactic acid compounds. United States Patent 9,062,201, issued June 23, 2015, assigned to PolyOne Corp.

[21] Zhu S, Avakian RW. Flame retardant polylactic acid compounds, United States Patent 9,534,116, issued January 3, 2017, assigned to PolyOne Corp.

[22] Pearson R. Enhanced Compost Rate and Simultaneous Toughening by Multifunctional Additives in Polylactic Acid. paper presented at SPE's ANTEC 2022, Charlotte, NC, USA.

[23] Krishnaswamy R. Impact Modification of PLA (Poly Lactic Acid) by Blending Small Amounts of Amorphous PHA (Polyhydroxyalkanoate) Copolymers. paper presented at SPE's ANTEC 2022, Charlotte, NC, USA.

[24] Kim M-Y, Kim T-Y, Hwang -J-J, Chung J-I. Poly(lactic acid) resin composition and molded product comprising same. United States Patent 11,118,051, assignee: SK Chemicals Co., Ltd., issued September 14, 2021.

[25] Cohn D, Hotovely-Salomo A.Biodegradable multiblock PEO/PLA thermoplastic elastomers: Molecular design and properties. Polymer 2005;46(7):2068–75.

[26] Avakian RW. Thermoplastic polyhydroxyalkanoate compounds, United States Patent 7,968,657, Assigned to PolyOne Corp. issued June 28, 2011.

[27] Babu RP, O'Connor K, Seeram R. Current progress on bio-based polymers and their future trends. Prog Biomater 2013;2(1):8. doi.org/10.1186/2194-0517-2-8/

[28] Donnelly Z. Polyolefin/polylactic acid blends, United States Patent Application 2012,0035,323 A1, assigned to Arkema Inc.

[29] 5 Bio-Degradable Plastic Companies for a Greener Future, Jan 19, 2018, https://www.goodnet.org/articles/5-biodegradable-plastic-companies-for-greener-future/.

[30] Jacques B, Pees B, Werth M. Polyamide-based composition for flexible pipes containing oil or gas, United States Patent 6,913,043, July 5, 2005, assigned to Arkema.

[31]   Longdon T, Law PW, Sillence K, Wetters DJ, Perez DA. United States Patent 9,062,186, "Bioresins", assigned to Biome Bioplastics Limited (UK), issued June 23, 2015.

[32]   Diyana Z, Jumaidin R, Selamat M, Ghazali I, Julmohammad N, Huda N, Ilyas R.Physical properties of thermoplastic starch derived from natural resources and its blends: A review. Polymers 2021;13:1393. doi.org/10.3390/polym13091393.

[33]   Andrew JJ, Dhakal HN. Sustainable biobased composites for advanced applications: Recent trends and future opportunities – a critical review. Composites Part C 2022;7:100220. https://doi.org/10.1016/j.jcomc.2021.100220.

[34]   Hottle TA, Bilec MM, Landis AE. Biopolymer production and end of life comparisons using life cycle assessment. Resour Conserv Recycl 2017;122:295–306. https://doi.org/10.1016/j.jclepro.2020.121158/.

[35]   Walker S, Rothman R. Life cycle assessment of bio-based and fossil-based plastic: A review. J Cleaner Prod 2020–07–10;261:121158. doi: 10.1016/j.jclepro.2020.121158.

[36]   De Armitt C. The Plastics Paradox: Facts for a Brighter Future, Terrace Park, OH, USA: Phantom Plastics LLC; ISBN 978-0-9978499-6-7.

[37]   Michael C, Dirk C, Harald K, Jan R, Joachim V. Policy paper on bio-based economy in the EU: level playing field for bio-based chemistry and materials, 2011, https://renewable-carbon.eu/publications/product/nova-paper-1-on-bio-based-economy-level-playing-field-for-bio-based-chemistry-and-materials-%E2%88%92-full-version/.

[38]   Mitchell G, Mahendra V, Pinheiro J, Sousa D, Abdulgahni S, Dias J, Faria P, Gaspar F, Mateus A. Challenges with biopolymers. presented at 8th World Congress on Biopolymers & Bioplastics June 28–29, 2018, Berlin, Germany, https://www.longdom.org/proceedings/challenges-with-biopolymers-19169.html/.

[39]   Bicerano J. A Practical Guide to Polymeric Compatibilizers for Polymer Blends, Composites and Laminates, SpecialChem, December 2005; http://polymerexpert.biz/blog/135-compatibilization-of-immiscible-polymers/.

Ashok Adur

Chapter 24
Bio-based fibers and plastics composites

24.1 Introduction

Bio-composites are compounds that contain a biopolymer and/or a reinforcing bio-fiber using some sort of a compounding equipment. Typically, reinforcements are added to plastics to improve their mechanical properties and thermal resistance. As is covered in the previous chapter on Bio-Compounds, bio-composites would also need the use of various additives to ensure that the biopolymer does not degrade during the compounding step, and can meet various end-use applications and handling requirements. They may contain one or more bio-based major ingredients, meaning either the matrix polymer, the reinforcement, or both could be bio-based.

These additives and the reinforcements are added during a compounding process step to overcome the inherent limitations of biopolymers and biofibers in composites, and to increase the range of performance characteristics required by customers in various applications. This is what drives the need, and hence the practice of using additives and reinforcements using the right compounding equipment and process to achieve target specification requirements for each application.

This chapter covers the various commonly used biofibers, and the several emerging and commercial bio-composites. So, this chapter will dive into what is common and what is different for biofibers versus non-bio-sourced fibers like glass or carbon fibers as well as for bio-composites versus conventional petro-based plastics composites. Challenges in bio-based plastic composite products in trying to make them compatible with industry practice, equipment, capital investment, etc., and how specific bio-composites have been successful and what factors led to their success will also be discussed. Many chapters of a 2007 book [1] covered many combinations of such composites but there has been a lot of progress and developments in bio-based plastics composites since then.

24.2 Processing to produce bio-based plastics composites

Similar to other polymers with high viscosity, compounding of bio-composites needs sufficiently high shear and temperature to bring down the viscosity of the polymer

Ashok Adur, Everest International Consulting (LLC), Jacksonville, FL32256, USA, e-mail: Aadur@Outlook.com

https://doi.org/10.1515/9783110791228-024

matrix to a range where adequate mixing is easy, irrespective of the additives and reinforcements to be mixed into the molten polymer are in liquid, powder, pellet, or fiber form. The process to add additives to make the polymers processible downstream and modify their properties into useful articles is called compounding. This includes adding reinforcements to polymers to produce plastics composites. When one or more reinforcements are incorporated in a polymer, it always forms two or more phases. It is more challenging when two polymers are blended because the reinforcement does not melt, but similar to other bio-sourced polymers, can degrade during processing.

When reinforcements are added to polymers using a compounding equipment, care must be taken to ensure that the reinforcing fiber is uniformly distributed in the polymer. This starts with ensuring adequate and uniform feeding of the fibers as well as other additives in liquid, pellet, or powder form into the compounding equipment. If the fibers are not fed at a constant rate that can keep up with the production rate, special accessory equipment may be needed to ensure that it is, irrespective of the fibers being bio-based or not. The larger the aspect ratio of the fibers, the more difficult it is to feed them at a consistent rate. Also, if the fibers stick to one another, the difficulty of feeding increases.

Similar to biopolymers, almost all biofiber reinforcements are very hygroscopic and absorb water, and hence need adequate drying to minimize the moisture content prior to the compounding process. The moisture absorption of biofibers has been discussed for each biofiber in the section on biofibers. Too much moisture can cause foaming, voids, and other issues. In a few end products, this sort of foaming and voids in the end product could be useful to reduce the specific gravity of the composite part, but this is more of an exception. Some of the moisture can be driven out through the hopper by a proper selection of the extruder screw configuration. Nevertheless, additional equipment is needed to control moisture to the right level needed for the application. Non-biofibers, like glass fiber and carbon fiber, do not have this issue at all since they are not hygroscopic and do not absorb large quantities of water. Hence, the small amount of water in these other fibers does not need a pre-drying step, like biofibers do, and gets converted to steam; some through back-venting through the feed hopper and/or the die.

The equipment used to produce bio-composites is similar to that of non-bio-based composites, and is not repeated here. However, the process parameters are usually a little different. Besides the additional equipment to ensure moisture control, the temperature and shear control need to be moderated to ensure the fiber does not degrade, both from mechanical chopping, which can reduce the fiber aspect ratio too low to affect mechanical properties and from thermal or oxidative degradation. Figure 24.1 shows the processing steps for bio-based plastics composites.

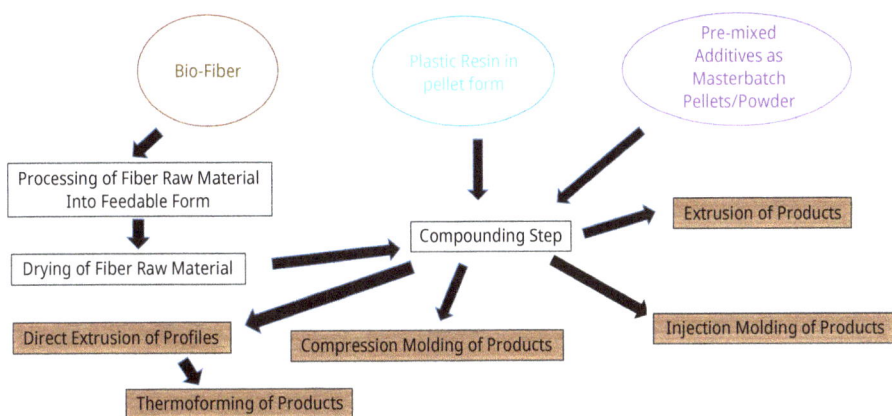

Figure 24.1: Processing steps for bio-fiber plastics composites.

In addition, the aspect ratio of the reinforcement needs to be considered. Usually, the reinforcing fiber has a higher length than its width or depth, like a rod, i.e., a higher aspect ratio compared to mineral reinforcements could be spherical, round, or flat in shape. A fiber is long only in one direction but in some cases, it is woven into a fabric or converted into a non-woven, from the natural fiber itself so it is larger in two dimensions than the third, it is also used as a reinforcement. During extrusion, most fibers orient in one direction, typically the direction of the extrudate flow. This leads to different properties in the extrusion direction versus the perpendicular or the transverse direction. This can happen a little during the compounding stage and a lot more at the subsequent extrusion stage or the injection molding stage. In the injection molding, the design of the gate determines the direction of the fiber orientation. In order to obtain the same properties in both directions, various downstream methods can be used or in some cases, rather than feed the fibers into the compounding equipment, a different set of processes are employed. Sometimes, depending upon the end application, the fiber is first converted into a cloth or a non-woven fabric, which is then processed using a compression molding or similar process to convert the cloth or fabric into a two-dimensional sheet of the fiber-reinforced composite. Compression or lamination molding and a variety of cast processes are more commonly used with thermoset type of plastics. Two-dimensional reinforcements are also used for end-use applications with thermoplastic composites, where the mechanical strength in more than one direction is insufficient. In some cases, the biofiber is powdered into flour, especially for wood plastics composites (WPC), which will be covered in detail later in this chapter.

In most cases, the reinforcement is chemically incompatible. Incompatible polymers can be made compatible by adding small amounts of compatibilizer as additives during compounding.

24.3 Unique challenges

Similar to biopolymers, biofibers have advantages due to their outstanding biodegradability, and if some of them are not degradable or compostable, at least they are renewable, and in many cases they can be produced into useful form starting from plant sources or even from waste products, like industrial or consumer waste. However, they generally possess poor mechanical properties, higher sensitivity to moisture and water absorption, a short fatigue life, low chemical resistance, poor long-term durability, and limited processing capability compared to petro-based plastics [2]. Most of these challenges and disadvantages of synthetic polymers and biopolymers can be overcome, at least partially, by compounding them with other polymers to form compounds and/or with reinforcing fillers or, in some cases, nanofillers to form composites [2]. A nanofiller is small enough that at least one of its dimensions is in the nanometer range or below 100 micrometers.

Similar to the inherent attractiveness of biopolymers, biofiber materials also have a positive environmental impact because they are low-cost, renewable, reusable, and biodegradable, and hence environmentally friendly and sustainable. A common disadvantage that almost all of these fibers share with biopolymers, which is covered in the previous chapter, is that the fibers are also hydrophilic polar polymers themselves; they absorb water and when wet, lose some of their mechanical strength, which is measured by the tensile strength, bending strength and especially rigidity, which is usually measured as flexural modulus.

Being biopolymers themselves, biofibers also have a fairly low degradation temperature, especially when compared to non-biofibers like glass fiber or carbon fiber. However, in general, biofibers are slower to degrade than biopolymers. Some of this is due to the presence of lignin in many biofibers as well as their inability to melt in most cases, and in some cases, it is due to their crystallinity.

Since all these biofibers are found in nature, there is a lot of inconsistency from one batch grown at one location to another grown elsewhere. This occurs due to differences in their growth due to differences in soil, temperature, and rainfall or the availability of water at each location. Even biofibers from the same plant or tree at the same location exhibit variability. So, consistency in mechanical performance is more difficult to maintain, compared to synthetic fibers, which are more precisely controlled through process control. On average, the standard deviation of the tensile strength of dry biofibers is 5–10 times higher than that of glass fibers. When biofibers are wet, they lose dimensional stability, and this range can increase up to ten times more. Most large scale producers of bio-composites have learned to handle this variability by ensuring that before the fibers go into their compounding equipment, their moisture content is below a threshold that their equipment can handle to still meet customer requirements; the producers also widen the specifications, compared to those without bio-based content. Some manufacturers also carry out post-blending of pellets after the compounding stage to minimize variability.

Another unique issue about biofibers is that due to the higher lignin levels in many biofibers, the color of the composite gets darker and, more specifically, more brown as a function of the heat applied to it during its processing, both during the production of the biofiber and during the processing of compounding it into biocomposites as well as during further conversion into the final product. The color darkens as a function of its heat history beyond a threshold temperature, which varies based on each type of biofiber. To prevent excessive darkening, monitoring and controlling the maximum temperature it sees during processing is even more critical. This is not a major issue with reinforcing fibers from non-bio-based sources.

Due to the higher stiffness and rigidity that reinforcing fibers provide to biocomposites, these composites typically are more brittle and less ductile than the polymer matrix on its own. This is true for non-bio-based fiber reinforced composites as well. In many cases, higher stiffness in composites, compared to unreinforced plastics, means lower impact properties as well. However, with adequate increase in the interfacial adhesion between the reinforcing fiber and polymer matrix, this see-saw effect of rigidity versus impact can be overcome and a synergistic combination of properties can be achieved.

In order to obtain better properties and minimize the see-saw effect, it may be necessary to either chemically bond or "couple" the biofiber to the bioplastic matrix [3]. Alternatively, the surface of the reinforcing fiber should be ozone or corona treated or chemically modified by a process called "functionalization" [4]. This type of surface modification and the use of coupling agents are also commonly used with non-biofibers such as glass fiber for petro-based thermoplastics. The choice of such functionalization is based on the specific chemistry of the matrix and the reinforcement, similar to conventional petro-based materials.

For example, polypropylene does not form good bonds with glass fiber. So, a polypropylene grafted with maleic anhydride is used as a coupling agent during the compounding process to bond to the amine groups on the amino-silane on the glass fiber surface. On the other side, the polypropylene chain on the coupling agent is able to co-crystallize with the polypropylene matrix to provide that critical interfacial adhesion between the glass fiber and the polypropylene; thus giving the composite significantly higher tensile strength, flexural modulus, impact strength, long term fatigue performance, and improved heat distortion temperature [5].

Another example is where during the production process of these specialized glass fibers, the hot glass fibers get dipped into a sizing solution containing alternating copolymers of ethylene and maleic anhydride. This incorporates anhydride groups on the glass surface, which during the subsequent compounding process downstream binds to the amine end groups of nylon. This reaction results in a covalent bonding between the glass fiber and the nylon matrix to improve the interfacial adhesion between the reinforcing fiber and the polymer matrix, and improve the mechanical properties, increase the heat distortion temperature, and the resistance to harsh chemicals like the radiator fluid used in automotive parts in the engine compartment of automobiles [6–8]. Before this technology of this type of surface treatment of the glass fiber to produce glass fiber-reinforced nylon-6,6 composites was developed, radiators used to leak frequently. In the last five years, such occurrences are very rare.

Similar technology can and has been used with biofibers to improve the interfacial adhesion between the biofiber and the polymer matrix. When such chemical bonding of the biofiber to the biopolymer occurs, it results in significant improvement in performance, opening up a wider range of applications, but makes biodegradability even more challenging. Yet, from a net carbon dioxide balance perspective, such bio-composites would be beneficial. Such bio-composites may be 100% bio-sourced but suffer from the drawback that life cycle analysis results show they consume a lot more of two precious resources, namely energy and water, than petro-based composites, and sometimes cost more, based on data from life cycle analysis!

So, rather than concentrating on biodegradability or compostability of bioplastics and their composites, this chapter will cover compounds made from biopolymers to turn them into useable bioplastics and bio-based composites to include combinations of polymers reinforced with fibers where at least one of the combining materials is bio-based.

24.4 Biofibers & other bio-reinforcements

24.4.1 Introduction

A biofiber is a fiber sourced from biological matter, and has either plant or animal origin. Sometimes, it is used naturally without further processing, or is more often processed further in some sort of a subsequent process. The main difference between a biofiber and a non-biofiber is the biological origin of the former. Modern technology has blurred this distinction somewhat, because the production of fibers with a chemical structure that is identical to those found in nature through chemical and non-chemical modification is now possible [9]. Unlike chemically modified fibers, biofibers are abundant in nature in fibrous materials, especially those in the cellulosic family. Hence, some prefer to call biofibers, "natural fibers"

The promise of biofibers is the preservation of natural resources with a focus on the utilization of renewable raw materials that, in many cases, are biodegradable at end of life. Besides the fact that they contribute to the consumption of carbon dioxide gas during their growth stage, this advantage of natural fibers is neutralized if at the end of their lives they are burned; the same amount of CO_2 emission is generated. The low abrasive nature of biofibers reduces the wear of the processing equipment and is hence greener compared to more abrasive carbon or glass fibers [10]. In spite of these benefits, finding applications that consume sufficiently large quantities of these materials to enable price reduction and allow biofibers to compete economically in the market remains a continuing challenge [9]. The major exception to this is the area of wood plastic composites (WPC) where wood scrap converted to wood flour from the wood industry is used as a relatively cheap fiber-reinforcing material, but mostly with synthetic petro-based polymers, which again are relatively cheaper than biopolymers.

From an environmental and sustainability perspective, biofibers are typically believed to have a life cycle that is typically measured from "cradle-to-grave" from an environmental impact perspective that includes sourcing, manufacturing, use, and end of life, when compared to non-bio-based polymer fibers like glass or carbon fibers. However, there are many key performance indicators in each aspect of the lifecycle. Two of the most often considered are greenhouse gas emissions in production and end-of-life. However, this type of analysis is somewhat flawed because it ignores other factors like resource usage, which is another important factor to consider, particularly when evaluating biofiber production. Biofibers typically require a huge amount of water for both irrigation and processing. In terms of environmental biodegradability, biofibers typically break down more quickly than petro-based polymers, which is a positive. But from a cost perspective, except for a few exceptions, biofibers have a disadvantage over petro-based polymer or glass fibers, due to the issue of scale, covered earlier, as well as the cost of the water and energy needed for irrigation, cleaning, processing, and drying [11].

There are many naturally occurring biofibers that are used in fiber-polymer composites. Among them are cellulose fiber, hemp fiber, kenaf, flax, bamboo, jute, coir, rice husk, corn husk, wood fiber, leaf fibers (like those from sisal, abaca and banana leaves), pineapple fiber, crop residues, etc. While a lot of these fibers are cellulosic in nature, there are other chemical compounds present, depending on the source of these fibers. The chemical and mechanical structure of each type of fiber gives each of these fibers unique set of performance characteristics when they are used in composites.

Most of these biofibers can be used after cutting them into the desired fiber length, keeping in mind that inherently during compounding and further process stages, the fiber length and hence the aspect ratio gets trimmed, a feature that is common to fiber compounding, bio-based or not.

In order to use natural fiber reinforcements and achieve isotropy in two directions, the fibers need to be converted into a textile preform, with many different structures to suit each end application, similar to man-made reinforcements. The most widely used structure is the non-woven preform, which is usually co-mingled (blended) with 50% wt. polypropylene so that it can be directly compression molded for making automotive parts. The polypropylene intertwined between the biofiber allows for good moldability in this process. Due to the random fiber orientation obtained in this process, high fiber packing and consequently high fiber volume fraction are achieved. Non-woven reinforcements can be cost effective and at the same time utilize hard fibers that otherwise would be difficult to spin. Also, they provide a porous structure with up to 30% void content, which is critical in achieving the levels of thermal insulation and vibration damping required by the transportation industry, and specifically by the automotive industry. A few of the important tier 1 suppliers of these type of composites to the automotive sector in the United States were FlexForm and Ecotechnilin in Europe, a couple of years ago [11]. Today, with the rapid adoption of bio-based polymers and biofibers, and of more diverse types of bio-fiber reinforced composites by many different industries, there are several more in many countries around the globe.

These natural fibers also include fibers from various parts of plants, such as leaves, seeds, grass, branches, and trunks of trees as well as from products like wood scrap, waste paper, and other bio-derived sources. Such biofibers are being used as alternative materials for fillers in the conventional petro-based polymer matrices and in bioplastics. Usually, fibers are made from easily available plants locally, such as bamboo, rice straw, sugar palm fiber, kenaf, flax, hemp, roselle, and pineapple. Other natural fibers that are derived from waste products, forest clippings, and remnants after a crop harvest is completed can also be used as a bio-sources for use in the production of bio-composites as well. Maintaining consistency in quality for such fibers is inherently difficult for routine production of composites because there is no guarantee that the fiber properties will be similar one day to the next or even within a batch of fiber. This makes producing composites with the same properties daily or even hourly very difficult [12]. There are several good reviews of these cellulose-based natural fibers available in literature [10–17], which cover many different perspectives – from sourcing through converting to various polymer composites to a myriad of end applications.

Figure 24.2: Various biofibers [from Ref. 11].

Figure 24.2 shows a few photographs of the various biofibers that are used to produce plastics composites in both biopolymers and petro-based polymers.

Table 24.1 shows the chemical composition of several common biofibers [11–15]. Cellulose is typically the largest component in biofibers, with varying amount of hemi-cellulose, lignin, and other minor components like waxes, pectin as well as inorganic and water-soluble compounds. While cotton fiber is listed in these tables, it is rarely used because of its demand for clothing and hence higher cost, in spite of a lot of research data available in literature.

Table 24.1: Composition of natural biofibers.

Fibers	Cellulose (wt.%)	Hemicellulose (wt.%)	Lignin (wt.%)	Waxes (wt.%)
Abaca	56–63	20–25	12–131	–
Bagasse	55.2	16.8	25.3	–
Bamboo	26–43	30	21–31	–
Banana	63–64	–	5–11	–
Coir	32–43	0.15–0.25	40–45	–
Corn Husk	31–39	34–41	14-Feb	–
Cotton	85–90	5.7	–	0.6
Flax	71	18.6–20.6	2.2–20.6	1.5–1.7
Hemp	68–74	15–22.4	3.5–10	0.8
Jute	61–71.5	13.6–20.4	12–13	0.5
Kenaf	45–72	20.3–21.5	8–13	–
Pineapple	80.5	17.5	8.3	–
Ramie	68.6–76.2	13–16	0.6–0.7	0.3
Rice husk	36–40	0–4	28–32	–
Sisal	65–78	10–14	9.9–14	2

Table 24.2: Mechanical properties of natural biofibers.

Fibers	Tensile (MPa)	Young's modulus (GPa)	Elongation at break (%)	Specific gravity (g/ml)
Abaca	400	12	3.0–10.0	1.50
Bagasse	350	22	5.8	0.89
Bamboo	290	17	1.5–3.0	1.25
Banana	529–914	27–32	5.9	1.35
Coir	220	6	15–25	1.25
Corn Husk	145–165	4–5	18–24	N/A
Cotton	400	11.8	3.0–10.0	1.51
Flax	800–1500	60–80	1.2–1.6	1.40
Hemp	550–900	70	1.6	1.48
Jute	410–780	26.5	1.9	1.48
Kenaf	930	53	1.6	1.20–1.45
Pineapple	413–1627	60–82	14.5	1.44
Ramie	500	44	2.0	1.50
Rice Husk	25–75	2.5 to 4	0.5–1.0	0.95
Sisal	610–720	10–40	2.0–3.0	1.3–1.5

Table 24.2 shows the mechanical and physical properties of these natural fibers. The values of these properties can be used to calculate the estimated values of composites produced from them, with relevant assumptions or, even better, microscopic measurements of fiber lengths in the final part.

Dimensions of fiber length and diameter for some of the biofibers are shown in Table 24.3.

Table 24.3: Fiber dimensions, growth cycle and performance benefits of biofibers.

Fiber type	Fiber length (mm)	Fiber diameter (µm)	Growth cycle (month)	Performance benefits
Bagasse	0.65–2.2	20–30	10–15	Good durability, low sp. gr.
Bamboo fiber	60–200	6–10	3–6	Antimicrobial, UV-resistant, high tensile strength
Banana fiber	60–80	15–25	10–15	Light weight
Brown coir	200–350	12–25	3–6	Resistance to microbes and salt water
Flax	90–125	18–22	2–4	Low growth cycle
Hemp	120–200	16–50	10–20	Low sp. gr., good thermal and acoustic insulation
Jute	1000–4000	15–20	4–5	Strong fiber, antistatic
Kenaf	2.5–4	15–25	3–7	Stiff fiber
Rice husk	5–7		4–6	High silica content
Sisal	1.0–2.0	20–25	24–36	Good strength and siffness
Wood fiber	60–200	6–14	12–24	Good flexibility

Both fiber length and diameter contribute to the mechanical properties of the fiber and of the composite. In the discussion of composites, the significance of the chemical composition, fiber length and diameter, cell structure, original polymer matrix attributes, interaction of the fiber with the matrix, and other characteristics, and their effect on composites will be better understood. The growth cycle is related to the economics of renewal and hence the raw cost of the biofiber, and ultimately of the bio-based plastic composite. The longer the growth cycle, obviously the more expensive the fiber becomes, assuming all the other factors remain the same.

Another factor that needs some discussion here is the challenge of providing sufficient volumes of both biopolymers and biofibers. Production of both materials, based on natural sources, is limited by factors like weather issues and the current bottles necks in supply chain, locally and globally. With respect to issues like scale up and the availability of supply, it varies significantly across each individual biofiber. Some are already produced in large quantities for other lower value applications. Also, the growth cycle of many common biofibers, shown in Table 24.3, is relatively shorter than the time needed to set up a manufacturing plant for biopolymers. Consequently, for the scale-up of composites produced, starting with biopolymers reinforced with bio-fibers, the time to develop large-scale manufacturing is determined by the speed at which the production of the biopolymer can be scaled up.

Table 24.4 shows the annual global production of different biofibers in terms of their use in composites. The global production of natural fibers was estimated to be around 25 million tons per year in the mid 2000s [15]. During the past decade, the global annual natural fiber production has ranged between 28 million tons and 35 million tons, with most of the variation from one year to the next caused by changes in yields linked to weather. This data collected by DNFI, in collaboration with the Food and Agriculture Organization of the United Nations (FAO), on 14 categories of natural fibers, ranging from abaca to wool and other animal fibers, is available for public viewing [18]. This shows that the production of such fibers has not kept up with growth in global GDP. This is likely because the large majority of these fibers are used for clothing and a relatively small fraction is used as reinforcement in bio-composites.

Table 24.4: Annual production and origin of biofibers.

Fiber	Annual production (tons)	Origin
Abaca	70,000	Leaf
Bamboo	10,000,000	Stem
Banana	200,000	Leaf/stem
Coir	100,000	Fruit
Cotton lint	18,500,000	Stem
Flax	810,000	Stem
Jute	2,500,000	Stem
Kenaf	770,000	Stem
Ramie	100,000	Stem
Rice and corn husk	Abundant	Fruit/grain
Sisal	4,500,000	Leaf
Wood	1,750,000,000	Stem

24.4.2 Cellulose fiber

Cellulose fibers used for reinforcement of composites are fibers made with some type of processing, whether it is derivatives of cellulose ethers or esters of cellulose, which can be obtained from the bark, wood or leaves of plants, or from other plant-based material. Such derivatives have been covered in other chapters in this book. Besides cellulose, the fibers typically also contain hemicellulose and lignin, with different percentages of these components. This variability alters the mechanical properties of the fibers, depending upon many factors such as the species of the plant or tree it came from, and how much processing was carried out to separate the cellulose from lignin, hemi-cellulose, and other extractables. While there are patents that cover cellulose fiber composites, in order to cover as many of the biofibers, rarely do they use pure cellulose for such composites when used with plastics. So, this chapter will cover as much as possible each of the major types of biofibers that have cellulose individually.

24.4.3 Jute fiber

Jute is a herbaceous plant that is cultivated with edible young shoots for jute fiber, and is primarily grown in India and Bangladesh. It has since been cultivated in many tropical countries. The name originates from the Bengali *jhuto* meaning "matted hair." Jute is a long, soft, shiny bast fiber, composed primarily of the plant materials cellulose and lignin. Jute is renewable, versatile, nonabrasive, porous, hydroscopic, viscoelastic, biodegradable, combustible, compostable, and reactive. The fiber has a high aspect ratio, high strength-to-weight ratio, is low in energy conversion, and has good insulation properties. Earlier, jute was used for low value products, such as gunny bags, twine, and carpet backing, considering jute as a low quality resource, viewing its biodegradability and combustibility as disadvantages. Biodegradability provides a means of predictable and programmable disposal, which is not easily achieved with other resources [19].

When wet, jute like other cellulosic fibers, absorbs water and swells but retains enough tensile strength to be used when twisted into rope. However, for most industrial applications, minimizing water is critical to ensure dimensional stability as a function of humidity. Using bio-based polymers that are hygroscopic to produce such composites is not practical for most industrial applications. For example, biodegradable jute fiber-biopolymer composites, such as those covered by S. R. T. Reddy et al. [20], are hygroscopic, and products produced from such composites are not very useful for many applications due to the high moisture absorption, poor dimensional stability, and susceptibility to insect and microbial attacks.

Fibers like jute can also be used after processing them into a cloth. The processes to make the fibers useful as cloth reinforcements or as other reinforcements include thermal lamination, compression molding, and a variety of cast processes.

24.4.4 Coir fiber

Coir is the fibrous material found between the hard, internal shell and the outer coat of a coconut and is sometimes called coconut fiber. Commercial coir is a natural fiber extracted from the outer husk of coconut. There are two types of coir fiber. The more commonly used brown coir is from mature coconuts while white coir is from unripe tender coconuts with finer fiber and low lignin content. Among vegetable fibers, brown coir fiber has one of the highest concentrations of lignin, making it stronger but less flexible than cotton, and unsuitable for dyeing. The tensile strength of coir is lower than that of abaca or bamboo, but it has good resistance to decomposition from microbial action and salt water damage, and needs no chemical treatment. This key advantage for coir is used for manufacturing durable products. Coir fibers measure up to 35 cm in length with a diameter of 12–25 μm. Coconut trees are grown on over 10 million hectares of land throughout the tropics. A coconut harvest occurs once in about 45 days. From 1000 coconuts, it would be possible to extract 10 kg of coir [21]. Coir is a fibrous material that is widely used to overcome the problem of erosion. This works based on weaving them into geotextiles and placed on areas in need of erosion control, where it promotes new vegetation by absorbing water and preventing top soil from drying out. Such geotextiles retain the natural ability of coir to retain moisture and protect from the Sun's radiation, just like natural soil. Unlike geo-synthetic materials, it provides good soil support for up to three years, allowing natural vegetation to become established.

Coconuts are typically grown by small-scale farmers, who use local mills for fiber extraction. Globally, around 650,000 tons of coir are produced annually, mainly in India and Sri Lanka, where the coir industry is fully developed and are also the main exporters, followed by Thailand, Indonesia, the Philippines, Malaysia, Vietnam, and Brazil. There are many other countries that grow coconuts but they have not moved to commercialize coir fiber. About 80 percent of the coir produced is exported in the form of raw fiber, while smaller quantities are exported as yarn, mats, matting, and rugs. Research and development efforts are continuing to focus on the use of coir in geotextiles and other new applications like coir plastics composites, as the market shows promising prospects. Coconuts are grown in more than 93 countries globally and hence there is considerable scope to develop coir industry in more countries [21].

24.4.5 Hemp fiber

Hemp fiber is derived from the *cannabis sativa* plant. This fiber is located in the stalks of hemp plants. The stalks are surrounded by thick skins, with bundles of fibers present directly under this outermost layer. Hemp is also one of the bast fibers known from ancient times. Recent excavations in Egypt have shown hemp to be one of the oldest cultivated fiber plants but is currently finding renewed interest holistically, including all parts of the plant's fibrous material. Thus, hemp is both historical and holistic! It has been widely cultivated as a source of bast fibers in many countries that have a mild climate with humid atmosphere and a decent rainfall – a rainfall of at least 65 cm per year. Hemp is an annual plant that can grow on a wide variety of soils, but a well-drained, nitrogen-rich, and non-acidic soil is essential for ample hemp cultivation.

The chemical composition of hemp stems shows a significant difference between the outer bast and the woody core. The chemical composition of the outer bast is 60–70% cellulose, 15–20% hemi-cellulose, less than 5% lignin, and much smaller amounts of pectin, fats, and waxes. So, only the bast is used for fiber production and not the hemp woody core, which is very similar to that of hardwood, with roughly 40% of both cellulose and hemi-cellulose and 20% lignin; and hence more difficult to process. The process for removing fibers from the hemp plant is similar to the process for flax fiber production [22]. Chapter 25 has more details on hemp.

Hemp has been grown for many end uses – textiles, matting, ropes, and sail cloth – for its fiber properties. Hemp has received a bad rap, especially in the United States, and was banned for most of the twentieth century because of its similarity to the marijuana plant. This negative reputation is undeserved. In the United States, we have recently started exploring the potential of the hemp plant once again. The CBD industry, for instance, would not exist without hemp. Industrial hemp generally has less than 1% THC, which is a psychoactive chemical. However, the two varieties appear very identical, which often causes confusion. Today, it is one of the fastest growing plants, as CBD usage has grown [23].

24.4.6 Bagasse fiber

Bagasse is the dry pulpy fibrous material that remains after crushing sugarcane or sorghum stalks to extract their juice. Chemical analysis of washed and dried bagasse shows it has 45–55% cellulose, 20–25% hemicellulose, 18–24% lignin, and smaller quantities of ash and waxes [24]. Bagasse fiber length is 0.65–2.17 mm and its width is 21–28 μm [25].

For every 100 tons of sugarcane that is crushed, a sugar factory produces nearly thirty tons of wet bagasse. It is challenging to directly use this byproduct as a fuel because of its high moisture content, typically 40–50 percent. Instead, bagasse is typically stored prior to further processing. Products from bagasse biodegrade quickly and can be composted without releasing any harmful chemicals. It is considered a waste product, meaning it requires no additional resources to produce, and the largest current application for bagasse is as fuel to generate electricity, but only after drying [26]. One company has used it to produce what is claimed to be the world's best disposable plates [27].

24.4.7 Bamboo fiber

Bamboo, (subfamily *Bambusoideae*), is part of tall tree-like grasses that belong to the *Poaceae* family, comprising more than 115 genera and 1,400 species. Bamboos grow in tropical and subtropical to mild temperate regions, with the heaviest concentration and largest number of species in East and Southeast Asia and on islands of the Indian and Pacific Oceans. A few species are native to the southern United States along river banks and in marshy areas. Bamboos are typically fast-growing perennials. Depending on the species, some grow as much as 30 cm per day. The smallest species of bamboos grow to just 10 to 15 cm (about 4 to 6 inches) in height while others grow to more than 40 m [28].

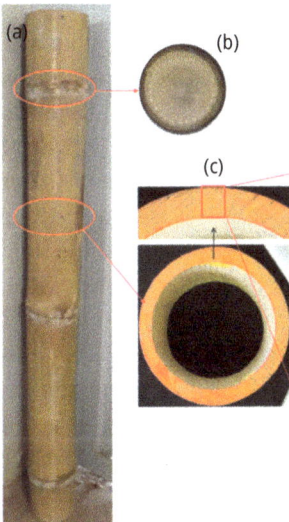

Figure 24.3: Bamboo and its nodes (adapted from [29]).

The woody ringed stems, known as culms, are typically hollow between the rings (nodes) and grow in branching clusters from a thick rhizome (underground stem). In Figure 24.3, (a) is the bamboo stem where the first red circle is the node, (b) shows the cross-section of the node, and (c) shows the cross-section of the hollow bamboo where there is no node [29].

Bamboo is usually extracted into fibers and this fiber is then used as a reinforced material in the polymer matrix. Various properties of bamboo-fiber-based composites have been investigated extensively, and will be covered in this chapter in the next section. The reason bamboo fiber is selected as a reinforcement is because of its good mechanical and thermal properties, facile extraction and fiber treatment, low cost, environment friendly nature, and ability to be used as a product in the industry. Bamboo fiber is identified to have strength and stiffness, and it contains microfibrillar angles that give it very high aspect ratio naturally, and its thick cell walls make bamboo fiber an ideal reinforcing fiber. However, these fibers themselves have poor compressive strength. Most researchers have separated the bamboo fibers from the nodes before using them. But the nodes were left in place by one researcher. This will be discussed later in the section on Bamboo Fiber Plastics Composites in this same Chapter 24 where most of the limitations of bamboo were mitigated and it was used in an application where the high tensile strength was useful. The water absorption reduced significantly and its poor compression set was compensated by using it in as a reinforcement in concrete, where the concrete's high compression strength was used for synergistic effect.

24.4.8 Kenaf fiber

Kenaf is a plant that is native to Southern Asia, though its exact origin is unknown. The name also applies to the fiber obtained from this plant. These fibers are in many ways similar to fibers of jute, and shows similar characteristics. It is an annual or biennial herbaceous plant (rarely a short-lived perennial), growing to a height of 1.5–3.5 m, with a woody base. The stems are 1–2 cm diameter, often but not always branched. The leaves are 10–15 cm long, variable in shape, with leaves near the base of the stems being deeply lobed with 3–7 lobes, while leaves near the top of the stem are shallowly lobed or unlobed lanceolate. The flowers are 8–15 cm diameter, white, yellow, or purple; when white or yellow, the center is still dark purple [30].

The fibers in kenaf are found in the bast (bark) and core (wood). The bast constitutes 40% of the plant. "Crude fiber" separated from the bast is multi-cellular, consisting of several individual cells stuck together [30]. The individual fiber cells are about 2–6 mm long and slender. The cell wall is thick (6.3 μm). The core is about 60% of the plant and has thick (≈38 μm) but short (0.5 mm) and thin-walled (3 μm) fiber cells [31].

The stems produce two types of fiber: a coarser fiber in the outer layer (bast fiber) and a finer fiber in the core. The bast fibers are used to make ropes. Kenaf matures in 100 to 200 days. First grown in Egypt over 3000 years ago, the leaves of the kenaf plant were a component of both human and animal diets, while the bast fiber was used for bags, cordage, and the sails for Egyptian boats. This crop was not introduced into southern Europe until the early 1900s. Today, while the principal farming areas are China and India, Kenaf is also cultivated for its fiber in Bangladesh, southern part of the United States of America, Indonesia, Malaysia, South Africa, Viet Nam, Thailand, parts of Africa, and to a small extent in southeast Europe [30, 31].

24.4.9 Rice & corn husk fiber

The rice husk, also called rice hull, is the dry outer covering of a grain of rice. Lab tests have shown that the husk contains about 40% cellulose-based fiber, 30% lignin group, and 20% silica. The last two of these are hard materials, which have the function to protect the seed during the growing season. Rice husk is the exterior shell of rice grain, which is always removed during the milling of rice. It is of no direct nutritional value as food for mankind, and in most mills, it is often discarded or allowed to rot away [32, 33]. Rice husk can absorb water ranging from 5% to 16% by weight, and the unit weight of rice husk is 83–125 kg/m^3 [34]. The rice husk is removed in the milling process, which accounts for 20% of the world's total rice production. With about 760 million tons of world paddy production in 2017, according to the FAO data [35] and data from knoema. com [36], rice husk can be calculated to be about 152 million tons available globally.

Similarly, corn husk is the outer covering of an ear of corn, also called maize. Commercially bought corn husks are commonly the whole husk and are dried in sunshine, or by air or in an oven. Corn husk fiber has low crystallinity and rough surface with hollow cross-sectional area. Cornhusk fiber shows less tensile strength and more elongation, compared to other popular lignocellulosic fibers like jute and cellulosic cotton, due to lower lignin, and higher cellulose and hemi-cellulose content. Despite wide availability in many countries around the world, the only applications appear to be in various culinary recipes and in tribal handicrafts; not much as a biofiber [37].

24.4.10 Wood fiber

Wood is a porous and fibrous structural tissue found in the stems and roots of trees and other woody plants. It is a natural biocomposite of cellulose fibers that is strong in tension and embedded in a matrix of mostly lignin and hemicellulose that together resists compression.

Wood fibers are usually cellulosic elements that are extracted from trees, and used to make materials, including paper and paperboard. This is one of the largest three traditional applications, the others being furniture and building construction. The end paper product dictates the species, or blends of two or more species are used, which are best suited to provide the desirable sheet characteristics of the paper or paperboard. This also dictates the required fiber processing. In North America, virgin wood fiber is primarily extracted from both hardwood and softwood trees for producing paper and paperboard. The wood fiber is either extracted as a primary product or collected during the milling of lumber [38, 39]. Wood fibers can also be recycled from used paper materials. The furniture and building construction industries also use the right species of wood that is cut at lumber mills. Most of the wood fiber used for wood plastics composites (WPC) are from lumber mill waste that comes mainly from these three industries. Depending upon the type and species of wood and what part of the tree it is derived from, the fiber can be 1 to 3 mm in length and 20–30 microns in diameter [39].

24.4.11 Leaf fiber, especially from sisal, abaca, and banana leaf

Various leaves from trees and plants can also be used, but those from sisal, abaca, and banana leaves are the most important. Sisal fiber has one of the largest cross-sectional diameters among natural, fibers which gives it a higher tensile strength and stiffness, making it useful for its current applications for cords, power transmission, marine ropes, baler, binder twine, sacks, paper filter, and other industrial uses. While some research work has been carried out to use sisal fiber as a reinforcement for plastic composites, which is covered later in this chapter, it has not been commercialized for such applications, at least in a large scale. Same is the case with fiber from abaca and banana leaves.

Abaca fibers, also called manila hemp, are extracted from the leaf sheath around the trunk of the abaca plant (*Musa textilis*). The plant is native to the Philippines and is widely found in the humid tropics close to the Equator. The process of harvesting abaca is labor intensive because each stalk must be cut into strips, which are scraped to remove the pulp, and the fibers are then washed and dried. The leaf fiber is composed of long slim cells that form part of the leaf's supporting structure. The high lignin content of 15% is the reason that abaca fiber is prized for its mechanical strength, resistance to saltwater damage, and long fiber length (up to 3 meters). The best grades of abaca are fine, lustrous, light beige in color, and very strong. It has been pulped to make sturdy manila envelopes. Today, it is still used to make twines, ropes, fishing lines and nets, as well as coarse cloth for producing sacks. There is also a flourishing niche market for abaca clothing, curtains, screens and furnishings, but paper-making is currently the main use of the fiber, especially in Indonesia [40].

Banana fibers, obviously from banana trees, are concentrated near the outer surface. The fibers are extracted by hand scraping, chemically, by retting, or by using raspadors. They can also be extracted by boiling the leaf sheaths in sodium hydroxide solution [41]. The use of these fibers for textiles has been proposed based on research in using them in composites, but no large commercial applications are known yet [42].

24.4.12 Fibers from crop residues and other biomass

There have been a lot of statements from many companies around the world about using crop residues and biomass; only a few large-scale productions have started using what is called a bio-refinery, except for generating bio-fuels. This consists of various conversion processes and equipment, which convert biomass into fuels, chemicals, and power [43]. Deep insight into chemistry, production, and conversion technology is required to transform biomass to useful products. If the process temperature is controlled sufficiently, the residue left over contains a lot of useful fiber, which could be used for producing composites. This is especially true if agricultural crop residues or waste products come from paper making or other industries, which essentially use natural bio-sources that have strong fibers.

Crop residues are the materials left over after a crop harvest in the form of leaves, stalks, stems, seeds, etc. The amount of crop residue depends on the type of the crop and its growth, as well as the methods of tillage and harvesting. Cereal crops produce large quantities of residues, which are useful in adding soil organic matter. They are also used as feed, fodder, fiber and construction material, and as fuel for energy [44]. Use of such biomass as feedstock for biopolymers has been covered in Chapters 2, 14 and 25. Post-harvest residues, which are used for soil amendments, like husk and bagasse, will not be covered here because they have already covered earlier. In Europe, brewer's spent grain, olive pomace and residual pulp from fruit juice production, and other agri food wastes that are produced annually in huge quantities, pose a serious problem economically and environmentally because landfilling biowaste is expensive and unsustainable in the long run. All such lignocellulosic matter can be converted into fibers along with forest residues, yard clippings, wood chips, and even municipal solid waste, as per the objectives of the circular economy [45]. Industrial biomass can also be grown from numerous types of plants, including miscanthus, switchgrass, wheat, hemp, corn, poplar, willow, sorghum, sugarcane, bamboo, and a variety of tree species, ranging from eucalyptus to palm oil. After valuable oils, other bio-chemicals and, in many cases, bio-fuels have been already extracted. The only issue with biomass, sometimes, is that besides carbon, hydrogen, oxygen, and nitrogen, other elements like alkali, alkali earth, and heavy metals are also present in such bio-fibers [45], which might restrict their use in some downstream applications.

24.5 Bio-based fiber plastic composites

24.5.1 Introduction

Fiber composites are compounds where a polymer is reinforced with a reinforcing agent like a fiber, and the fiber provides reinforcement and rigidity to the composite. As such, composite structures are generally a combination of two or more materials at the macroscopic level, and typically have an interface layer between the two main phases [3, 46, 47]. This section covers composites, irrespective of the polymer and/or the reinforcement being bio-based in the composite. In some cases, the carrier polymer used is bio-based; in some composites, biofibers were used to reinforce non-biopolymers; and in others, both the reinforcing fiber and the polymer matrix are bio-based. Natural fiber is used in these composites as a reinforcement material. It is embedded in a polymer matrix where the polymer is either a thermoplastic or a thermoset. Composites that contain natural biofiber and polymers have been called many different names, like natural fiber composites (NFC) [48, 49], natural-fiber-reinforced polymer composites (NFPs) [50], etc.

A bio-based composite's properties are influenced by a number of major variables, including the fiber type, environmental conditions (where the plant fibers are sourced), processing methods, and any modification of the fiber. Ultimately, the properties of the composite are determined by the individual properties of the reinforcing fiber and those of the polymer matrix, as well as the interfacial interaction between the two phases. The properties of the composite are dependent on the following factors:

- aspect ratio, measured by the ratio of the length of the fiber to its diameter,
- length of the fiber,
- mechanical strength of the fiber,
- volume ratio of the fiber to the matrix,
- distribution and dispersion of the fiber in the matrix,
- orientation of the fiber within the matrix,
- interfacial adhesion of the fiber to the matrix,
- mechanical strength of the matrix, which itself is dependent on its crystallinity, its orientation, and its molecular weight, among other factors.

Most natural bio-based polymers and bio-based fibers are polar, hygroscopic, and have high enough water absorption capability. This factor restricts their use without either chemical modification of those polar groups and/or sufficient coverage by non-polar polymers, almost all of which are petro-based, like polyethylene and polypropylene.

One of the greatest challenges in working with biofiber-reinforced plastic composites, either experimentally in trials or in large-scale production, is the large variation in properties and characteristics of biofibers [51]. Natural fibers vary a lot compared to synthetic fibers that consist of metal, glass, or synthetic polymer – as measured by standard deviation.

Over the past few decades, there has been a significant increase in the production of a wide range of products that use natural-fiber-reinforced polymer composites. These composites are one of the best alternatives to produce environmentally friendly materials by combining bio-based and non-bio-based polymers and natural fibers in various products for many different applications [48–52]. Using these natural fibers has a high impact on the manufacturing industry, as these materials are readily available, cost less, are easy to design, and increase productivity [50]. Natural fiber sources are increasingly gaining attention and acceptance for use in fiber-reinforced polymer composites, such as with biopolymer matrixes like:

- polyhydroxyalkanoates (PHA) [53],
- polylactic acid (PLA) [54],
- chitosan [55],
- polycaprolactone, [56], and
- thermoplastic starch [57].

Kuciel, et al. published their results on short fiber-reinforced bio-composites produced using polyamides from castor oil, produced from renewable sources in 2012 [58]. Jumaidin, et al. reported on their work on processing and characterization of banana leaf fiber-reinforced thermoplastic cassava starch composites [59]. Another detailed review was published in a book entitled "Natural polymers, biopolymers, biomaterials, and their composites, blends, and IPNs" in 2021 [60] and by Plotz [9]. While a few applications of banana fiber-reinforced epoxy composites in automotive and aircraft parts have been pursued, there is no confirmation of commercialization available in literature.

Numerous applications exist currently for bio-based fiber polymer composites in packaging, automotive parts, furniture, building and construction, medical devices, and other end-use markets. For example, Mercedes Benz has used a composite produced from abaca yarn in polypropylene rein matrix for molding automobile body parts. Replacing glass fibers by natural fibers can reduce the weight of automotive parts and facilitates more environmentally friendly production and recycling of the parts, typically using maleic anhydride-grafted polypropylene as the coupling agent. Due to the high mechanical strength of the fiber as well as its length, application of abaca, even in highly stressed components, offers great potential for this and other industrial applications [61].

24.5.2 Cellulose fiber composites & cellulose nano composites

Cellulose fiber has been used as reinforcement for polymer composites with many different plastics matrices, for example, with dimer fatty acid-based polyamides [62]. At the Society of Plastics Engineers (SPE) meeting hosted by the Automotive Composites Conference (ACCE) in Novi, Michigan in September 2015, Dr. Alper Kiziltas, Lead Research Scientist at the Research & Innovation Center of Ford Motor Co. discussed the blends of bio-based polyamide (PA-6,10 and 10,10) reinforced with cellulose fibers sourced from sustainable forestry [63].

What is not well known is that the strongest part of a tree lies not in its trunk or its sprawling roots, but in the walls of its microscopic cells. A single wood cell wall is constructed from fibers of cellulose (nature's most abundant polymer), the main structural component of all plants and algae. Within each fiber, there are reinforcing cellulose nanocrystals (CNCs), which are chains of cellulosic polymers arranged in nearly perfect crystal arrays or patterns. At the nanoscale, CNCs are stronger and stiffer than Kevlar™. The idea was that if the crystals could be worked into materials in significant fractions, CNCs would provide a route to stronger, more sustainable, naturally derived plastics composites. Recently, a team at the Massachusetts Institute of Technology in the USA developed a composite, made mostly from cellulose nanocrystals mixed with a bit of synthetic polymer. The bio-crystals make up about 60 to 90% of the material – the highest fraction of CNCs achieved in a composite to date.

Researchers found that the cellulose-based composite is stronger and tougher than some types of bone, and harder than typical aluminum alloys. They found that this engineered material has a brick-and-mortar microstructure that resembles nacre, the hard inner shell lining of some mollusks [64].

24.5.3 Kenaf fiber polymer composites

Kenaf fibers have also been used as reinforcements in various plastics composites. Examples are Anand Sanadi's work with polypropylene as far back as 1995 [65] and a review covering kenaf fiber-reinforced plastics composites with many polymers by Akil et al. in 2011 [66]. A Japanese patent developed kenaf fiber-reinforced biodegradable resin composition for molded articles such as parts for electrical and electronic equipment [67]. However, it is not known if any commercial breakthroughs have happened with kenaf fiber-based polymer composites.

24.5.4 Bagasse fiber polymer composites

Bagasse fibers are low cost waste products from sugar production after sugar cane juice is extracted from it. These fibers have been evaluated as a reinforcing filler to enhance mechanical properties, reduce product weight, and enhance environmentally friendly characteristics of a wide range of thermoplastics and thermosetting matrices, such as polypropylene, polyesters, polyvinyl chloride, polyurethane, epoxy, and high density polyethylene (HDPE) to form composites.

Bagasse has also been used to produce polymer composites with significant performance characteristics using gamma radiation to produce *in situ* thermoset composites using unsaturated polyester and styrene monomers as well as styrene-acrylonitrile monomers [68].

Bagasse fiber has been used along with another waste product from coal-powered power plants, called fly ash, to produce bagasse-fly ash polymer composites using the same radiation processing method, and a patent was granted [69]. The combination of bagasse, a low-cost raw material, and fly ash, which power plants pay companies to haul away, actually has a negative cost. This enables products made from this combination and used for composites as an amazing synergy of performance and cost for producing products with excellent performance-to-cost ratio.

More recently, a review was published on bagasse-fiber-composites with various polymers by Verma et al. [70]

24.5.5 Jute fiber polymer composites

Using jute fiber for composites has many advantages. Jute is renewable, versatile, nonabrasive, porous, hydroscopic, viscoelastic, biodegradable, combustible, compostable, and reactive. The fiber has a high aspect ratio, high strength-to-weight ratio, is low in energy conversion, and has good insulation properties.

A comparative analysis of jute fiber polymer composites with similar composites using jute and bamboo was published in 2017 [71].

Production of surface-modified bio-composite jute fiber hybrid, and differing polylactide and PCL fractions has been recorded with improved rigidity and durability. This type of bio-composites was fabricated by the hot pressed solvent-impregnated prepregs process. The addition of PCL led to an improvement in the biodegradation rate of bio-composites; thus making them more environmentally friendly [72]. The effect of compatibilizer to improve properties of short jute fiber-reinforced polypropylene composites has also been studied [73]. However, such composites are no longer biodegradable.

24.5.6 Bamboo polymer composites

The fundamentals and the processes of bamboo-based thermoset composites have been reviewed recently [74]. Most of these have been with bamboo fibers, and some in combination with wood fibers. Many of the initial commercial applications used phenol-formaldehyde, polyurethane, polyvinyl acetate, and epoxy resins, mostly to produce oriented strand board for housing and construction applications.

Both thermoset and thermoplastic bamboo fiber-reinforced composites have been reviewed recently. These references cover properties, fabrication, preparation, and chemical modification, including methods for improving bonding of the bamboo, impregnation methods used for thermoset polymers, process details, and applications [75–79].

Bamboo fibers (BF) were compounded with polypropylene in a twin screw extruder and injection molded into test specimen for determining mechanical and other properties by Lee et al. [80] They found that properties like tensile modulus, flexural modulus, and water absorption were enhanced by increasing the bamboo fiber loading. They also determined that the addition of tetramethoxy orthosilicate (TMOS) and aminopropyltrimethoxysilane (AS) after the alkali pretreatment for the fibers increased tensile, flexural, impact strength, water desorption, and other properties of the resultant composites, due to improved interfacial adhesion between the biofibers and the PP matrix. The melting temperature, melting enthalpy, crystallization enthalpy, and crystallinity were decreased by the increase in fiber loading and the "AS + TMOS" treatments. On the other hand, the crystalline temperature was increased by the addition of BF, AS, and TMOS. This data proved that AS and TMOS are effective coupling agents for the bamboo-polypropylene thermoplastic composite systems.

A 1944 patent describes the first impregnation of bamboo fibers; it was carried out with synthetic group of resins like phenolic aldehyde resins and urea aldehyde resins to form thermoset composites [81]. The first use of radiation processing to prepare bamboo-polymer composites was reported by Wang [82, 83]. Some preliminary work on the preparation of bamboo-polymer composites using radiation processing was also published in a paper by Mani et. al [84].

Several years ago, a unique method was developed using sections of bamboo instead of first separating the fibers and throwing away the rigid nodes. In this method, the long bamboo stems were instead cut along the length of the bamboo, into sections, with the nodes left in place. The sections were placed in a specially designed chamber to remove moisture and air to a vacuum of 4 mm of mercury for 4 h. Liquid monomers were added into the impregnation container, so the bamboo could soak the monomer mixture overnight. Two monomer mixtures were used: styrene-acrylonitrile-benzene hexachloride (St-AN) and unsaturated polyester-styrene-chlorinated paraffin (St-PE) in ratios of 60:40:3 and 20:80:30, respectively, in separate set of experiments. Chlorinated paraffin and the benzene hexachloride were used as free radical chain propagators to accelerate the polymerization and minimize the radiation dosage needed. Chlorinated paraffin also helped to reduce the viscosity of the unsaturated polyester to improve the impregnation into the bamboo. The container was then slowly restored to ambient pressure. This technique to get fibrous material to incorporate monomers or other liquids is called vacuum impregnation [85–87]. The soaked samples were then packed in polyethylene bags, then sealed, and irradiated with cobalt-60 gamma radiation at a dose rate of 0.11 MRad per hour. From prior work, the exact radiation dosage was determined to achieve full polymerization. Total dosage used for the St-AN mixture was 2.12 MRads and for the St-PE mixture was 2.49 MRads, based on preliminary work to determine the dosage needed to get to the required gel ratio. The microstructure of the composites showed that the polymer covered the cellulosic fiber of the bamboo using phase-contrast photo-microscopy and electron microscopy [85, 88].

Test specimens were prepared using machine-cut strip pieces of bamboo that were 50 cm in length and having a cross-sectional area of 1 sq. cm. The test results showed improvements in tensile strength, rigidity, modulus of elasticity, and reduction of water absorption even after 24 h and 3 months [85, 89]. Huge improvements were observed in resistance to attack by fungus, bacteria, and insects without using biocides over a 6-month period by which time, the untreated bamboo fell apart in indoor open air [85].

At a polymer loading of 30% on three different commercially available bamboo species, water absorption was reduced to the 1–2% range. The dimensional stability in water was adequate for the intended application for these bamboo-polymer composites was replacing steel rebar with these composites in concrete for building structures [85, 86], [90]. The properties of these composites were such that they had a higher strength-to-weight ratio compared to steel rebar, which is used for reinforcement of concrete. Testing of these composites was carried out by a large housing government corporation and found to be viable, based on higher bonding strength to concrete and other properties, at a time when there was a huge shortage of steel in India [90]. However, the economics fell apart with the tripling of monomer prices in 1972–74. The significance of this research is that, with these bamboo plastics composites, several of the limitations of bamboo, like high water absorption, poor dimensional stability, poor resistance to rotting, etc. were mitigated and bamboos were used in an application where the high tensile strength of the bamboo was synergized with concrete's high compression strength. The poor compression set of bamboo was not a factor in using it as reinforcement in concrete. Moreover, water absorption reduced significantly, which helped to ensure bamboo's other disadvantages were overcome in these unique composites [90].

In the US, Cali-Bamboo has been selling its BamDeck™ composite, made from 60% recycled bamboo fibers and 40% recycled plastics for decking, fencing, poles, edging, flooring, etc. [91, 92].

24.5.7 Wood polymer composites (WPC)

Wood fiber can either be extracted as a primary product or collected during the milling of lumber. Wood fibers can also be recycled from used paper materials. Most of the wood fiber used for wood plastics composites (WPC) are from lumber mill waste. Of all of the biofiber-based composites, this is the one with the largest sales volume and revenue globally and in most countries around the world. A total of 10,866 patents have been granted, just in the USA, related to wood polymer composites, also called wood plastic composites, which just shows how much interest there is in this area. There are several good reviews available on this topic [93, 94].

Wood plastic composites can be divided into three types:
(1) Traditional WPC such as particle board, fiber-based board, plywood, oriented strand board (USB), etc. in which the wood particles, fibers, and veneers are held together with a thermoset adhesive, used at 4–15% [60, 94].
(2) Non-adhesive-based WPC covers both wood that is impregnated with monomer systems and polymerized *in situ* to form thermoset WPC [95, 107].
(3) Non-adhesive-based WPC, are produced from either wood fiber, powder/flour or even wood micro powder or wood-based source, is recovered from waste materials [96].

Both Type (1) and Type (2) are reviewed in this eBook on "Wood Polymer Composites" [93]. Type (2) are materials in which wood is impregnated with monomers that are later polymerized in the wood to tailor the material for special applications. This polymerization technique is called "*in situ* polymerization." The resulting performance properties of these materials, like low specific gravity, boosted mechanical properties, and superior sustainability, has created a rising number of applications in building and construction as well as in automotive and other transportation industries. This eBook review covers a wide variety of topics such as the manufacture of wood-polymer composites, how their properties can be evaluated and enhanced and their range of uses, as well as key aspects of manufacture, including raw materials, manufacturing technologies, and interactions between wood and synthetic polymers. The remaining chapters discuss the mechanical and other properties, such as durability, creep behavior, and processing performance [93].

Prof. Li's book reviewed well over 33 different publications, covering various monomers polymerized *in situ* with peroxides and other initiators, from 1968 to 2011 [94].

Many of the initial WPC products were produced without using too many additives or optimized for long-term applications. They had problems like resistance to moisture, and exhibited swelling after contact with water. As a result, loss of properties was observed when the composite parts got wet, later resulting in rotting due to insect and micro-organism attack, among other issues. Over time, many of these were mitigated or eliminated by pretreating the wood or adding additives [95].

Some of the initial work on high-energy radiation processing of WPC with cobalt-60 gamma radiation was carried out in Taiwan and the USA with methyl methacrylate, styrene, and other monomers after impregnating these monomers into the wood with vacuum impregnation [96, 97]. Later on, several papers were presented and published based on the work carried out on WPC using vacuum impregnation, followed by radiation, at Bhabha Atomic Research Center in India [98–100].

Most current commercial production of WPC by several companies around the world is by using thermoplastic plastics like high density polyethylene (HDPE), low density polyethylene (LDPE), linear low density polyethylene (LLDPE), or a mixture of these polyethylenes, or polyvinyl chloride (PVC). Recycled HDPE from blow-molded milk bottles is used, sometimes mixed with virgin HDPE and some LLDPE to improve impact performance, when it is needed for some specific end applications. For some end indoor applications, recycled LDPE, usually from post-consumer film waste, works if the application is such that it is not likely to see temperatures above 25 °C. In most cases, the process used is shown in Figure 24.4, which shows a typical WPC compounding process to produce WPC pellets.

The main carrier resin (or mixture) is added in the main hopper in pellet form while an additive master batch (concentrate containing a coupling agent, antioxidants, flow aids, lubricants, and biocides) to ensure resistance to rotting, especially for decking, fencing, boat docks, and other products that can get wet. The wood fiber, ideally in pellet form, is also added to the main hopper, but if feeding is slower than the required production rate, it is sometimes added through a secondary feed through two separate feeds or with a mechanical stuffer. The wood fiber in powder/flour form is added downstream to ensure that it is added once the plastic pellets melt. Typically, the ratio of plastic to wood is 50:50 by weight, which ensures that the volume ratio of the plastic is higher than that of the wood due to the former's lower specific gravity.

The wood fiber needs to be sufficiently dry to ensure that voids are not created, which can reduce the tensile, impact, and bending strength of the WPC. The remaining moisture in the wood is removed using a dryer and vacuum pump system as shown in the figure.

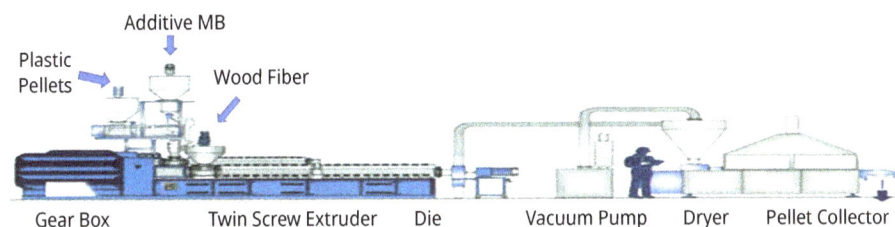

Figure 24.4: Typical compounding process to produce WPC pellets.

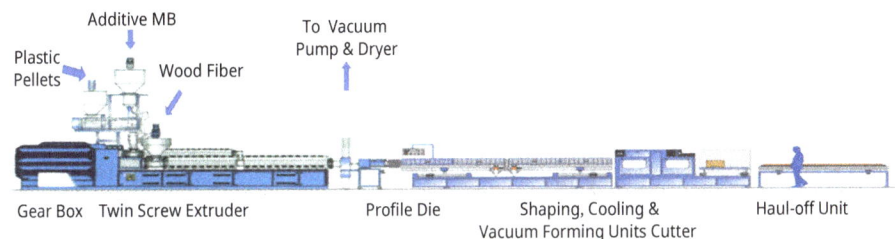

Figure 24.5: Typical compounding process to produce WPC profiles.

In cases where WPC pellets are produced, they are converted by the customer into whatever products the customer needs in a secondary process, which was discussed in section 2 of this chapter. In most cases, large companies that produce profiles in large volume for decking, boat docks, or along water fronts, extrude the profiles directly in a single step. Instead of a pelletizing die as shown in Figure 24.4, they have a profile die and a cooling system that can also indent specific designs or print colors or designs as shown in Figure 24.5.

Many WPC companies use a combination of virgin and recycled plastics to produce their products with many different proprietary designs, especially when viewed from a cross-sectional perspective, as well as various colors, surface designs, and grooves.

Based on a patent granted, Trex Co., a leader in the WPC industry, uses a different method to produce their polyethylene-based WPC. Rather than use polyethylene pellets, they use specialized equipment to first compress post-consumer polyethylenes in films, bottles, and other forms. These are then stuffed into a thermoplastic component, with a particle size of less than about 3/4 inch (19 mm.). The process also involves grinding the wood to a particle size of less than about 600 microns, proportioning the wood component and the first thermoplastic component in a weight ratio of about 65/35 to about 40/60 to form a wood-thermoplastic mixture. This pelletizes the said wood-thermoplastic mixture into a first stable non-separating high bulk density feedstock, wherein the said first feedstock is capable of being readily air transported. These steps are repeated to form a second stable non-separating high bulk density feedstock, after which the second high bulk density feedstock comprising a second thermoplastic component having different physical characteristics from the said first thermoplastic component is then mixed with the first and the second feedstocks to form a blended high bulk density feedstock [101]. This ensures the use of recycled materials with wood-based powder to produce profiles that have among the highest recycled and bio-sourced materials in the market.

Other WPC companies like Timbertech™ (now part of Azek), Fiberon™ (now owned by Fortune Brands) use more traditional method of production, similar to what is shown in Figure 24.5, So, rather than producing pellets, the molten WPC is directly converted using a profile die into decking boards or fence posts or other profiles. These products are sold with a 25- to 50-year warranty. In some products, HDPE or PVC is used. Anderson Windows & Doors sells Fibrex™ windows produced from PVC and reclaimed wood.

The combination of using recycled wood from construction and demolished structures with recycled or even virgin plastics forms wood-polymer composites (WPC) have a very wide scope of usage. Such recycled composites have advantages such as very low environmental impact in terms of abiotic potential, global warming potential, and greenhouse potential. Another significant benefit is the flexibility because WPCs can be compounded and processed – they can be easily modified to meet customers' predetermined strength values needed for each end application. If the WPC is thermoplastics, conventional polymer composite manufacturing techniques, such as injection molding and extrusion, can be used, but for thermoset WPCs, such processes are not feasible without precisely controlling the process parameters. Many rheological characterization techniques need to be determined to ensure the quality of final WPCs, based on the effect of formulation and process parameters [102].

Even though wood-plastic composites can reduce the plastic waste mass in landfills and can improve the physical and mechanical properties of wood, they also create new problems in the lifecycle when WPC products are disposed off. They increase the volume of plastic waste because of including the volume of wood, also making it difficult to recycle, especially with WPC that is produced based on thermoset polymers [103]. At least, scrap thermoplastic-based WPC can be recycled at percentages below 25% but such products acquire a darker color than the original WPC.

However, modern WPC products can last 30–50 years, depending on the brand, and many large producers provide multi-decade warranties, compared to treated wood, which is only good for 6–10 years. WPC profiles from reputable companies are also maintenance-free, whereas treated wood for decks needs to be re-stained every year after the first 2 years. This is the value that WPC products provide and this value is why these type of bio-based fiber plastic composites have grown so large all over the world. The amount of WPC produced keeps increasing. The wood plastic composite market is projected by researchandmarkets.com to grow at a CAGR of 5.75%, to reach a market size of $7.533 billion by 2026 from $5.092 billion in 2019 [104].

The growing demand is primarily in the building and construction end market, with applications like decking, molding and siding, and fencing. In the furniture market, WPCs outperform wood, which has been the main traditional furniture-making material for centuries. These composites are resistant to rotting, moisture, and infestation, and can be extruded or molded into a variety of forms, as required. Over time, the volume of recycled materials (combination of biofiber and plastics) has gradually increased. In April 2019, a WPC was produced with up to 95 % recycled material by weight – including building and demolition waste – by Conenor, a Finnish business specializing in composite extrusion techniques, based on the technology developed by Technalia, a prominent R&D center in Spain [104]. Why is this sort of progress important? The ratio of renewable/recycled to non-renewable resources has a direct influence on the environmental impact of WPCs and the added benefit of reducing the quantity of materials going into landfills. This shows that the use of recycled materials is growing with WPC, both by using recycled plastics and recycled wood fibers. Fossil fuel-based polymers, which were widely utilized in many such applications a couple of decades ago, based on non-renewable raw materials and typical plastics used for these WPC applications, are non-biodegradable and considered undesirable, whereas wood used before that time would need maintenance each year and replacement every five to ten years.

However, cost is a challenge where the consumer needs to be educated because surface-treated wood and recycled plastic profiles are much lower in cost while WPC has a significantly high initial cost [104].

Another factor that needs to be considered is that during installation, material thermal expansion and shrinkage happen. Traditional wood planks used in these applications expand and contract due to their ability to absorb water, since composites are a blend of wood fibers and plastics, and are not completely natural. Many projects cause the thermal expansion of the floor to curve since there is no gap between the boards. Moreover, since a few plastics have greater chemical heat content and can melt, some WPC compositions have more fire hazard qualities than wood alone, unless fire-retardants are used in the compounding of the WPC.

24.5.8 Other biofiber composites

Besides the biofiber composites discussed in previous sections (6.2 to 6.7), other biofibers have also been used to prepare and, in a few cases, produce, biofiber plastics composites. For example, biodegradable polymer nanocomposites have been prepared and are found to be biodegradable [105]. Another example is that of a German patent

[106] that disclosed reinforced biologically degradable polymer compositions, based on thermoplastic starch or a polymer mixture containing thermoplastic starch and at least one hydrophobic biologically degradable polymer. This was reinforced by natural fibers, which are incorporated in the polymer. The polymer mixture furthermore contains a phase mediator for the molecular coupling of the starch phase with the hydrophobic polymer phase. The reinforcing natural fiber is at least one selected from the group consisting of ramie, cotton, jute, hemp, sisal, flax, linen, silk, and abaca. The data shows that the substantially biologically degradable polymer composite is prevented from at least partially losing its biological degradability when reinforced with natural fibers in particular, such as sisal or ramie fibers.

Recycled newspaper fibers were used to produce fiber-polypropylene thermoplastic composites [107].

Thermoset composites, based on polyester reinforced with corn husk fibers, have been studied by a group of researchers from Indonesia, Malaysia, and India [108].

Hemp fiber composites are starting to get commercialized in Europe and the USA. US-based startup, The Hemp Plastic Company, produces a 100% plant-based PLA composite, reinforced with 25% hemp content. The material substitutes conventional plastics without significant capital or operating expense, making the transition easier in injection molding applications [109]. A Canadian company called INCA Renewtech™ has developed some proprietary technology and has made inroads into using their thermoset hemp-based composites for applications in wind energy and in the automobile industries [110].

Several researchers have also studied hybrid biofiber plastic composites in which natural fibers, in combination with artificial fibers, show interesting properties –sufficiently high enough – over using only biofibers to make them suitable for certain engineering applications [111].

There are several publications that summarize biofiber-reinforced plastics composites for different end-use market applications:

–   Ballistic applications are reviewed by Odesanya and his co-writers [112].
–   A comprehensive review on recent developments in biopolymer-based composites and their use in drug delivery and bio-medical applications was published in 2017 [113].
–   Medical applications of chitosan-based biofiber composites are among the several topics covered in this review by researchers at the M.G. University in Kottayam, Kerala, India [114].
–   Biopolymers in electronics are reviewed in a 2017 book, which addressed the challenges posed by the increased use of biopolymeric materials in electronic applications for applications such as sensors, actuators, optics, fuel cells, photovoltaics, dielectrics, electromagnetic shielding, piezoelectrics, flexible displays, and microwave absorbers [115].
–   At Texas Tech University, a novel composite material, using a new composite material formulation of a compound combining two biopolymers polycaprolactone

(PCL) and polyglycolide (PGA), was used to produce miscible 50/50 PCL-PGA blended electro-spun fibers. These biofibers were compounded into a PCL matrix to form these unique bio-based plastics composites. This bio-based composite was developed for fabricating temporary implanted orthopedic devices. The composite can be produced from biopolymer composites that biodegrade and can be replaced by natural tissues. In order to further increase their strength, single-walled carbon nanotubes (SWNTs) were purified and wrapped with double stranded deoxyribo-nucleic acid (dsDNA) and introduced in the fibers. This design utilizes the long degradation rate of PCL while acquiring the strength of PGA along with increasing the interfacial bonding between the fibers and the matrix. The dsDNA is used to improve the dispersion of the SWNTs in the fibers, which enables the mechanical properties of the composite to be enhanced. These devices can be used for permanent applications such as total knee replacements where the device is essential for the lifetime of the patient, or for temporary applications such as bone fractures where the device is no longer needed once the patient has healed, by a modification of the formulation and the process [116]. There are numerous and growing examples of such bio-based fiber-reinforced composites in the bio-medical industry, and are discussed in a book published in 2018 [117].

– Many of the medical industry applications of a new product called Fibertuff™ are covered in Chapter 26 by Robert Joyce in this book.

24.6 Limitations of bio-based plastics composites and challenges for growth

Research into bio-based polymers has been active, as evidenced by the large number of patents, technical publications, and new products. The rising environmental and sustainability consciousness has motivated efforts for developing many different bio-based composite materials for different end-use applications, and as a novel and more sustainable alternative to conventional non-renewable synthetic fibers such as glass and carbon reinforced composites.

On the other hand, bio-composite materials are not problem-free direct substitutes; they have some drawbacks, such as poor moisture resistance (hydrophilicity), fiber/matrix incompatibility, low thermal stability, flammability, poor electrical properties, extraction, processing, surface modification, machining, manufacturing, supply logistic issues, and characterization-associated challenges, along with highly anisotropic properties [118]. However, due to challenges related to the properties of these materials, such as their resistance to chemicals or weather, compared to synthetic polymers, their applicability and usage has not dramatically increased yet in many end-use markets, as was originally forecasted. For demanding applications as well as for those where durability is essential, such drawbacks will preclude biopolymers, bioplastics, and bio-fibers from displacing incumbent non-bio-based products.

As stated earlier, a common disadvantage almost all of these fibers have is that they are hydrophilic polar polymers and absorb water. When the fibers get wet, they lose some of their mechanical strength, especially tensile strength and rigidity – measured as flexural modulus. Another issue is that each of these has a fairly low degradation temperature, especially when compared to non-biofibers like glass fiber or carbon fiber. A third issue is that since these are all found in nature, there is a lot of inconsistency from one batch grown at one location to another grown elsewhere. This occurs due to differences in their growth due to differences in soil, temperature, and rainfall, or the water available at each location. So, consistency in mechanical performance is more difficult to maintain, compared to synthetic fibers. Some of these issues can be mitigated by innovative compounding techniques that have been discussed in this chapter as well as by improving the quality control of the incoming raw materials and the outgoing quality assurance of products.

An unappreciated complexity that Prof. Midani points out is that there are few in the composites industry with enough experience and knowledge of natural fiber reinforcements to work with them confidently. He says that there is a large gap between the different stages of the value chain. For example, processors of fibers often operate with comparatively low-tech equipment, while composites manufacturers are relatively high tech. Given this difference, there is a compelling need to bridge this gap between the different value chain actors: at the one end, to educate the farmers, fiber processors, and fabric makers on this new end-use of natural fibers, and at the other end, to educate composites manufacturers of composites on this new type of reinforcement and how it differs from synthetic fibers, whether they are petro-based or glass [118]. We can assume that over time this issue will resolve by itself.

In the fiber-reinforced composite industry, in most cases, glass and carbon fibers have poor adhesion and chemical compatibility to the plastics polymer matrix and hence chemical modification of these reinforcing fibers, either through surface modification or by the use of coupling agents during production processes, are carried out to provide adhesion and compatibility between the two phases. For using biofibers, several projects targeting the improvement of compatibility of natural fiber reinforcements to various matrixes (bio-based and non-bio-based) have resulted in achieving the goal of providing off-the-shelf reinforcements, similar to glass and carbon fiber. Examples of such projects include using bio-fiber reinforced plastics composites using low-twist rovings, unidirectional tapes, woven fabrics, spread-tow fabrics, and prepregs, and most of these have involved flax fibers provided by Bcomp Ltd. (based in Fribourg, Switzerland) and Composites Evolution (based in Chesterfield, U.K.). Another approach to improve compatibility of the biofiber with the resin matrix is the water-based acrylic resin, Acrodur, developed by BASF in the USA. Midani points out that there is also a project called

QualiFlax, which is responsible for specifically developing to control and ensure consistent technical performance, reliable supply, and pricing of natural fiber reinforcements, targeted primarily at flax. In addition, other projects like Fibragen in Spain aim to develop genetically selected flax, with improved and consistent properties [118].

Table 24.5: Fiber production by country.

Country	Share (%)
China	23%
India	21%
USA	16%
Brazil	10%
Bangladesh	6%
Sri Lanka	5%
Turkey	3%
Uzbekistan	2%
Australia	2%
Mexico	2%
Pakistan	1%
Turkmenistan	1%
Others	8%

Many of the supply chain issues on biofibers have been covered, except for one that has not been addressed so far. The latest data available for global fiber production by country is from the 2018 data of the Discover Natural Fibres Initiative [119], along with data from other sources. This data is compiled and shown in Table 24.5. As Midani has pointed out, natural fiber reinforcements are mostly derived from plants grown mostly in China, India, Brazil, Bangladesh, and Sri Lanka, while till a few years ago, most of the larger manufacturers of composites were mostly located in Western Europe, the United States, and Japan. Such a geographic barrier made it difficult for composites manufacturers to easily access natural fiber reinforcements for large-scale production, and even for research. Moreover, the wide ban on growing hemp in certain countries significantly limits its availability, both for trials and production. Hence, composite manufacturers had to develop composites based on biofibers available locally, such as flax, which is mostly grown in France and Belgium – close to regions where many manufacturers of composites are located. In fact, the easy and quick access to flax in Western Europe has significantly increased its development as a viable biofiber for trials, and now flax has become the most widely used natural fiber reinforcement [118]. Recently, the use of natural fiber composites based on locally availability has been growing in India, China, Brazil, and SE Asia. The world production of natural fibers is estimated at 33.7 million tons in 2022 by Discover Natural Fibres Initiative [119].

24.7 Conclusions

While biopolymers and bio-fibers have several limitations, it is clear from this chapter that interest in them is very high, not just related to research and creative ways to overcome some of these limitations, but also to move forward with scaling up these innovations to commercial products. Issues related to moisture absorption and dimensional stability are handled by encasing the biofiber in a plastic material that provides a moisture barrier to protect the fiber. Another limitation of poor interfacial adhesion is overcome with the use of coupling agents and compatibilizers using techniques covered in a couple of online courses and a blog [3, 120].

Biofiber reinforcements have many advantages over glass fibers. For example, biofibers have low density, high specific properties, are biodegradable, are derived from renewable resources, have a small carbon footprint, and provide good thermal and acoustical insulation. These relative advantages must be emphasized when developing new products from bio-based fiber plastics composites. This is the main reason why the automotive industry is at the forefront of adoption of such composites these days. OEMs of transportation vehicles are taking advantage of such benefits as more and more specification requirements for modern-day vehicles are demanding such adoptions because of the growth of electric vehicles and the ever-stricter environmental regulations [11].

Furthermore, there are many enthusiastic researchers, engineers, architects, and designers who are exploring new horizons for bio-based plastics composites. For example, in the Netherlands, an entire façade of a building was constructed using a biocomposite using bio-based epoxy resin as the matrix and with hemp as the reinforcing fiber. Also, in the summer of 2021, two composite pavilions were built from bio-based bioplastic composites in Germany. The first one, produced using flax-reinforced bio-based epoxy, uses a mold-less filament winding process, called Livmat™. The second was the Biomat™ composite pavilion, which was constructed with flax and hemp-pultruded rods. Other interesting applications of biofiber-reinforced composites include:

- Hardcase travel bags by a Danish company called Projektkin, available from their online store for US$450.
- Applications for printed circuit boards are being pursued by another UK-based company. Jiva, using fully sustainable substrates.
- Greenboats, a German-based company, which is famous for its iconic flax 27 Daysailer™, 2021 installed the first NFC nacelle in the Netherlands in summer in a test program to study the performance of such new materials in the wind industry [9].
- A specific application of hemp-fiber-based thermoset plastic composites is also covered in the chapter by Saltman, with successful commercialization for wind blade and automotive applications [109].

As more and more bio-based polymers and biofibers as well as compounds and composites are being adopted, applications will grow and the total volume of such biomaterials should see an attractive annual increase, estimated by several market reports to grow at CAGRs in the range of 7 to 12%.

References

[1]   Stoyko F, Debes B. Handbook of Engineering Biopolymers – Homopolymers, Blends and Composites, Cincinnati, OH, USA: Hanser Gardner Publications, Inc.; ISBN: 1-68015-262-9, 1-56990-405-7, 2007.

[2]   Díez-Pascual AM, Cinelli P. Synthesis and Applications of Biopolymer Composites, Basel, Switzerland: MDPI; ISBN 3039211323 & 9783039211326 2019.

[3]   Adur A. Compatibilization of polymer alloys blends and composites. online course taught on specialchem.com, 2002-2004, available from SpecialChem, Paris, France.

[4]   Marechal E. Surface Modification of Natural Fibers: Chemical Aspects", Chapter 2 in Book, "Handbook of Engineering Biopolymers – Homopolymers, Blends and Composites, Cincinnati, OH, USA: Hanser Gardner Publications, Inc.; pp. 49–77ISBN: 1-68015-262-9. 1-56990-405-7.

[5]   Constable RC, Adur AM. Chemical Coupling of Glass-Fiber Reinforced Polypropylene using Acid or Anhydride-Modified Polypropylenes, SPE 49th ANTEC"91 at New York City, NY, April 2-5, 1991, Preprints 1892–96.

[6]   Adur AM. Surface modification of glass fibers & other reinforcements using ethylene-maleic anhydride alternating copolymers. InnoPlast's Polymer Compounding for Innovations in Plastics Industry, Hilton Airport Hotel, Atlanta, GA, USA Sep. 23-25, 2014; https://www.prlog.org/12348262.

[7]   Adur AM. Using zemac® copolymers to modify surface chemistry of glass fibers to improve performance of glass-reinforced polyamides. 2017 International Congress on Glass Annual Meeting, Istanbul, Turkey, Oct. 22 to 25 2017.

[8]   Adur AM. Modifying surface chemistry to improve performance of polyamides reinforced with glass fibers and other functional additives for tailoring polyamide properties to meet automotive industry requirements. In: AMI's 2nd Performance Polyamides USA Conference, 2018, Pittsburgh, PA, USA, Nov. 6-7, 2018.

[9]   Plotz C. What is a biofiber? Int Fiber J 2021;1:40–42. https://fiberjournal.com/what-is-a-biofiber/.

[10] Bledzki AK, Gassan J. Composites reinforced with cellulose-based fibres. Prog Polym Sci 1999;24:221–74. https://doi.org/10.1016/S0079-6700(98)00018-5/.

[11] Midani M, Elseify LA. Natural fiber composites – a practical guide for industrial utilization. Int Fiber J 2022;(1):https://fiberjournal.com/natural-fiber-composites-a-practical-guide-for-industrial-utilization/.

[12] Ngo T-D. Natural Fibers for Sustainable Bio-Composites. In: Günay E, editor. Book "Natural and Artificial Fiber-Reinforced Composites as Renewable Sources, London, UK: IntechOpen; editor 2018. ISBN:978-1-78923-061-1/.

[13] Ilyas RA, Sapuan SM, Ibrahim R, Abral H, Ishak M, Zainudin E, Asrofi M, Atikah MSN, Huzaifah MRM, Radzi AM, Azammi AMN, Shaharuzaman MA, Nurazz NM, Syafri E, Sari NH, Norrrahim MNF, Jumaidin R. Sugar palm cellulosic fiber hierarchy: A comprehensive approach from macro to nano scale. J Mater Res Technol 2019;8:2753–66. https://doi.org/10.1016/j.jmrt.2019.04.011/.

[14] Luhar S, Suntharalingam T, Navaratnam S, Luhar I, Thamboo J, Poologanathan K, Gatheeshgar P. Sustainable and renewable bio-based natural fibers and its application for 3d printed concrete: A review. Sustainability 2020;12:10485. https://doi.org/10.3390/su122410485/.

[15] Latif R, Wakeel S, Khan NZ, Siddiquee AN, Verma SL, Khan ZA. Chemical composition of natural fibers" in "surface treatments of plant fibers and their effects on mechanical properties of fiber-

reinforced composites: A review. J Reinf Plast Compos 2019;38:15–30. https://doi.org/10.1177/0731684418802022.

[16] Gholampour A, Ozbakkaloglu T. A review of natural fiber composites: properties, modification and processing techniques, characterization, applications. J Mater Sci 2020;55:829–92. https://doi.org/10.1007/s10853-019-03990-y/.

[17] Taj S, Munawar MA, Khan S. Natural fiber-reinforced polymer composites. Carbon N Y 2007;44:129–44. Google Scholar.

[18] Townsend T. Natural Fibers and the World Economy, July 2019, https://dnfi.org/coir/natural-fibers-and-the-world-economy-july-2019_18043/.

[19] Rowell RM. Potentials for jute based composites. In: Paper presented at Jute India '97 Conference & Expo, New Delhi, India, October 20-22, 1997; https://www.fpl.fs.fed.us/products/publications/specific_pub.php?posting_id=13212/.

[20] Reddy SRT, Ratna Prasad AV, Ramanaiah K. Tensile and flexural properties of biodegradable jute fiber reinforced poly lactic acid composites. In: Materials Today: Proceedings, Vol. 44, Part 1, December 2020, pp 917–21.

[21] Future Fibres. https://www.fao.org/economic/futurefibres/fibres/coir/en/.

[22] Hemp Fiber: Properties, Processing and Uses. https://www.textileblog.com/hemp-fiber-properties-processing-and-uses/.

[23] What is hemp fiber? https://secretnaturecbd.com/blogs/cbd/what-is-hemp-fiber/.

[24] https://en.wikipedia.org/wiki/Bagasse.

[25] http://www.paperpulping.com/application/bagasse-pulp-making.html\.

[26] Biofuel and Electricity Cogeneration from Sugarcane, https://tractorexport.com/biofuel-and-electricity-cogeneration-from-sugar-cane/.

[27] Rathore R. Bagasse FAQ. June 30, 2020 https://www.biogreenchoice.com/blogs/blog/bagasse-faq/.

[28] https://www.britannica.com/plant/bamboo/.

[29] Kadivar M, Gauss C, Ghavami K, Savastano H Jr. Densification of bamboo: State of the art. Materials 2020;13(19):4346. Densification of Bamboo: State of the Art – PubMed (nih.gov)/.

[30] https://en.wikipedia.org/wiki/Kenaf/.

[31] Nanko H, Button A, Hillman D. The World of Market Pulp, Appleton, WI, USA: WOMP, LLC; 2005, p. 258. ISBN 0-615-13013-5.

[32] Rice Husk. https://www.dairyknowledge.in/article/rice-husk/.

[33] Phonphuak N, Chindaprasirt K. Chapter 6 – Types of Waste, Properties, and Durability of Pore-Forming Waste-Based Fired Masonry Bricks. In: Pacheco-Torgal F, Lourenço PB, Labrincha JA, Kumar S, Chindaprasirt P Editors, Eco-Efficient Masonry Bricks and Blocks. Sawston, Cambridge, UK: Woodhead Publishing; 2015, pp. 103–27. ISBN 9781782423058.

[34] Mansaray KG, Ghaly AE. Thermo gravimetric analysis of rice husks in an air atmosphere. Energy Source 1998;20:653–63.

[35] FAO, vol. XXI Issue no. 1, April 2018, https://www.fao.org/markets-and-trade/commodities/rice/en/.

[36] https://knoema.com/atlas/World/topics/Agriculture/Crops-Production-Quantity-tonnes/Rice-paddy-production.

[37] Corn Husk. https://www.specialtyproduce.com/produce/Corn_Husk_473.php#.

[38] Philip Joseph Burton PJ. Towards Sustainable Management of the Boreal Forest, New York, USA: NRC Research Press, Great River, 2003, p. 759. ISBN 978-0-660-18762-4.

[39] Rowell RM, editor. "Handbook of Wood Chemistry and Wood Composites', Boca Raton, FL, USA: CRC Press; 2005.

[40] Abaca, https://www.fao.org/economic/futurefibres/fibres/abaca0/en/.

[41] Chand N, Fahim M. Tribology of Natural Fiber Polymer Composites, 2nd edition, 2020, Elsevier Ltd. ISBN 978-0-12-818983-2.

[42] Debnath S. Chapter 3 – "Sustainable Production of Bast Fibres", in "Sustainable Fibres and Textiles". In: Muthu SS, editors, The Textile Institute Book Series, Sawston, Cambridge, UK: Woodhead Publishing; 2017, pp. 69–85. ISBN 9780081020418.

[43] Zafar S. Biorefinery as a Source of Advanced Biofuels. August 29, 2020, www.cleantechloops.com/biorefinery/.

[44] https://www.farmpractices.com/crop-residues-types-management-uses/.

[45] Biomass- Types, Uses, Merits and Demerits of Biomass, https://byjus.com/biology/biomass/.

[46] Huang Y, Petermann J. Interface layers of fiber reinforced composites with transcrystalline morphology. Polymer Bulletin 1996;36:517–24.

[47] Go KL, Aswathi MK, De Silva RT, Sabu T editors. Interfaces in Particle and Fibre Reinforced Composites, Amsterdam, The Netherlands: Elsevier Ltd; 2020.

[48] Lau AK-T, Cheung KHY. 1 – Natural Fiber-Reinforced Polymer-Based Composites. In: Lau AK-T, Hung AP-Y, editors. Book "Natural Fiber-Reinforced Biodegradable and Bioresorbable Polymer Composites. Sawston, Cambridge, UK: Woodhead Publishing; 2017, pp. 1–18. ISBN 9780081006566.

[49] Faruk O, Bledzki AK, Fink H-P, Saini M. Biocomposites reinforced with natural fibers, 2000-2010. Prog Polym Sci 2012;37(11):1552–96. https://doi.org/10.1016/j.progpolymsci.2012.04.003/.

[50] Ilyas RA, Sapuan SM. Biopolymers and biocomposites: Chemistry and technology. Curr anal Chem 2020;16:500–03. https://do.org/10.2174/1573411016052006003095311.

[51] Rajak DK, Pagar DD, Menezes PL, Linul E. Fiber-reinforced polymer composites: Manufacturing, properties, and applications. Polymers (basel) 2019;11(10):1667. https://doi.org/10.3390/polym11101667.

[52] Saheb DN, Jog JP. Natural fiber polymer composites: A review. Adv Polym Technol 1999;18(Issue 4):351–63.

[53] Gómez-Gast N, Cuellar MDRL, Vergara-Porras B, Vieyra H. Biopackaging potential alternatives: Bioplastic composites of polyhydroxyalkanoates and vegetal fibers. Polymers (basel) 2022 Mar 10;14 (6):1114. https://doi.org/10.3390/polym14061114.

[54] Ilyas RA, Zuhri MYM, Aisyah HA, Asyraf MRM, Hassan SA, Zainudin ES, Sapuan SM, Sharma S, Bangar SP, Jumaidin R, Nawab Y, Faudzi AAM, Abral H, Asrofi M, Syafri E, Sari NH. Natural fiber-reinforced polylactic acid, polylactic acid blends and their composites for advanced applications. Polymers 2022;14:202. https://doi.org/10.3390/polym14010202/.

[55] Ilyas RA, Aisyah HA, Nordin AH, Ngadi N, Zuhri MYM, Asyraf MRM, Sapuan SM, Zainudin ES, Sharma S, Abral H, Asrofi M, Syafri E, Sari NH, Rafidah M, Zakaria SZS, Razman MR, Majid NA, Ramli Z, Azmi A, Bangar SP, Ibrahim S. Natural-fiber-reinforced chitosan, chitosan blends and their nanocomposites for various advanced applications. Polymers 2022;14:874. https://doi.org/10.3390/polym14050874.

[56] Ilyas RA, Zuhri MYM, Norrrahim MNF, Misenan MSM, Jenol MA, Samsudin SA, Nurazzi NM, Asyraf MRM, Supian ABM, Bangar SP, Nadlene R, Sharma S, Omran AAB. Natural fiber-reinforced polycaprolactone green and hybrid biocomposites for various advanced applications. Polymers 2022;14:182. https://doi.org/10.3390/polym14010182/.

[57] Mohammed AABA, Omran AAB, Hasan Z, Ilyas RA, Sapuan SM. Wheat biocomposite extraction, structure, properties and characterization: A review. Polymers 2021;3:3624. https://doi.org/10.3390/polym13213624.

[58] Kuciel S, Romanska P, Liber-Knec A. Polyamides from renewable sources as matrices of short fiber reinforced biocomposites. Polimery 2012;57:627–34. https://www.researchgate.net/publication/274882725/.

[59] Jumaidin R, Diah N, Ilyas R, Alamjuri R, Yusof F. Processing and characterization of banana leaf fiber reinforced thermoplastic cassava starch composites. Polymers 2021;13:1420. https://doi.org/10.3390/polym13091420.

[60] Natural Polymers, Biopolymers, Biomaterials, and Their Composites, Blends, and IPNs. In: Thomas S, Ninan N, Mohan S, Francis E editors, Print Book, Toronto, Canada: Apple Academic Press; 2021.

[61] Abaca, https://www.fao.org/economic/futurefibres/fibres/abaca0/en/.

[62] Hablot E, Matadi R, Ahzi S, Averous L. Renewable bio-composites of dimer fatty acid-based polyamides with cellulose fibers: thermal, physical and mechanical properties. Compos Sci Technol 2010;70:504–09. https://doi.org/10.1016/j.compscitech.2009.12.001.

[63] Kiziltas A. Bio-based polyamides reinforced with cellulose nanofibers-processing and characterization. In: Paper presented at SPE's Automotive Composites Conference & Exhibition (ACCE), Novi, Michigan, USA. Sept. 2015; https://www.researchgate.net/publication/283505156/.

[64] (a) MIT team engineers plant-derived composite with potential for stronger, tougher applications, March 4, 2022, https://www.compositesworld.com/news/mit-team-engineers-plant-derived-composite-with-potential-for-stronger-tougher-applications/ (b) Rao, A, Divoux, T, Owens, CE, Hart, AJ. Printable, castable, nanocrystalline cellulose-epoxy composites exhibiting hierarchical nacre-like toughening. Cellulose 2022;29:2387–2398. https://doi.org/10.1007/s10570-021-04384-7/(c) https://www.goodnewsnetwork.org/new-plant-derived-sustainable-plastic-is-tough-as-bone-mit/.

[65] Sanadi AR, Caulfield DF, Jacobson RE, Rowell RM. Renewable agricultural fibers as reinforcing fillers in plastics: Mechanical properties of kenaf fiber-polypropylene composites. Ind Eng Chem Res 1995;34(5):1889–96. https://doi.org/10.1021/ie00044a041/.

[66] Akil HM, Omar MF, Mazuki AAM, Safiee S, Ishak ZAM, Abu Bakar A. Kenaf fiber-reinforced composites: A review. Mater Des 2011;32:4107–21. https://doi.org/10.1016/j.matdes.2011.04.008/.

[67] United States Patent 7,445,835. Serizawa S, Inoue K, Iji M. Kenaf-fiber-reinforced Resin Composition, Tokyo, Japan: assigned to NEC Corporation, issued November 4 2008.

[68] Adur AM, Majali AB, Iya VK. Development of Radiation Processed Wood-Polymer and Bagasse-Polymer Composites. In: Paper presented at the High Polymer Symposium, IIT, Kanpur, U.P., India, January 21-23, 1972, Preprints, p. 135.

[69] Indian Patent 139,260. Iya VK, Adur AM, Majali AB. Bagasse-flyash polymer composites. V. K. Iya, A. B. Majali & A. M. Adur Chem Abstr 1980;92:77703.

[70] Verma D, Gope PC, Maheshwari, Sharma RK. Bagasse fiber composites-a review. J Mater Environ Sci 2012;3(6):1079–92. https://www.researchgate.net/publication/284625466/.

[71] Bansal S, Ramachandran M, Raichurkar P. Comparative analysis of bamboo using jute and coir fiber reinforced polymeric composites. Mater Today: Proc Issue 2, Part A 2017;4:3182–87. https://www.sciencedirect.com/science/article/pii/S2214785317304121O/.

[72] Goriparthi BK, Suman KNS, Mohan Rao N. Effect of fiber surface treatments on mechanical and abrasive wear performance of polylactide/jute composites. Compos Part A Appl Sci Manuf 2012;43 (10):1800–08. https://www.sciencedirect.com/science/article/abs/pii/S1359835X12001649/.

[73] Rana AK, Mandal A, Mitra BC, Jacobson R. Short jute fiber-reinforced polypropylene composites: Effect of compatibilizer. J Appl Polym Sci 1998.

[74] Nkeuwa WN, Zhang J, Semple KE, Chen M, Xia Y, Dai C. Bamboo-based composites: a review on fundamentals and processes of bamboo bonding. Compos Part B Eng 2022;235:109776. https://www.researchgate.net%2fpublication%2f359018108/.

[75] Tahir PM, Ahmed AB, SaifulAzry S, Ahmed A. Retting process of some bast plant fibres and its effect on fiber quality: A review" (PDF). Bioresources 2011;6(4):5260–81.

[76] Popat TV, Patil AY. A review on bamboo fiber composites. Iconic Res Engg J 2017;1(2):pp. 54–72. IRE Journals, 1700041.

[77] Kaur N, Saxena S, Gaur H, Goyal P. A review on bamboo fiber composites and its applications. In: Paper presented at 2017 International Conference on Infocom Technologies and Unmanned Systems (Trends and Future Directions) (ICTUS), https://doi.org/10.1109/ICTUS.2017.8286123/.

[78] Muhammad A, Rahman M, Hamdan S, Sanaullah. Recent developments in bamboo fiber-based composites: A review. Polym Bull 2019;76:2655–82. https://doi.org/10.1007/s00289-018-2493-9/.

[79] Radzi AM, Sheikh Ahmad Zaki SA, Hassan MZ, Ilyas RA, Jamaludin KR, Daud MYM, Aziz SA.
      Bamboo-fiber-reinforced thermoset and thermoplastic polymer composites: A review of properties,
      fabrication, and potential applications. Polymers (basel) 2022;14(7):1387. https://doi.org/10.3390/
      polym14071387;https://www.ncbi.nlm.nih.gov/pmc/articles/PMC9003382/.
[80] Lee S-Y, Chun S-J, Doh G-H, Kang I-A, Lee S, Paik K-H. Influence of chemical modification and filler
      loading on fundamental properties of bamboo fibers reinforced polypropylene composites. J
      Compos Mater 2009;43:1639–57. https://doi.org/10.1177/0021998309339352/.
[81] Shannon. Method of impregnating bamboo with synthetic resin. United States Patent 2352740,
      Issued May 14, 1944, assigned to Bakelite Corp., New York City, NY, USA.
[82] Wang UP. Papers Presented at Study Group Meeting on "Impregnated Fibrous Materials", held at
      Bangkok, Thailand. Proceedings, I.A.E.A., Vienna, Austria, 1968, pp. 35; pp. 293.
[83] Hao PLC, Wang UP. Report on I. A E. A. contract # 565/RI/RB, I.A.E.A., Vienna, Austria, 1970.
[84] Mani RS, Jayaraman MT. Preparation of bamboo-plastics composites. Indian J Technol 1969;7
      (1):16–22.
[85] Adur AM. Studies on the Radiation Processing of Bamboo- Polymer Composites, Part IV of M.Sc.
      Thesis, Univ. of Bombay, 1974.
[86] Adur AM, Nigam SK. Studies on cobalt-60 gamma radiation processing of bamboo polymer
      composites. Isotpenpraxis 1975;11(1):21–24. https://www.osti.gov/biblio/4219041-cobalt-gamma-
      radiation-processing-bamboo-polymer-composites/.
[87] Adur AM. Gamma radiation processed bamboo-polymer composites I. preparation, process
      parameters and product microstructure. J Radiat Curing 1977;4(4):2–17.
[88] Adur AM. Gamma radiation processed bamboo-polymer composites II. Mechanical properties,
      moisture content and water absorption characteristics. J Radiat Curing 1978;5(2):4–12.
      https://www.osti.gov/biblio/5242471.
[89] Adur AM. Gamma radiation processed bamboo-polymer composites III. Possible applications for
      tensile reinforcement of concrete. J Radiat Curing 1978;5(4):9–16. https://www.osti.gov/biblio/
      5378997.
[90] CaliBamboo Introduces First Compoisite Deck Made from Bamboo. https://www.multihousingnews.
      com/cali-bamboo-introduces-first-composite-deck-made-from-bamboo/.
[91] https://www.calibamboo.com/composite-decking/bamdeck-3g.html/.
[92] Oksman K, Bengtsson M. Wood Fiber Thermoplastic Composites: Processing, Properties, and Future
      Developments. In: Stoyko F, Bhattacharyya D, editors. Handbook of Engineering Biopolymers –
      Homopolymers, Blends and Composites, Munich, Germany: Hanser Publishers; 2007, pp. 655–71.
[93] Niska KO, Sain M, editors. Wood-Polymer Composites. 1st edition, Sawston, Cambridge, U.K.: eBook,
      Woodhead Publishing; 2008. ISBN: 9781845694579.
[94] Li Y. Wood-Polymer Composites. In book "Advances in Composite Materials – Analysis of Natural
      and Man-Made Materials, September 2011, org/10.5772/17579/.
[95] Ergun Baysa E, Kemal MY, Altinok M, Sonmez A, Peker H, Colak M. Some physical, biological,
      mechanical, and fire properties of wood polymer composites (WPC) pretreated with boric acid and
      borax mixture. Const Build Mater 2007;21:1879–85. https://www.academia.edu/20153251/.
[96] Iannazzi FD, Levins PL, Perry FG, Jr, Lindstrom RS. Technical and Economic Consideration for an
      Irradiated Wood-plastic Material, United States Atomic Energy Commission, TID-21434.
[97] Narayanamurti D, Jayaraman MT, Thacker R, Anantanarayanan S. Note on radiation processed
      wood-plastic materials. Wood Sci Technol 1970;4:226–36. https://doi.org/10.1007/BF00571857/.
[98] Iya VK, Majali AB, Adur AM, Nigam SK. Radiation processed wood polymer composites. In: Paper
      presented at Symposium on Wood Panel Products at Bangalore, India, 1974.
[99] Adur AM, Majali AB, Thacker RC. Radiation Processing of WPC and Drierite Plastic Combinations.
      In: Paper Presented at Chemistry Symposium; Madras, India (25 Nov 1970) of Proceedings of the

Chemistry Symposium. Vol. II. Dept. of Atomic Energy (1970); 1970. pp 31–41. https://www.osti.gov/biblio/4655925.

[100] (a) Adur AM, Nigam SK, Patil ND. On the electrical properties of radiation processed wood polymer composites. In: Paper presented at DAE Chemistry Symposium, Aligarh, U.P., India, December 21-23, 1972, Preprints, Vol. I, pp. 333–39. (b) Adur AM, Nigam, SK. Electrical Conductivity of Radiation Processed Wood Polymer Composites. J. Radiat Curing 1979;5 (1):18–24. https://www.osti.gov/biblio/5343202.

[101] Gustafsson K-A, Muller JJ, Wittenberg RA. Method of producing a wood-thermoplastic composite material, United States Patent 5,746,958, issued May 5, 1998 assigned to Trex Co.

[102] Ramesh M, Rajeshkumar D, Sasikala G, Balaji D, Saravanakumar A, Bhuvaneswari V, Bhoopathi R. A critical review on wood-based polymer composites: processing, properties, and prospects. Polymers 2022;14(3):589.

[103] Taifor A. A review of wood plastic composites effect on the environment. J Univ Babylon Pure Appl Sci 2016;25(2):360–67. https://www.researchgate.net/publication/317344750.

[104] Wood plastic composite market – forecasts from 2021 to 2026. market report by ResearchAndMarkets.com, https://www.globenewswire.com/news-release/2022/03/04/2397099/28124/en/Global-Wood-Plastic-Composite-Market-2021-to-2026-Rising-Demand-from-the-Construction-Industry-is-Driving-Growth.htm/.

[105] Mclauchlin AR, Thomas NL. Biodegradable Polymer Nanocomposites. In: Advances in Polymer Nanocomposites, Sawston, Cambridge, UK: Woodhead Publishing; 2012, pp. 398–430. https://scholar.google.com/citations?view_op=view_citation&hl=en&user=wMwpY_8AAAAJ&pagesize=80&sortby=pubdate&citation_for_view=wMwpY_8AAAAJ:35N4QoGY0k4C/.

[106] Tomka I. Reinforced biodegradable polymer. United States Patent 5,663,216, assigned to Bio-Tec Biologische Naturverpackungen GmbH (Germany) issued September 2, 1997.

[107] Sanadi AR, Young RA, Clemons C, Rowell RM. Recycled newspaper fibers as reinforcing fillers in thermoplastics: Part i-analysis of tensile and impact properties in polypropylene. J Reinf Plast Compos 1994;13(1):54–67. https://www.researchgate.net/publication/237535647/.

[108] Sari NH, Pruncu CI, Sapuan SM, Ilyas RA, Catur AD, Suteja S, Sutaryono YA, Pullen G. The effect of water immersion and fiber content on properties of corn husk fibers reinforced thermoset polyester composite. Polym Test 2020;91:106751. https://doi.org/10.1016/j.polymertesting.2020.106751/.

[109] Innovative biocomposite and bioplastic plastic alternatives that can reduce your carbon footprint and plastic use, https://hempplastic.com/products/.

[110] Kistaiah N, Chavan UK, Reddy GR, Rao MS. Mechanical characterization of hybrid composites: A review. J Reinf Plast Compos 2014;33(14):1364–72. https://doi.org/10.1177/0731684413513050/.

[111] Odesanya KO, Ahmad R, Jawaid M, Bingol S, Adebayo GO, Wong YH. Natural fiber-reinforced composite for ballistic applications: A review. J Polym Environ 2021;29:3795–812. https://doi.org/10.1007/s10924-021-02169-4/.

[112] Jana S, Maiti S, Jana S. Biopolymer-based Composites: Drug Delivery and Biomedical Applications eBook. Duxford, UK: Woodhead Publishing, an imprint of Elsevier; 2017.

[113] Thomas S, Ninan N, Mohan S, Frances E. editors. Natural Polymers, Biopolymers, Biomaterials, and Their Composites, Blends, and IPNs, Toronto, Canada: Apple Academic Press; Book 2021.

[114] Sadasivuni KK, Ponnamma D, Kim J-W, AlMaadeed MA, Cabibihan -J-J. editors. Biopolymer Composites in Electronics, New York, USA: Elsevier; Print Book 2017.

[115] Swain Spearman S. Fabrication of biodegradable biopolymer composites for orthopedic applications. Ph.D. Thesis Dissertation, Texas Tech University, https://ttu-ir.tdl.org/2016-06-20T20:01:08Z/.

[116] Shimpi NG, editor. Biodegradable and Biocompatible Polymer Composites: Processing, Properties and Applications, Duxford, United Kingdom: Woodhead Publishing, an imprint of Elsevier; Print Book 2018.

[117] Andrew JJ, Dhakal HN. Sustainable bio-based composites for advanced applications: Recent trends and future – a critical review. Compos Part C Open Access 2022;7:100220. https://doi.org/10.1016/j.jcomc.2021.100220/.

[118] Midani M. Natural fiber composites: What's holding them back? https://www.compositesworld.com/articles/natural-fiber-composites-whats-holding-them-back/.

[119] (a) Data from various publications from Discover Natural Fibres Initiative (DNFI). https://dnfi.org/. (b) "World Natural Fibre Update", August 2022, https://dnfi.org/statistics/dnfi-world-natural-fibre-update_34610/ (c) Townsend T. 1B – World Natural Fibre Production and Employment. In: book, 'Handbook of Natural Fibres", 2nd edition, 2020, Woodhead Publishing Series in Textiles, pp. 15-36; https://doi.org/10.1016/B978-0-12-818398-4.00002-5/.

[120] Bicerano J. A Practical Guide to Polymeric Compatibilizers for Polymer Blends, Composites and Laminates. Special Chem, December 2005; http://polymerexpert.biz/blog/135-compatibilization-of-immiscible-polymers/.

Camille Sobrian Saltman, David Saltman

Chapter 25
Turning biomass into business™

25.1 Introduction

Powerful market forces are leading industry to adopt more environmentally friendly materials in their products. These include government regulations to reduce plastics pollution, sequester carbon, and achieve higher lifecycle performance. They also include investor focus on environmental, social, and governance – ESG initiatives and supply chain challenges. These metrics are now being used to measure investment risk and determine the value of corporations. Equally important is the growing consumer demand for less toxic and more sustainable products. Consumer purchasing represents 70% of the US economy, and as customers change their buying patterns, major brands are listening.

These powerful market trends have led to significant innovations in bio-based polymer chemistry and natural fiber-reinforced materials; however, there are two significant barriers to wide scale industrial adoption: price and performance.

Many biopolymers, including polyhydroxyalkanoates (PHA) and polylactic acid (PLA), are more expensive than petroleum-derived incumbents such as polypropylene PP and polyethylene PE. This is mainly due to scale. New materials have to compete with the enormous, vertically integrated petrochemical companies that have invested billions of dollars in manufacturing infrastructure. Price is impeding the formation of the major industrial product development partnerships necessary to develop high-performance alternatives. Until bio-based materials can compete on price they will remain niche products.

INCA Renewable Technologies is a Canadian-based corporation with the mission to provide our industrial partners with natural fiber composites that are stronger, lighter, cost competitive, and far more sustainable, and thereby solve this pain point in the marketplace. Instead of procuring processed fiber offshore, as competitors do, INCA is purchasing hemp biomass, a renewable agricultural resource, directly from farmers. The company will process the biomass in-house and then directly transform this low-cost, renewable resource into a set of proprietary high-value products for major industrial customers. There will be a seamless integration of INCA's operation – from farm gate to factory floor – and INCA's products are being developed to meet customers' price and performance specifications.

Camille Sobrian Saltman, 250-300-5254, e-mail: csaltman@incarenewtech.com
David Saltman, 250-300-5253, e-mail: dsaltman@incarenewtech.com

https://doi.org/10.1515/9783110791228-025

Just as the petroleum industry leverages its expertise in exploration, extraction, and chemicals manufacturing to maximize financial return on every barrel of oil, INCA will leverage the company's expertise in agronomics, fiber processing, composites innovation, and manufacturing to maximize return on every bale of agricultural biomass. INCA's ability to directly process biomass and utilize this renewable agricultural resource rather than petroleum, wood, or fiberglass alternatives will provide INCA's industrial customers with higher performing, lower cost, and far more sustainable material solutions that help reduce plastics pollution, deforestation, and greenhouse gas emissions.

25.1.1 Vertical integration

To achieve this vertical integration, INCA will deploy two factories, one in Vegreville, Alberta, within a 150-mile radius of hemp cultivation, and the second in Indiana, within 150 miles of customers in the automotive, RV, and consumer products industries. Although separated by 1,600 miles, these two factories function as a single coordinated operation.

Hemp has been legally cultivated in the Canadian Prairies for over 20 years where it is primarily grown for plant-based protein. The remaining stalks contain some of the strongest natural fiber on earth. If properly refined and manufactured, hemp fiber can function as a direct replacement for glass fiber in composites.

INCA will acquire this "waste" biomass in the form of baled straw and decorticate it, a process that mechanically separates the long outer bast fiber from the short inner core (called hurd). The bast fiber will be further refined to create the ultra-clean material required for its products. In Alberta, the hurd will be used to manufacture INCA BioBalsa™, a direct replacement for the balsa wood used in boats and wind turbine blades and BioPlastics (ensure TM is used) for the consumer products industry. The refined long bast fiber will be sent *via* rail to the company's second factory in Indiana. There, it will be manufactured into two additional products: INCA BioPanels™ for the RV industry and INCA PrePregs™ for the automotive industry.

Domestically sourcing and processing hemp lowers raw material and logistics costs and reduces supply chain challenges versus procuring and importing expensive jute or kenaf from Southeast Asia, as other bio-composite manufacturers do. Locating the fiber processing operation close to the hemp fields enables quality control of raw material. Sending only the refined long fiber from Canada to the US reduces transportation costs by 75% versus shipping whole hemp bales. Siting composites manufacturing close to customers again reduces logistics costs.

25.1.1.1 A brief history of hemp

Hemp (Cannabis sativa L.) was one of the first crops cultivated by man. Recorded use dates back over 8,000 years. Hemp cultivation began in what is now China and spread quickly across Asia, Europe, Africa, and later South America. It was such a strategic crop that in 1533, King Henry VIII fined English farmers if they did not raise hemp. Literally, Britannia could not have ruled the waves without hemp rope and canvas. Until the late 1800s, most of the paper in the world was produced from hemp pulp.

Hemp was originally introduced to North America in 1606 where it was grown for papermaking, cordage, lamp fuel, and clothing. However, at the turn of the twentieth century, powerful forces set out to delegitimize the cultivar. It was seen as a direct competitor by the pulp and paper industry, and by DuPont, which invented Nylon as a petroleum-based substitute for hemp textiles. Backed by powerful industry lobbyists, hemp was included in the 1938 Opium and Narcotic Drug Act in Canada, and in the 1937 Marijuana Tax Act in the US. This legislation essentially killed the hemp industry.

The good news is that hemp is now experiencing a major rebirth in North America. In 1998, it became legal to grow it for commercial production. Canada's Industrial Hemp Regulations, part of the Cannabis Act 2018, allows for whole plant utilization of the crop. Health Canada controls the production, processing, transportation, and sale of industrial hemp. In 2020, there were approximately 53,000 acres of hemp being cultivated, and this is expected to expand dramatically as the market for plant-based protein and fiber grow.

In 2009, the Government of Alberta made a strategic investment in the nascent hemp industry by building a large fiber processing plant at the InnoTech Alberta research facility in Vegreville. INCA is partnering with InnoTech to further advance hemp genetics and agronomics, focusing on cultivators that can provide both seed and fiber yield.

The Alberta government is investing in the hemp industry for the following reasons:

- Hemp matures in as little as 90 days, and can be grown with low water and chemical inputs;
- As a rotational crop, it breaks disease cycles and has a long tap root that aerates the soil;
- It yields 4X the biomass of a forest in one season, compared to 25 years of tree growth; and,
- Every portion of the plant can be commercialized, creating multiple revenue streams.

Thanks to this early government support as well as the long hours of sunshine, which the Canadian Prairie affords, Alberta currently grows 40% of Canada's hemp. INCA has partnered with companies that are breeding high yield hemp seed. By providing seed and signing long-term offtake agreements with the farmers for the remaining biomass leftover after protein is harvested, the dual income will make hemp one of the most profitable rotational crops in the region.

25.2 Advanced materials from hemp carbohydrates

Today, many of the same products that are synthesized from hydrocarbon molecules derived from oil and natural gas can be produced from carbohydrates derived from agriculturally grown plants. This includes plastics, textiles, fertilizers, building materials, furniture, papermaking, packaging, industrial fillers in paints, plastics and adhesives, pharmaceuticals, nutraceuticals, and cosmetics.

The potential benefits of bio-based materials are very promising. Petroleum-based products are energy-intensive to refine, generate harmful byproducts in the process, and are often difficult to recover and recycle at the end of their useful life-cycles. Plant-based products sequester carbon, generate no toxic byproducts, and are much more easily recovered, recycled, or even composted. INCA has chosen a product development strategy focused on producing materials that can meet its customer's price, performance, and environmental objectives. This work has led to the development of four proprietary products:

1. INCA BioBalsa™ – a direct replacement for the balsa wood and plastics used in boats and wind turbine blades;
2. INCA BioPanels™ – a direct replacement for hardwood plywood used in the RV industry;
3. INCA PrePregs™ – a replacement for natural and glass fiber-reinforced plastics for auto interiors;
4. INCA BioPlastics™ – a replacement for glass-reinforced plastics in the consumer products industry.

25.2.1 INCA product lines

25.2.1.1 INCA BioPanels™

Since the 1970s, the recreational vehicle industry has been dependent upon rainforest hardwood plywood from Southeast Asia to construct side walls and roofing systems. Hardwood plywood's strength, light weight, and smooth surface make it a product in high demand not only for RVs but also for transport trailers, furniture, decorative paneling, and even Hollywood movie sets.

In 2021, the RV industry alone used 620 million board feet of lauan plywood. Unfortunately, the popularity of this material has led to devastation of primordial rain forests throughout Indonesia, Malaysia, Borneo, and The Philippines. Entire ecosystems are being lost, species endangered, and indigenous communities displaced. Plywood prices have skyrocketed, quality has fallen, and manufacturers are actively seeking alternatives.

To construct an RV, sheets of 4′ x 8′ and 5′ x 10′ plywood are seamed together like stick frame housing. Exterior wall panels are laminated with aluminum or fiberglass, and interior walls with paper or vinyl. Urethane foam insulation is sandwiched between the inner and outer walls. However, plywood presents significant performance problems for RV owners. It off-gasses toxic formaldehyde, absorbs moisture, and burns easily. After just a few years, seams telegraph through the exterior walls, and the strength and visual appearance of the vehicles are compromised.

Oriented strand board (OSB) is being used as a plywood substitute, but it is much heavier, and it too absorbs water and off-gasses formaldehyde. Fiberglass-reinforced plastic (FRP) panels of various types are being used in higher-end RVs. Manufacturers such as Crane Composites and Vixen Composites create multi-layered wall systems with

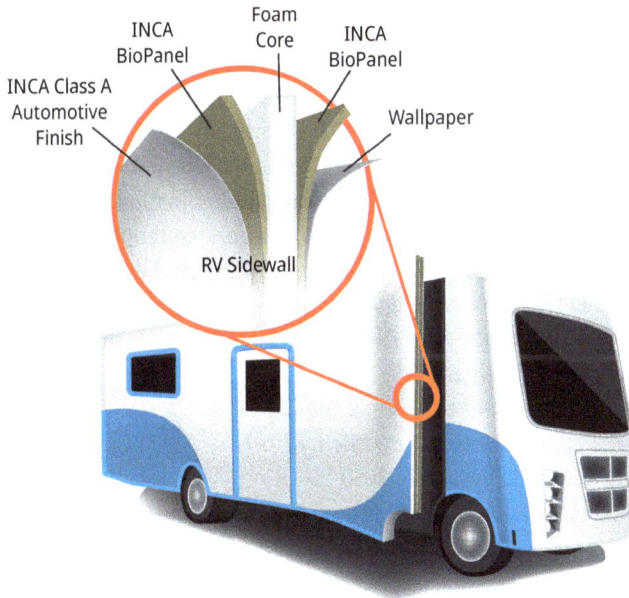

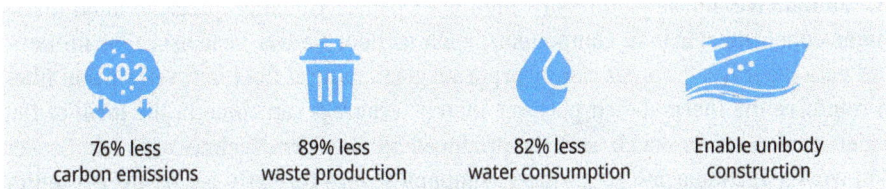

Figure 25.1: INCA BioPanelsTM versus rainforest plywood (© 2022 INCA Renewtech).

expanded polystyrene (EPS) cores, FRP skins, and aluminum framing. In addition to being much heavier, more expensive and labor intensive, FRP panels offer their own performance challenges, including delamination, warping, and poor insulation properties.

INCA has signed an exclusive sales agreement with Elkhart-based Genesis Products, Inc., which designs, engineers, and manufactures a range of products for the RV, construction, transportation, and furniture markets. Their customers include Thor Industries, Winnebago, and Forest River. Consolidation is the name of the game in the RV industry. Forest River purchased Coachman, which was in turn acquired by Berkshire Hathaway. Thor Industries now owns Airstream, Heartland, Jayco, and Keystone. Winnebago owns Newmar. These public companies have implemented aggressive ESG targets that include sourcing alternatives for rainforest plywood.

INCA has developed a revolutionary panel product made from a blend of hemp fiber and a proprietary thermoset resin system. Panels will be produced on a twin-belt press capable of making boards up to 10 feet wide and 45 feet long. Rather than having to seam together sheets of plywood, BioPanels™ will save significant labor costs and create lighter, stronger, and more durable unibody vehicles. INCA BioPanels™ will be fire rated, moisture resistant, lighter in weight, and available with Class A automotive finishes, saving manufacturers the significant cost of painting.

Genesis supplies one-third of the plywood used in the North American RV industry. The company currently operates multiple hot and cold lamination lines in nine factories, with over one million square feet of production, assembly, and warehousing space. As the exclusive distributor of INCA BioPanels™, Genesis has facilitated a product development agreement with Winnebago for the Company's entire panel production capacity. Winnebago is partnering with INCA to help them achieve their stated environmental sustainability goals. These include:
- Achieving net-zero greenhouse gas emissions by 2050
- Diverting 90% of all manufacturing waste by 2030
- Reducing fresh water use by 30% in 2050
- Including eco-friendly options on all products by 2025

25.2.1.2 INCA PrePregs™

The automotive industry utilizes resin-infused panels, called prepregs, to mold three-dimensional interior trim components, such as door panels, seat backs, headliners, and package trays. Prepregs can incorporate glass, natural fiber, or even carbon fiber to reinforce the thermal-melt polymer matrix. Prepregs can come in the form of flat panels or non-wovens such as those produced by FlexForm Technologies and Carver Non-woven Technologies. (Both these companies were formerly led by INCA's senior management.) The automobile industry was an early adopter of these materials due to their need for lower cost and high strength-to-weight ratio, which resulted in better fuel mileage and customer safety.

Manufacturing non-wovens involves blending natural fiber and polymer fiber such as polypropylene, and then depositing this mat on a moving belt. The material is consolidated, carded, cross-lapped, and sent through a needle-punch system to interlace the final product. Automotive manufacturers heat these prepregs in hot presses, surface them with cover stock, and compression mold them into final 3D components. While non-wovens have proven to be of great value to the automotive industry, they depend upon expensive natural fiber, such as jute or kenaf imported from Southeast Asia. They also use a high percentage of polymer fiber, adding considerable cost.

INCA has developed a methodology to deposit multi-layered mats of hemp and polymer fiber, and consolidate the material on our twin-belt press to produce thermoformable prepregs for Toyota North America. These prepregs are stronger, improving side impact resistance. They reduce vehicle weight, improving mileage. They sequester carbon and are made of a higher percentage of renewable resources. Finally, they can be easily recycled back into new products once vehicles have reached the end of their useful lifecycles. These factors are of great importance to Toyota, which has engaged INCA to develop proprietary prepregs for their 2026 model year electric vehicles.

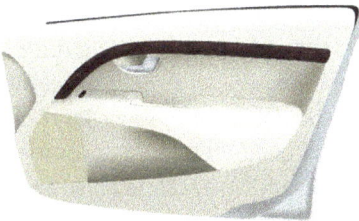

Figure 25.2: INCA PrePregs™ will be molded by Toyota into 3D interior trim components (© 2022 INCA Renewtech).

25.2.1.3 INCA BioPlastics™

Plastics are a multi-billion-dollar industry led by the world's largest petrochemical companies. Thermoplastics such as polypropylene (PP), high-density polyethylene (HDPE), and polyvinyl chloride (PVC) are used in packaging, furniture, flooring, roofing membranes, automobile interiors, medical devices, and the countless consumer products that surround us. Their relatively low cost, high moisture resistance, and ability to be heated and shaped using injection, compression, or extrusion molding machines make plastics ubiquitous in the modern world. The top polymer producers include Dow Chemical, Lyondell Basell, Exxon Mobil, SABIC, INEOS, and BASF.

Most polymers are derived from petroleum or natural gas. Their extraction and refinement generates millions of tons of CO_{2e}. They often contain toxic chemicals that bio-accumulate in humans and animals. As only nine percent of plastic products are recycled, they cover the earth and fill the seas with pollution. Often, thermoplastics are compounded with glass fibers to reinforce the polymer matrix and enhance structural properties. However, these materials increase the cost and weight of the resulting products. Glass fiber is extraordinarily energy-intensive to manufacture. It has high tensile strength but low impact strength. Glass is also abrasive on processing equipment and difficult to recycle, once products have reached the end of their useful life cycles.

Bio-based polymers represent a small percentage of the millions of tons of plastics produced annually. However, the global bioplastics and biopolymers market is growing rapidly. Market drivers include industry and consumer demand for sustainable products, increasingly stringent regulatory mandates, and growing public concern regarding plastics pollution.

Europe has been the major hub for bio-based innovation, and currently produces 55% of all biopolymers, but capacity now is expanding rapidly in Asia and North America. Many of the companies leading the charge are agricultural bioscience companies, or petrochemical companies making the transition to plant-based chemistry.

Danimer Scientific produces PHA, a starch-based polymer. Total Corbion operates a 75,000 ton PLA production facility in Thailand and has plans to build a second PLA factory in France. NatureWorks, owned by Cargill and PTT Global Chemical, has a refinery in Minnesota capable of producing 300 million pounds of PLA per year from corn feedstock.

While these polymers show great promise, they tend to be significantly more expensive than incumbent petroleum-based polymers and lower in structural performance. INCA has developed a novel manufacturing methodology to replace glass fiber with natural hemp fiber as a reinforcement in polymeric composites. In INCA's process, hemp fiber is refined to precise geometries and then blended with various polymers and additives under relatively low heat and pressure. This preserves the structural properties of the natural fiber as well as that of the polymer. Product benefits of INCA BioPlastics™ include cost and weight reduction, improved stiffness and impact strength, and lower wear on processing equipment. Unlike glass fiber, refined hemp fiber is more amenable to multiple recycling episodes without significant loss in structural properties.

INCA formulates materials to meet customer specifications. For example, one company may want to utilize recovered polypropylene to lower product costs and increase recycled content, while a second customer may specify a bio-based polymer such as PLA or PHA to create 100% bio-based products. These formulations are brought together in the form of compounded "master batch" pellets that can be shipped directly to manufacturers for use in their injection or compression molding operations with little or no modifications to their existing lines. Test results demonstrate that compounding polymers with refined natural fiber can double structural properties. It not only increases strength, it decreases the cost, enabling customers to make stronger, lighter, and less expensive final products.

25.2.1.4 INCA Biobalsa™

The balsa tree (Ochroma pyramidale) is a large, fast-growing species, native to the rainforest regions – from southern Mexico to southern Brazil. Although classified as a hardwood, balsa is exceptionally soft and light weight. This is because the trunk of the tree has large cells that store water. When cut and dried, these empty cells retain their structure, giving the resulting lumber exceptionally low density and high compressive strength. As a result of these properties balsa is widely used as a core material in automotive, aerospace, and marine applications.

Recently, balsa has become one of the most sought-after materials in the construction of wind turbine blades where it is used to reduce blade weight and increase blade stiffness, particularly in the root section where the mechanical forces are most demanding. This is accomplished by placing shaped balsa in the core of the blade and encapsulating it with styrene-resonated glass fiber to form stressed skin panels. Blade skins are put into tension on one side and compression on the opposite side, and held in opposition by the balsa core to resist collapse.

Historically, most balsa wood came from the rainforests of Ecuador, but the species has been largely clear-cut, with devastating impacts on ecosystems and indigenous communities. Sixty percent of balsa wood is now plantation grown. However, without the many years required to grow mature trees large enough to produce solid lumber, the industry has adopted end-grain balsa – essentially slicing logs across the grain, milling rectangular pieces, and gluing them back together to form board stock. This has added considerable cost to the product, reduced consistency, and lowered technical properties. As a result, many blade manufacturers are turning to high-density polyethylene terephthalate (PET) type of closed cell thermoplastic structural foam – for the tips and mid-sections of the blades. Unfortunately PET foam offers significantly less compressive and sheer strength, is more expensive and is petroleum-based and therefore not sustainable.

INCA engineers have developed a novel methodology to transform refined hemp hurd into a balsa-like material with high, uniform compressive strength, and a density of 10 pounds per cubic foot – the ideal weight for turbine blades. Unlike balsa wood, it can be formulated to increase moisture resistance, and "tuned" to be compatible with specific resin systems. These characteristics provide significant advantages to engineers who are no longer forced to compensate for variations in the material properties of plantation-grown trees.

INCA has signed a strategic development agreement with Gurit to develop and manufacture BioBalsa™. Gurit develops and manufactures advanced composite solutions for the wind energy, aerospace, and marine industries, and their portfolio of structural core materials, prepregs, adhesives, and engineering services is recognized world-wide. The industries Gurit serves are actively seeking an alternative core material that can deliver the compressive and shear strength of balsa as well as meet their aggressive sustainability goals.

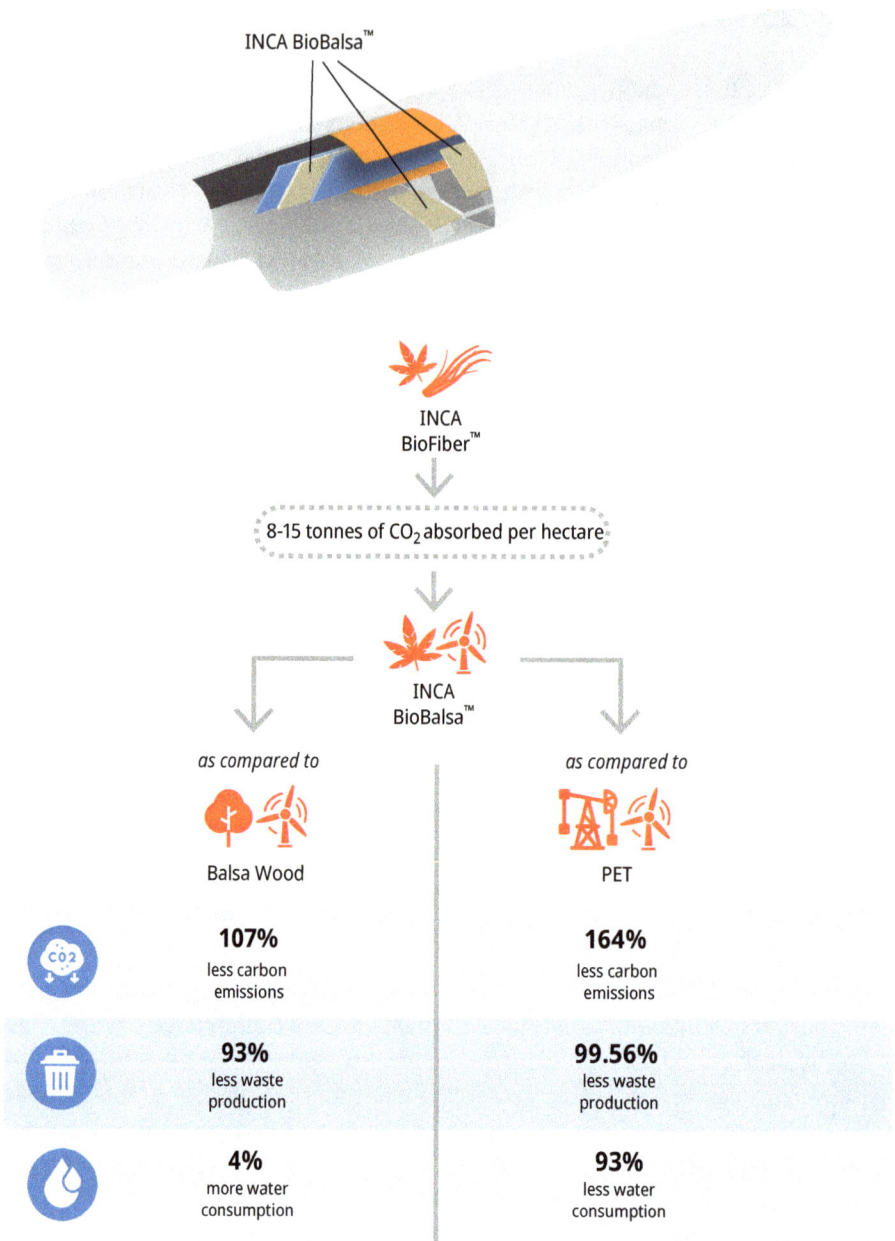

INCA BioBalsa™

INCA BioFiber™

8-15 tonnes of CO_2 absorbed per hectare

INCA BioBalsa™

as compared to

Balsa Wood

107%
less carbon
emissions

93%
less waste
production

4%
more water
consumption

as compared to

PET

164%
less carbon
emissions

99.56%
less waste
production

93%
less water
consumption

Figure 25.3: INCA BioBalsa™ versus conventional alternatives (© 2022 INCA Renewtech).

Constructed from hemp hurd cellulose, INCA BioBalsa™ presents a product of comparable density but with higher compressive strength, thanks to the randomness of fiber placement – which delivers uniform properties in all directions. Unlike balsa, it can be formulated to reject moisture and attract specific binding resins. It has the shear strength required for installation in all sections of the wind turbine blades, including the root section.

LIFECYCLE ANALYSIS

INCA commissioned GreenStep Solutions, an independent research and certification organization, to undertake a Product Life Cycle Analysis of the company's products versus conventional alternatives. This study utilized the Greenhouse Gas Protocol Product Life Cycle Accounting and Reporting Standard. GreenStep is a certified B Corporation that has helped more than 3,000 private, public, non-profit, and academic institutions to understand the full impact of their operations along the entire lifecycle.

The company's in-depth reports[1] analyze the impacts of industrial hemp farming, product formulations, raw material extraction and processing, energy and water inputs, product manufacturing, and fuel consumption at each stage of the process, including downstream transportation and distribution. Their findings demonstrate that:

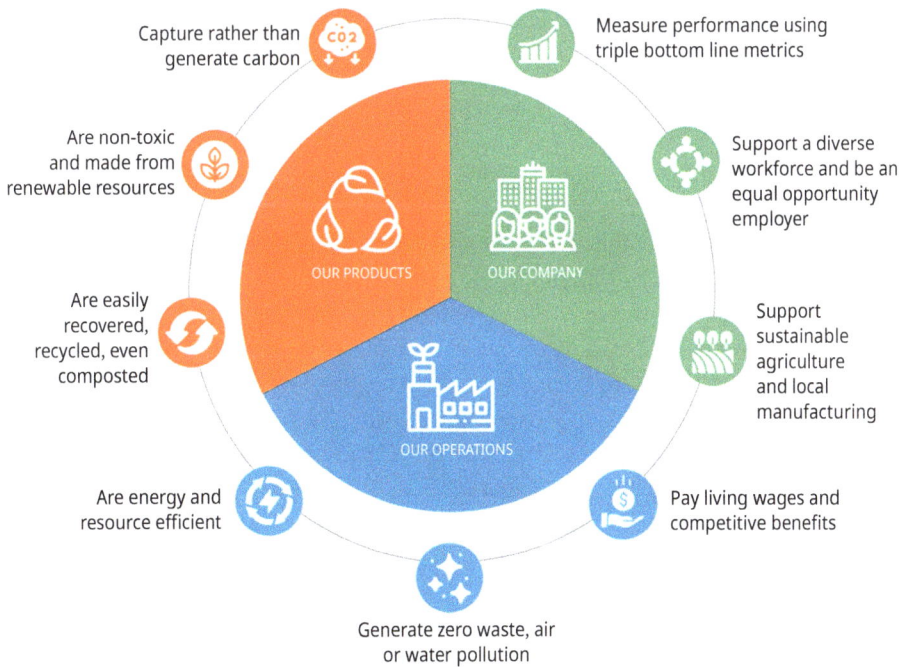

Figure 25.4: INCA Renewtech ESG Commitment (© 2022 INCA Renewtech).

- **INCA BioPanels**™ reduce carbon emissions by 76%, waste production by 89%, and water consumption by 82%, compared to lauan rainforest plywood. One square meter sequesters 4.27 kg of CO_{2e}.
- **INCA BioPlastics**™, when compounded with recycled polypropylene, reduces carbon emissions by 91%, waste production by 64%, and water consumption by 59%, compared to glass fiber-reinforced plastics (GRP). One cubic meter absorbs 517.96 kg of CO_{2e}.

The findings in their report on INCA BioCore[2] demonstrate that:

- **INCA BioBalsa**™ has 164% less carbon impact than PET and 107% less carbon impact than balsa wood. Production results in 93% less waste than balsa and 99.56% less than PET. It also means 93% less water consumption during manufacturing. One cubic meter of BioBalsa absorbs 262.31 kg of CO_{2e}.
- **INCA PrePregs**™ (needs to be superscript) has 27% less carbon impact, produces 30% less waste and saves 68% more water.

25.3 Manufacturing operations

25.3.1 Fiber processing

The 164,000 square foot Alberta fiber processing facility is being built on a 26-acre site, adequate for three months of raw material storage. It will be adjacent to a major highway for incoming bales of hemp, and on a rail spur for outgoing products. At full capacity, five decortication lines will have the ability to process 55,000 metric tons of hemp biomass per year. The facility will have sophisticated temperature and humidity control, and a state-of-the-art dust collection system to reduce fire danger and assure exceptional indoor air quality for INCA's employees. While INCA's products and processing techniques are novel, the equipment used in the company's factories will be purchased from world-leading equipment suppliers with decades of experience.

1. GreenStep Solutions Life Cycle Assessment Final Report, INCA BioPlastics and Bio-Panels for INCA Renewable Technologies, July 8, 2022
2. GreenStep Solutions Life Cycle Assessment Final Report INCA BioBalsa, for INCA Renewable Technologies, September 12, 2022
3. GreenStep Solutions Life Cycle Assessment Final Report INCA PrePregs, for INCA Renewable Technologies, November 1, 2022

During Phase 1 of operation, two large customized DILOTemafa decortication lines will be installed, each capable of processing 2,500 kilos of biomass per hour. This will yield approximately 500 kilos of clean bast fiber for composites and 1,250 kilos of clean hurd for BioBalsa. The remaining 750 kilos represents excess moisture and unusable organic material. The moisture will be captured and used in our humidity

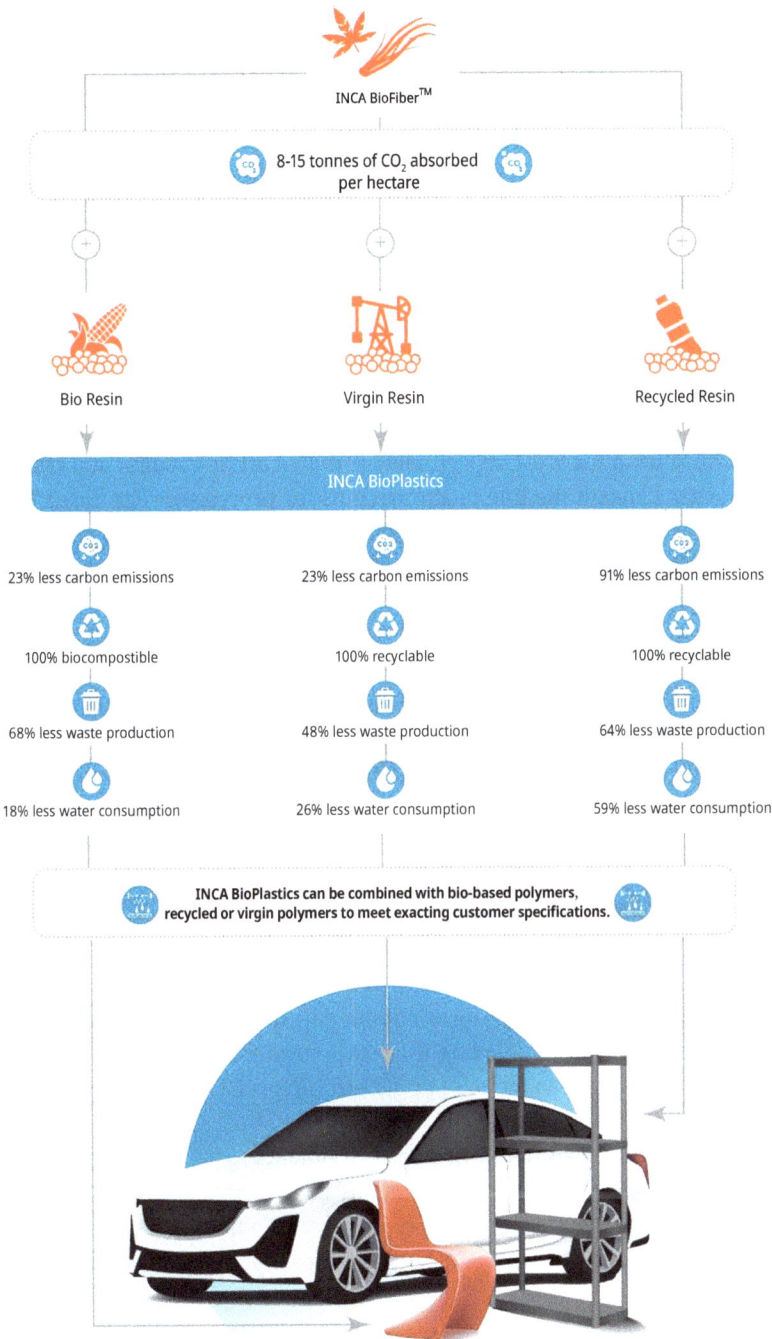

Figure 25.5: INCA BioPlastics combined with bio-polymers, recycled polymers and virgin polymers versus conventional glass reinforced petroleum based plastics. (© 2022 INCA Renewtech).

control system. The organic material will be composted on site. Our goal will be to be a zero-waste facility. During Phase 2 expansion, three additional lines will be installed. The following is a description of the fiber manufacturing processes:

Stage 1
Whole hemp bales are loaded onto a bale breaker, binding straps removed, and bundled stalks decompressed. They are fed into a hammer mill that crushes the stalks to begin separating the outer fiber from the inner hurd. At this point in the line, metal detectors and weight scales ensure ferrous and non-ferrous materials, such as rocks and stones, are removed to prevent equipment damage. The material is then fed continuously into a skrutching turbine, which performs further fiber separation.

Stage 2
The low-density hurd from the working cylinders is pulled off by vacuum and transported by a conveyor beneath the line to storage bins for processing. The bast fiber goes through a series of aggressive step openers and wire drums, opening the vascular fiber bundles to achieve fiber that is 99% clean – a specification essential for successful composites manufacturing.

Stage 3
The final short fiber will be conveyed to a bulk storage bin and long fiber products are conveyed to dedicated baling systems that compress the product into 200-kilogram blocks, and wrap them in polypropylene film and tie straps for rail transport to the Indiana factory.

25.3.2 INCA BioBalsa™ manufacturing

The 99% clean hurd is diverted to a separate processing line. The material is hydro-pulped using advanced thermo-mechanical equipment. Using steam injection, the fiber bundles are opened while preserving their structural integrity. Various ingredients can be added at this stage to increase shear and compressive strength, improve dimensional stability, and adjust for resin absorption.

The resulting material is dewatered in a radio frequency dryer, and then deposited onto a moving belt press to create a continuous board stock. Radio frequencies affect only the water molecules, leaving the properties of the fiber intact. Panels are edge-trimmed, cut to length, stacked, and palleted for shipping to customers and distributors.

25.3.3 Bioplastics needs to be changed to BioPlasticsTM (with TM as superscript) manufacturing

Natural fiber for injection moldable composites will require clean bast fiber 4–7 milli-meters in length, with aspect ratios of approximately 15:1. They will be decompressed and dehumidified using a radio frequency dryer, and then combined with various polymers and additives per customer specifications into customized formulations. For example, one customer may want to use recycled polypropylene (PP) to increase their recycled content and lower costs, while another company may want to use a bio-based polymer such as polylactic acid (PLA) to create 100% bio-based, bio-degradable products.

Dry formulations of fiber, polymer, and additives are fed into feed towers, which mix the material at relatively low heat and pressure, extrude the dimensioned pellets through a rotary die, and cut them to length. This low temperature process is essential to preserving the integrity of the natural fiber and polymer, as additional heat cycles degrade the structural properties of the final products. Pellets are cleaned and cooled in a large holding tank, conveyed to a bagging machine where they are weighed into supersacks, and palletized for shipment.

INCA will initially install three compounding lines, each having the capacity to produce 3,000 pounds of pellets per hour. The value of this pelletized product is that it can be easily fed into customers' existing manufacturing lines, where additional additives such as colorants, fillers, flame retardants, and antioxidants can be added prior to injection or extrusion molding. Without pelletization, companies would have to invest millions of dollars in air filtration systems, RF dryers, unique feeding systems, and other modifications to use natural fiber.

25.3.4 INCA Biopanels needs to be BioPanels™ and INCA prepregs needs to be PrePregs™ manufacturing

Indiana, is the epicenter of the recreational vehicle manufacturing industry in the US and close to Detroit which is the center of manufacturing for the automotive industry. The Great Lakes region is also a major manufacturing hub for the consumer products and commercial furniture industries.

During Phase 1 operations, INCA is planning to acquire an existing composites manufacturing operation in Indiana and upgrade the facility. With these renovations INCA will have the technical capabilities to produce thermal-set panels for the RV industry and thermal-melt prepregs for the auto industry on a 60″ wide twin belt press built by IPCO in Germany. IPCO is the world's leading manufacturer of double-belt presses and precision scattering systems. The press line will have the technical capabilities to produce thermal set panels for the RV industry and thermal melt prepregs for the auto industry. It will be similar in design to the much larger 120″ press line that will be installed during the Phase 2 expansion.

Stage 1

Bales of refined hemp fiber arriving *via* rail from Alberta will be placed on an apron belt where they are decompressed, opened in a willow card system, and sent to six scattering towers located above the press line.

Stage 2

Each pair of scattering towers will have bulk storage and a blending box, called a mix master. These units precisely deposit multiple layers of natural fiber onto a continuously moving scrim. The scrim provides a surface for transportation of the fiber mat through the needle punch and press lines. It remains on the product as a surface for glue retention during RV or automotive component manufacturing.

Stage 3

A proprietary resin injection system converts liquid resin into a low-density foam, which is then injected into the mat, infusing the entire fiber matrix with polymer. During the liquid phase, INCA will have the option to blend in various additives, such as flame retardants or biocides. All liquids are metered into the formulation *via* a process logic controller and flow metering valves. The accuracy of foam placement and precise control of additives will be unmatched in the industry.

Stage 4

Prior to entering the IPCO press, INCA will have the option to deploy a needle loom system. The purpose of this equipment is to further consolidate the fiber mat, lending additional strength for the automotive prepregs, which must pass rigid impact tests. The needle loom will be bypassed for thermoset panels for the RV industry.

Stage 5

INCA and IPCO engineers have designed a proprietary "camel back" twin-steel-belt thermal compression line, consisting of a single lower belt and a two-section upper steel belt. Operation of the lower belt is consistent with standard press configurations, with a first stage heating zone and a final cooling zone. The first stage of the upper belt covers six meters of heating and an optional two meters of cooling. This configuration allows the line to process thermosets at maximum line speed while enabling finished panels to freely release from the steel belt.

A space between the two upper belts is novel and allows for the application of color films or scattering of powder finishes prior to entering the cooling zone. This allows films or powders to bond to the underlying panel and then be chilled for mold release before exiting the press line.

Stage 6

At the end of the line, there are three saws, one for edge trimming and the second for dimensional cross cutting. When the larger 120″ press line is installed, 100% of its capacity will be utilized for motor home sidewalls. The first 60″ press line will be entirely devoted to production of thermal-melt prepregs for the auto industry. The company will also install a finishing line capable of putting Class A automotive finishes on the 3800 GSM exterior walls.

25.3.5 Environmental production

INCA's manufacturing facilities will be optimized in terms of location, cleanliness, safety, labor, and efficiency. Both factories will recycle all waste back into the production system, and the company anticipates yield losses to be very low. In Alberta INCA will use geothermal heating and cooling to save energy.

Safety procedures will be put in place, and staff trained to maintain a safe and healthy work environment. Compliance with ISO 9000 (quality) and ISO 14,000 (sustainability) standards will be verified. Our ability to displace significant volumes of petrochemical plastics, reduce deforestation and greenhouse gas emissions, and contribute to a more sustainable future is at the heart of the company's mission and vision.

25.4 Commercialization strategy

The Company's commercialization strategy is entirely focused upon developing products that meet customer's price, performance, and environmental specifications. This customer centrism has led to the commercialization partnerships with clients Winnebago, Toyota, and Gurit. Early market traction has also been largely due to relationships that INCA's management team have built over decades of work in natural fiber composites.

Chairman & Chief Executive Officer, David Saltman, has built a number of successful cleantech companies. He and Chief Technology Officer Garry Balthes founded Flexform Technologies, the first natural fiber composites facility in North America. They developed programs that include door panels, package trays, truck liners, and load floors for Honda, Toyota, GM, Ford, Chrysler, Mercedes, and Tesla. Saltman is the recipient of the Award for Excellence in Innovation (ACE) from the American Composites Manufacturing Association. In 2010, Balthes received the Toyota Global Award for Sustainable Composite Technology. He went on to design and build Carver Non-Woven Technologies, the most advanced composites manufacturing facility.

Chief Marketing and Sustainability Officer, Camille Sobrian Saltman, is the former President and Chief Marketing Officer of Malama Composites, manufacturers of the first rigid urethane foams made from bio-based polymers. INCA President, Paul Wybo, brings 30 years of manufacturing management expertise, most recently as the former Chief Operating Officer of ABC Industries. INCA's President of Canadian Operations, Paul Bullock, has been involved in the modern hemp and cannabis industries. From a fourth-generation agricultural family, Bullock's knowledge of plant genomics and field operations, and his direct relationships with hemp farmers are mission critical to acquiring a steady, high-quality supply of raw material.

25.5 External market drivers

Powerful trends are driving the development and commercialization of new bio-based materials. Among these external market forces are included:

25.5.1 Green purchasing guidelines

The US government is the single largest "customer" in the world, spending more than $550 billion annually in goods and services. To create jobs in rural communities, add value to agricultural commodities, and reduce America's dependence upon foreign imports, the US Department of Agriculture has created the BioPreferred Program. This program directs government agencies and their suppliers to utilize domestically produced bio-based products, wherever possible.

Canada's green procurement program requires that federal and provincial governments incorporate environmentally responsible guidelines in their purchasing decisions while maintaining compliance with all legislative, regulatory and policy obligations. Both the US and Canada define environmentally preferable products as those that reduce greenhouse gases and other toxic emissions, improve resource conservation, enable reuse and recycling, and protect biodiversity. Europe is several years ahead of both the US and Canada in this regard.

25.5.2 LEED certification

The Leadership in Energy and Environmental Design (LEED) program was originally created by the Natural Resource Defense Council and the US Green Building Council to provide independent third-party verification that homes and buildings are designed and built to achieve higher energy and resource efficiencies and better indoor air quality. LEED has become the *de facto* standard for the built environment, influencing not only design and construction methodologies, but also material selection.

25.5.3 ESG investing

While government policies are crucial to supporting more sustainable products and services, the single most powerful force influencing industry trends are the capital markets, and the single most important shift among money managers and investors is their move toward environmental, social and governance (ESG) criteria in making investment decisions. According to Forbes, ESG considerations now account for over $20 trillion in capital, or approximately a quarter of all professionally managed assets.

Investors recognize that ESG metrics are financially relevant to managing risk and understanding corporate purpose, strategy, and management objectives. This shift is driving a major increase in carbon prices. The origins of this trend can be traced back to work done by the United Nations in conjunction with the International Finance Corporation (IFC) and the Swiss government, which produced a report entitled "Who Cares Wins." This report argued that incorporating ESG considerations in evaluating investment decisions makes sound business sense. The report formed the background for the launch of Principles for Responsible Investment at the New York Stock Exchange.

25.5.4 Consumer demand

While government regulations and investment criteria are critical to bringing about more sustainable societies, the real market makers are consumers. Their spending accounts for roughly 70% of the US economy, and they are voting with their dollars. A recent NYU Stern Center for Sustainable Business study found that over half of all consumer-packaged goods growth from 2013 to 2018 came from products marketed as sustainable. The category expanded at five times the rate of products that did not have a legitimate environmental message.

25.6 Conclusion

Economic factors are now playing a powerful role in the adoption of bio-based materials. However, cost and performance continue to be an impediment to wide-spread adoption. In spite of the goal of decoupling materials use from economic growth, legislation and regulation can only come into play as a driver of change when policy makers perceive a solution is within reach from a commercialization perspective. INCA Renewtech has recognized that to achieve market penetration and compete with incumbents, the company requires a business model that can deliver price and performance along with circularity and carbon neutrality.

Robert C. Joyce

Chapter 26
BioProducts and biomaterials that drive sustainability in a medical market

26.1 Background

During the 1980s, the oil crisis accelerated the development of sustainable materials to replace fossil fuels. These sustainable materials, based on cellulosic materials, included wood fibers, hemp, flax, and sisal that were mixed or compounded with various thermosets or thermoplastics polymers. The innovative blends or compounds developed in the next decades helped define biocomposite materials of interest in diverse end uses, including medical applications. These new bio products, made with sustainable biomaterials, have replaced old and outdated technologies; however, innovation has not been easy to commercialize. A personal journey in this field is highlighted in this chapter.

26.1.1 BioProducts and biomaterials: a preface

Robert Joyce started to investigate the compounding of wood fiber with thermoplastics in circa 2003 with a twin-screw extruder he purchased through Ford Motor Company. The twin-screw Leistritz extruder had a 54" sheet die, with a downstream to calendar sheet for thermoforming automotive parts – package trays. Joyce was successful in the production of bio products and a wood fiber calendar sheet, and had built a 14,000 sq. ft. facility in Dundee, Mi, to support production. However, after product development and several letters in communications with the automotive giant, it became evident that the compounding of natural fibers to produce sheets was a commodity business with small margins of 15% or less. Joyce's company was helping drive prices lower for wood fiber composite sheets in the production of package trays. It was further realized there was no possibility to increase product margins for the bio product unless there was innovation behind the production. Hence, the quest for innovating with natural fibers and thermoplastics in compounding and molding bio products began for markets such as the automotive, building, construction, and furniture. He was met with challenges that included product costs, recycle ability, mechanical performance, offensive smells or odor, natural fiber bleaching, and fungus contamination or microbial attack. These technological challenges were overcome through a series of experiments and research to build a commercialization platform. Joyce first started investigating the injection molding process with

Robert C. Joyce, Innovative Plastics and Molding, Inc, Toledo, OH, USA, e-mail: robert@fibertuff.us

https://doi.org/10.1515/9783110791228-026

bio composites and with gas assist. Nevertheless, the barriers to market were difficult to injection mold biopolymers. The obstacles that Joyce experienced included a breakdown of understanding business acumen and pushback from established manufacturers.

The bioproducts business experience for Joyce helped grow his competency in entrepreneurship and progress technically to generate new biomaterial innovation in the medical market. In 2015, Joyce would pivot his business towards the medical market after his bio composite compound passed a basic cell study performed by a university in Northwest Ohio, USA. Joyce refined the biopolymer compound, processing natural polymers as in cellulose fiber with thermoplastics. Through Joyce's knowledge and know-how, innovative filaments and powders for 3D printing were developed for the medical market. These cellulose fiber thermoplastic compounds met the medical challenges in both mechanical and biological performance. Further, Joyce's innovative biomaterial products, identified as novel and patented for 3D printing, could replace inconsistent, less efficient, and more expensive products. The commercialization process to help get the 3D printed medical grade filaments and powders to market were most challenging, especially finding funding for support of the innovation. The local community innovation fund got millions of dollars and failed to share grant funding with the surrounding towns. For over two years, Joyce's efforts in commercialization failed to generate any community support for a loan or grant. Joyce would not give up and eventually achieved sales revenue from his inventions. He continued down the path to innovate biomaterials for 3D printing. The innovative technology developed by Joyce was trademarked Fibre-Tuff. Joyce also applied for patents and had several patents referred to as FibreTuff technology [1–5]. The technology – composition, process, and products – is evolving, showing relevance and profitability in the medical market. FibreTuff has global filament customers that print anatomical models, phantoms, and soon-to-be-released surgical guides and bone grafts. These customers include hospitals, medical centers, physicians, and universities. These third parties are significant in helping provide necessary product feedback, data, and justification in the practice of commercialization and further research.

26.1.1.1 Entrepreneurship and small business innovation with bioproducts

An entrepreneur as well as a business can be short-lived, depending on the person or person's perseverance, persistence, passion, drive, risk tolerance, and results. The author believes a successful entrepreneur must have vision for the short- and the long-term to run a sustainable business. Also, the successful entrepreneur must embrace the responsibility to manage daily events and make decisions, positive or negative, with minimal data or input. In many cases, the entrepreneur must generate their own data or through independent third parties for a sustainable business proof of concept. The data and entrepreneurial vision will drive decisions that are part of the overall strategy to achieve positive results. Moreover, positive results are necessary elements of the entrepreneurial vision – a talent necessary to achieve or generate funding and/

or profitability to advance the journey. He or she cannot prepare for the entrepreneurial journey, solely by reading books. The entrepreneur needs theoretical information and vision to pick a good partner, a productive employee, managing personnel, or even the ingredients for novel compounds to make bio products. In addition, because bio products made from biomaterials is a nascent technology without much history, there are many possibilities for the entrepreneur for innovation and to think outside the box. However, thinking outside the box can be a daunting task for entrepreneurs and they must possess the ability to be curious, and communicate to others their vision. Furthermore, a passion for the vision must be demonstrated and the time being spent should be worth pursuing, hopefully, bearing fruit. The entrepreneur must have a goal to make money but also understand there is a high level of volatility in the work. Lastly, making sacrifices to achieve one or more goals may include hardships on the personal front, including family and acquaintances.

The entrepreneurship experience for the company founder, Robert Joyce of Innovative Plastics and Molding, started through education and life experiences. He acquired practical and theoretical knowledge through books, college, and the experiences gleaned from family and his job. Joyce's entrepreneurial spirit was realized in his early thirties, working for a large industrial controls company in sales and in applications of plastics. Joyce believes there is a moment when the entrepreneurial spirit rises, curiosity and impatience is realized, and he does not fit into the role of an employee. Joyce felt his idea to reuse and recycle industrial controls could be profitable, fulfilling, fun, but risky. Joyce left his job and proceeded to start his first business based on the innovative idea of reuse and recycle. In six months, Joyce became financially challenged in entrepreneurship after leaving his former employer. He found a way to fix and sell higher value plastic machinery and products as a way to survive. His entrepreneurial spirit would eventually evolve to operate plastic machinery, extruding, and melt blending wood fiber with thermoplastics for producing bioproducts.

26.2 BioProducts manufacturing

In circa 2003, Robert Joyce of Innovative Plastics and Molding purchased from Ford Motor Company a twin-screw compounding extruder with a 54" sheet die and a downstream calendar sheet for thermoforming automotive parts – package trays. Joyce proceeded to install the twin-screw extruder calendar sheet in a 14,000 sq. ft. facility he helped build in Dundee, MI. Joyce wasn't experienced in the operation of the extrusion line but found a way to connect with the former extrusion operator and chemist who retired from Ford Motor. After several unsuccessful compounding and sheet production trials with the chemist, Joyce made several adjustments to the formulation and operation to produce a 54" wood composite sheet for thermoforming. Joyce proceeded to produce the package trays to show customers for possible partnerships and collaboration. He would identify a few possible automotive customers, including an OEM. After a few discussions and quotations, it was determined that a 15% margin to supply package trays for the automotive interior application was not a possibility to get a contract.

In 2005, Innovative Plastics and Molding began to explore innovation for reducing the bio product costs, increase mechanical performance, and create value. Joyce would invent a molded article that utilized his bio composite and gas-assist technology. The gas-assist process would use nitrogen – a gas with a low water content. It was introduced through gas pins in an injection molded tool. Further, the bio composite and nitrogen gas would be mixed to produce one or more cavities and porosity in thick-walled parts, creating an impervious skin. Depending on the processing technique, heat, and pressure, you could get different results. While Joyce was producing the bio composite, he decided to mix two types of synthetic polymers – polyamide and polyolefin – with wood fibers to make a bio composite compound for injection molding. This bio composite compound could then be sold to produce thick-walled bio products to the automotive, furniture, and building industries.

Innovative Plastics and Molding would shortly thereafter partner with a compounding entity to help commercialize the wood fiber gas-assist technology. Joyce would sign confidential agreement and contract, submit a provisional patent application, and share his knowledge and ideas for bio composite compounding and a commercialization strategy. The partnership worked for approximately twelve months and the commercialization of bio composite compounds showed profitability in several industries. Joyce's entrepreneurial spirit and innovation created value, recognized by the business partner. Unfortunately, the contractual agreement would be breached and the commercialization strategy would be derailed.

The two years of litigation would slow Joyce's work related to bio products. He did manage to continue testing and validating the novelty of his bio products and biomaterial compound utilizing natural fibers. Based on customer requirements and necessity, Joyce began to investigate replacing natural cellulosic fibers with cellulose fiber to blend with thermoplastics. This innovative cellulose fiber thermoplastic composition developed by Joyce focused on reducing impurities, such as lignin and hemicellulose.

26.3 BioProducts evolving to biomaterials, and new innovations

Innovative Plastics and Molding continued its plan to demonstrate sustainability and value with bio composite and bio products made with cellulosic fibers. Joyce would pursue opportunities to injection mold his cellulosic fiber thermoplastic bio products with automotive car companies and suppliers. Joyce's bio composite compound was injection molded for car interior applications. The two significant challenges were ability to color and smell or part odor. Joyce would reformulate the biocomposite compound, replacing the cellulosic fibers with cellulose fibers blended with thermoplastics. Joyce would also change some of the additives to pass the interior car part material requirements. Joyce innovation led to the patented technology – US Patent 7,994,241 [2] – for Innovative Plastics and Molding – a low moisture composite pellet for molding.

Joyce would again try partnering with a compounding company in Michigan and an automotive car company manufacturer. Innovative Plastics and Molding would execute confidential agreements with both entities in the fall of 2012. Shortly thereafter, the automotive car company would successfully injection mold an instrument panel with Joyce's participation and his bio composite. See Figure 26.1. The injection molded instrument panel had no cellulose fiber agglomeration, no offensive odor, and the cycle time for production was 20% less than the production part. The automotive car company would require more product testing and qualification of the bio composite. Joyce would accommodate, spending the necessary effort and funds for bio composite qualification. Many meetings between Joyce and the automotive car company would take place over the next 16 months, discussing product qualifications for purchase. In one meeting, the car company engineers informed Joyce his technology was considerably

Figure 26.1: Cellulose fiber thermoplastic instrumentation panel.

more innovative than Thrive, a bio composite produced by a large wood products company. Unfortunately, Joyce would be told by the automotive car company engineers that Innovative Plastics and Molding was too small a company for purchasing his composite. Later, Joyce found out that the automotive engineers he was meeting with to qualify his bio composites were developing a similar bio composite and patenting a bioproduct.

In parallel to the automotive application development, Joyce would be investigating collaboration with universities and nonprofits to get non-dilutive funding or grants. A university in Northwest Ohio would test the FibreTuff bio composition for possible uses in the medical industry. The university performed a basic cell study on Joyce's bio composite and the testing would come back favorable. As a matter of fact, the university tested Joyce's compound and parts twice to verify the unexpected results. These results at the university helped identify possibilities and set a goal to replace PEEK – polyether ether ketone – for implants. The bar was set extremely high for Joyce to produce the bio composition to be identified as biocompatible – mechanical qualities to be equal to or better than PEEK. The term biocompatible was defined as the acceptance of the cells and tissues without any adverse effects. As for the biomaterial characterization of the composition, this was not determined until actual *in vivo* study was performed. After the basic cell study, the focus of Innovative Plastics and Molding turned more to the gathering of data points for the validation of FibreTuff's mechanical attributes. Joyce would engage with the bioengineering professors and participate with students from the university in Northwest Ohio. These university students would test and compare the properties of FibreTuff with other biomaterials. Joyce had got a profile sample with aligned fibers to be compared with PEEK's mechanical properties. Several data points were generated with significance and Joyce set out to invent new products with his newly recognized, FibreTuff, a cellulose fiber thermoplastic composition. His direction for new product development with FibreTuff would be in 3D printing the cellulose fiber thermoplastic filament. See figure 26.2.

In a short period of time, FibreTuff would be printed in complex geometries, having bone-like performance and evolving to pass stringent tests for biocompatibility – meeting expectations for mechanical strengths and flexibility to print. Joyce also discovered his printed FibreTuff biocompatible composition had radiopacity or radio density. The radio density of FibreTuff could be scaled in Hounsfield Units from a computerized tomography or CT scan. The radiopacity was a differentiator compared to many other polymers that were radiolucent, and metals that were radiopaque. The concept of Fibre-Tuff, to be used in an implant, would be most beneficial, for determining location and bone bridging. Actual bone would be brighter than printed FibreTuff and data could be collected for determining how quickly the patient was healing. Once repeatable and accurate data is generated, artificial intelligence (AI) is a real possibility for FibreTuff biomaterials. In the process to drive the printing of the FibreTuff biomaterial technology forward, Joyce began to explore various means to produce filament for 3D printing but not without challenges. His company would conduct several small-scale printing experiments with FibreTuff filaments. One of the first test examples included a printing

Figure 26.2: FibreTuff filament for 3D printing.

manufacturer and a high school. Unfortunately, the filament testing did not fare well due to a combination of errors, and the company needed to regroup. The process of regrouping would include a 3rd party who supported the printer manufacturer, Ulti-maker, and would fix the printer to print bone models. Once the acceptance grew from the printer manufacturer, a soft launch was established for commercialization. In the commercialization process, Joyce had found not all printers could print FibreTuff fila-ments equally well, as in accuracy and resolution; even the machine control software had some challenges. Furthermore, the Fused Filament Fabrication (FFF) or fused depo-sition modeling (FDM) printers were priced accordingly, according to their sophistica-tion and complexity, from $300 to $250,000. Still, the FibreTuff filaments were designed for the higher-priced printer and the advanced professional. The FibreTuff filament had competition, mostly nylon, where the printed anatomical bone models had a very plastic appearance in nature and no similarities to the real bone, on grounds of compo-sition and mechanical properties. Thus, surgeons did not get the chance to plan a bone-mimicking estimation of operational parameters for the actual surgery.

The need for pre surgical planning models has been recognized by American Medical Association (AMA). In 2021, AMA has assigned reimbursement codes. The al-ternatives to printed bone models are few, as a matter of fact, the printed FibreTuff bone models could be an alternative to cadaver bones. The 3D printing of FibreTuff produced bone-like anatomical models that were functional, felt like real bone, with a very similar appearance, with the capabilities to have good screw retention, cutting, and sawing ability. The printed FibreTuff was determined to be a powerful tool for pre surgical assessment as it allows for the consideration of patient- and defect-specific anatomy in the pre-surgical planning models.

As the FibreTuff filament sales would grow globally to larger customers that included universities, hospitals, medical centers, and physicians, they would print bone-like parts, such as skulls, mandibles, foot and ankle bones, metatarsals, hands, and fingers. This soft launch in commercialization continued to refine the composition to produce filaments as the customers became more sophisticated. Examples of sophisticated users that were early adopters include Dr. Edward Strong, Professor and Vice Chairman, Department of Otolaryngology at the University of California Davis, Dr. Michael Harmon, a cranio maxillofacial surgeon, founder and owner of Digital Provisionalization Technologies in Mechanicsburg, PA, and Dr. Ross Salary of the Marshall University, Department of Mechanical Engineering. For instance, Marshall University professor Dr. Ross Salary would write about 3D fabrication of bone structures using FibreTuff. A most important commercial challenge for FibreTuff filaments occurred in the late summer of 2019 when 8 printed skulls were requested by Dr. Edward Strong for a cranio maxillofacial event held in Tampa, FL in November. Figure 26.3 this event would have 50 surgeons attend an educational and training seminar where the printed Fibre-Tuff skulls would demonstrate an orbital placement. Fortunately, the exercise would prove to be most beneficial in helping promote FibreTuff creditably. The second example of an early adopter includes Dr. Michael Hartman, who began printing prototype surgical models with FibreTuff filaments. These prototype surgical guides were for mandible reconstruction. This application would help FibreTuff filaments define suggestive uses for point-of-care. Another early adopter of FibreTuff filaments included Dr. Ross

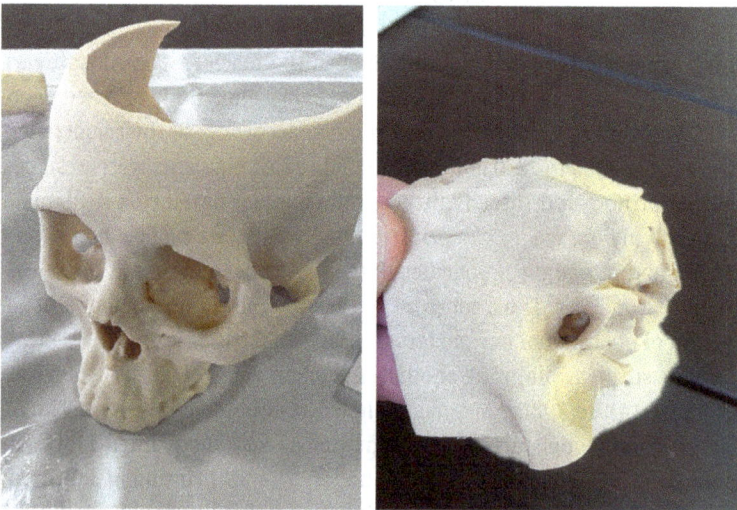

Figure 26.3: A 3D printed skull and temporal bone made with FibreTuff.

Salary at the Marshall University. Dr. Salary would investigate printed FibreTuff, producing posters and publications.

3D Fabrication of Bone Structures, Based on FibreTuff

Research Group: Dr. Ross Salary, Division of Engineering, Marshall University, Huntington, WV, 25,504

FibreTuff polymer filaments were used for the fabrication of biocompatible bone structures using material extrusion (fused filament fabrication) manufacturing technique. A multitude of process parameters were optimized with the aim to obtain strong, dimensionally accurate, and repeatable bone structures. It was experimentally observed that nozzle size, bed temperature, oven temperature, cooling rate, and print speed would significantly affect the quality as well as the performance of the fabricated bones.

Dr. Ross Salary would continue to investigate the properties of FibreTuff, and submitted an abstract to ASTM titled Investigation of the Mechanical Properties and Bioactivity of Additively Manufactured Bone Tissue Scaffolds, Composed of Polyamide, Polyolefin, and Cellulose Fibers (See Fig 26.4).

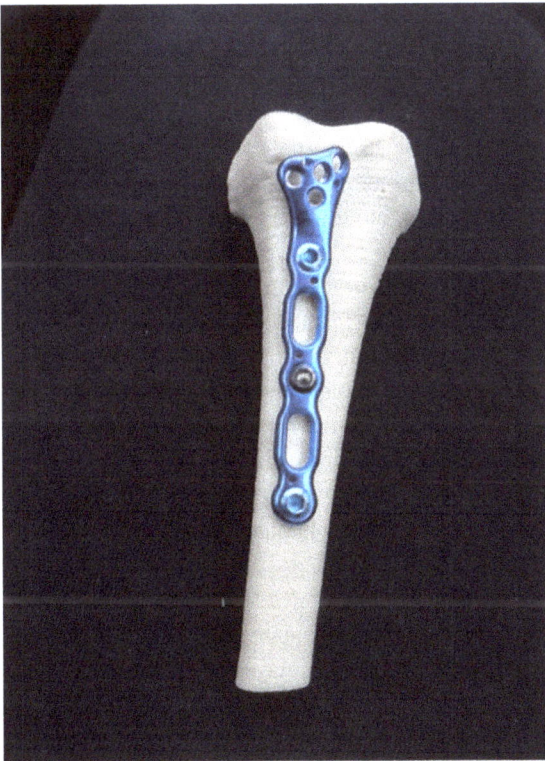

Figure 26.4: A 3D-printed bone, composed of polyamide, polyolefin and cellulose fiber (FibreTuff, Toledo, OH).

The abstract evolved into a published ASTM paper [6] in July 2022; the authors cited include; Paavana Krishna Mandava, Department of Mechanical Engineering, CECS, Marshall University, Huntington, WV 25,755, USA James B. Day, MD, Department of Orthopedic Surgery, School of Medicine, Marshall University, Huntington, WV 25,701, USA Robert Joyce FibreTuff Biotechnology Company, Toledo, OH 43,615, USA Roozbeh (Ross) Salary, PhD, Department's of Mechanical & Biomedical Engineering, Marshall University, Huntington, WV 25,755, USA. The overarching goal of this research work had been to fabricate mechanically robust, porous, and biocompatible bone scaffolds, with textured surfaces (allowing for cell/tissue adhesion) for the treatment of osseous fractures, defects, and diseases. In pursuit of this goal, the objective of the work is to investigate the mechanical properties of several triply periodic minimal surface (TPMS) bone scaffolds, fabricated using fused deposition modeling (FDM) additive manufacturing process, based on a medical-grade composite composed of polyamide, polyolefin, and cellulose fibers. FDM is a material from extrusion-based additive manufacturing process; it has emerged as a high-resolution method for the fabrication of a broad spectrum of biological tissues and constructs.

In late 2021, further mechanical validation and testing on FibreTuff would be performed at the Drexel University's Implant Research Center by Abigail Tetteh, Dr. Daniel MacDonald, and Dr. Steven Kurtz. They had written an abstract [7], Anisotropy in Fused Filament Fabrication (FFF) Printed FibreTuff for Synthetic bone models. The testing confirmed that the alignment of cellulose fibers would increase the compressive strengths in printing bone-like models. The abstract validated that the compressive strength was higher in the z-direction, which could be because of directionally arranged cellulose fibers in the material to increase its strength. We found the compressive modulus of the FFF-printed FibreTuff, tested in both directions, to be in range of the cancellous bone, while the compressive strength was significantly higher. Figure 26.5 below.

It should be noted that compression testing was performed with a mechanical testing machine, which was not equipped with an extensometer, and the strain values of the experiment were calculated from recorded time values. Therefore, the reported values are apparent compressive modulus compressive strength. Further, 3D printed samples with 100% infill had a significant amount of porosity, which may have affected the performance during mechanical testing. Nevertheless, given the results of this study and the printability of this material, FibreTuff remains an attractive material for patient-specific bone modeling.

After introducing the bone-like printed FibreTuff models, Joyce would continue to develop new products to include phantoms for radiographic imagery. Phantoms, by definition, are specially designed objects that are scanned or imaged in the field of medical imaging to evaluate, analyze, and tune the performance of various imaging devices.

Joyce would discover how to produce a printed phantom in his kitchen one evening when a customer complained there was no bone ring to mimic a real bone. Through ingenuity and a little investigation with glue, Joyce found a way to mimic real bone models, having produced a bone ring with about 3000 HU Hounsfield units. The production

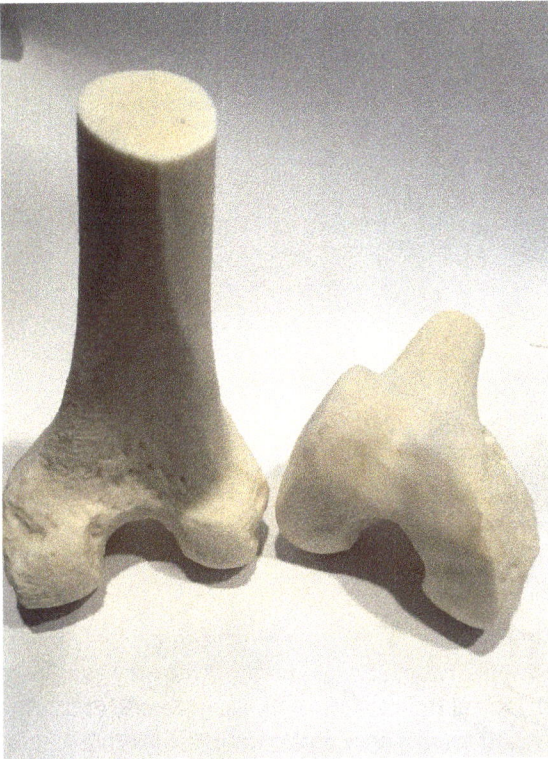

Figure 26.5: A 3D-printed femur bone made with FibreTuff having anisotropic behavior.

of the printed phantoms with FibreTuff would be evaluated by radiologists in hospitals, see Figure 26.6 below. The printed FibreTuff phantom demonstrated that there is a less expensive technology to replace many of the existing phantom products, including a temporal bone's priced at $600. Joyce' s success with printing FibreTuff phantoms would lead to printing surgical guides. Surgical guides are used by surgeons to properly place an implant in the right locations. FibreTuff would be printed as a surgical guide for a cranio maxillofacial surgeon to be used in the construction of a dental implant. The printed FibreTuff would be an effective tool in the replacement of printed nylon, commonly used in printing surgical guides. Unlike the FibreTuff, nylons do have processing constraints and limitations, such as hole size, cutting ability, and limited instrument use. Overall, the FibreTuff biocompatible composition is starting to be data driven, toward advanced bone replacement. The printed FibreTuff could be used to develop implantable bone grafts.

These attractive mechanical qualities and performance attributes helped initiate biocompatibility testing. The biocompatibility testing performed on the printed Fibre-Tuff, included NAMSA, a regulatory body located in Northwood OH. NAMSA performed an animal study that showed the formulation passed certification for USP Class VI. The

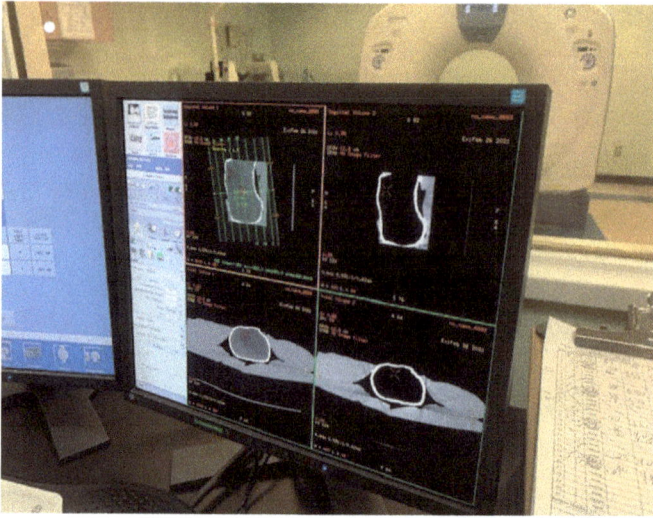

Figure 26.6: FiberTuff phantom product.

USP Class VI is a certification for implantable of less than 30 days in the body. Additional biological investigations on printed FibreTuff were performed by the Northeast Ohio Medical University (NEOMed), located in Rootstown, OH. Fayez Safadi, PhD., conducted two biological tests on several 3D printed discs made with the FibreTuff composition. The investigation for 22 days produced data to showed that osteoblast (bone forming cells) did attach to the FibreTuff. It was also identified where FibreTuff might induce bone matrix mineralization than maturation, i.e. it is great to have a matrix that will enhance the mineralization of osteoblast matrix. See Figure 26.7 below

In November 2021, further biological work was performed on FDM-printed Fibre-Tuff micropore specimen. This biological work was performed by Dr. Prabaha Sikder, Ph.D. Assistant Professor Mechanical Engineering Department Washkewicz College of Engineering Cleveland State University; Bharath Tej Challa Graduate Student, Mechanical Engineering Department, Cleveland State University Cleveland State University. They titled the paper Results of a bioactive study using a 3D printed cellulose fiber advanced composite. Dr. Sikder sterilized and immersed the printed FibreTuff micropore cellular structure in simulated body fluids (SBF). The study concluded there was evidence to show cell and tissue adhesion for bone regeneration without closure of the printed micropores for cell proliferation. See Figure 26.8. The high magnification SEM micrograph shows the formation of calcium-deficient apatite structures on the FiberTuff specimens after they were immersed in DMEM + SBF for 7 days. The calcium-deficient apatite is characterized by the formation of nano-sized well-defined globules, covering the strands of the porous FiberTuff specimens. The formation of such calcium-deficient apatite is an indication that FiberTuff is inherently bioactive in nature.

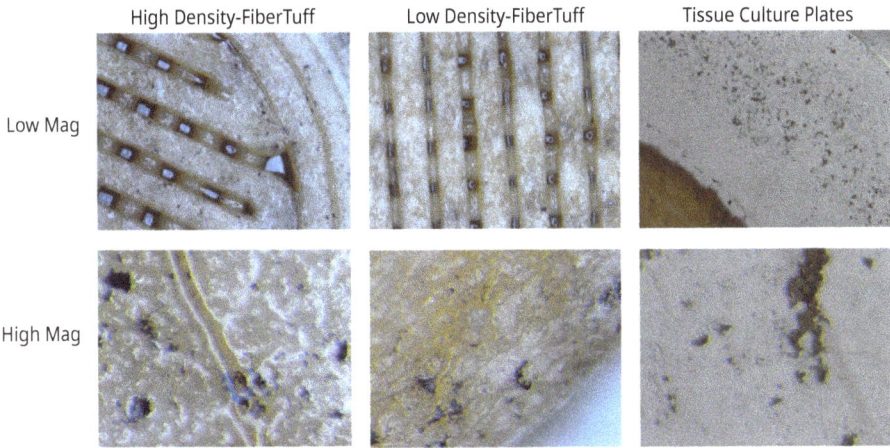

Figure 26.7: Printed FibreTuff filament disk showing osteoblast formation.

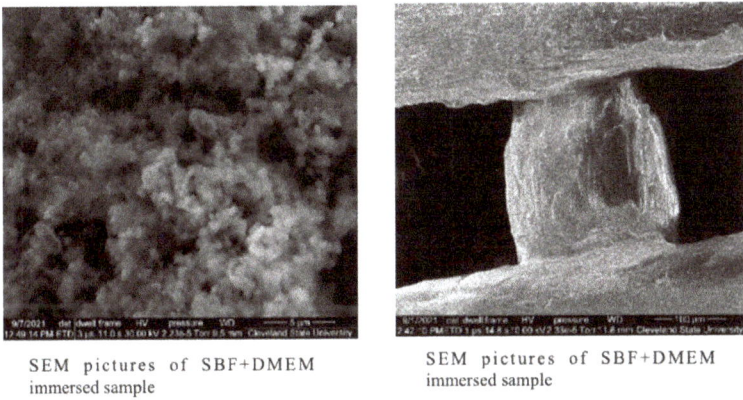

SEM pictures of SBF+DMEM immersed sample

SEM pictures of SBF+DMEM immersed sample

Figure 26.8: SEM micrographs of the printed porous made with FibreTuff.

Innovative Plastics and Molding's commercialization process is ongoing and will include approvals from the Food and Drug Administration's printing experts. The FDA will be able to evaluate FibreTuff for a master file. A FibreTuff master file with the FDA approvals would enable any device or implant manufacturer to utilize FibreTuff in a 510k for device or implant construction. Furthermore, the FDA would require a predicate to show equal to or better performance with the device construction for 510k approval. Joyce had identified a bioactive HaPEEK as the predicate. To date, the bioactivity claims of ingredients used for printing include bioactive glass, hydroxyapatite (Ha), HaPEEK, and, recently, FibreTuff.

The bioactive HaPeek will consist of a hydroxyapatite (Ha) with polyether ether ketone (PEEK). This biomaterial will only be bioactive after to 30 to 40% hydroxyapatite content is compounded and or blended. The HaPeek biomaterial is radiolucent and is considered resorbable with a resorption rate of 5–15% per year. Furthermore, by adding 30% of Ha to PEEK, there is a loss in mechanical properties of approximately 20–25%, or lower than 50MPa, which is borderline cortical bone. The printing of HaPeek has been primarily selective laser sintered (SLS), which produces stiff and brittle printed structure. In some cases, printed porosity is produced to enable cell and tissue growth for scaffold use in mimicking bone. The FibreTuff tensile strength of 45MPa can be designed and printed with more mass and fiber alignment to surpass the scaffold strength of HaPeek. Furthermore, biomaterial relies on the principles of real bone, consisting of inorganic and organic components. The inorganic ingredients are mineral, playing a vital role in defining the cell identity, and driving tissue-specific functions. The organic component comprises a protein backbone for cell and tissue proliferation. This is unlike any other printed bioactive materials, to date.

Other differentiators from the above bioactive biomaterials include the following: 1) FibreTuff has been printed in a very small micropore structure – 60–80 micron, conducive to bone ingrowth, unlike PEEK, which is printed on a macro pore structure of >120 micron. FibreTuff includes cellulose, which does not conduct heat; aiding faster cooling, when printed. 2) FibreTuff has radiopacity, showing an implant location and bone bridging (Real bone would be brighter than FibreTuff and HaPEEK is radiolucent). Additionally, FibreTuff generates healing data for artificial intelligence to support a patient's progress. 3) FibreTuff has hydrophilic ingredients with absorbent qualities to reduce brittleness and improve flexibility and impact.

26.4 The commercialization of biomaterials for the medical market

The six years of research and development of FibreTuff to commercialize filaments and powders for 3D printing, evolved into a directive or plan. For this disruptive technology to be successful, the soft launch for printing anatomical models was necessary. The customer feedback was sometimes discouraging, but also enlightening. We continued to innovate based on customer input; hence, the printing and post processing of customized phantoms. We also collaborated with surgeons in production of surgical guides. The last hurdle for a complete commercialization strategy, and the most profitable, will include printed implantable bone graft. The commercialization strategy will now include one printer method Fused Deposition Modeling (FDM) with the printer manufacturer having a high-resolution printer. Overall, the printing of FibreTuff filaments since 2018 has created a roadmap to achieve and identify value and customer success. This roadmap for point-of-care (POC) is designed for healthcare facilities (HCF) to improve customer satisfaction, improve quality of care, and reduce costs. This roadmap is the commercial strategy for products printed with FibreTuff filaments. See Figure 26.9

Roadmap For Point of Care

Pre surgical planning to implantable grafts

Anatomical Bone Models	Phantoms	Surgical Guides	Permanent Implants
Versatility to print bone with tissue	X-Ray and CT Scan	Machinability with push pin	Non re sorbable having Micropore structure
Cortical bone with Cancellus bone	Superior bone like images - bone ring and tissues	Machine ability having small threads	More consistent and effective to replace costlier grafts
Excellent screw retention with gluing capability	Eco friendly coating	High speed drills	Customization will help reduce time in OR
Radiopacity w Laser cutting, sawing like bone	Improved visualization	Small hole size 120 micron	Bioactive properties
Standard sterilization methods - Autoclave	Quick turnaround	Laser cutting	Eliminate variability identified in Allograft and Bio re sorbables
Look and appearance of bone	Customization	No special drills / tooling required	Adheres to collagen

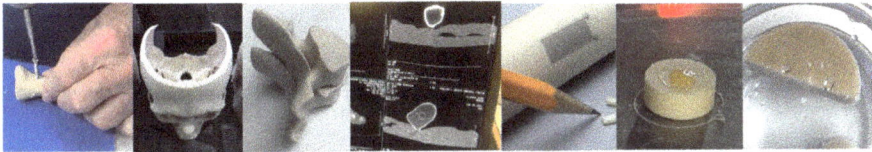

Figure 26.9: Printed FibreTuff filament having four silos for point-of-care (POC) in hospitals.

below. The silo # 1 includes printing FibreTuff to include the following "functional" bone-like qualities: good screw retention, cutting and sawing like bone, gluing ability, and fiber alignment. The silo # 2 is printing FibreTuff to produce customized phantoms for radiographic imaging. The printed FibreTuff has radiopacity for radiographic applications, including customized phantoms in hospitals. The silo # 3 is the determination that printed FibreTuff is an excellent surgical guide because it has passed biocompatibility testing. This biological discovery prompted investigation for prototyping a bone-like implantable graft. Hence, the silo # 4 is printing bone grafts with FibreTuff –the last hurdle to complete the POC roadmap. For printed FibreTuff to enter the bone graft market and be commercially viable, implementation must be guided through ISO 10993 regulations. The FDA classifies bone graft as a Class II or Class III implant. The bone graft printed with FibreTuff will require a 510 K, as per FDA regulations. Once the *in vivo* studies have passed ISO 10993–1 protocols, the first in-man study will be initiated to provide a clearer path for commercialization. The entire process for commercialization of the implantable bone grafts is expected to take 14–16 months.

The company is positioned globally to provide a four-silo commercialized printing solution for generating revenue as defined in the point-of-care (POC) roadmap. Of these four silos, FibreTuff has been able to generate revenues from two silos. The third and fourth silos can be revenue generators with the completion of the ISO 10993 studies. The expense for the healthcare facility (HCF) to implement printing FibreTuff with the designed POC roadmap can be \$90 – 120 K USD expense. The healthcare facility (HCF) can justify the expense in a 6–12 month period.

26.5 Summary

The Bioproducts investigation of early 2000's grew into the research and development of the novel medical grade biomaterials branded FibreTuff. The journey is long and sometimes challenging but continues towards commercialization of life saving devices and implants. The success of the devices and implants will introduce more collaborators and partners to help health care patients achieve better outcomes. The company and its founder wish to thank all that have contributed in making this published paper part of the journey.

References

[1]   Joyce R. Molded Article, U.S. patent 7,214,240.
[2]   Joyce R. Wood Composite Alloy Composition having a compatibilizer that involves the ability to process and compress cellulosic fiber, U.S Patent 7,994,241.
[3]   Joyce R. Composition of Matter for thermoplastic composition biopolymer "FiberTuff", U.S. patent 8,546,470.
[4]   Joyce R. Composition of Matter for thermoplastic biopolymer: FiberTuff", U.S Patent 9,109,118.
[5]   Joyce R. Cellulose Fiber thermoplastic composition having a cosmetic appearance and molding thereof, U.S. patent 10,233,309.
[6]   Mandaava KR, Day JB, Joyce R, Salary R. Investigation of the mechanical properties and bioactivity of additively manufactured bone tissue scaffolds, composed oy polyamide, polyolefin, and cellulose fiber, MSEC and ASME, June 27, 2022.
[7]   Chaffina A, Yu M, Claudio PP, Day JB, Salary R. Investigation of the functional properties of additively fabricated Triply Periodic Minimal surface-based bone scaffolds for the treatment of Osseous fractures, Aug 4, 2021 – American Society of Mechanical Engineers ASME.

About the author

Chapter 2

Ram is a subject matter expert in synthetic polymer chemistry and an entrepreneur. He developed novel functional materials and demonstrated their application in encapsulation & controlled release of chemical and biological actives. Successful new products designed in the last five years are smart coatings for corrosion detection and self-healing, stain-repellent and antimicrobial coatings for textiles, and icephobic coatings for power cables. Ram founded Aries Science & Technology and developed novel bio-based functional materials for the coatings industry. He has 41 granted US patents, nine pending patent applications, and more than 25 publications. He received four R&D 100 awards for his innovative research and successful commercialization.

Chapter 3

Michael A. Schultz is a chemical engineer by training and enjoys working on challenging scaleup problems to commercialize sustainable technology. He is currently consulting under PTI Global Solutions, and has prior industrial roles at LanzaTech, UOP, Battelle Science and Technology Malaysia, and Vertellus.

Chapter 4

John McArdle Chemical Engineer with MBA in Finance. Current Position: Principal Partner, Redwood Innovation Partners. Previous Positions: Battelle Research Institute, Koch Industries, and Honeywell Corporation. Provides consulting services to identify, evaluate, and accelerate the commercial introduction of innovative technologies focused on water, energy, advanced materials, and environmental control and remediation processes.

Chapter 5, 6

Dr. Herman P. Benecke was an employee at Battelle Memorial Institute between 1980 and 2016, and he is a named inventor on 32 US patents while twice being awarded Battelle's "Inventor of the Year" awards. Much of his recent work involved development of processes for the conversion of bioproducts to products previously derived from fossil fuels.

https://doi.org/10.1515/9783110791228-027

Chapter 7

Dr. Srinivas Nunna is an associate research fellow in carbon fiber and composite research at Carbon Nexus, Deakin University. He has eight years of experience working on university–industry collaborative projects for the automotive, wind turbine, and nanomaterial manufacturing industries where his focus has been upon creating enhanced carbon fiber. Currently, he is leading the precursor fiber development research at Carbon Nexus and his research interests include sustainable alternative precursors for the low cost carbon fibers, tailor the chemical composition of polyacrylonitrile precursors to assist rapid thermal treatment process, high compression strength, and multifunctional carbon fibers.

Dr. Claudia Creighton is a Senior Research Fellow for Carbon Fibre and Composites at Carbon Nexus, Deakin University (Australia). She has over 15 years of experience working with the aerospace, automotive and wind turbine manufacturing industries. Her current research interests include sustainable manufacture of carbon fiber and the fundamental understanding of the effect of the carbon fiber production process on fiber morphology and resulting properties.

Dr. Huma Khan is currently a postdoctoral researcher at Deakin University, Australia. Her Ph.D. research focused on developing bicomponent and non-circular precursor fiber and their conversion into carbon fiber. Dr. Khan's current research focuses on producing sustainable carbon fibers from bio-based precursor materials.

Dr. Nguyen Duc Le received his B.S. in Chemistry from Viet Nam National University, Hanoi-University of Science in 2014, and his Ph.D. in Material Engineering from Deakin University. His current research focuses on the development of bio-based precursors for carbon fiber, fiber spinning and characterization, chemical, mechanical, and microstructural analysis of carbon fibers.

Cai Li Song is a researcher at Group Research and Technology for PETRONAS, where she covers projects related to Specialty Chemicals Technology, mainly focusing on driving sustainability agenda of the corporate. She holds MEng and Ph.D. in chemical engineering degrees from Imperial College London, with a strong research background in physical chemistry, especially on material characterization and processing, advanced spectroscopy, and molecular interactions. As an early career researcher, she has been recognized for her excellence in scientific achievement, creative thinking and/or engagement between academia and industry through the 2022 Malcolm McIvor Prize and 2020 Dudley Newitt Prize for Experimental Excellence.

Prof. Russell J. Varley earned his B.Sc. (Hons) in Physical and Inorganic Chemistry from The University of Adelaide, South Australia in 1987. After graduation, he worked for 1.5 years at Yorkshire Chemicals in Melbourne, Victoria, a manufacturer of chemicals for the textile industry, before moving to CSIRO in 1989. In 1998 he received his Ph.D. in Materials Engineering from Monash University and worked for CSIRO until 2016 where he left to join the Institute for Frontier Materials at Deakin University. His primary research interests include the materials chemistry of network polymers, carbon fiber synthesis, self-healing materials and processable thermoplastic composites.

Chapter 8

Dr Nur Liyana Ismail graduated from Imperial College London with a PhD in Chemistry focusing on biomass pretreatment in the protic ionic liquids. Currently resides in PETRONAS Research Sdn Bhd as a senior reseacher, her main research interests revolve around the development of biobased and specialty chemicals. A detail-oriented and naturally curious chemist who specialized in the synthesis and product characterization. She is also passionate about solving the chemical industry's major issues with an emphasis on sustainability and circular economy.

Siti Fatihah Salleh graduated from Universiti Teknologi Petronas in 2011 with a B.Eng (Hons) in Chemical Engineering. In 2015, she completed a fast-track PhD programme in Bioprocess Engineering at Universiti Sains Malaysia. She then worked as a postdoctoral researcher at the Institute of Sustainable Energy and the Institute of Energy Policy and Research (IEPRe) for four years, where she merged her technical engineering knowledge with regulatory and business insight. She joined Petronas Research Sdn Bhd in 2019 to gain industrial experience, and she is thrilled to develop cost-effective environmentally friendly products for a global customer base.

Jofry Othman graduated from Loughborough University with a B.Eng in chemical engineering, MBA from INTI International University, MA in management from University of Hertfordshire and progressing for PhD in chemical engineering. With 19 years of R&D and industrial engineering experience in specialty chemical, petrochemicals and refining processes. Technical publications and presentations in hydrogenolysis, process engineering/optimization, CO2 conversion, ethylene oxide ethylene glycol, butyl glycol ether, olefins hydrogenation, ethanolamine, catalysis, and biomass studies.

Sara Shahruddin obtained her Ph.D. in Chemical Engineering from Imperial College London focusing on Specialty Chemicals Design for Inhibiting Solid Deposition in the oil and gas pipeline. She is currently a Manager (Technology Program) & Staff Scientist at PETRONAS Research Sdn Bhd with more than 15 years of experience in applied research. Her research is focused on developing sustainable specialty chemicals and materials for various end-use. She provides technical leadership on the execution of R&D projects ensuring business relevance within PETRONAS, especially in functional specialty chemicals design, processing strategy, economic analysis, and application of sustainable chemistry principles. Her core skills include synthesis, detailed product characterization, and presently exploring in-silico methods to design specialty chemicals.

Chapter 9

Mohamad Fakhrul Ridhwan Samsudin received his MSc and Ph.D. degrees in Chemical Engineering from Universiti Teknologi PETRONAS, Malaysia. He currently served as an Executive (Researcher) at PETRONAS Research Sdn Bhd. His research interests focus on the development of novel photocatalysts for the photo(electro)catalytic hydrogen production and wastewater treatment application. Additionally, his research also focused on the catalytic conversion of biomass into specialty chemicals products for multifarious applications.

Nur Amalina Samsudin obtained her master's degree in Chemical Engineering at Imperial College London, United Kingdom in 2018. At present, Amalina serves a Researcher at PETRONAS Research Sdn Bhd. Her research is currently focused on sustainability-based projects including the development of emollients for personal care applications, catalytic biomass conversion into specialty chemicals, and the development of bio-based polymers. She was shortlisted as the finalist for the Institution of Chemical Engineers (IChemE) Malaysia for the Sustainability Award in 2020 and the Young Researcher Award in 2021. Amalina is also in the top 2 percentile in the world for intelligence level as verified by Mensa Society.

Dr. Wong Mee Kee obtained her bachelor's and Ph.D. degrees in Chemical Engineering from Universiti Teknologi PETRONAS (UTP). Her academic career began with Curtin University as a lecturer before joining PETRONAS Research Sdn Bhd in 2018. Her research interests include renewable chemicals, process intensification, and CO_2 capture.

Mohammad Syamzari B. Rafeen is a Principal Scientist and Head Delivery for Specialty Chemicals Technology in PETRONAS specializing in the area of process technology specifically in separation and purification, chemical reaction engineering, and process scale-up. He has over 19 years of experience in the oil and gas industry, especially in R&D and technology development. He is a chemical engineer by background, and he pursued his Ph.D. in Chemistry focusing on mercury removal adsorbent research. His contribution to mercury removal technology was acknowledged by the Institute of Chemical Engineering – IChemE and the Don Nicklin Medal awards in 2013 where he is one of the key researchers in technology development. He is now actively involved in sustainability-based research and development projects that include the conversion of biomass into chemicals and the development of emollients for personal care applications.

Chapter 6, 10

Daniel Garbark is a synthetic chemist with 20 years of experience in chemical formulation, product characterization, process development, and product development. He has created novel approaches in the focus areas of bio-based products leading to 29 granted and 18 pending US patents. His main areas of focus have been in the development of polyols for polyurethane applications, lubricants, laundry builders, bio-based direct coal liquefaction, coalescing solvents, and surfactants from a variety of bio-based feedstocks.

Chapter 11

Michelle N. Young received her Ph.D. in Environmental Engineering from Arizona State University in 2018. While a research scientist at Arizona State University, her research focus on nutrient and energy recovery from wastewater systems, including efficient production of hydrogen peroxide using microbial electrochemical cells and improving methane recovery from municipal solid waste streams using anaerobic digestion at water resource recovery facilities (WRRFs). She currently works with Carollo Engineers, Inc. as a Lead Technologist on WRRF projects.

Rachel Yoho received her Ph.D. in Biological Design from Arizona State University in 2016. Her research focused on understanding electron transport pathways in microorganisms such as *Geobacter sulfurreducens* and *Geoalkalibacter ferrihydriticus*. She currently holds a faculty position at George Mason University in Fairfax, Virginia, USA.

Chapter 12

Sunil Chandran leads Amyris research and development including collaborations as Chief Science Officer and Head of Research & Development. Previously, he was appointed Senior Vice President of Research and Development in 2019. With more than 15 years of experience in industrial biotechnology, he has a proven track record of bringing multiple biotech products to market. Since joining Amyris in 2006, Sunil has led multiple metabolic engineering focused projects, and was the primary architect of our world-class Automated Strain Engineering (ASE) platform. Notably in 2015, Sunil was the principal investigator who led an ambitious project that earned Amyris a $35 million Technology Investment Agreement from the US Defense Advanced Research Projects Agency (DARPA). Prior to joining Amyris, Sunil was a scientist at Kosan Biosciences, where he studied the ability of polyketide synthases to make novel pharmaceutical drug candidates. He has more than 20 published scientific papers, book chapters, and patents and has been an invited speaker at numerous international conferences on biotechnology. He holds a B.S. in chemistry from the University of Mumbai, an M.S. in chemistry from the Indian Institute of Technology, Mumbai, and a Ph.D. in chemistry from Michigan State University. Sunil also completed a post-doctoral fellowship in chemistry and biochemistry at the University of Wisconsin-Madison.

Beth Bannerman, Amyris' Chief Engagement & Sustainability Officer, is a principal strategist and advisor to the C-Suite and executive team responsible for setting the strategic direction for corporate communications, partnerships and all aspects of internal and external communications delivery including the publication of the company's annual ESG report. She leads Amyris' ESG Council to ensure the company is capitalizing on opportunities to strategically advance sustainability and the bioeconomy. Previously, Beth led international communications at The Royal Bank of Scotland Group, PLC, later named NatWest Group, advising the COO, CFO and other C-level executives on communications strategy, crisis management, policy, investor relations and media relations. Beth is a founding member of CHIEF, a network for executive women, and serves on the Speaker Committee for UNICEF Northwest USA Board of Directors.

Chapter 13

Dr. Crispinus Omumasaba, a Researcher at UCDI, joined the company from the Research Institute of Innovative Technology for the Earth (RITE/Kyoto). He earned his Ph.D. from Kagoshima University following stints in academia and the food processing industry in Nairobi. He boasts a number of scientific papers, book chapters and patents to his credit.

Dr. Alain A. Vertés, Business Strategy Adviser, UCDI, is Managing Director at NxR Biotechnologies, a global boutique consulting firm based in Basel, Switzerland. Dr. Vertés has extensive, worldwide experience in the pharmaceutical (Lilly, Pfizer, Roche), industrial biotechnology (Mitsubishi Chemical Corporation), research (Institut Pasteur, RITE/Kyoto, Battelle Memorial Institute, PPD/BioDuro) and consulting (Australian Strategic Policy Institute) sectors with award-winning accomplishments (Scrip Award 2008) and commercial successes. With an M.Sc. from the University of Illinois at Urbana-Champaign, a Ph.D. from the University of Lille Flandres Artois and a Sloan Fellowship from the London Business School (MBA-M.Sc.), Dr. Vertés has a long publication record that spans the fields of science and business.

Dr. F. Blaine Metting, Jr., a Consultant in Microbial Biotechnology and Carbon Sequestration, recently retired from Pacific Northwest National Laboratory (PNNL) where he managed the Biological and Environmental Science Product Line and commercial accounts for the Fundamental and Computational Sciences organization, conceived and implemented initiatives and oversaw joint programs with other US government research institutes. He is currently Science Advisor to UCDI and has consulted for GEI, RITE, Dainippon Ink & Chemical, the Australian Renewable Energy Agency, and the US Department of Energy, and chaired the International Energy Agency (IEA) Microalgae Biofixation Network. Dr. Metting co-founded a microalgae company that secured small business innovation research grants from the US government to work on environmental restoration, contributed to the Battelle-Malaysia Renewable Energy Laboratory project, and was a founding member of the Industrial and Environmental Section at the Biotechnology Industry Association (BIO) for which he has perennially served on program committees. Dr. Metting earned his Ph.D. from Washington State University where he studied systematics, ecology and technical potential of soil microalgae and cyanobacteria and has over 30 refereed scientific articles and book chapters.

Dr. Hideaki Yukawa, Founder and CEO of UCDI Company, earned his Ph.D. in molecular biology, microbial applications, and enzyme chemistry from the University of Tokyo and is former Research Fellow at Mitsubishi Chemical Corporation. His dogged pursuit of a low-carbon future society and advancement of biotechnology-based solutions to global food security issues is reflected in his pivotal accomplishments such as the establishment of the Microbiology Research Group at the Research Institute of Innovative Technology for the Earth (RITE), founding of the Green Earth Institute (GEI) Company, which have been recognized through awards from the Japan Bioindustry, the Tsukuba Foundation for Chemical and Biotechnology, the Japan Society for Bioscience, Biotechnology and Agrochemistry, and the Nikkei Global Environmental Awards. A Fellow of the Society of Industrial Microbiology and Biotechnology, Dr. Yukawa is a widely published author of scientific papers, books, and patents and is a mentor to many scientists from diverse cultures.

Chapter 14

Dr. Vikas Kumar is an assistant professor in fish nutrition and nutrigenomics at the University of Idaho, USA. Prior to joining the University of Idaho, Dr. Kumar worked as an assistant professor in Fish Nutrition at Kentucky State University, USA. He has more than a decade of research and teaching experience in aquaculture and nutritional biochemistry. Dr. Kumar is currently serving as an editor for *Animal Feed Science and Technology* (Elsevier), an editorial board member of *Scientific Reports* (Nature Publishing Group), and a leading associate editor for the *Journal of the World Aquaculture Society* (Wiley). He has authored more than 116 papers in peer-reviewed journals, 19 book chapters, 32 magazine articles, and a book entitled *Enzymes in Human and Animal Nutrition: Principles and Perspectives* with Elsevier, Academic Press. He received undergraduate degree in fisheries science and Master's degree in fish nutrition and biochemistry from India. To expand his research desire into reality, he moved to the University of Hohenheim, Germany, to complete his doctoral study followed by postdoc from the Ohio State University, USA.

Dr. Anisa Mitra is a Ph.D. graduate from the Fisheries and Aquaculture Division, Department of Zoology at the University of Calcutta. She completed her M.Phil. in fisheries management from the University of Calcutta. She started her teaching career as a lecturer in Zoology and thereafter continuing her teaching as Assistant Professor in Zoology in Sundarban Hazi Desarat College affiliated to the University of Calcutta, and she is in active research at the Department of Zoology in Fisheries and Aquaculture Division, University of Calcutta. She has published many scientific articles in various esteemed national and international journals, book chapters, and in magazines. She has authored a book on Biology and culture potential of *Chitala chitala*. She is the awardee of young scientist award in poster presentation in Prof. Hiralal Chaudhuri Commemorative Conference in 2016, organized by The Zoological Society, Kolkata, in collaboration with the University of Calcutta, and D. N. Ganguly Award (organized by The Zoological Society, Kolkata) in 2007. She is the Life Member of the Zoological society of India, Kolkata, and Aquatic Biodiversity Conservation Society, NBFGR, India. She is a regular reviewer of internationally reputed journals.

Her total research experience is around 12 years and teaching experience is more than 11 years.

Her area of interest is fish nutritional and developmental biology, ecotoxicology, sustainability, and aquaculture.

Chapter 15

Dan Derr, Ph.D., has spent his career in renewable technologies, initially in energy, then in fuel, and now in chemicals. Dan received his Ph.D. in Inorganic Chemistry from Colorado State University and B.S. in Chemistry from Ithaca College.

Chapter 16

Dr. Jacyr Quadros Jr. is a chemical engineer from the University of São Paulo (1988), with postgraduate studies in business and economics at FGV (1998) and a Ph.D. from the University of S. Paulo (2015). He co-founded Nexoleum Bioderivados S. A., now Innoleics USA Corp., and focuses on bio-based innovation. He holds two commercially successful international patents for renewable PVC plasticizers.

Chapter 17

Gunnar Lynum, President, CEO and Founder of SMD Products Company, is a serial innovator who prior to founding his own company worked in government and private industries in the field of chemistry and bio-technologies. Gunnar is an industry renowned figure who helped generate record sales of soybeans to the Japanese market with a farm direct program and was featured in the *Wall Street Journal*.

Steven Lynum, Corporate Officer and Co-head of the Diagnostics & Lifesciences Domain of PHC Holdings Corporation (publicly traded on the Tokyo Stock Exchange), currently serves as President of Epredia Holdings and is an unpaid advisor to SMD Products Company.

Chapter 18

Jonathan M. Cristiani is a clean fuels specialist and technology manager at Black & Veatch with nearly two decades of experience in a host of renewable and alternative energy technologies. His duties include low-carbon-fuel technology expertise, front-end project development/consulting, and engineering/project management support. Mr. Cristiani has significant experience with the conversion of bio-based feedstocks into energy products as well as with the production, storage, and utilization of hydrogen for numerous end use applications.

Justin Distler, P.E. is a professional Mechanical Engineer at Black & Veatch with a background in thermal performance, power generation, renewable energy, and hydrogen systems. Distler graduated with his B.S.M.E. from the University of Missouri – Columbia and M.S.M.E. from the University of Missouri – Kansas City. Distler is a life-long learner and aspires to be a catalyst in decarbonization and energy sustainability in a rapidly evolving power generation space.

Andrew Doerflinger is a Process Engineer at Black & Veatch focusing on decarbonization projects. Within this role, Andrew primarily supports front-end conceptual project development for various hydrogen applications, including blending and co-firing hydrogen with natural gas for power generation, fuel cell electric vehicles (FCEV) for heavy-duty and light-duty vehicles, fuel cells for stationary power generation, hydrogen liquefaction, and hydrogen derivative production (i.e., ammonia, methanol). Andrew also supports consulting efforts, with experience examining hydrogen and clean fuel integration for decarbonization studies and performing independent assessments of novel electrolyzer technologies. Andrew is a licensed Professional Engineer in Kansas.

Vincent Mazzoni is a Process Engineer at Black & Veatch and is active on projects that range from feasibility studies to detailed engineering design for clients in the Energy & Process Industries. These projects include developing solutions for hydrogen energy storage, hydrogen/natural gas blending, renewable natural gas (methanation), ammonia, and mobility end use applications. Vincent works closely with technology providers, including electrolyzer, fuel cell, compressor, storage, and hydrogen dispensing original equipment manufacturers to develop project solutions and technology evaluations for clients. Vincent is a licensed Chemical Engineer in California.

Chapter 20

Pramod Kumbhar holds a Ph.D. in Chemical Engineering and is President and Chief Technology Officer for last 10 years at Praj Matrix – R&D, focusing on driving innovations in biotechnology to make biofuels and biochemicals. He is Fellow of Maharashtra Academy of Sciences and received the ICI Process Development Award from Indian Institute of Chemical Engineers. He has more than 30 international patents and publications in scientific journals.

Dr. Anand Ghosalkar is currently working as Chief Scientist with Praj Matrix-R&D center (a division of Praj Industries Ltd). He has more than 15 years of experience in the Industrial Biotechnology R&D sector. Dr. Anand holds PhD in Biochemical Engineering from Indian Institute of Technology Delhi and M.Tech in Chemical Engineering from IIT Kharagpur.

Dr. Anand's research interests include metabolic engineering of bacteria and yeast species for biofuels & biopolymers production and development of bioprocess from laboratory to pilot scale. Dr. Anand holds 3 Indian patents, published 13 articles in international peer reviewed journals, co-authored 2 book chapters and has presented in several national and international conferences.

Dr. Yogesh Ramesh Nevare is a Staff Scientist at Praj Industries Limited, Pune. He received his M.Sc. (Industrial Chemistry, 2011) and M. Tech. (Polymer Technology, 2013) from Kavayitri Bahinabai Chaudhari North Maharashtra University, Jalgaon, India, and PhD (Chemical Sciences, 2021) from CSIR-National Chemical Laboratory (CSIR-NCL), Pune, India, under the supervision of Dr. S. Sivaram, Dr. Prakash P. Wadgaonkar, and Dr. Harshawardhan V. Pol. His current research focuses on exploring synthetic approaches to aliphatic polyesters, polymers from renewable resources, polymer synthesis involving chain and step-growth polymerization, and the structure-property relationship in polymers.

Chapter 21

Sandeep Kulkarni earned a Ph.D. in Polymer Chemistry, following which he worked in the packaging and consumer packaged goods industry for over two decades. Dr. Kulkarni is the founder and President of KoolEarth Solutions, Inc., providing product and business development consultancy as well as educational services in the field of sustainable packaging. He is also a team member of Ubuntoo, an innovative environmental solutions platform, as well as a staff member at the Association of Plastic Recyclers (APR). Dr. Kulkarni is an advisor for several startup companies and also a member of the Executive Committee of the Sustainable Packaging Coalition.

Chapters 19, 23, 24

Ashok M. Adur has a Ph.D. in Polymer Science and Engineering and trained in Executive Entrepreneurship. With over 45 years of success in R&D, business development, marketing, and general management at chemical, plastics, and paper companies, some working on recycling, bio-based fibers and sustainability, he has published 85 papers in professional journals and presentations at professional conferences and filed 77 patent applications, resulting in business of over $2.5 billion/year. As Consultant at Everest International Consulting (LLC), he is considered an expert in polymers, compatibilization of multi-component systems, and developing strategies for intellectual property issues in polymers and plastics.

Chapter 22

Dr. Ray Bergstra has a B.Sc. in Applied Chemistry from McMaster University and a Ph.D. in Organic Chemistry from the University of Alberta. Ray has 13 years of international experience in the lubricants industry having held positions as in synthetic lubricant research and development, technical marketing, and sales management. In 2003, Ray established MTN Consulting Associates, a market research, technology commercialization, and project management consulting firm, specializing in supporting new initiatives and ventures in bio-based fuels, chemicals, and industrial materials. Ray is currently Director of Technology and Business Development for TerraVerdae Bioworks, a Canadian-based bioplastic technology and production company.

Chapter 25

David Saltman is a cleantech entrepreneur and leading advocate of environmental innovation and sustainable business development. He is currently Chairman and CEO of INCA Renewable Technologies, a world leader in the development, commercialization, and manufacturing of innovative bio-composite materials for a wide range of industrial applications.

Camille Saltman is Chief Marketing and Sales Officer of INCA Renewtech, a serial entrepreneur, angel impact investor and a pioneer in both the US and Canada in commercializing clean technologies and advocating for a circular economy.

Chapter 26

Robert Joyce is the President and Founder of FibreTuff a developer of cellulose fiber with thermoplastic compositions for 3D Printing, Molding and Extrusion. FibreTuff has commercialized 3D printing biomaterials for the medical industry. These advanced materials in both FDM and SLS methods has produced functional "bone like" anatomical models. FibreTuff customers have included Universities, Medical Centers, Hospitals and physicians. The FibreTuff technology has grown to include phantoms, surgical guides and prototypes of implantable products such a bone grafts.Robert Joyce has two college degrees. An associates from Macomb College and bachelors from Ohio University. He was recently feature on a podcast presented by Talking Additive. He has contributed to two research papers both published by the American Society of Mechanical Engineer (ASME) regarding the Investigation of the Mechanical properties of porous bone scaffolds, composed of polyamide, polyolefin and cellulose fibers. His work includes 7 North American patents for composition, processing and products and he has been a speaker at biopolymer conferences to include International BioComposites Symposium, Society of Plastic Engineers, Society of Manufacturing Engineers and Forest Products Society.

Editor/Chapter 1 and 2

Dr. Bhima Vijayendran has wide ranging experience over his 45-year career in senior leadership roles at Fortune 500 companies and at Battelle Memorial Institute, World's largest non-profit, independent R&D organization. Since his retirement in 2012, Dr. Vijayendran has been active as a consultant and advisor in the areas of renewable technologies and advanced materials and their commercialization. After retirement, co-founded Redwood Innovation Partners and in 2017 co-founded California Biobased Innovation Center (CBIC).

Dr. Vijayendran's accomplishments span a wide-ranging area including results oriented R&D management, Global R&D development, Strategy and Value Creation of IP Investments, Product and Process innovation and commercialization, Strategic Alliance and Partnerships and mentoring and STEM initiatives to train next-generation scientists and engineers. He has over 100 patents, 50 technical articles and is widely recognized with many awards such as the Presidential Green Chemistry Award, 10 R&D Awards, American Soybean Association and Battelle CEO awards. Since his retirement, he has been focusing his activities as consultant and advisor to startups and established companies.

Index

https://doi.org/10.1515/9783110791228-028

www.ingramcontent.com/pod-product-compliance
Lightning Source LLC
Chambersburg PA
CBHW060943210326
41598CB00031B/4706